U.S. Department of Transportation
National Highway Traffic Safety Administration

www.nhtsa.dot.gov

people saving people

DOT HS 809 833

October 2004
Technical Report

Lives Saved by the Federal Motor Vehicle Safety Standards and Other Vehicle Safety Technologies, 1960-2002

Passenger Cars and Light Trucks

With a Review of 19 FMVSS and their Effectiveness in Reducing Fatalities, Injuries and Crashes

1. Report No. DOT HS 809 833	2. Government Accession No.	3. Recipient's Catalog No.
4. Title and Subtitle Lives Saved by the Federal Motor Vehicle Safety Standards and Other Vehicle Safety Technologies, 1960-2002 – Passenger Cars and Light Trucks – With a Review of 19 FMVSS and their Effectiveness in Reducing Fatalities, Injuries and Crashes		5. Report Date October 2004
		6. Performing Organization Code
7. Author(s) Charles J. Kahane, Ph.D.		8. Performing Organization Report No.
9. Performing Organization Name and Address Evaluation Division; National Center for Statistics and Analysis National Highway Traffic Safety Administration Washington, DC 20590		10. Work Unit No. (TRAIS)
		11. Contract or Grant No.
12. Sponsoring Agency Name and Address Department of Transportation National Highway Traffic Safety Administration Washington, DC 20590		13. Type of Report and Period Covered NHTSA Technical Report
		14. Sponsoring Agency Code
15. Supplementary Notes		

16. Abstract

The National Highway Traffic Safety Administration (NHTSA) began to evaluate its Federal Motor Vehicle Safety Standards (FMVSS) in 1975. By October 2004, NHTSA had evaluated the effectiveness of virtually all the life-saving technologies introduced in passenger cars, pickup trucks, sport utility vehicles and vans from about 1960 up through the later 1990's. A statistical model estimates the number of lives saved from 1960 to 2002 by the combination of these life-saving technologies. Fatality Analysis Reporting System (FARS) data for 1975-2002 document the actual crash fatalities in vehicles that, especially in recent years, include many safety technologies. Using NHTSA's published effectiveness estimates, the model estimates how many people would have died if the vehicles had not been equipped with any of the safety technologies. In addition to equipment meeting specific FMVSS, the model tallies lives saved by installations in advance of the FMVSS, back to 1960, and by non-compulsory improvements, such as the redesign of mid and lower instrument panels. FARS data have been available since 1975, but an extension of the model allows estimates of lives saved in 1960-1974.

Vehicle safety technologies saved an estimated 328,551 lives from 1960 through 2002. The annual number of lives saved grew quite steadily from 115 in 1960, when a small number of people used lap belts, to 24,561 in 2002, when most cars and light trucks were equipped with numerous modern safety technologies and belt use on the road achieved 75 percent.

17. Key Words FARS; statistical analysis; evaluation; benefits; effectiveness; fatality reduction; injury reduction; crashworthiness; crash avoidance	18. Distribution Statement Document is available to the public through the National Technical Information Service, Springfield, Virginia 22161	
19. Security Classif. (Of this report) Unclassified	20. Security Classif. (Of this page) Unclassified	21. No. of Pages 422	22. Price

Form DOT F 1700.7 (8-72)   Reproduction of completed page authorized

TABLE OF CONTENTS

LIST OF ABBREVIATIONS

ABS   Antilock brake system

ACIR   Automotive Crash Injury Research, a crash data file of the 1950's and 60's

ACTS   Automotive Coalition for Traffic Safety (before 1999, American Coalition for Traffic Safety)

AIS   Abbreviated Injury Scale

AMC   American Motors Corporation

ANPRM   Advance Notice of Proposed Rulemaking

ANSI   American National Standards Institute

BMW   Bayerische Motoren Werke

CATMOD   Categorical models procedure in SAS

CHMSL   Center high mounted stop lamp

CRASH   Calspan reconstruction of accident speeds on the highway

CY   Calendar year

DMV   Department of Motor Vehicles

ESC   Electronic stability control

FARS   Fatality Analysis Reporting System (a census of fatal crashes in the United States since 1975)

FHWA   Federal Highway Administration

FMCSA   Federal Motor Carrier Safety Administration

FMH   Free-Motion Headform for testing upper interior components

FMVSS   Federal Motor Vehicle Safety Standard

GM   General Motors

GSA	General Services Administration of the Federal government
GVWR	Gross Vehicle Weight Rating (specified by the manufacturer, equals the vehicle's curb weight plus maximum recommended loading)
HIC	Head Injury Criterion
HPR	High penetration resistant windshield
HSL	Highway Safety Literature, an on-line literature database that is a subfile of the automated Transportation Research Information Service (TRIS) file, accessible at http://199.79.179.82/sundev/search.cfm
IIHS	Insurance Institute for Highway Safety
LATCH	Lower anchors and tethers for children
LTV	Light trucks and vans (includes pickup trucks, SUVs, minivans and full-sized vans)
MCOD	Multiple Cause of Death file, a supplement to FARS in 1987-98, listing causes of death from the occupant's death certificate
MDAI	Multidisciplinary Accident Investigations (a file of in-depth crash investigations conducted by NHTSA and others, 1967-78)
MDB	Moving deformable barrier
mph	Miles per hour
MVMA2D	Motor Vehicle Manufacturers' Association's 2-dimensional computer simulation of the occupant's motion in a frontal crash
MY	Model year
NASS	National Automotive Sampling System (a probability sample of police-reported crashes in the United States since 1979, investigated in detail)
NCAP	New Car Assessment Program (consumer information supplied by NHTSA on the safety of new cars and LTVs, based on test results, since 1979)
NCSA	National Center for Statistics and Analysis, NHTSA
NCSS	National Crash Severity Study (a probability sample of police-reported towaway crashes in seven multi-county areas, 1977-79, investigated in detail)

NHTSA	National Highway Traffic Safety Administration
NOPUS	National Occupant Protection Use Survey (statistics for the United States, since 1994, from a national observational survey based on a probability sample)
NPRM	Notice of Proposed Rulemaking
NTSB	National Transportation Safety Board
RF	Right-front
RSEP	Restraint Systems Evaluation Project (a probability sample of police-reported towaway crashes involvements of model year 1973-75 cars in five urban or multi-county areas, 1974-75, investigated in detail)
RWAL	Rear-wheel antilock brake system
SAE	Society of Automotive Engineers
SAS	Statistical analysis software produced by SAS Institute, Inc.
SCI	Special Crash Investigations, NHTSA National Center for Statistics and Analysis
SID	Side impact dummy
SSF	Static stability factor (Half of the vehicle's track width divided by the height of its center of gravity)
SUV	Sport utility vehicle
TTI	Thoracic Trauma Index
TTI(d)	Thoracic Trauma Index for the dummy in a side-impact test
UMTRI	University of Michigan Transportation Research Institute
VIN	Vehicle Identification Number
VMT	Vehicle Miles of Travel
VW	Volkswagen

EXECUTIVE SUMMARY

The National Highway Traffic Safety Administration (NHTSA) began to evaluate the effectiveness of its Federal Motor Vehicle Safety Standards (FMVSS) in 1975. By October 2004, NHTSA had evaluated virtually all the life-saving technologies introduced in passenger cars and in LTVs (light trucks and vans – i.e., pickup trucks, sport utility vehicles, minivans and full-size vans) from about 1960 up through the later 1990's. The agency is now ready to estimate the number of lives saved from 1960 to 2002, year-by-year, by the combination of all these life-saving technologies, and by each individual technology.

Past evaluation reports estimated the *effectiveness* of a safety technology – a percentage reduction of fatalities – by statistically analyzing crash data on vehicles produced just before vs. just after a make-model received that technology. Effectiveness, if accurately estimated, should not change much over time. But the *benefits* of a technology – the absolute number of lives saved in a year – readily change from year to year depending on the number of vehicles equipped with the technology, their mileage and the crash-involvement rate of the driving population (exposure). This report will:

- Review the effectiveness estimates in past evaluations of safety technologies for cars and LTVs, describing how they work, and the history of the FMVSS that regulate them.

- Develop a model that uses Fatality Analysis Reporting System (FARS) data and these past effectiveness estimates to calculate how many lives the following technologies have saved, individually and in combination, in each year from 1960 to 2002:

FMVSS: Safety Technologies	Cars	LTVs	Heavy Trucks
105: Dual master cylinders & front disc brakes[1]	X	X	
108: Conspicuity tape for heavy trailers			X[2]
(201) Voluntary mid/lower instrument panel improvements	X	X	
203/204: Energy-absorbing steering assemblies	X	X	
206: Improved door locks	X	X	
208: Lap belts	X	X	
3-point belts	X	X	
2-point automatic belts	X		
Voluntary NCAP-related improvements for belted occs.	X		
Frontal air bags	X	X	
212: Adhesive windshield bonding	X	X	
213: Child safety seats	X	X	
214: Side door beams	X	X	
Voluntary (pre-1994) side impact protection in 2-door cars	X		
216: Roof crush strength (eliminate true hardtops)	X		

---

[1] Applied to cars and LTVs, but also saves pedestrians, bicyclists and motorcyclists not hit by these cars and LTVs.
[2] Applied to heavy trailers, but also saves occupants of cars and LTVs that avoid collisions with these trailers.

In addition to safety equipment installed to meet specific FMVSS, the model tallies lives saved by installations in advance of the FMVSS, and by non-compulsory improvements, as shown in the preceding list, such as the redesign of mid and lower instrument panels and modifications to improve performance on the New Car Assessment Program. The model includes car/LTV occupants saved by car/LTV technologies or child safety seats (99 percent of the total), plus pedestrians/bicyclists/ motorcyclists saved by car/LTV brake improvements, and car/LTV occupants saved by conspicuity tape on heavy trailers.

The model does not include technologies so recent that NHTSA has not yet evaluated them based on statistical analysis of crash data, such as the dynamic-test standard for side impact protection (1994-97 phase-in), or head air bags. The study is limited to technologies in cars and LTVs, or that save lives of car/LTV occupants; for example, motorcycle helmets are not included. It is limited to vehicle technologies. It does not estimate the effects of behavioral safety programs (such as the reduction of impaired driving) – except, of course, to the extent that programs to increase belt use have contributed greatly to the number of lives saved by belts; roadway and traffic engineering improvements; and shifts in the vehicle fleet – e.g., from large to small cars, or from cars to LTVs. The model is limited to estimating fatality reduction by the safety technologies: NHTSA does not have enough "building blocks" (evaluation results) to develop estimates for the numbers of nonfatal injuries prevented over the years.

How the model works Consider 1,000 cases of driver fatalities in directly frontal multivehicle crashes in cars with 1960 technology: no energy-absorbing steering columns, all drivers unbelted, no air bags. A NHTSA evaluation estimates that energy-absorbing columns reduce fatalities of drivers in frontal crashes by 12.1 percent. Thus, if these cars had been equipped with them, there would have been only 879 fatalities, a saving of 121 lives. Another evaluation estimates that 3-point belts, in cars with energy-absorbing columns, reduce drivers' fatality risk by 42 percent in these types of crashes. If the cars had been equipped with 3-point belts in addition to energy-absorbing columns, and the drivers had buckled up, the 879 fatalities would have diminished to 510, saving another 369 lives. A third evaluation estimates that air bags reduce fatality risk by 25.3 percent for belted drivers in these types of crashes, in cars with energy-absorbing columns. Air bags would have cut the 510 fatalities down to 381, saving another 129 lives.

The model uses 1975-2002 FARS data and performs the same calculations in reverse order: e.g., there might be 381 actual FARS cases of 3-point-belted driver fatalities in directly frontal multivehicle crashes in model year 1999 cars, all of which are equipped with air bags and energy-absorbing columns. If air bags, the most recent (1990's) safety technology, had been removed from the cars, fatalities would have increased to 510. In other words, there must have been 129 potentially fatal collisions in these model year 1999 cars that did not become FARS cases because air bags saved the driver's life. If the 3-point belts, a 1970's technology, had also been removed from the cars, and the drivers had been unbelted, the fatalities would have increased to 879. Finally, if the energy-absorbing columns, a 1960's technology, had been replaced by rigid columns, degrading these cars all the way back to a 1960 level of safety, fatalities would have increased to 1,000. The three technologies, in combination, saved 619 lives: 129 by air bags, 369 by 3-point belts and 121 by energy-absorbing columns. In summary, FARS cases of fatalities in vehicles equipped with modern safety technologies constitute evidence of an even larger number of fatalities that would have occurred without those technologies. This approach, based on "reverse chronological order" is not the only one that could have been used in the model; however, alternative approaches would have generated the same estimate of overall lives saved in 1960-2002, differing only in how they allocated that total among the individual safety technologies.

FARS data have been available since 1975, but the FMVSS date back to January 1, 1968, and some technologies were introduced before that. An extension of the model allows estimates of lives saved in 1960-1974.

<u>Lives saved in 1960-2002</u> Safety technologies saved an estimated 328,551 lives from 1960 through 2002. Table 1 shows that the annual number of lives saved grew quite steadily from 115 in 1960, when a small number of people used lap belts, to 24,561 in 2002, when most cars and LTVs were equipped with numerous modern safety technologies and belt use on the road achieved 75 percent. (Safety belt use continued to increase after 2002, and reached 80 percent in 2004.)

Figure 1 tracks the benefits of vehicle safety technologies. Fewer than 1,000 lives per year were saved in 1960-67. Starting in 1968, vehicles incorporating most of the safety improvements of the 1960's superseded older vehicles; lives saved reached 4,000 in 1978, but remained at that level for 6 years as belt use temporarily declined. The greatest increase, from 4,835 in 1984 to 11,265 in 1988, came with buckle-up laws. Since 1988, continued increases in belt use, air bags and other recent technologies, and a steadily escalating "base" of more vehicles and more VMT (vehicle miles of travel) have helped the fatality reduction grow steadily, exceeding 15,000 in 1994 and 20,000 in 2000, reaching 24,561 in 2002.

TABLE 1: LIVES SAVED BY VEHICLE SAFETY TECHNOLOGIES, 1960-2002

(Car and LTV occupants saved,
plus non-occupants and motorcyclists saved by car/LTV brake improvements)

CY	LIVES SAVED
1960	115
1961	117
1962	135
1963	160
1964	203
1965	251
1966	339
1967	509
1968	816
1969	1,179
1970	1,447
1971	1,774
1972	2,226
1973	2,576
1974	2,518
1975	3,058
1976	3,240
1977	3,671
1978	4,040
1979	4,299
1980	4,539
1981	4,455
1982	4,057
1983	4,248
1984	4,835
1985	6,389
1986	8,523
1987	9,973
1988	11,265
1989	11,487
1990	11,711
1991	12,194
1992	12,483
1993	13,796
1994	15,154
1995	16,117
1996	17,813
1997	18,560
1998	19,380
1999	19,942
2000	21,789
2001	22,605
2002	24,561
	============
	328,551

FIGURE 1: LIVES SAVED BY VEHICLE SAFETY TECHNOLOGIES, 1960-2002

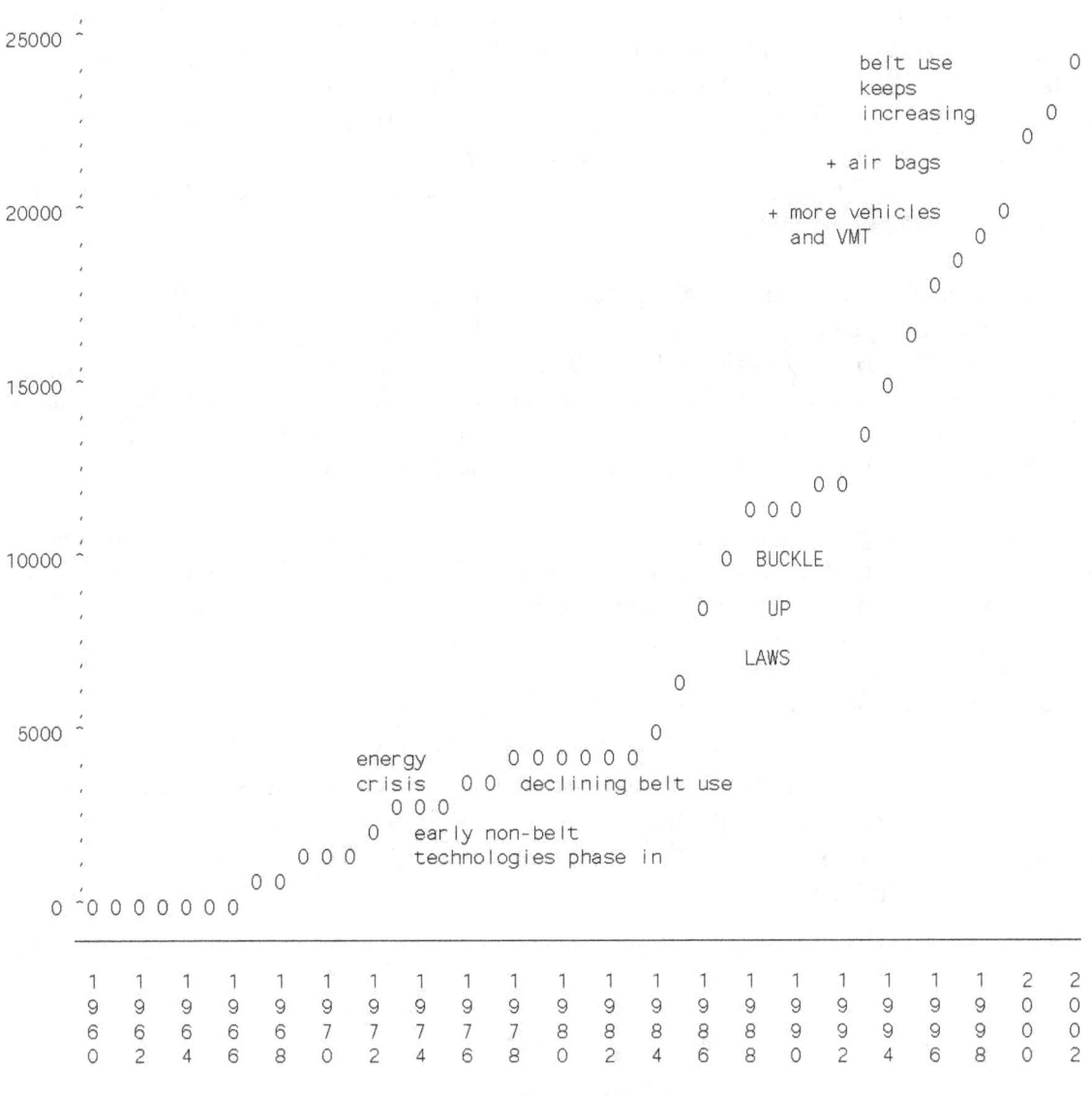

CALENDAR YEAR

Car/LTV occupants: actual fatalities, potential fatalities and percent saved Among the 328,551 lives saved in 1960-2002, 326,371 were occupants of cars and LTVs. (The remaining 2,180 were pedestrians, bicyclists and motorcyclists who avoided fatal impacts by cars or LTVs because dual master cylinders or front disc brakes improved the car or LTV's braking performance.) The sum of the actual fatalities and the lives saved is the number of fatalities that potentially would have happened if cars and LTVs still had 1960 safety technology and nobody used safety belts. Table 2 shows 1,443,030 actual car/LTV occupant fatalities in 1960-2002; without the 326,371 lives saved, there would have been 1,796,401 potential fatalities. Actual car and LTV occupant fatalities only increased from 28,183 in 1960 to 32,737 in 2002. Without the vehicle safety technologies and increases in belt use, they would have more than doubled, from 28,298 in 1960 to 57,242 in 2002.

Figure 2 compares the trends in actual and potential fatalities. Up to the early 1980's, both trend lines were fairly close together, and both moved up or down in response to baby boomers starting to drive (1960's), energy crisis (1970's) and recession (early 1980's). From the mid 1980's, vehicle safety made a big difference. Potential fatalities kept rising as registered vehicles and VMT increased in an affluent society. But increased belt use, air bags and other vehicle safety technologies held the line on actual fatalities at about 32,000 a year.

The overall, combined *effectiveness* of the vehicle safety technologies is the percent of potential fatalities that were saved, as shown in the right column of Table 2. The effectiveness grew in *every* year from 1960 to 2002, from a humble 0.40 percent in 1960 to a very substantial 42.81 percent fatality reduction in 2002. Figure 3 charts the trend, showing:

- Not much effect before the FMVSS.
- Steady growth in the early-to-mid 1970's as the early FMVSS phased in.
- A slowdown in 1978-82, when belt use declined prior to national buckle-up campaigns.
- The largest gains came with the buckle-up laws in the mid-to-late 1980's.
- Steady progress since the late 1980's thanks to continued increases in belt use, air bags and other recent FMVSS.

TABLE 2: ACTUAL OCCUPANT FATALITIES, POTENTIAL FATALITIES WITHOUT THE VEHICLE SAFETY TECHNOLOGIES, AND LIVES SAVED IN CARS/LTVs

CAR+LTV OCCUPANT FATALITIES

CY	ACTUAL	W/O SAFETY TECHS.	LIVES SAVED	PERCENT SAVED
1960	28,183	28,298	115	0.40
1961	28,087	28,204	117	0.41
1962	30,544	30,679	135	0.44
1963	32,664	32,823	159	0.49
1964	35,603	35,805	202	0.56
1965	36,518	36,767	249	0.68
1966	39,130	39,465	334	0.85
1967	39,327	39,826	499	1.25
1968	41,019	41,818	799	1.91
1969	42,117	43,273	1,156	2.67
1970	39,556	40,972	1,415	3.45
1971	38,916	40,651	1,735	4.27
1972	40,103	42,281	2,178	5.15
1973	38,739	41,258	2,520	6.11
1974	31,145	33,608	2,463	7.33
1975	31,361	34,355	2,995	8.72
1976	32,222	35,398	3,176	8.97
1977	33,173	36,772	3,599	9.79
1978	34,988	38,951	3,964	10.18
1979	35,108	39,325	4,217	10.72
1980	35,097	39,554	4,456	11.27
1981	33,911	38,284	4,373	11.42
1982	29,855	33,834	3,979	11.76
1983	29,209	33,384	4,176	12.51
1984	30,177	34,935	4,758	13.62
1985	30,044	36,357	6,314	17.37
1986	32,380	40,827	8,447	20.69
1987	33,306	43,203	9,898	22.91
1988	34,217	45,407	11,190	24.64
1989	33,709	45,127	11,418	25.30
1990	32,830	44,470	11,640	26.18
1991	30,928	43,060	12,131	28.17
1992	29,542	41,966	12,424	29.60
1993	30,182	43,917	13,735	31.27
1994	30,979	46,075	15,096	32.76
1995	32,057	48,113	16,056	33.37
1996	32,534	50,289	17,755	35.31
1997	32,501	51,003	18,502	36.28
1998	31,940	51,263	19,323	37.69
1999	32,151	52,038	19,887	38.22
2000	32,234	53,968	21,734	40.27
2001	32,009	54,558	22,548	41.33
2002	32,737	57,242	24,506	42.81
	===========	===========	===========	
	1,443,030	1,769,401	326,371	

FIGURE 2: ACTUAL VS. POTENTIAL CAR/LTV OCCUPANT FATALITIES

("A" = actual fatalities; "P" = potential fatalities without the vehicle safety technologies)

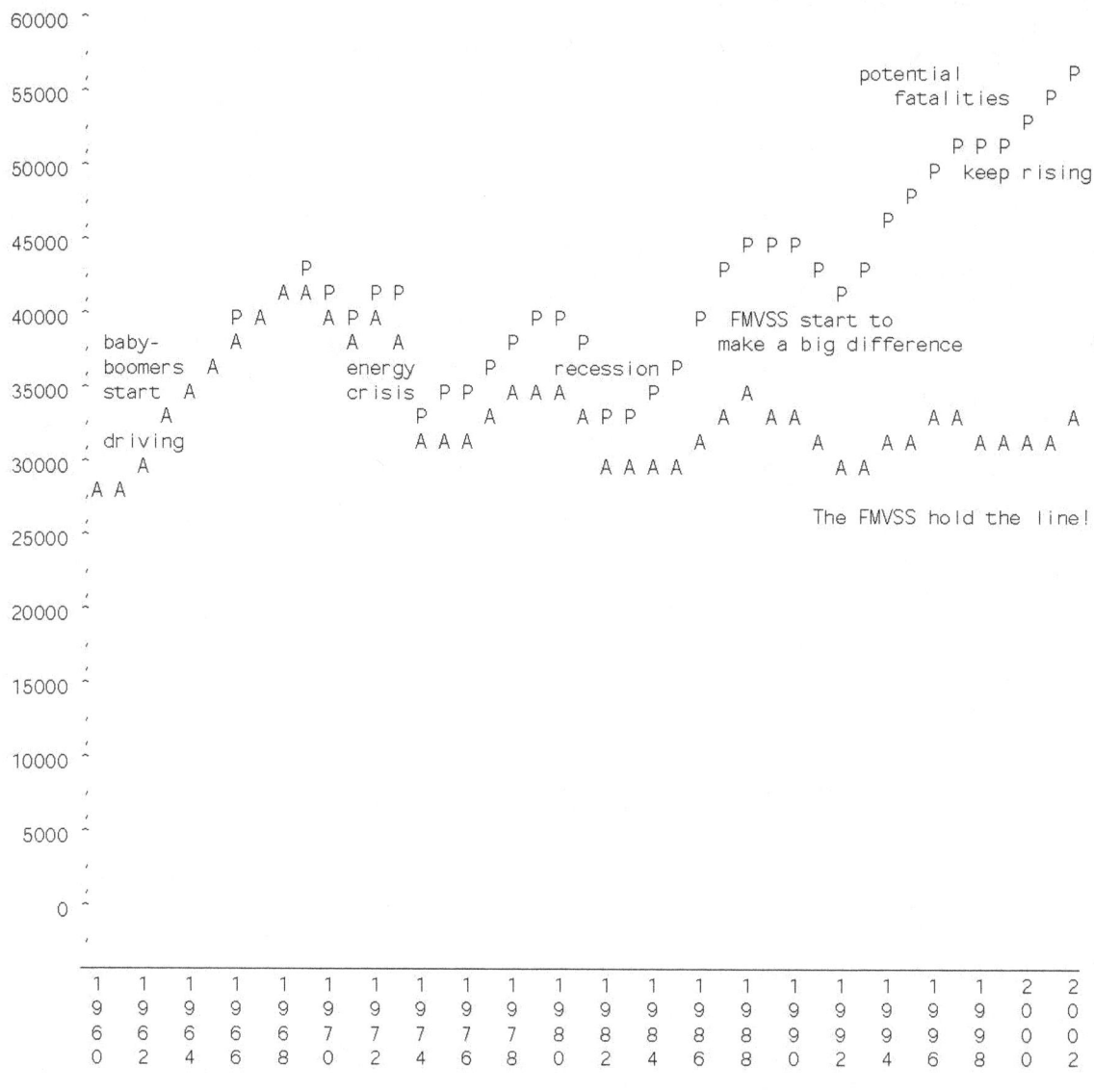

FIGURE 3: PERCENT OF POTENTIAL FATALITIES SAVED BY VEHICLE SAFETY
TECHNOLOGIES, 1960-2002

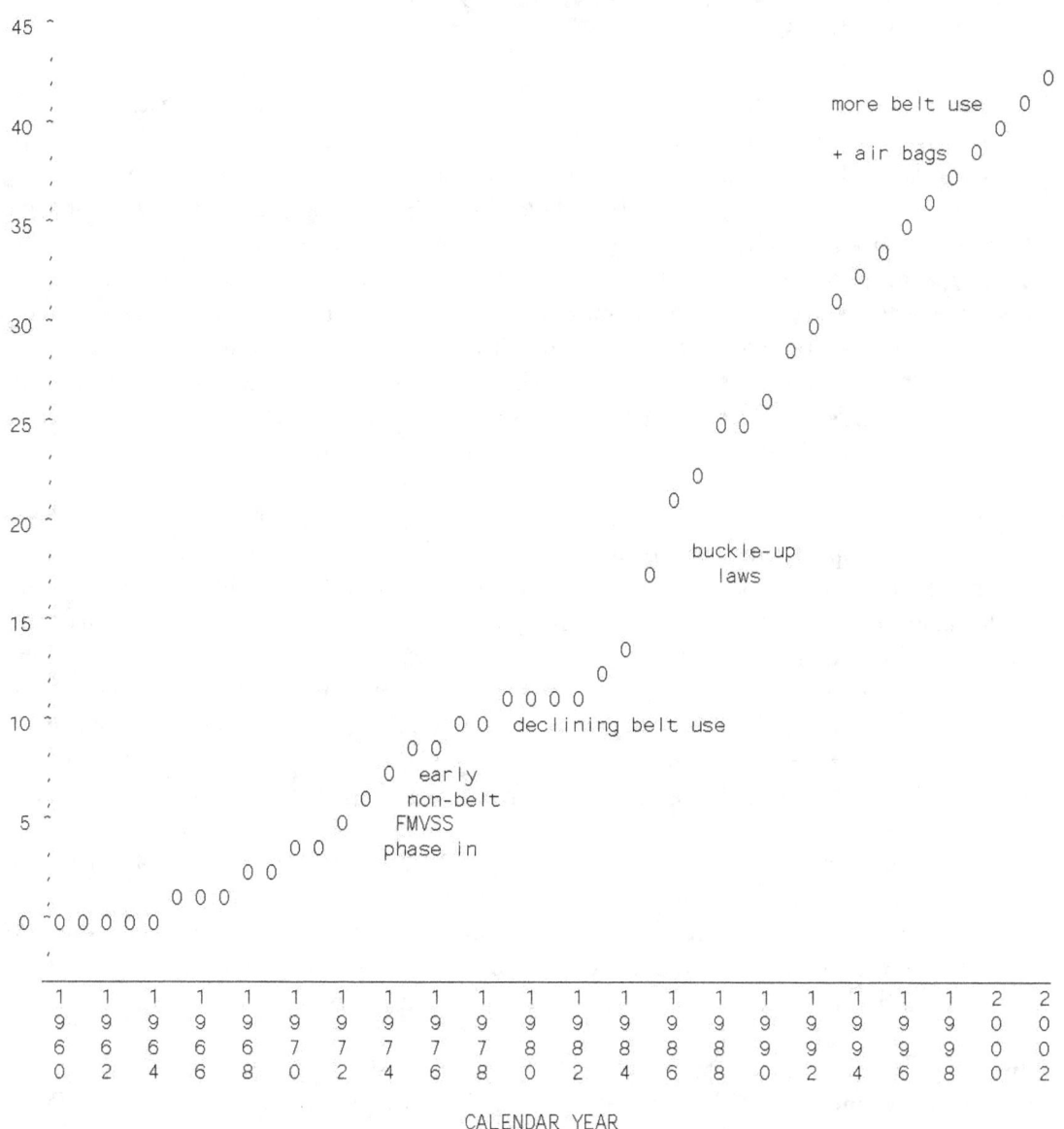

<u>Estimates of lives saved by each FMVSS</u> Car/LTV safety technologies saved an estimated 24,561 lives in 2002, comprising 14,175 car occupants, 10,331 LTV occupants and 56 pedestrians, bicyclists and motorcyclists saved by car/LTV braking improvements. Table 3 shows how many lives were saved by the individual FMVSS – i.e., by the technologies associated with each FMVSS:

- Safety belts are by far the most important occupant protection, saving an estimated 14,570 lives[3]: over half the total of 24,561. The estimate includes safety belts of all types (3-point, lap-only, automatic), at all designated seat positions. Safety belts are designed to keep occupants within the vehicle and close to their original seating position, provide "ride-down" by gradually decelerating the occupant as the vehicle deforms and absorbs energy, and, if possible, prevent occupants from contacting harmful interior surfaces or one another (however, NHTSA recommends correctly installed, age-appropriate safety or booster seats for child passengers until they are at least 8 years old, unless they are at least 4'9" tall). Safety belts are especially important in LTVs, where a large proportion of unrestrained fatalities are ejections and/or rollover crashes; belts saved 6,872 lives in LTVs, nearly two-thirds of the 10,331 LTV occupants saved.

- Frontal air bags saved 2,473 lives in 2002[4], when 63 percent of cars and LTVs on the road were equipped with driver or dual air bags. Benefits can be expected to grow in future years as the on-road fleet approaches 100 percent air bag equipped. Air bags have significant benefits in frontal and partially frontal impacts for nearly all occupants age 13 and older, including the oldest drivers and passengers, by providing energy absorption and ride-down and by preventing head contacts with the windshield or windshield header. Risk from air bags to child passengers age 12 and younger can be eliminated by riding in the back seat, correctly restrained – or by turning off the on-off switch in pickup trucks where children cannot ride in a back seat correctly restrained.

- Energy-absorbing steering assemblies meeting FMVSS 203 and 204 are an important "built-in" safety technology that saved an estimated 2,657 lives in 2002. In the 1960's, they were the first basic protection for drivers in frontal crashes, designed to cushion their impact into the steering assembly. Today, the combination of energy-absorbing columns, safety belts and air bags provides far better protection for the driver in frontal crashes.

- Improvements to door locks, latches and hinges, generally implemented by manufacturers in the 1960's and regulated by industry standards subsequently incorporated into FMVSS 206, saved 1,398 lives in 2002. They reduce the risk of occupant ejection by keeping doors closed in rollover crashes.

---

[3] NHTSA's official estimate in *Traffic Safety Facts 2002 – Occupant Protection*, is 14,164 lives saved by safety belts. This report uses slightly different computational procedures as it estimates the lives saved by all vehicle safety technologies, not just belts, air bags and safety seats.

[4] NHTSA's official estimate in *Traffic Safety Facts 2002 – Occupant Protection*, is 2,248 lives saved by air bags.

TABLE 3: ESTIMATES OF LIVES SAVED BY SAFETY TECHNOLOGIES IN 2002

FMVSS & Safety Technology	Car Occupants	LTV Occupants	Pedestrians Bicyclists Motorcyclists	TOTAL
105: Dual master cylinders & front disc brakes	288	194	56	538
108: Conspicuity tape for heavy trailers	91	68		159
201: Voluntary mid/lower instrument panel improvements	631	299		930
203/204: Energy-absorbing steering assemblies	1,660	997		2,657
206: Improved door locks	704	694		1,398
208: Safety belts – all types, all seat positions	7,699	6,872		14,570 *
208: Frontal air bags	1,642	831		2,473 *
212: Adhesive windshield bonding	229	118		347
213: Child safety seats	223	112		335 *
214: Side door beams & voluntary (pre-1994) TTI(d) reductions	848	146		994
216: Roof crush strength (eliminate true hardtops)	161			161
TOTAL	14,175	10,331	56	24,561

* NHTSA's official estimates of lives saved by safety belts (14,164), air bags (2,248) and child restraints (376), published in *Traffic Safety Facts 2002 – Occupant Protection*, were computed by a similar method. This report's estimates are not identical; slightly different computational procedures were used to estimate the lives saved by all vehicle safety technologies, not just belts, air bags and safety seats.

- Side door beams in cars and LTVs meeting the original static crush test of FMVSS 214, plus modifications to 2-door cars in anticipation of the dynamic test requirement later added to FMVSS 214 saved 994 lives in side impacts. Side door beams are primarily effective in side impacts with fixed objects, such as trees or poles.

- Improvements to mid- and lower instrument panels, not actually required by FMVSS 201 but historically and functionally associated with that standard to some extent, saved an estimated 930 right-front passengers in frontal crashes in 2002. Instrument panels were redesigned, using energy-absorbing materials, to decelerate occupants at a safe rate and keep them in an upright position during frontal crashes.

- Car/LTV braking improvements directly or indirectly associated with FMVSS 105 include dual master cylinders and front disc brakes. By eliminating brake failure or helping cars and LTVs stop more effectively, they saved 538 lives in 2002, including 56 pedestrians, bicyclists or motorcyclists

- Adhesive windshield bonding saved 347 lives in 2002 by keeping the windshield attached to the vehicle in severe impacts and preventing occupant ejection via the windshield portal. FMVSS 212 regulates windshield mounting.

- Child safety seats meeting FMVSS 213 saved an estimated 335 young passengers in 2002.[1] Child safety seats are the basic protection system for passengers who are too small to obtain full benefits from safety belts. Newborns should start with rear-facing infant seats, graduate to forward-facing toddler seats, booster seats and, finally, when they are at least 9 years old or 4'9" tall, to adult safety belts.

- FMVSS 216, roof crush strength, is associated with the redesign of true hardtops as pillared hardtops or sedans during the 1970's. True hardtops had no B-pillars to support the roof, making it more susceptible to crush in a rollover. If cars were still built that way there would have been 161 additional fatalities in 2002.

- FMVSS 108 requires red-and-white conspicuity tape on heavy truck trailers. The tape reflects another vehicle's headlights strongly and it is highly visible in the dark. Although this device is furnished on heavy trailers, not cars or LTVs, it is the occupants of cars and LTVs who primarily benefit by avoiding collisions with the trailers. The tape saved an estimated 159 car and LTV occupants in 2002.

Table 4 shows cumulative lives saved from 1960 through 2002: 232,255 car occupants and 94,117 LTV occupants, plus 2,180 pedestrians, bicyclists and motorcyclists saved by car/LTV braking improvements, for an estimated total of 328,551. Safety belts (168,524) accounted for more than half the total. Air bags, one of the most recent technologies, had saved 12,074 lives by the end of 2002, and child safety seats, 5,954. The "built in" non-belt technologies regulated by the remaining nine FMVSS in Table 4 (105, 108, 201, 203/204, 206, 212, 214 and 216) add

---

[1] NHTSA's official estimate in *Traffic Safety Facts 2002 – Occupant Protection*, is 376 lives saved by child restraints.

TABLE 4: ESTIMATES OF LIVES SAVED BY SAFETY TECHNOLOGIES IN 1960-2002

FMVSS & Safety Technology	Car Occupants	LTV Occupants	Pedestrians Bicyclists Motorcyclists	TOTAL
105: Dual master cylinders & front disc brakes	7,993	2,880	2,180	13,053
108: Conspicuity tape for heavy trailers	683	422		1,105
201: Voluntary mid/lower instrument panel improvements	16,670	4,373		21,043
203/204: Energy-absorbing steering assemblies	41,545	11,472		53,017
206: Improved door locks	19,504	9,398		28,902
208: Safety belts – all types, all seat positions	109,519	59,004		168,524
208: Frontal air bags	8,770	3,304		12,074
212: Adhesive windshield bonding	5,248	1,462		6,710
213: Child safety seats	4,854	1,100		5,954
214: Side door beams & voluntary (pre-1994) TTI(d) reductions	14,002	701		14,703
216: Roof crush strength (eliminate true hardtops)	3,466	—	—	3,466
TOTAL	232,255	94,117	2,180	328,551

up to 142,000 lives saved; energy-absorbing steering assemblies, improved door locks, and voluntary instrument panel improvements saved the most lives.

<u>Comments on some assumptions in the model</u> The effectiveness estimates used in the model derive from past NHTSA evaluations. Estimates were based on statistical analyses of crash data, comparing fatality risk in vehicles built just before and just after make-models were equipped with the technology; the reductions were statistically significant, and the analyses attempted to statistically control for factors other than the technology by using double-pair comparison, control groups, logistic regression, or other techniques. However, in the preparation of this report, the estimates in past evaluations were generally not updated with data that subsequently had become available.

The basic assumption of the model is that FARS fatality cases with a safety technology are evidence of additional crashes where that technology saved lives: if there are 100 belted fatalities in a type of crash where statistical analysis shows 50 percent belt effectiveness, there must have been another 100 people in potentially fatal crashes who were saved by the belt. This is a leap of faith to the extent that we cannot identify those 100 specific crashes were occupants were "saved by the belt" – we assume they must exist, based on our effectiveness estimate.

The model simulates "removing" safety equipment from a modern vehicle one piece at a time, starting with the most recent technology and working backward. Some of these technologies were introduced at about the same time, and it is not always obvious which was first: for some of the earliest ones, there is limited written information, and the people who worked on them have long since retired. A case could be made for changing the order of "removing" the technologies. The model would still produce the same estimate of overall lives saved, but the allocation among the FMVSS could change.

The model assumes that the belt use of *fatally injured* occupants (not survivors) on FARS is accurately reported. NHTSA has long believed this to be true, based on statistical analyses comparing FARS data with belt use observed in surveys. In the future, conceivably, event data recorders could provide more direct evidence on belt use in crash data files.

Finally, when the model says vehicle safety technology saved 328,551 lives, it means there would have been that many additional fatalities in 1960-2002 if everything else had stayed the same: the same increase in VMT from 1960-2002, the same driving behaviors. It is somewhat of a paper estimate. If safety belts and the other modern vehicle safety technologies had never been invented, and if occupant fatalities had continued climbing toward 57,000 instead of remaining near 32,000, as shown in Table 2, the public might have demanded much stronger regulation of drivers (e.g., licensing) or the infrastructure (e.g., speed limits). Consumers might purchase a different mix of vehicles (e.g., larger cars) and some people might be more reluctant to travel during the riskiest hours (e.g., weekend nights). Those measures might have prevented at least some of the additional 328,551 fatalities – but surely not as efficiently, and with as little impairment of driving enjoyment and mobility as the vehicle safety technologies.

LIVES SAVED BY THE FEDERAL MOTOR VEHICLE SAFETY STANDARDS AND OTHER VEHICLE SAFETY TECHNOLOGIES, 1960-2002

FRAMEWORK FOR THE ANALYSIS

The National Highway Traffic Safety Administration (NHTSA) began to evaluate the effectiveness of the Federal Motor Vehicle Safety Standards (FMVSS) in 1975, well before Executive Order 12291 (February 1981), Executive Order 12866 (October 1993) or the Government Performance and Results Act of 1993 required Federal agencies to evaluate their existing regulations. By October 2004, NHTSA had issued 48 evaluations of individual safety standards, programs or technologies.[1]

A typical evaluation estimates the *effectiveness* of a safety technology – a percentage reduction of fatalities, injuries and/or crashes – by statistically analyzing crash data on vehicles produced just before vs. just after receiving the technology. It estimates the *benefits* of that technology – absolute numbers of lives saved, injuries avoided, or crashes avoided per year – by applying effectiveness estimates to baseline numbers of annual fatalities, injuries or crashes. "Baselines" have typically been the year that a report was written.

NHTSA has evaluated all the major crashworthiness standards and several crash avoidance standards for passenger cars and LTVs (light trucks and vans – i.e., pickup trucks, sport utility vehicles, minivans and full-size vans). The agency has studied consumer information on vehicle safety such as the New Car Assessment Program and safety technologies that are not mandatory for cars under Federal regulations, such as the redesign of mid and lower instrument panels.

By now, the agency has evaluated virtually all the life-saving technologies introduced in cars and LTVs from about 1960 up through the later 1990's. Having estimated the lives saved by each individual standard, we are now ready to assess the overall effect of vehicle safety improvements by essentially adding up the individual estimates. "Building up an estimate one standard at a time" is the most empirical and defensible way to estimate how many lives are saved by all the vehicle safety technologies. It is preferable to a complex statistical analysis of the long-term reduction in overall fatality rates per 100,000,000 vehicle miles of travel (VMT) that attempts to tease out the relative effects of vehicle, behavioral, roadway and demographic factors.

Estimating the combined net lives saved by the vehicle safety technologies, as well as the lives saved by each individual technology in each year, has been a long-term goal for NHTSA. The agency's evaluation plan of 1998 proposed to start it in the near future, and the current evaluation plan of 2004-2007 says it is underway and describes the proposed approach.[2]

---

[1] "Executive Order 12291 – Federal Regulation," *Federal Register* 46 (February 19, 1981): 13193; "Executive Order 12866 – Regulatory Planning and Review," *Federal Register* 58 (October 4, 1993): 51735; *Government Performance and Results Act of 1993*, Public Law 103-62, August 3, 1993; *National Highway Traffic Safety Administration Evaluation Program Plan, Calendar Years 2004-2007*, NHTSA Report No. DOT HS 809 699, Washington, 2004, pp. 37-47; www.nhtsa.dot.gov/cars/rules/regrev/evaluate/Summaries.html .

[2] *Evaluation Program Plan, 1998-2002*, NHTSA Report No. DOT HS 808 709, Washington, 1998, p. 12; *NHTSA Evaluation Program Plan, Calendar Years 2004-2007*, pp. 2-3.

Basic analysis method

We will rely on the individual effectiveness estimates (percentage reductions) developed in past NHTSA evaluations. But it is not as simple as merely adding up past reports' estimates of lives saved per year. The absolute estimates in the various reports are not directly comparable, and are no longer accurate today, because they involve many different, past baselines.

Instead, a process is needed that applies the effectiveness estimates in a consistent manner and to a consistent "baseline" number of fatalities. The Fatality Analysis Reporting System (FARS) data serve as the starting point, indicating the actual number of fatalities during 1975-2002 in the fleet of cars and LTVs that was on the road. It is a mix of vehicles, some built recently and meeting many of the FMVSS, others quite old and pre-FMVSS.

Each 100 actual fatality cases on FARS represent a potentially even greater number of fatalities that could have happened if the vehicles had not met any of the FMVSS. The process begins with the actual FARS fatality cases and computes how many additional fatalities there would have been if the vehicles had not been equipped with any safety technologies. The computations rely on the effectiveness estimates from past evaluations. For example, given that 3-point belts reduce fatality risk by 45 percent in cars, 100 belted FARS fatality cases are equivalent to $100/(1 - .45) = 182$ fatalities without belts – i.e., there must have been 182 belted occupants involved in crashes that would have been potentially fatal without belts, but 82 of them did not become FARS cases, because the belts saved the occupant's life. The process is repeated for other FMVSS and safety technologies until all of them have been "removed" from the vehicle – until the vehicle has been degraded to a level of safety performance characteristic of the 1950's rather than its actual model year. The technologies are removed in the reverse chronological order that they were historically introduced into vehicles. At each step into the past, the model tallies the lives saved by the latest safety technology – i.e., the additional fatalities that would have occurred if that technology had been removed. This is the process that NHTSA already uses to estimate the number of lives saved by air bags and safety belts, but expanded to also count the benefits of the other FMVSS.[3] "Reverse chronological order" is not the only approach that could have been used in the model; alternative approaches are considered in Part 2 of this report (Summary of the Estimation Method). However, the various techniques would have generated the same estimate of overall lives saved in 1960-2002, differing only in how they allocated that total among the individual safety technologies.

The model produces unbiased estimates of the lives saved by the various technologies, and it is not an exercise in double counting, because the effectiveness estimates in past evaluations are based on analyses of vehicles produced *just after* vs. *just before* the FMVSS in question. They estimate the *incremental* effect of that FMVSS on a vehicle that already meets all the earlier FMVSS. For example, NHTSA's evaluation of air bags was a study of cars, some without air bags and some with air bags, but all equipped with 3-point belts and energy-absorbing columns. The evaluation of 3-point belts was based on older cars equipped with energy-absorbing columns but not yet with air bags. The evaluation of energy-absorbing columns was based on even older cars without air bags or 3-point belts. These effectiveness estimates are incremental, and they

---

[3] *Traffic Safety Facts 2002 – Occupant Protection*, NHTSA Publication No. DOT HS 809 610, Washington, 2003.

may be applied in sequence to estimate the total fatality reduction for the combination of the three technologies.[4]

What is included and what is excluded?

This will be a study of the lives saved in 1960-2002 by vehicle safety technologies that were implemented in cars or LTVs from approximately 1960 until the later 1990's, or that were implemented in other vehicles but benefited occupants of cars and LTVs. The short explanation for limiting the study to vehicle safety technologies in general, and to these vehicles and timeframe in particular, is that they are the technologies that have been evaluated by NHTSA[5] – inclusively enough to add up the lives saved by the individual technologies and say, "Here is the net impact of the vehicle program."

The benefits of roadway improvements or behavioral safety programs, such as the effort to prevent drunk driving are not explicitly included here. One exception: the benefits of safety belts, a vehicle technology, would not be anywhere near what they are today without all the behavioral programs that have increased belt use; the benefits of the behavioral "occupant protection program" are implicitly and inseparably part of the benefit of safety belts. Unlike the FMVSS, there are generally no easy statistical methods to estimate the effectiveness of individual behavioral or roadway programs. NHTSA does not have a comprehensive set of effectiveness estimates for behavioral or roadway programs, based directly on statistical analysis of crash data, corresponding to what it has for the vehicle programs.

NHTSA has thoroughly evaluated the life-saving benefits of FMVSS and other technologies for passenger cars. The set of estimates for LTVs is almost as complete, and where there are some gaps, estimates can in most cases be plausibly inferred from the results for cars. The list of evaluations for motorcycles, heavy trucks and buses is not so complete (although this is a future priority for NHTSA).

The timeframe of vehicle technologies is as up-to-date as feasible. However, some of the rules or technologies introduced in the mid-to-late 1990's, and anything more recent than that cannot be included because NHTSA is only now acquiring, or has not yet acquired enough crash data to evaluate their effectiveness in production vehicles.

For the beginning of the timeframe, it seems most logical to start with the technologies more or less directly regulated by the initial FMVSS of January 1, 1968. Many of these technologies, however, were actually introduced some years before 1968. For a full picture of the benefits of the FMVSS-era technologies, it makes sense to take the analysis back to 1960, as long as we keep separate accounts of lives saved in pre-standard and post-standard vehicles. The oldest technologies more or less directly regulated by the initial FMVSS include lap belts, introduced in

---

[4] Kahane (1996), pp. 7-9; Kahane, C.J., *Fatality Reduction by Safety Belts for Front-Seat Occupants of Cars and Light Trucks*, NHTSA Technical Report No. DOT HS 809 199, Washington, 2000, pp. 5-10; Kahane, C.J., *An Evaluation of Federal Motor Vehicle Safety Standards for Passenger Car Steering Assemblies*, NHTSA Technical Report No. DOT HS 805 705, Washington, 1981, pp. 197-203.

[5] *NHTSA Evaluation Program Plan, Calendar Years 2004-2007*, pp. 37-47; www.nhtsa.dot.gov/cars/rules/regrev/evaluate/Summaries.html .

the late 1950's and early 1960's; improvements to door locks throughout the 1960's; and many effective devices introduced in 1965-67. There does not seem to be much point in going back before 1960 or attempting to predict how many fatalities there would be today if cars still had, say, 1905 technology; in any case, NHTSA has not evaluated safety improvements that long precede the FMVSS era, such as enclosed, metal car bodies, hydraulic brakes, safety glass, or electric headlamps.

One feature of the estimation model in Part 2 of this report is that estimates for the later technologies (such as air bags or 3-point belts) are unaffected by the inclusion or exclusion of any technology that preceded them. Thus, readers have the option of just subtracting the benefits for the earliest technologies on the list (e.g., lap belts or the 1960's improvements to door locks) if, in their opinion, they ought not have been included in this report.

NHTSA is also preparing a summary report of its past evaluations of the cost of individual FMVSS, and estimates of the cost per car or LTV of all the FMVSS, by model year.[6] It will be a companion to this report. A supplement to these two reports will compare overall lives saved and costs on a substantial "core" group of FMVSS for which NHTSA has evaluated effectiveness as well as costs.

List of FMVSS, safety technologies and effectiveness evaluations

Part 1 of this report is a review of 19 FMVSS, plus the New Car Assessment Program (NCAP) that provides consumers with information about vehicle safety performance. Part 1 is grouped into 16 chapters. These FMVSS either regulate cars and/or LTVs, or they regulate other vehicles/equipment but result in benefits to occupants of cars and LTVs. Part 1 reviews 31 individual safety technologies directly or indirectly associated with FMVSS/NCAP that NHTSA has evaluated based on statistical comparisons of the crash experience of vehicles built before and after the introduction of those technologies.

Historically, the FMVSS have had numbers. The 100-series are crash avoidance standards; the 200-series, crashworthiness; and the 300-series, post-crash fire prevention. Within each series, the numbering is chronological.

Each chapter of Part 1 presents the rationale for a FMVSS, the safety problem it addresses, and its regulatory history, including major *Federal Register* citations. For each safety technology, Part 1 summarizes what was added or changed in vehicles and how this equipment works, when it was introduced and by whom, and why it was expected to reduce fatalities, injuries or crashes. The data and statistical methods of NHTSA's evaluations are summarized, with examples if possible, and so are the principal findings on effectiveness, benefits, and side effects (if any). For technologies that require some action by drivers or other occupants (e.g., safety belts, on-off switches for air bags, ABS, head restraints), Part 1 describes how to use them most effectively.

---

[6] *NHTSA Evaluation Program Plan, Calendar Years 2004-2007*, pp. 3-4.

Table 1-1 lists the safety technologies reviewed in Part 1, grouped by chapter (FMVSS). It summarizes the effectiveness of each technology in reducing fatalities, injuries or crashes (100-series only) of cars and LTVs:

- Yes = NHTSA's evaluation found a statistically significant reduction
- No = the evaluation did not find a significant reduction, despite ample data.
- (Yes), (No) = LTV results inferred by analogy from the passenger car results
- Mixed results = significant reduction on some crash types, significant increase on others.
- Unknown = NHTSA has not performed an evaluation
- N/A = the safety technology was not installed on this type of vehicle.

NHTSA's evaluations demonstrated significant benefits of some type – if not a fatality reduction then at least a reduction of injuries, crashes or fires – for 28 of the 31 technologies: all except rear window defoggers, seat back locks and glass-plastic windshields (and of these three, only seat back locks are required by a FMVSS).

Almost all the effectiveness estimates in Part 1 may be found in published NHTSA reports. Those NHTSA publications are cited in footnotes and listed in the References at the end of this report. Moreover, Appendix B provides capsule summaries of 48 evaluations published as of October 2004, in chronological order. The only new analyses performed especially for this report are: fatality reduction by front-seat lap belts, injury reduction by air bags, and fatality reduction for riding in the back seat.

For many technologies, NHTSA has published only one evaluation, statistically comparing vehicles just before vs. just after that technology was introduced, based on crash data for a number of calendar years after introduction. The report assumes the effectiveness, in specific crash types, has stayed about the same in subsequent years. It can be difficult to estimate the effectiveness of a technology in later model-year vehicles. However, NHTSA has completed or plans follow-up evaluations of:

- Technologies that have changed over the years, such as frontal air bags or child safety seats.
- Front-seat 3-point belts, because of their exceptional importance.
- Crash avoidance technologies whose effectiveness may change over time as a result of drivers' response, such as ABS or Center High Mounted Stop Lamps.

TABLE 1-1: SAFETY TECHNOLOGIES EVALUATED BY NHTSA

FMVSS / SAFETY TECHNOLOGY	Effectiveness					
	Cars			LTVs		
	Fatals	Injuries	Crashes	Fatals	Injuries	Crashes
103: Windshield defrosting and defogging						
Rear-window defoggers	Unknown	Unknown	No	Unknown	Unknown	Unknown
105: Hydraulic brake systems						
Dual master cylinders	Yes	Yes	Yes	Yes	Yes	Yes
Front disc brakes	Yes	Yes	Yes	Yes	Yes	Yes
Rear-wheel ABS for LTVs		N/A			Mixed results	
4-wheel antilock brake systems (ABS)		Mixed results			Mixed results	
Electronic stability control (ESC)[1]	Yes	(Yes)	Yes	Yes	(Yes)	Yes
108: Lamps, reflective devices						
Side marker lamps	No	Yes	Yes	No	Yes	Yes
Center High Mounted Stop Lamps	No	Yes	Yes	No	Yes	Yes
Retroreflective tape on heavy trailers[2]	Yes	Yes	Yes	Yes	Yes	Yes

---

[1] Preliminary evaluation; injury reduction inferred from the crash reduction and the fatality reduction.
[2] Tape installed on heavy trailers is effective in preventing cars and LTVs from hitting those trailers.

TABLE 1-1 (continued): SAFETY TECHNOLOGIES EVALUATED BY NHTSA

FMVSS	SAFETY TECHNOLOGY	Effectiveness			
		Cars		LTVs	
		Fatals	Injuries	Fatals	Injuries
201: Occupant protection in interior impact					
	Voluntary mid and lower instrument panel improvements	Yes	Yes	Yes	Yes
202: Head restraints					
	Head restraints for front-outboard occupants	No	Yes	(No)[3]	Yes
203: Impact protection from the steering control					
204: Steering control rearward displacement					
	Energy-absorbing steering assemblies	Yes	Yes	Yes	Unknown
205: Glazing materials					
	High-penetration resistant (HPR) windshields	No	Yes	(No)	(Yes)
	Glass-plastic windshields	No	No	N/A	N/A
206: Door locks					
	Improved locks, latches and hinges for side doors[4]	Yes	(Yes)	(Yes)	(Yes)

---

[3] Result for LTVs inferred from the evaluation of cars.
[4] Injury reduction and LTV fatality reduction inferred from the fatality reduction in cars.

TABLE 1-1 (continued): SAFETY TECHNOLOGIES EVALUATED BY NHTSA

FMVSS	SAFETY TECHNOLOGY	Effectiveness			
		Cars		LTVs	
		Fatals	Injuries	Fatals	Injuries
207: Seating systems	Seatback locks for 2-door cars	No	No	N/A	N/A
208: Occupant crash protection					
209: Seat belt assemblies					
210: Seat belt assembly anchorages	Lap belts for front-seat occupants	Yes	Yes	Yes	(Yes)[5]
	Lap belts for back-seat occupants	Yes	Yes	Yes	(Yes)
	Manual 3-point belts for front-seat occupants	Yes	Yes	Yes	(Yes)
	3-point belts for back-seat occupants	Yes	Yes	Yes	(Yes)
	Automatic safety belts	Yes	Unknown	N/A	N/A
	Frontal air bags	Yes[6]	Yes	Yes	Yes
	On-off switches for passenger air bags in pickup trucks	N/A	N/A	Yes	Unknown

[5] "(Yes)" indicates result for LTV injuries inferred from the results for passenger car injuries and LTV fatalities.
[6] Except that fatalities increased for certain groups of child passengers in cars and LTVs.

8

TABLE 1-1 (continued): SAFETY TECHNOLOGIES EVALUATED BY NHTSA

		Effectiveness			
		Cars		LTVs	
FMVSS	SAFETY TECHNOLOGY	Fatals	Injuries	Fatals	Injuries
212: Windshield mounting					
	Adhesive windshield bonding	Yes	Yes	(Yes)[7]	(Yes)
213: Child restraint systems					
225: Child restraint anchorage systems					
	Child safety seats	Yes	Yes	Yes	Unknown
	Riding in the back seat	Yes	Unknown	Unknown	Unknown
214: Side impact protection					
	Side door beams	Yes	Yes	Yes	Unknown
	Voluntary (pre-1994) TTI(d) improvements in 2-door cars	Yes	Unknown	N/A	N/A
216: Roof crush resistance					
	Redesign of true hardtops as pillared hardtops or sedans	Yes	Unknown	N/A	N/A

---

[7] "(Yes)" indicates results for LTVs inferred from corresponding results for cars.

9

TABLE 1-1 (concluded): SAFETY TECHNOLOGIES EVALUATED BY NHTSA

		Effectiveness			
		Cars		LTVs	
FMVSS	SAFETY TECHNOLOGY	Fatals	Injuries	Fatals	Injuries
301: Fuel system integrity					
	1975-77 upgrade: rollover, rear- and lateral-impact tests	No	Yes[8]	No	No
NCAP: New Car Assessment Program					
	Frontal NCAP-related improvements, vehicles w/o air bags	Yes	Unknown	Unknown	Unknown

---

[8] Significant reduction of crashes with post-crash fires.

Part 1 also summarizes the vehicle modifications and rationale for 12 safety technologies that are already available in production vehicles cars and LTVs, but are still too new for NHTSA to have completed an evaluation based on statistical analysis of crash data. The agency's evaluation plan of 2004-2007[1] outlines these ongoing and future studies:

- FMVSS 201 – head impact protection upgrade
- Head air bags
- Pretensioners and load limiters for safety belts
- 1998-99 redesign of air bags
- Advanced air bags
- Upper tethers and anchorages for child safety seats
- LATCH (lower anchors and tethers for children)
- FMVSS 214 – side impact protection, dynamic test standard for passenger cars
- Side air bags
- Frontal NCAP-related modifications in vehicles with air bags
- Side NCAP-related modifications
- Rollover resistance NCAP-related modifications

The agency's evaluation plan of 2004-2007 also outlines two ongoing, but not completed studies of safety technologies for heavy trucks that will likely prevent collisions with cars/LTVs or reduce injury risk to the car/LTV occupants (but the technologies are not discussed in Part 1 of this report):

- FMVSS 121 – Antilock Brake Systems for vehicles with air brakes
- FMVSS 223/224 – Rear impact guards for heavy trailers

In addition to the 19 FMVSS listed in Table 1-1, there are 24 other FMVSS (in effect as of April 2004) that regulate new cars, new LTVs, or car/LTV components that have not been evaluated by NHTSA. The following FMVSS definitely or quite possibly resulted in tangible changes to vehicles, but were not evaluated because existing or potentially available data do not adequately identify what vehicles were modified; or because the type of crashes/injuries mitigated by the FMVSS cannot be singled out in available data (or cannot be identified at all); or because there is no hope of obtaining enough data for a statistically meaningful analysis of the limited effect, if any, that could reasonably be expected for that FMVSS:

- 114: Theft protection
- 116: Motor vehicle brake fluids
- 118: Power-operated windows
- 124: Accelerator control systems
- 125: Warning devices
- 129: New non-pneumatic tires for passenger cars
- 219: Windshield zone intrusion
- 302: Flammability of interior materials

---

[1] *NHTSA Evaluation Program Plan, Calendar Years 2004-2007, op. cit.*

The following standards have not been evaluated even though they may regulate vehicle subsystems that are important for safety (e.g., tires, mirrors). The agency believes they by and large did not result in extensive changes to those subsystems, or does not know if they have resulted in changes. In many cases, the FMVSS may have largely incorporated other organizations' standards or industry-wide practices that vehicles had already been meeting for quite some time before 1968:

- 101: Controls and displays
- 102: Transmission shift lever sequence
- 104: Windshield wiping and washing systems
- 106: Brake hoses
- 109: New pneumatic tires
- 110: Tire selection and rims
- 111: Rearview mirrors
- 113: Hood latch systems
- 117: Retreaded pneumatic tires
- 119: New pneumatic tires for vehicles other than passenger cars
- 120: Tire selection and rims for vehicles other than passenger cars

The following standards went into effect recently. NHTSA has not yet begun an evaluation, or proposed one in its 2004-2007 evaluation plan:

- 135: Passenger car brake systems
- 303: Fuel system integrity of compressed natural gas vehicles
- 304: Compressed natural gas fuel container integrity
- 305: Electric-powered vehicles: electrolyte spillage and electrical shock protection
- 401: Interior trunk release

Part 1 ends with Tables 1-2, 1-3 and 1-4 summarizing the effectiveness of safety technologies: their estimated percentage reductions in fatalities, nonfatal injuries and crashes (always specifying to what group of crashes/injuries these percentages apply). Tables 1-3 and 1-4 also summarize the estimated annual benefits of nonfatal injury reduction and crash avoidance, as stated in NHTSA's evaluation reports. However, those estimates of benefits, in general, are not directly comparable and do not add up to the overall crash avoidance and injury reduction by all the FMVSS in, say, 2002. NHTSA does not have enough "building blocks" to develop models for overall crash avoidance and injury reduction comparable to the analysis of fatal crashes in Part 2 of this report.

Estimating lives saved by safety technologies, 1960-2002

Part 2 of this report focuses on the safety technologies that have significantly reduced fatality risk. The individual effectiveness estimates and the basic analysis method, described above, are applied to FARS data to estimate how many lives were saved in 1960-2002. The tables in Part 2 also estimate how many lives were saved:

- In each calendar year
- By each individual safety technology, and grouped by FMVSS
- By vehicle type:
  - Car occupants
  - LTV occupants
  - Pedestrians and motorcyclists saved by car/LTV braking improvements
- Distinguishing between lives saved by a FMVSS in effect and "voluntary" saves such as:
  - Improvements introduced before the effective date of a FMVSS
  - Technologies not required for meeting any FMVSS, although perhaps indirectly associated with a FMVSS because they address the same general safety problem

Part 2 compares the actual number of fatalities in 1960-2002, or in any specific year to the number that potentially would have occurred, given the same driving exposure, if none of the cars and LTVs had been equipped with any of the safety technologies. It computes the percentage of the potential fatalities that were saved by the technologies. Part 2 also compares the trends in fatalities per 100,000,000 VMT – with and without the vehicle safety technologies.

Every life-saving technology in Table 1-1 is included in Part 2 except electronic stability control (ESC) systems and "riding in the back seat." ESC systems began to appear in the late 1990's, and NHTSA has so far only conducted a preliminary evaluation. The back seat is not a new technology of the FMVSS era, but has been available since the early days of the automobile. The FMVSS 213 chapter of Part 1 only discusses the fatality reduction for riding in the back seat because it is highly relevant to the child passenger safety issue. Just as Part 2 does not count lives saved by the recent shift from front to back seats among the "benefits of the FMVSS," it does not count the effects of other market shifts between existing vehicle types, such as between:

- 2-door cars and 4-door cars
- Large cars and small cars
- Passenger cars and LTVs

While these shifts can and do affect the number of fatalities, they cannot be considered benefits of new safety technologies of the FMVSS era. Part 2 considers only their implicit effects on the year-to-year changes in actual and potential fatalities.

PART 1

*REVIEW OF 19 FEDERAL MOTOR VEHICLE SAFETY STANDARDS
AND THE NEW CAR ASSESSMENT PROGRAM*

COMPRISING 43 SAFETY TECHNOLOGIES

*AND THEIR EFFECTIVENESS IN REDUCING FATALITIES, INJURIES AND
CRASHES FOR PASSENGER CARS AND LIGHT TRUCKS*

Following the review of the FMVSS and NCAP, Tables 1-2, 1-3 and 1-4 summarize the fatality-, injury- and crash-reducing effectiveness (percentage reductions) of the safety technologies. Tables 1-3 and 1-4 also summarize annual benefits of individual technologies: injuries and crashes avoided per year. Part 2 of this report estimates the annual fatality reduction, for each technology individually and for all of them together.

FMVSS 103: Windshield defrosting and defogging systems

A vehicle modification whose safety benefits have been evaluated by NHTSA is grouped with this standard merely because the functions are similar:

- *Rear window defrosting and defogging systems*

FMVSS 103 only regulates windshield defrosting and defogging. One of NHTSA's initial safety standards, effective on January 1, 1968, it required passenger cars and SUVs to have windshield defroster/defoggers and it set performance requirements for them, incorporating SAE Recommended Practices dating back to 1964. Cars and LTVs had windshield defroster/defoggers well before 1968. They remained unchanged during the mid-to-late 1960's.[1]

FMVSS 103 has never required, or proposed to require rear-window defoggers. Their development has been voluntary on the part of the industry, in response to customer demand. Drivers obviously want a clear rear window, and they like a device that clears it for them automatically, so they don't have to wipe or scrape it repeatedly.

History of rear window defoggers *Ward's Automotive Yearbooks* (Ward's Publications, Detroit) began to include rear window defoggers in MY 1973 among their statistics for factory-installed equipment in domestic cars by make-model. In that year, 16 percent of new cars were equipped with them; presumably, they were offered in smaller numbers some years before that. Installations grew steadily in the 1970's and 1980's. By 1992, over 90 percent of new cars were equipped with them. By MY 2001, they were standard equipment on most cars, SUVs and minivans, but not pickup trucks or full-size vans; 94 percent of new cars were equipped with them.[2]

How they work Current rear-window defoggers are grids of electric wires attached to the rear window. The wires are thin enough not to obstruct vision. Controlled by a switch on the instrument panel, they heat up to evaporate condensation or melt ice and snow. The switch automatically turns off after a certain number of minutes, in order to save wear and tear on the system, and the driver has to turn it back on if the window is not clear. During the 1970's some defoggers consisted of an electric heater and blower-motor. That type was gone by 1982.

Two things must happen for rear-window defoggers to "work": (1) There has to be some environmental factor such as rain, snow or cold that fogs or ices up the window; (2) The driver has to be aware of the problem and turn on the switch. In warm, dry, sunny conditions, windows are normally clear and defoggers are not needed. The situations that might put condensation, snow or ice on the window include any kind of precipitation; early morning hours when water vapor in the outside air condenses as dew; and very cold weather that can make water vapor inside the vehicle condense on windows. Whereas rear-window defoggers rapidly dispel condensation, they cannot melt large amounts of ice, but they make it easier to scrape off.

---

[1] *Code of Federal Regulations*, Title 49, Government Printing Office, Washington, 2002, Part 571.103; *Federal Register* 32 (February 3, 1967): 2414.

[2] Morgan, C., *Evaluation of Rear Window Defrosting and Defogging Systems*, NHTSA Technical Report No. DOT HS 809 724, Washington, 2004, pp. 1-4.

Expected benefits Rear-window defoggers might help prevent collisions by allowing a driver to see through the back window, either directly or via the inside rear-view mirror – but only when there are environmental conditions that are fogging or icing up the windows, and when the driver has activated the defoggers. Specifically, there are two maneuvers where a driver must know, or at least ought to know what is behind the vehicle. The maneuvers could result in rear-impact collisions if performed in an unsafe manner:

- Backing up: obviously, the driver needs to know what is directly behind the vehicle, and that is much easier if he or she can look through a rear window that has been cleared by a defogger. Without a defogger, the driver's alternative strategies would be relying on outside mirrors, getting out of the car to look around/wipe the window, or just backing up a short distance very slowly and hoping for the best.
- Changing lanes: good, defensive driving includes periodic scanning to the rear of the vehicle to know if it will be safe to change lanes should it suddenly become necessary. Obviously, this is easiest with a correctly positioned inside rear-view mirror and a clear window. Without them, the driver would have to rely more on outside mirrors, change lanes with less confidence that it is safe, or even forego lane changes unless they are absolutely necessary.

Analysis of crash avoidance – passenger cars NHTSA's study, published in 2004, used crash data from Michigan (1981-91) and Florida (1986-99). The basic analysis method was to compare the number of rear-impacts involving either of the pre-crash maneuvers where rear-window defoggers might be effective – backing up or changing lanes – to the number of rear-impact involvements where the vehicle had been standing still for some time and was hit by somebody else. The latter are a control group, because the drivers did nothing to cause the crash, and rear-window defoggers would not have prevented it. Did the ratio of impacts while backing up or changing lanes to impacts while standing still decrease with rear window defoggers – in the types of environmental conditions when defoggers might be in use (precipitation, early morning, winter)?[3]

This is a difficult statistical analysis. Neither the police-reported data elements nor the VIN included on the Michigan and Florida files indicate if any specific vehicle was equipped with rear-window defoggers. The VIN only identifies the make-model and the model year. *Ward's Automotive Yearbooks* indicate the percent of vehicles with rear-window defoggers by make-model and MY. However, unlike most other safety technologies that went from 0 to 100 or near-100 percent of vehicles from one MY to the next, rear-window defoggers increased their market share of nearly all make-models gradually, a few percent more each year. That precluded a relatively simple 3-way contingency table analysis (yes/no defogger, impact type, environmental condition) and necessitated logistic regressions on each State's data. The dependent variable was the type of rear impact (backing-up or changing-lanes vs. control group). The independent variables were the proportion of vehicles of that make-model and MY equipped with defoggers, the environmental conditions (precipitation, early morning or winter vs. none of the above), vehicle age, driver age/gender, calendar year, and the specific make-model. However, the key independent variable was the interaction term between rear-window defogger and adverse

---

[3] *Ibid.*, pp. 9-14.

environmental conditions. If defoggers are effective, there should be a significant reduction in backing-up/changing-lanes involvements under adverse environmental conditions.[4]

The statistical analyses showed little or no effect for defoggers in either State. The evaluation was unable to conclude that rear-window defoggers reduce police-reported crashes. It may be that the complexity of the analysis, necessitated by the data, made it harder to identify an effect. It may also be that rear-window defoggers are not that essential to bottom-line safety. As explained in "Expected Benefits," drivers usually have alternative strategies to compensate for the absence of a defogger: more reliance on outside mirrors, getting out of the car to look around or wipe the window when that can be done safely, or simply backing up and changing lanes less frequently/more gradually. These alternatives are, at best, inconvenient and sometimes unnerving. Therefore, we would expect most drivers will continue wanting to have rear-window defoggers on their vehicles.[5]

---

[4] *Ibid.*, pp. 4-8 and 14-17.
[5] *Ibid.*, pp. v-vi and 18-27.

FMVSS 105: Hydraulic and electric brake systems

NHTSA has evaluated five innovative braking technologies for passenger vehicles:

- *Dual master cylinders*
- *Front disc brakes*
- *Rear-wheel antilock (RWAL) brake systems for light trucks*
- *Four-wheel antilock brake systems (ABS) for passenger cars and light trucks*
- *Electronic stability control (ESC) systems (preliminary evaluation)*

The goal of dual master cylinders is to provide dual hydraulic circuits, so that a fault in one hydraulic system will not lead to a catastrophic loss of all braking power. They have been standard equipment on most passenger cars since model year 1967. Front disc brakes enhance a driver's control by providing a more linear pedal "feel" than drum brakes. They are also less prone to "fade" from overheating or exposure to water. They have been standard on most cars since model year 1977. Rear-wheel antilock (RWAL) brake systems are designed to prevent rear-wheel lockup and loss of control during braking. Four-wheel antilock brake systems (ABS) do what RWAL does and additionally allow drivers to keep control of a vehicle and steer it during severe braking. Since the mid-1990's, antilock brakes have been installed as standard or optional equipment on approximately 60 percent of new cars (4-wheel) and light trucks (4-wheel or rear-wheel). Electronic stability control (ESC) systems detect when a vehicle is about to go out of control and automatically apply the brakes to individual wheels and/or reduce engine torque to help the driver stay on course. ESC systems were first offered on selected model year 1997 cars, and they were installed on 7.4 percent of vehicles in model year 2003.

FMVSS 105 regulated hydraulic brake systems until 2000-2002, when it was superseded by FMVSS 135.[1] FMVSS 105 is a performance standard, specifying stopping distances or deceleration rates for a series of stopping tests under various conditions. It does not prescribe technologies. However, dual master cylinders satisfied the FMVSS 105 requirement for a dual or split braking system. Front disc brakes helped vehicles pass the fade and water-recovery tests added to FMVSS 105, effective September 1, 1975. Antilock brakes have never been required for vehicles with a GVWR less than 10,000 pounds, but Congress asked NHTSA in 1991 to consider an ABS requirement in FMVSS 105. Electronic stability control systems, likewise, have never been required on motor vehicles. NHTSA's evaluations of these five technologies will now be discussed, one-by-one.

Dual master cylinders

Regulatory history Before it was superseded by FMVSS 135 in 2000-2002, FMVSS 105 applied to all motor vehicles with hydraulic brakes. The first version of FMVSS 105 was one of NHTSA's initial safety standards, with an effective date of January 1, 1968 for passenger cars. To a large extent, it incorporated SAE Recommended Practices dating back to 1966. FMVSS 105 allows two alternatives for hydraulic brakes: (1) Vehicles may have a "split service brake

---

[1] *Code of Federal Regulations*, Title 49, Government Printing Office, Washington, 2002, Parts 571.105 and 571.135. However, after 2002, FMVSS 105 continued to apply to vehicles with a GVWR of 7,716 pounds or more.

system" – i.e., "a brake system consisting of two or more subsystems actuated by a single control, designed so that a single failure in any subsystem (such as a leakage-type failure of a pressure component of a hydraulic subsystem...) does not impair the operation of any other subsystem" that enables the vehicle to stop within a specified distance even "in the event of any one rupture or leakage type of failure in any component of the service brake system." Dual or tandem master cylinders met the requirement. (2) Alternatively, vehicles may have a single hydraulic circuit if they can meet even more stringent stopping tests in the partial failure mode.[2] Dual master cylinders were installed on 9 percent of model year 1962 and 1963 cars, 7 percent of model year 1964 and 1965 cars, and most new cars from model year 1967 onwards. It is likely that dual master cylinders were also installed in LTVs at about the same time (although FMVSS 105 was only extended to LTVs effective September 1, 1983).[3]

How dual master cylinders work Without dual master cylinders, a significant loss of pressure in the hydraulic system due to a leak can result in a complete, catastrophic loss of braking power. Moreover, when there is a slow leak that will eventually lead to a loss of pressure, the vehicle does not send an early warning that can be easily understood by the average driver. The Indiana Tri-Level Study of the Causes of Accidents suggested that approximately 2 percent of all accidents in cars with 1960's technology were caused by some type of catastrophic brake failure, most commonly hydraulic failures.[4]

A dual braking system consists of two separate hydraulic circuits – typically split front-and-rear in rear-wheel-drive cars, or diagonally in front-wheel-drive cars. Both circuits are activated by the brake pedal through one master cylinder that has two chambers, called a "dual master cylinder." If one of the circuits fails, the other is still available. The car can be stopped from high speeds, although of course not as quickly as when both circuits are intact. A second important feature is that the dual master cylinder has a sensor to detect the relative pressure in the chambers. If there is an imbalance, it activates a warning light on the instrument panel. It is an unambiguous early warning that the brake system needs repair.[5]

Expected benefits Dual master cylinders should prevent many of the crashes involving catastrophic brake failure, specifically those due to failures in the hydraulic system. Possible exceptions could include certain types of failure within the master cylinder or cases where the driver ignores the warning light. Since catastrophic brake failures can occur in fatal and injury crashes, dual master cylinders ought to have an effect on fatalities and injuries. Moreover, when brake failure in one vehicle results in a collision with another vehicle or pedestrian, preventing the brake failure in that one vehicle will prevent the fatalities or injuries of every crash partner, including the other vehicles' occupants and pedestrians.

---

[2] *Ibid.*, Part 571.105 S4, S5.1.2.1 and S5.1.2.2; *Federal Register* 32 (February 3, 1967): 2414; *1967 SAE Handbook*, Society of Automotive Engineers, New York, 1967, pp. 856-857.

[3] Kahane, C.J., *A Preliminary Evaluation of Two Braking Improvements for Passenger Cars*, NHTSA Technical Report No. DOT HS 806 359, Washington, 1983, pp. 1-7; *Federal Register* 46 (January 2, 1981): 55.

[4] Treat, J.R. et al., *A Tri-Level Study of the Causes of Traffic Accidents*, Vol. 1, NHTSA Technical Report No. DOT HS 805 085, Washington, 1977.

[5] The sensor in some vehicles also detects low fluid level in the master cylinder reservoirs.

Crash avoidance – passenger cars NHTSA's evaluation was published in 1983. It is based on crash data from North Carolina (1971-81) and Texas (1972,74,77).[6] The data allow comparison of the proportion of crash involvements in which "defective brakes" are a "contributing factor" for cars with single vs. dual master cylinders. However, it is important to control for vehicle age, since the rate of brake defects increases strongly as a vehicle ages. The analysis method is linear regressions of the proportion of crash involvements (in a given model year and calendar year) in which defective brakes are a contributing factor, by percentage of cars with dual master cylinders, vehicle age, and other control variables. The regressions attributed an unequivocal, statistically significant reduction of brake defects upon the introduction of dual master cylinders. The effectiveness estimate is an 0.7 percent reduction of all police-reported crashes involving at least one car, when all cars are equipped with dual master cylinders.[7] The incidence of "defective brakes" was reduced by approximately 35 percent (0.7/2.0). Using 1979-80 as the "baseline" years, the evaluation estimated that dual master cylinders would prevent 40,000 police-reported crashes per year when all cars on the road were equipped with them.[8]

Injury reduction – passenger cars Estimates of injury reduction were obtained by limiting the preceding analyses of North Carolina and Texas data to injury-producing crashes – i.e., crashes where at least one occupant in any of the involved vehicles (not necessarily the vehicle equipped with a dual master cylinder) was injured.[9] The regression showed virtually the same effectiveness in injury-producing crashes as in property-damage crashes. The evaluation estimated that dual master cylinders prevent 0.7 percent of injury-producing crashes, resulting in a reduction of 24,000 injuries per year.[10]

Fatality reduction – passenger cars The preceding regression analyses were also performed with 1975-81 FARS data.[11] They showed a statistically significant reduction for dual master cylinders, proportionately about the same as in the nonfatal crashes. The evaluation assumed that brake defects are a contributing factor in similar percentages of fatal and nonfatal crashes (i.e., 2 percent of crashes, in cars without dual master cylinders). It estimated that dual master cylinders prevent 0.7 percent of fatal crashes that involve passenger cars, saving 260 lives per year.[12]

Light trucks The 1983 evaluation addressed only passenger cars. However, FMVSS 105 also applies to light trucks. Both types of vehicles use fairly similar hydraulic braking technology. It seems likely that dual master cylinders could have about the same effect in LTVs as in cars: a 0.7 percent reduction of crashes, injuries and fatalities.

---

[6] Kahane (1983 Brakes), pp. 15-38.
[7] *Ibid.*, p. 43; 90% confidence bounds: 0.58 to 0.82 percent.
[8] *Ibid.*, pp. 55-57; 90% confidence bounds: 33,000 to 47,000 crashes prevented.
[9] *Ibid.*, pp. 28-29.
[10] *Ibid.*, pp. 55-57; 90% confidence bounds: 19,000 to 28,000 injuries prevented.
[11] *Ibid.*, pp. 38-40.
[12] *Ibid.*, pp. 55-57; 90% confidence bounds: 220 to 310 lives saved.

Front disc brakes

Regulatory history Front disc brakes began to appear on domestic passenger cars in model year 1965. Consumers welcomed the new technology. It jumped from 13 percent of new cars in 1968 to 86 percent by 1973. In 1970, NHTSA proposed FMVSS 105a, incorporating more stringent stopping distance, fade- and water-recovery tests than the original FMVSS 105, with a proposed effective date of October 1, 1972. After several revisions, this regulation took effect for passenger cars on January 1, 1976 and LTVs on September 1, 1983. Although some cars with four-wheel drum brakes could and did pass the new tests, it was easier to meet them with front disc brakes. Furthermore, the superior self-adjusting characteristics of disc brakes allowed for increased vehicle stability during the high-speed stopping tests in FMVSS 105. By model year 1978, most cars and LTVs produced for sale in the United States were equipped with front disc brakes. (FMVSS 105 or 135 have never been explicitly required disc brakes.)[13]

How front disc brakes work Disc brakes are calipers equipped with abrasive pads that squeeze rotors, metal plates parallel to the wheels that rotate with the wheels. Drum brakes are shoes with abrasive linings that press against the insides of drums that rotate with the wheels. Whereas drum brakes readily heat up from overuse or fill up with water upon immersion, resulting in losses of braking ability, disc brakes are ventilated to dissipate heat or shed water quickly. Furthermore, drum brakes have a "self-energizing capability" (friction building up more rapidly than pedal pressure) that makes drivers prone to lock the wheels. Disc brakes have a more linear relationship between pedal pressure and vehicle deceleration, making it easier for drivers to deliver just the right amount of pressure short of locking the wheels.[14]

Expected benefits Front disc brakes might prevent some crashes involving catastrophic brake failure, especially on mountain roads, where brakes could fade from overuse on the downhills, or on flooded roads, where water gets into the wheels. If drivers use the more linear pedal feel to advantage, that could manifest itself in two ways in the crash data: if locking the wheels is reported as "defective brakes," a reduction in lockup could result in fewer reported brake defects. If drivers can better optimize their deceleration rate short of lockup, they will stop in a shorter distance, possibly reducing the risk of frontally hitting other vehicles in the rear. Since brake failures/defects can occur in fatal and injury crashes, front disc brakes could have an effect on fatalities and injuries. Moreover, when brake failure in one vehicle results in a collision with another vehicle or pedestrian, preventing the brake failure in that one vehicle will prevent the fatalities or injuries of every crash partner, including the other vehicles' occupants and pedestrians.

Crash avoidance – passenger cars NHTSA's 1983 evaluation has several analyses of the proportion of crash involvements in which "defective brakes" are a "contributing factor" for cars with drum vs. front disc brakes. The most reliable one is based on North Carolina data (1971-

---

[13] *Ibid.*, p. 7; *Federal Register* 35 (November 11, 1970): 17345, 37 (September 2, 1972): 17970, 38 (May 18, 1973): 13017, 39 (February 22, 1974): 6708, 40 (June 9, 1975): 24525, 46 (January 2, 1981): 55; *Ward's Automotive Yearbook*, Ward's Communications, Detroit, 1975-79.

[14] *Ibid.*, pp. 6-10; Ballard, C. and Andrade, D., *Systems and Hardware Effects of FMVSS 105-75*, Paper No. 760216, Society of Automotive Engineers, Warrendale, PA, 1976; Kahane, C.J. and Ichter, K.D., *Statistical Evaluation of Brake Safety Improvements for Passenger Cars*, Paper No. 841236, Society of Automotive Engineers, Warrendale, PA, 1984.

81). It subdivides the data by make-model as well as model year, since front disc brakes were introduced earlier in some models than in others.[15] The analysis method is a linear regression of the proportion of crash involvements (in a given make-model, model year and calendar year) in which defective brakes are a contributing factor, by percentage of cars with front disc brakes, vehicle age, and other control variables. The regressions attributed a statistically significant reduction of brake defects upon the introduction of front disc brakes, equivalent to an 0.17 percent reduction of all police-reported crashes involving at least one car.[16] Using 1979-80 as the "baseline" years, the evaluation estimated that front disc brakes would prevent 9,800 police-reported crashes per year when all cars on the road were equipped with them.[17] Other analyses of North Carolina data showed reductions of "defective brake" crashes on hilly and wet roads with front disc brakes, but not necessarily greater reductions than on other roads.[18] The analysis of front-to-rear crashes attributed to front disc brakes a 0.2 percent reduction in the likelihood of being the striking car, but this was not statistically significant.[19]

Injury reduction – passenger cars Estimates of injury reduction were obtained by limiting the preceding analysis of North Carolina data (by make-model) to injury-producing crashes.[20] The regression showed virtually the same effectiveness in injury-producing crashes as in property-damage crashes. The evaluation estimated that front disc brakes prevent 0.17 percent of injury-producing crashes, resulting in a reduction of 5,700 injuries per year.[21]

Fatality reduction – passenger cars Regression analyses of 1975-81 FARS data (not subdivided by make-model) attributed a statistically significant 0.55 percent reduction of "defective brake" involvements to front disc brakes.[22] However, the evaluation estimated, more conservatively, that front disc brakes would have proportionately the same effect in fatal as in nonfatal crashes: a 0.17 percent reduction of fatal crashes that involve passenger cars, saving 64 lives per year.[23]

Light trucks The 1983 evaluation addressed only passenger cars. However, front disc brakes have superseded front-wheel drum brakes in light trucks as well as in passenger cars. Both types of vehicles use fairly similar hydraulic braking technology. It seems likely that front disc brakes could have about the same effect in light trucks as in cars: a 0.17 percent reduction of crashes, injuries and fatalities.

---

[15] Kahane (1983 Brakes), pp. 26-28.
[16] *Ibid.*, p. 44; 90% confidence bounds: 0.10 to 0.24 percent.
[17] *Ibid.*, pp. 55-57; 90% confidence bounds: 5,800 to 13,800 crashes prevented.
[18] *Ibid.*, pp. 29-31.
[19] *Ibid.*, pp. 45-47.
[20] *Ibid.*, pp. 28-29.
[21] *Ibid.*, pp. 55-57; 90% confidence bounds: 3,400 to 8,100 injuries prevented.
[22] *Ibid.*, pp. 38-40.
[23] *Ibid.*, pp. 55-57; 90% confidence bounds: 38 to 90 lives saved.

Rear-wheel antilock (RWAL) brake systems for light trucks

Regulatory history and voluntary installations No type of antilock brake system (ABS) has ever been required for passenger vehicles with GVWR less than 10,000 pounds. However, the Highway Safety Act of 1991, Section 2507 instructed NHTSA to consider requiring ABS in passenger vehicles. NHTSA published an ANPRM at the beginning of 1994 asking for information about the effectiveness and potential benefits of ABS technologies.[24] Based on responses to the 1994 ANPRM, including statistical studies by NHTSA and others that failed to show significant net benefits for voluntarily installed ABS, NHTSA issued a second ANPRM in 1996 deferring indefinitely the ABS requirement.[25]

Rear-wheel antilocks were installed as standard equipment in 1987 Ford F-Series pickup trucks, Bronco and Bronco II. In 1988-90, they were phased into most domestic pickup trucks and SUVs as well as many vans, as standard equipment. During the 1990's, RWAL was increasingly superseded by 4-wheel ABS; by model year 2001, only 8 percent of new domestic LTVs were equipped with RWAL.[26]

How RWAL works The fundamental safety problem addressed by any ABS system is that few drivers are able to modulate pressure on the brake pedal optimally, given a sudden emergency situation or unexpectedly slippery surface. If excess pedal pressure locks the rear wheels, the vehicle can lose control. Light trucks are especially prone to rear-wheel lockup when they are not heavily loaded, and especially crash-prone once they are out of control. RWAL senses if any of the rear wheels have locked, and if so, quickly releases the brakes on that wheel and lets it start rolling again. Cycles of releasing, holding and reapplying brakes are repeated many times per second. RWAL, however, will not prevent front-wheel lockup or assure steering control during braking. If the front wheels lock while the rear wheels turn, the truck will just slow down on a straight line, without yawing. NHTSA conducted stopping tests on five surfaces with RWAL-equipped trucks, and also for the same trucks with the RWAL disabled. RWAL substantially but not completely reduced the frequency and severity of yawing in spike stops. That benefit, however, was offset by a slight (6 to 18 percent) increase in stopping distances on four of the surfaces.[27]

Expected benefits If RWAL prevents or substantially reduces yawing, it will help keep trucks from running off a straight road. By keeping the truck on the road, it may prevent the rollovers or impacts with fixed objects or pedestrians that can occur after a vehicle has left the road. Side impacts with fixed objects are especially characteristic of vehicles that have yawed out of control. RWAL would be less effective if the driving task requires steering as well as braking (e.g., on a curved road). It will not affect yawing caused by reasons other than brake-induced

---

[24] *Federal Register* 59 (January 4, 1994): 281.

[25] *Federal Register* 61 (July 12, 1996): 36698.

[26] Kahane, C.J., *Preliminary Evaluation of the Effectiveness of Rear-Wheel Antilock Brake Systems for Light Trucks*, NHTSA Docket No. 70-27-GR-026, Washington, 1993, p. 15; *2002 Ward's Automotive Yearbook*, Ward's Communications, Detroit, 2002, p. 333.

[27] Kahane (1993 RWAL), pp. 7-14; Hiltner, E., Arehart, C. and Radlinski, R., *Light Vehicle ABS Performance Evaluation*, NHTSA Technical Report No. DOT HS 807 813, Washington, 1991; Arehart, C., Radlinski, R. and Hiltner, E., *Light Vehicle ABS Performance Evaluation – Phase II*, NHTSA Technical Report No. DOT HS 807 924, Washington, 1992.

rear-wheel lockup (e.g., going around a curve too quickly). To the extent that RWAL is associated with a slight increase in stopping distances, there could be an adverse effect on multivehicle crashes that require a truck to stop in time to avoid hitting another vehicle.

Crash avoidance NHTSA's preliminary evaluation was published in 1993. It is based on crash data from Florida (1990-91), Michigan (1990-91), and Pennsylvania (1989-91). These data show consistent, statistically significant reductions of nonfatal rollovers and side impacts with fixed objects in trucks equipped with RWAL.[28] Involvements in rollovers or fixed-object collisions are compared to a control group: crash involvements in which the truck, standing still or moving 5 mph or less, was struck by another vehicle (where the truck's RWAL could not have been a factor in the crash). For example, the results for pickup trucks in Michigan indicate a 42 percent reduction of nonfatal rollovers[29]:

MICHIGAN, 1990-91 PICKUP TRUCKS	Rollovers	Control Group	Ratio of Rollover/Control
Last 2 model years without RWAL	1,095	3,634	.301
First 2 model years with RWAL	737	4,215	.175

On the average, the analyses showed about a 30 percent reduction of nonfatal rollovers and a 20 percent reduction of nonfatal side impacts with fixed objects.[30]

A similar analysis investigated the effect of RWAL in multivehicle crashes. Involvements as a frontally impacting truck in a multivehicle collision, traveling faster than 5 mph, were compared to the control group of vehicles struck while standing still or moving 5 mph or less. The ratio of striking to control-group involvements was about the same for trucks with and without RWAL.[31]

Fatality reduction Statistical analyses of 1989 – mid-1992 FARS data did not show an overall reduction of fatal rollovers or side impacts with fixed objects, relative to fatal control-group involvements.[32] The FARS analysis of fatal involvements as a frontally impacting truck in a multivehicle collision (traveling faster than 5 mph) showed mixed/negative results: little or no effect for RWAL when the data were limited to trucks of the first model year with RWAL vs. the last model year before RWAL; an increase in collisions for trucks of the first two years with RWAL vs. the last two years before RWAL.[33] Pedestrian crashes were the only type that the data hinted at a reduction with RWAL, possibly on the order of 10 percent.[34]

---

[28] Kahane (1993 RWAL), pp. 19-44.
[29] *Ibid.*, p. 23.
[30] *Ibid.*, p. 108.
[31] *Ibid.*, pp. 100-109.
[32] *Ibid.*, pp. 42-49 and 100-109.
[33] *Ibid.*, pp. 78-85.
[34] *Ibid.*, pp. 93-99 and 108.

NHTSA performed another statistical analysis of 1989-93 FARS data, using logistic regression of the ratio of relevant crashes to control-group crashes. This study, likewise, associated RWAL with a statistically significant increase of fatal involvements as the striking vehicle in a multivehicle collision. It did not find a reduction for RWAL in fatal rollovers or side impacts with fixed objects. Pedestrian crashes were not considered in that study.[35]

RWAL is apparently effective in preventing yaw in the relatively low-speed situations where rear-wheel lockup would have caused the truck to run off the road with nonfatal consequences. It may be less effective in high-speed scenarios that precipitate fatal crashes. Also, many of these fatal crashes do not involve pre-crash braking – e.g., when the driver is impaired. Since RWAL usually does not reduce, and may even increase stopping distances, the absence of benefits in multivehicle crashes is not surprising.

Four-wheel antilock *brake systems (ABS) for passenger cars and light trucks*

Regulatory history and voluntary installations As described above (regulatory history of RWAL), NHTSA considered, but eventually deferred any requirement for ABS on passenger vehicles with GVWR less than 10,000 pounds.

Modern 4-wheel ABS was first offered as standard equipment in 1985 on some lines of BMW, Lincoln and Mercedes and in 1986 on Chevrolet Corvette. Availability of ABS increased gradually in 1987-90 and dramatically in 1991-92, when it became standard on the majority of GM cars. Throughout 1994-2001, about 60-65 percent of new passenger cars were equipped with ABS. In general, ABS is usually standard on the larger and more expensive cars, optional and not too frequently sold on small economy cars. Four-wheel ABS installations for light trucks began in 1989 on some GM Astro/Safari minivans, Jeep Cherokee and Jeep Wagoneer. The market share for 4-wheel ABS in new light trucks steadily increased during the 1990's, as RWAL was phased out, and had reached about 80 percent by 2001.[36]

How ABS works Few drivers are able to modulate pressure on the brake pedal optimally, given a sudden emergency situation or unexpectedly slippery surface. If excess pedal pressure locks only the front wheels, the vehicle will continue in a straight path, but the driver will be unable to steer it and avoid obstacles. If it locks the rear wheels, the vehicle can lose control. ABS senses if any of the four wheels have locked, and if so, quickly releases the brakes on that wheel and lets it start rolling again. Cycles of releasing, holding and reapplying brakes are repeated many times per second.

ABS activation causes noise and pedal vibration in some vehicles, and steering may not be as easy as in normal operation. Nevertheless, a reasonably alert driver familiar with the system will maintain pressure on the pedal, stop in a minimum distance and be able to steer the vehicle throughout the crash avoidance maneuver.

---

[35] Hertz, E., Hilton, J. and Johnson, D.M., *An Analysis of the Crash Experience of Light Trucks Equipped with Antilock Braking Systems*, NHTSA Technical Report No. DOT HS 808 278, Washington, 1995.

[36] Kahane, C.J., *Preliminary Evaluation of the Effectiveness of Antilock Brake Systems for Passenger Cars*, NHTSA Technical Report No. DOT HS 808 206, Washington, 1994, pp. 9 and 119-128; *2002 Ward's Automotive Yearbooks*, Ward's Communications, Detroit; Kahane (1993 RWAL), pp. 15.

NHTSA conducted stopping tests on straight and curved paths and various surfaces with ABS-equipped cars, and also for the same cars with the ABS disabled. Results were impressive. Yawing was nearly eliminated and steering control maintained. Furthermore, on most surfaces, especially wet pavements, cars stopped in a shorter distance with ABS activated and steady pedal pressure than with ABS disabled and the trained test driver attempting to modulate and optimize the pedal pressure. The principal exception was a gravel surface, where ABS, although maintaining directional stability and control, took 28 percent longer to stop than just slamming on the brakes and sliding.[37]

Expected benefits The experience on the test track suggested that ABS could have safety benefits in many crash situations. Maintaining steering control and cutting stopping distances, especially on wet roads, could reduce frontal impacts into other vehicles and collisions with pedestrians. By preventing yaw and preserving steering control, ABS can help drivers keep their vehicle on a straight or curving road, and could prevent run-off-road crashes such as rollovers or fixed-object impacts. However, there could be some adverse impact due to longer stopping distances on gravel and other loose surfaces away from the road.

Crash avoidance NHTSA's preliminary evaluation of 4-wheel ABS for passenger cars was published in 1994. It is based on 1990-92 crash data from Florida, Missouri and Pennsylvania. Crash involvements as a frontally impacting car in a multivehicle collision, traveling faster than 5 mph, were compared to the control group of cars struck while standing still or moving 5 mph or less (where the car's ABS could not have been a factor in the crash).[38] On wet roads, cars with ABS experienced a statistically significant, 28 percent reduction of frontal impacts, relative to the control group and relative to cars of the same make-models without ABS[39]:

WET ROADS FL-MO-PA 1990-92	Multivehicle Frontals	Control Group	Ratio of Frontals/Control
Last 2 model years without ABS	1,220	922	1.323
First 2 model years with ABS	841	879	.957

On dry roads, the reduction was 5 percent, not statistically significant. For all road conditions (wet, dry, snowy, icy), the overall reduction in multivehicle frontal impacts with ABS was a statistically significant 9 percent.

Run-off-road crashes, however, increased rather than decreased with ABS. Relative to the control group, rollovers and fixed-object impacts increased overall by a statistically significant 19 percent in the ABS-equipped cars. The rate of increase was about the same on dry, wet and

---

[37] Kahane (1994 ABS), pp. 4-9; Hiltner et al., *Op. cit.*; Arehart et al., *Op. cit.*
[38] Kahane (1994 ABS), pp. 10-57 and 106, especially 51, 53 and 55.
[39] *Ibid.*, p. 53.

snowy/icy roads.[40] The issue of ABS performance in run-off-road crashes will be discussed in the next section.

Fatality reduction The analysis was based on 1989-93 FARS data. In multivehicle crashes on dry roads, the ratio of involvements as a frontally impacting car to involvements as a car that was struck in the rear or side, or while standing still, was nearly the same with or without ABS. That suggests ABS has, at most, a small effect on multivehicle crashes on dry roads. For the remaining analyses, multivehicle crash involvements on dry roads (regardless of impact type) serve as the control group.[41]

Multivehicle crash involvements on wet roads were reduced by a statistically significant 24 percent with ABS[42]:

FARS 1989-93 MULTIVEHICLE CRASHES	Wet Roads	Dry Roads	Ratio of Wet/Dry
Last 2 model years without ABS	246	1,021	.241
First 2 model years with ABS	158	858	.184

Multivehicle crashes on snowy/icy roads decreased by 13 percent (not significant). Fatal collisions with pedestrians, bicyclists, trains or animals, on all road surfaces, were reduced by a statistically significant 27 percent.[43]

On the other hand, fatal run-off-road crashes (rollovers and impacts with fixed objects) increased by a statistically significant 28 percent on all road surfaces, with ABS[44]:

FARS 1989-93	Rollover + Fixed Object	Dry Road Multivehicle	Ratio
Last 2 model years without ABS	431	1,021	.422
First 2 model years with ABS	463	858	.540

The increases were about equally large on dry, wet and snowy/icy roads. In particular, side impacts with fixed objects, a crash type typically preceded by loss of directional control, increased by a statistically significant 57 percent. The increase in run-off-road crashes nearly

---

[40] *Ibid.*, p. 84-92.
[41] *Ibid.*, p. 57-62.
[42] *Ibid.*, p. 65.
[43] *Ibid.*, p. 63-68.
[44] *Ibid.*, p. 95.

offset the reductions in pedestrian and wet-road multivehicle crashes, resulting in a near-zero net effect on overall fatal-crash risk.[45]

NHTSA performed another statistical analysis of ABS in passenger cars, based on 1989-93 FARS data as well as 1989-92/93 data from four States, using logistic regression of the ratio of relevant crashes to control-group crashes. This study confirmed the statistically significant increases of rollovers and side impacts with fixed objects in the ABS-equipped cars, in both fatal and nonfatal crashes. It also confirmed the benefits of ABS in fatal and nonfatal multivehicle crashes on wet roads.[46] Furthermore, NHTSA's initial analysis of 4-wheel ABS in light trucks showed similar increases (although not statistically significant) of rollovers and side impacts with fixed objects in the light trucks with the 4-wheel ABS.[47] A 1996 study by the Insurance Institute for Highway Safety likewise observed a 28 percent increase in run-off-road crashes with ABS.[48] Since all these analyses of United States data are based on more or less the same make-models (and could conceivably have the same biases), it is noteworthy that a Swedish study, based on a quite different vehicle mix, also showed a substantial increase of run-off-road crashes, relative to multivehicle crashes on dry roads, with ABS.[49]

Hypotheses for the observed increases in run-off-road crashes included:

- Driver inexperience/lack of knowledge about ABS. Drivers might remove their foot from the pedal in response to the [unanticipated] noise and vibration of ABS, or try to pump the pedal as with conventional brakes.
- More aggressive driving due to a misperception of how much ABS reduces stopping distances or enhances control.
- Longer stopping distances with ABS on the loose surfaces vehicles encounter after they leave the road.
- The enhanced steering control while braking with ABS could allow unsafe panic steering maneuvers.
- Possible flaws in ABS performance in certain maneuvers or on some roadway surfaces.

NHTSA undertook a multiyear research program to test out these and other hypotheses.[50] NHTSA, the manufacturers and suppliers and the insurance industry also developed media to educate vehicle owners about how ABS responds when activated, and how to use it properly ("Don't let up [on the brakes]"). The research, in fact, has not identified *any* significant problems with ABS other than owners' initial lack of knowledge and experience with the systems.

---

[45] *Ibid.*, p. 115-117.

[46] Hertz, E., Hilton, J. and Johnson, D.M., *An Analysis of the Crash Experience of Passenger Cars Equipped with Antilock Braking Systems*, NHTSA Technical Report No. DOT HS 808 279, Washington, 1995.

[47] Hertz et al. (1995 Light Truck ABS).

[48] Farmer, C.M., Lund, A.K., Trempel, R.E. and Braver, E.R., *Fatal Crashes of Passenger Vehicles Before and After Adding Antilock Brake Systems*, Insurance Institute for Highway Safety, Arlington, VA, 1996.

[49] Kullgren, A., Lie A. and Tingvall, C., "The Effectiveness of ABS in Real Life Accidents," *Proceedings of the Fourteenth International Technical Conference on the Enhanced Safety of Vehicles*, NHTSA, Washington, 1994, Paper No. 94-S4-O-07.

[50] Kahane (1994 ABS), pp. v-viii.

Three reports published in 2000-02 suggests that the increase in run-off-road crashes may have been largely or entirely temporary. NHTSA's follow-up analysis based on 1995-96 FARS and State data showed that the increase in fatal run-off-road crashes was of lower magnitude than in the earlier years. In fact, ABS did not have a statistically significant effect on any type of fatal crash, except for an increase in LTV rollovers, whereas several types of nonfatal crash involvements were significantly reduced with ABS.[51] The IIHS follow-up, while reconfirming the increase in run-off-road crashes of MY 1992 vehicles with ABS during CY 1993-95, found that these same MY 1992 vehicles did not experience any increase in run-off-road crashes during CY 1996-98, when they were 4-6 years old. However, this study did not find a statistically significant overall effect for ABS on fatal crashes in 1996-98; it did not analyze nonfatal crashes.[52] Harless and Hoffer confirmed the IIHS results and, moreover, demonstrated that the increase in run-off-road crashes during the earlier years was virtually confined to drinking drivers. It was especially prevalent in young, drinking drivers or in drinking drivers with a history of high-risk driving behavior.[53]

These recent results suggest that the problems with ABS may have been largely due to owners' inexperience and lack of knowledge about the systems. Thanks to the efforts to educate the public about ABS, and as owners simply accumulated years of experience with their vehicles, the increase in run-off-road crashes may have faded away. Moreover, the Harless paper suggests this wasn't everybody's problem, but was concentrated among drinking drivers. The combination of impairment with a sudden, unexpected noise and vibration may have led to inappropriate panic reactions. Furthermore, ABS technology has evolved over the years with improvements in the response and feedback to the driver, or with added features such as "brake assist" that amplifies input from the driver.

ABS may be due for a fresh look. If the effect on run-off-road crashes continues to be neutral, even if there is no significant overall effect on fatal crashes, the reductions of various types of nonfatal crashes could mark ABS as a technology with net safety benefits.[54]

Electronic stability control (ESC) systems (preliminary evaluation)

Voluntary installation history Mercedes-Benz first offered electronic stability control systems in 1997 as standard equipment on the top-of-the-line S600 and SL600 and as an option on some other luxury cars. The next year, ESC was standard on Cadillac DeVille Concours, BMW 700-series and a few additional Mercedes models. By 2000, it was standard on most BMW and Mercedes cars, Cadillac Seville and a few other GM luxury models, Lexus LS and GS and Acura RL. Among SUVs, ESC was standard on Mercedes in 1999, on Lexus LX in 2000, followed by Toyota 4Runner and Landcruiser and Lexus RX in 2001. ESC was installed on 7.4 percent of

---

[51] Hertz, E., *Analysis of the Crash Experience of Vehicles Equipped with All Wheel Antilock Braking Systems (ABS) – A Second Update Including Vehicles with Optional ABS*, NHTSA Technical Report No. DOT HS 809 144, Washington, 2000.

[52] Farmer, C.M., "New Evidence Concerning Fatal Crashes of Passenger Vehicles Before and After Adding Antilock Braking Systems," *Accident Analysis and Prevention*, Vol. 33, 2001, pp. 361-369.

[53] Harless, D.W. and Hoffer, G.E., "The Antilock Braking System Anomaly: A Drinking Driver Problem?," *Accident Analysis and Prevention*, Vol. 34, 2002, pp. 333-341.

[54] *National Highway Traffic Safety Administration Evaluation Program Plan, Calendar Years 2004-2007*, NHTSA Report No. DOT HS 809 699, Washington, 2004, pp. 26-27.

vehicles sold in model year 2003. No FMVSS requires ESC systems. However, NHTSA believes ESC is a promising safety technology; since 2001, the agency's annual *Buying a Safer Car* brochures have been informing the public what make-models are equipped with ESC.

How ESC works ESC systems detect and automatically assist drivers in situations that lead to loss of control (e.g. understeer and oversteer) and occur especially in unfavorable conditions (rain, snow, sleet, ice). Sensors monitor the speed of each wheel, the steering wheel angle, and the yaw rate and lateral acceleration of the vehicle. The yaw rate is the rate of change of the vehicle's heading. The system compares the measured yaw rate to the intended rate of change of heading consistent with the speed and lateral acceleration of the vehicle. For example, a yaw rate measurement greater than that consistent with the speed and lateral acceleration of the vehicle indicates oversteer. ESC would make a rapid automatic intervention to correct the vehicle heading by applying the brakes to individual wheels and possibly reducing engine torque to help the driver stay on the road. If the vehicle was experiencing the onset of oversteer in a left curve, ESC would momentarily apply the brake to the right front wheel to counteract the excessive yaw rate and stabilize the vehicle. Some systems may then slow down the vehicle to a speed more appropriate for conditions. All current vehicles with ESC are also equipped with ABS and traction control systems; the ESC to a large extent builds on ABS technology and shares ABS components.[55]

Expected benefits ESC would appear to be most effective in preventing single-vehicle run-off-road crashes including rollovers and collisions with fixed objects. That is because many of them involve a loss of directional stability or control during the pre-crash sequence of events. For example, if adverse conditions on the road trigger a skid, ESC could stabilize the vehicle and prevent the entire off-road excursion. Even if a careless driver has already drifted off the road with two wheels, ESC may help maintain directional control and stability, making it easier for the driver to get back on the road and avoid a crash. Presumably, ESC would have much less effect on collision involvements on the road with other vehicles, pedestrians or animals, because a smaller proportion of these crashes are preceded by a loss of directional stability. A unique feature of ESC that could make it especially effective is that it can activate without any action by a driver (unlike, for example, ABS, which cannot activate unless the driver applies the brakes).

Crash avoidance – passenger cars NHTSA's preliminary evaluation was published in 2004.[56] It is based on calendar year 1997-2002 crash data from Florida, Illinois, Maryland, Missouri and Utah for model year 1997-2002 cars of make-models that received ESC as standard equipment at some point during those model years. Involvements in single-vehicle crashes (excluding collisions with pedestrians/bicyclists, animals or trains) were compared to a control group of involvements in multivehicle crashes. Cars with ESC experienced a statistically significant 35 percent reduction of single-vehicle crashes, relative to the control group and relative to cars of the same make-models without ESC[57]:

---

[55] Dang, J.N., *Preliminary Results Analyzing the Effectiveness of Electronic Stability Control (ESC) Systems*, NHTSA Evaluation Note No. DOT HS 809 790, Washington, 2004.
[56] *Ibid.*
[57] 95% confidence bounds: 29 to 41 percent.

FL-IL-MD-MO-UT 1997-2002	Single-Vehicle Involvements	Multivehicle Involvements	Ratio of Single/Multi
Cars without ESC	1,483	19,044	.0779
Cars with ESC (same make-models)	699	14,090	.0496

Fatality reduction – passenger cars A similar analysis of fatal single- vs. multi-vehicle crash involvements in 1997-2003 FARS data showed a statistically significant 30 percent reduction of fatal single-vehicle crash involvements with ESC.[58]

Crash avoidance – light trucks In model years 1997-2002, the only LTV make-models equipped with ESC were SUVs. The analysis of single- vs. multi-vehicle crash involvements by SUVs in five States (a database about ¼ as large as the car sample) showed a statistically significant 67 percent reduction of single-vehicle crash involvements with ESC.[59]

Fatality reduction – light trucks Fatal single-vehicle crash involvements of SUVs were reduced by a statistically significant 63 percent with ESC.[60]

Comments on the findings Logistic regression analyses on the proportion of car crash involvements that are single-vehicle, controlling for vehicle age and make-model group, driver age and gender, attributed nearly the same crash reduction and fatality reduction to ESC as the preceding contingency table analyses.

Multivehicle crash involvements are not necessarily an ideal control group, as they, too, could be influenced up to a point by ESC. Because nonfatal multivehicle crashes are far more common than single-vehicle crashes, it is at least conceivable that a small relative increase in multivehicle crashes could offset the reduction in single vehicle crashes (not that there is any intuitive basis for anticipating such an increase). However, linear regression analyses of multivehicle crash involvement rates per 100,000 vehicle registration years, controlling for vehicle age, attributed significant or borderline-significant <u>reductions</u> in those rates to ESC. In other words, the preliminary evaluation did not see a negative side effect for ESC in multivehicle crashes, and possibly even a benefit.

However, a caveat on these preliminary findings is that the database consisted primarily of luxury make-models; BMW and Mercedes-Benz constituted 61 percent of the passenger car sample. A single manufacturer, Toyota/Lexus, contributed 78 percent of the SUVs in the database. NHTSA will feel more confident about the overall effectiveness of ESC when the data comprise a more representative cross-section of the fleet including non-luxury vehicles and a wider variety of manufacturers.

---

[58] 95% confidence bounds: 10 to 50 percent.
[59] 95% confidence bounds: 60 to 74 percent.
[60] 95% confidence bounds: 44 to 81 percent.

FMVSS 108: Lamps, reflective devices, and associated equipment

This standard regulates three innovative safety technologies that have been evaluated by NHTSA:

- *Side marker lamps*
- *Center High Mounted Stop Lamps (CHMSL)*
- *Retroreflective tape on heavy trailers*

The goal of side marker lamps is to make the sides of vehicles more visible in the dark, and to prevent nighttime angle collisions. They have been standard equipment on all cars and light trucks since model year 1970. CHMSL are designed to enhance the effect of stop lamps and reduce front-to-rear collisions, by day and night. They have been standard on cars since model year 1986 and were phased into light trucks during model years 1991-94. Retroreflective tape makes heavy trailers more visible from the side and the rear, in the dark, so other vehicles won't hit them. It has been standard on new trailers over 10,000 pounds GVWR since December 1, 1993 and also has to be on older trailers still on the road as of June 1, 2001.

The second paragraph of FMVSS 108 states, "The purpose of this standard is to reduce traffic accidents ... by providing adequate illumination of the roadway and by enhancing the conspicuity of motor vehicles ... so that their presence is perceived and their signals understood, both in daylight and in darkness."[1] FMVSS 108 regulates many lamps, reflectors, bulbs and flashers on the vehicle exterior, and the systems for controlling them. Most of those devices existed well before NHTSA, were specified in SAE standards subsequently incorporated into FMVSS 108, and remained unchanged or at most, underwent gradual or subtle changes during 1968-99. Side marker lamps, CHMSL and retroreflective tape are three exceptions where a fundamentally new safety device was required. NHTSA's evaluations of these three technologies will now be discussed, one-by-one.

Side marker lamps

Regulatory history FMVSS 108, one of NHTSA's initial safety standards, took effect for vehicles over 80 inches wide (large trucks and buses) on January 1, 1968. It was extended to passenger cars and light trucks, effective January 1, 1969. Side marker lamps and/or reflectors were required from the start. Vehicles must have two combination lamp/reflectors on each side, an amber one as close as possible to the front and a red one close as possible to the rear. (Vehicles more than 30 feet long require an additional amber lamp/reflector at the midpoint.) However, to give manufacturers of cars and LTVs some more lead time, the standard allowed use of just a lamp, or just a reflector, in lieu of a lamp/reflector, during calendar year 1969. Since January 1, 1970, all cars and light trucks must have four lamp/reflectors. The side marker lamp requirement primarily incorporates SAE Recommended Practice J592. Although issued by the SAE in 1964, it did not immediately put side marker lamps on all domestic cars and light trucks, because it was only a Recommended Practice, not a Standard. Instead, the proportion of new cars and light trucks with side marker lamps increased gradually from 5 percent in model

---

[1] *Code of Federal Regulations*, Title 49, Government Printing Office, Washington, 2002, Part 571.108.

year 1964 to 13 percent in 1967, then jumped to 88 percent in 1968 and reached 100 percent in 1970.[2]

How side marker lamps work   Before side marker lamps, a car or light truck was very hard to see from the side, in the dark. A vehicle entering a right-angle intersection, or pulling out of a driveway onto unlit or weakly illuminated roads was virtually invisible to cross traffic, or at least did not provide any visual cues to catch the attention of other drivers. Headlamps, in particular, focus their beams straight ahead and may barely be visible from the side (unless the approaching vehicle is still a substantial distance to the left or right, and there are no vision obstructions such as trees or buildings).

Side marker lamps make cars visible from the side, in the dark, and enable other drivers to avoid collisions. Moreover, the use of amber in the front and red in the back sends a visual cue: "You are approaching the side of a vehicle – here's how big it is, and it's not necessarily getting out of your way." The two colors distinguish the front of a car (white or amber light) from its rear (red light). Side marker lamps are on whenever the parking lights are on, and when they aren't (e.g., on a parked vehicle), the reflectors throw back some of the light from other vehicles' headlamps.

Two systems have been used to meet the standard. One makes the existing parking lamps and tail lamps visible from the side by designing their lens/reflectors to "wrap around" the sides of the vehicle. The other approach is to install dedicated side marker lamps, separate from the parking lamps and tail lamps. Both methods are acceptable under FMVSS 108 and are believed to be essentially equivalent. FMVSS 108, like SAE Recommended Practice J592, requires at least 0.62 candela for amber side marker lamps, and 0.25 candela for red ones, as measured from 9 specified observation points. That makes them considerably less bright than, for example, tail lamps, let alone stop lamps. Side marker lamp/reflectors are too dim to make vehicles more conspicuous by day, and perhaps even to be seen from a long enough distance, in the dark, to avoid high-speed, potentially fatal collisions.[3]

Expected benefits   Side marker lamps ought to reduce the number of angle collisions in reduced lighting conditions: dark not lighted, dark lighted, dawn or dusk. The definition of an "angle" collision varies but generally include crashes where two vehicles approach one another at an angle, typically 90 degrees, or where the front of one vehicle hits the side of the other. The lamps should have little or no effect on daytime crashes. When two vehicles approach at an angle in the dark, there could be benefits if either one is equipped with the lamps, and doubly so if both are equipped: lamps on vehicle 1 can help driver 2 avoid a crash, and lamps on vehicle 2 can help driver 1 avoid it. When the lamps prevent a crash, they not only save the lamp-equipped vehicle from damage, but also the other vehicle that would have been in the crash. Furthermore, if any occupant of either vehicle would have been injured, these injuries are prevented. As stated above, it is doubtful if the lamps would have much effect on fatal crashes.

---

[2] Kahane, C.J., *An Evaluation of Side Marker Lamps for Cars, Trucks and Buses*, NHTSA Technical Report No. DOT HS 806 430, Washington, 1983, pp. 1-7 and 29-32; *Federal Register* 32 (February 3, 1967): 2414, 32 (December 16, 1967): 18033; *1983 SAE Handbook*, Vol. 2, Society of Automotive Engineers, Warrendale, PA, 1983.

[3] Kahane (1983 Side Marker Lamps), pp. 1-7 and 29-32; Cole, B.L., Dain, S.J. and Fisher, A.J., *Study of Motor Vehicle Signal Systems*, HSL Publication No. DOT HS 022 690, Road Safety Information Service, Melbourne, 1977.

Crash avoidance NHTSA's evaluation was published in 1983. It is based on crash data from North Carolina (1971-80) and Texas (1972-74).[4] The data allow a tally of vehicle involvements in angle collisions under reduced-light conditions. They also allow counting a control group of involvements in daytime angle collisions. Tallies are compared, for example, in model year 1967, when only 13 percent of cars, trucks and buses had side marker lamps, and model year 1968, when 88 percent of cars, trucks and buses were equipped with the lamps:

	Angle Collisions		
	Night[5]	Day	Ratio of Night/Day
NORTH CAROLINA			
Model year 1967 (13% with side marker lamps)	5,971	20,755	.288
Model year 1968 (88% with side marker lamps)	7,075	26,351	.268
TEXAS			
Model year 1967 (13% with side marker lamps)	7,738	30,324	.255
Model year 1968 (88% with side marker lamps)	8,715	36,618	.238

In both states, the ratio of nighttime to daytime crashes is a statistically significant 6.7 percent lower in the 1968 vehicles. The arithmetic in the preceding tables works out to 8.8 percent fewer nighttime angle collision when one of the two vehicles has the lamps than when neither has the lamps.[6] That means 16.8 percent fewer nighttime angle collisions when *both* vehicles have the lamps than when neither have them.[7] Nearly identical effectiveness estimates were obtained when the data were expanded to two, three, or four model years before and after the 1968 shift to side marker lamps, or in regression analyses of the ratio of nighttime to daytime angle-collision involvements by presence of side marker lamps and other factors such as vehicle age.[8] The evaluation concluded that the best estimate of side marker lamp effectiveness, based on one of these regressions, is a 16 percent reduction of nighttime angle collisions when both vehicles have the lamps.[9] Using 1979-80 as the "baseline" years, the evaluation estimated that side marker lamps would prevent 106,000 police-reported nighttime angle collisions per year when all cars, trucks and buses on the road were equipped with the lamps.[10]

Injury reduction Estimates of injury reduction were obtained by limiting the preceding analyses of North Carolina and Texas data to injury-producing crashes – i.e., crashes where at least one occupant in any of the involved vehicles (not necessarily the vehicle equipped with side marker lamps) was injured. All of these analyses show a somewhat stronger reduction of injury-producing crashes than property-damage-only crashes. The evaluation's best estimate of side

---

[4] Kahane (1983 Side Marker Lamps), pp. 39-48.
[5] Includes dark not lighted, dark lighted, dawn and dusk.
[6] $(1 - .067) = [.88(1-.088) + .12] / [.13(1-.088) + .87]$
[7] $.168 = 1 - [(1-.088)(1-.088)]$
[8] *Ibid.*, pp. 69-85.
[9] *Ibid.*, pp. 120-122; 90% confidence bounds: 10 to 22 percent.
[10] *Ibid.*, pp. 136-139; 90% confidence bounds: 65,000 to 149,000 crashes prevented.

marker lamp effectiveness is a 21 percent reduction of injury-producing nighttime angle collisions when both vehicles have the lamps.[11] The evaluation estimated that side marker lamps would prevent 93,000 injuries in nighttime angle collisions per year.[12]

Fatality reduction When the preceding contingency-table and regression analyses were performed with 1975-81 FARS data, they did not show any reduction of fatal angle collisions, in the dark, with side marker lamps. A case-by-case review of Multidisciplinary Accident Investigations (MDAI) showed that the great majority of fatal nighttime angle collisions (but not nearly so many nonfatal collisions) involved at least one of these two factors that would make the lamps unlikely to help prevent a crash: (1) a vehicle traveling at least 50 mph, too fast for the driver to see the relatively dim lamps in time to take evasive action; (2) two vehicles approaching at right angles at relatively similar speeds, where drivers ought to see each others' headlamps more easily than the side marker lamps. In addition, many nighttime fatal crashes involve alcohol-impaired drivers who are not likely to be alert to the lamps. The evaluation concluded that side marker lamps, although highly effective in reducing nonfatal crashes and their injuries, have little effect in fatal crashes.[13]

Center High Mounted Stop Lamps (CHMSL)

Regulatory history After much experimental research (1974-79), on-the-road testing (1976-79) and regulatory analysis, NHTSA amended FMVSS 108 in October 1983 to require CHMSL on new passenger cars, effective September 1, 1985. CHMSL were standard equipment on all model year 1986 cars; also on 1985 Cadillacs. In addition, consumers welcomed the lamps, purchased about 4,000,000 retrofit kits and installed them on their pre-1986 cars. An April 1991 amendment extended the CHMSL requirement to pickup trucks, vans and SUVs manufactured on or after September 1, 1993. All light trucks had the lamps in model year 1994, some make-models as early as 1991.[14]

How CHMSL work Stop lamps are a basic tool for preventing rear-impact collisions, daytime and nighttime. When a driver applies the brake pedal, the lamps alert the following driver to slow down immediately, maintain a safe distance, and be ready to stop. Conventional stop lamps are less than optimal for accomplishing their mission. They are low and off toward the sides of the vehicle, not where the following driver is looking most of the time. During quick scanning, they might be confused with adjacent taillights and turn signals, and misinterpreted to mean a car is proceeding or turning, rather than slowing down.

NHTSA and the insurance industry tested various configurations of experimental stop lamps, in the laboratory and on the road, and concluded that the best system combined the two existing stop lamps with a third red lamp, mounted in the center of the vehicle, and if possible higher than

---

[11] *Ibid.*, pp. 48-53, 86-89, and 123; 90% confidence bounds: 12 to 29 percent.

[12] *Ibid.*, pp. 141-143; 90% confidence bounds: 51,000 to 132,000 injuries prevented.

[13] *Ibid.*, pp. 62-66, 100-112 and 125-130; Cole, Dain, and Fisher, *Op. Cit.*

[14] Kahane, C.J. and Hertz, E., *The Long-Term Effectiveness of Center High Mounted Stop Lamps in Passenger Cars and Light Trucks*, NHTSA Technical Report No. DOT HS 808 696, Washington, 1998, pp. 1-2, 9 and 62-63; *Federal Register* 48 (October 18, 1983): 48235, 56 (April 19, 1991): 16015; *Final Regulatory Impact Analysis, Federal Motor Vehicle Safety Standard 108, Center High-Mounted Stop Lamps*, NHTSA Docket No. 81-02-N02-001, 1983.

the other two. The central and raised location of CHMSL puts them in the area where the following driver most often glances. It separates them from tail lamps and turn signals and minimizes confusion. The high mounting might make it visible through the windows of a following vehicle and enable the driver of the third vehicle in a chain to react to the first vehicle's braking. The three lamps form a triangle that might be an additional cue to get a driver's attention and a warning to keep a safe following distance. As a consequence, CHMSL significantly reduced the average time for drivers to apply their brakes in response to the signal from the vehicle in front of them.[15]

Road tests with fleets of taxicabs and telephone-company cars were very encouraging, showing a 35 percent reduction of rear-impact collisions in the CHMSL-equipped cars relative to control-group cars with conventional stop lamps. They paved the way for the CHMSL regulation by suggesting the lamps would be highly beneficial and cost-effective.[16]

Two questions arose during the CHMSL rulemaking in the early 1980's. One was if the high effectiveness in the fleet tests would persist after the lamps became standard equipment on millions of vehicles, or if drivers would gradually become acclimatized to the lamps and less conscious of them. This would be examined by repeated evaluations of CHMSL for about 10 years after they became standard on passenger cars. The other was if CHMSL would be effective on pickup trucks, where they would likely be on top of the cab, far away from all other lamps, or on full-sized vans, where they might be above the following driver's usual line of vision. As a consequence, CHMSL were only required on passenger cars as of September 1, 1985, and the requirement was extended to light trucks as of September 1, 1993 only after additional research satisfied NHTSA that they would be effective there.[17]

Expected benefits CHMSL ought to reduce the probability of being struck in the rear by another vehicle, daytime or nighttime.[18] That will help prevent simple two-vehicle front-to-rear collisions and chain collisions involving three or more vehicles. When the lamps prevent a crash, they not only save the struck, CHMSL-equipped vehicle from damage, but also the striking vehicle. Furthermore, if any occupant of either vehicle would have been injured, these injuries are prevented. To the extent that fatal rear-impact crashes involve much higher speeds and different crash conditions from the nonfatal ones, it is not clear if the lamps would have much effect, if any, on fatal crashes.

---

[15] Kahane and Hertz, pp. 1-2; Digges, K.H., Nicholson, R.M. and Rouse, E.J., *The Technical Basis for the Center High Mounted Stoplamp*, Paper No. 851240, Society of Automotive Engineers, Warrendale, PA, 1985.

[16] Kahane and Hertz, pp. 2-3; Kohl, J.S. and Baker, C., *Field Test Evaluation of Rear Lighting Systems*, NHTSA Technical Report No. DOT HS 803 467, Washington, 1978; Rausch, A., Wong, Jackson I. and Kirkpatrick, M., *A Field Test of Two Single Center High-Mounted Brake Light Systems*, Submission to NHTSA Docket No. 81-02-N01-031, Insurance Institute for Highway Safety, Arlington, VA, 1981; Reilly, R.E., Kurke, D.S. and Buckenmaier, C.C., Jr., *Validation of the Reduction of Rear End Collisions by a High Mounted Auxiliary Stoplamp*, NHTSA Technical Report No. DOT HS 805 360, Washington, 1980.

[17] *Evaluation Plan for Center High-Mounted Stop Lamps*, NHTSA Docket No. 81-02-N02-002, 1983; *Final Regulatory Impact Analysis, Amendment to FMVSS 108 to Require Center High-Mounted Stop Lamps on Light Trucks and Buses*, NHTSA Docket No. 81-02-N10-001, 1991.

[18] Unlike side marker lamps in angle collisions, CHMSL are beneficial only on the struck vehicle. When two vehicles approach at a 90 degree angle, either driver may see the other vehicle's side marker lamps, but when two vehicles travel in the same direction, only the following driver can see the other vehicle's CHMSL.

Crash avoidance – passenger cars NHTSA evaluated CHMSL in 1987, 1989 and 1998; only the last evaluation, which superseded the earlier results, will be discussed here.[19] It is based on 1986-95 crash data from eight states: Florida, Indiana, Maryland, Missouri, Pennsylvania, Texas, Utah and Virginia. These state files identify each crash-involved car's model year and impact site (rear impact vs. other). The basic contingency tables compare the last model year(s) before CHMSL became standard equipment to the first model year(s) they were standard. They tally how many cars were struck in the rear, and how many cars were involved in any other kind of impact (the control group). The following example is based on Florida data for calendar year 1987, comparing the crash experience of model year 1985 and 1986 cars:

| | Collision Involvements | | |
Florida, CY 1987 crashes	Rear Impacts	Others	Ratio of Rear/Other
Model year 1985 (10% with CHMSL[20])	6,773	22,959	.295
Model year 1986 (100% with CHMSL)	7,161	25,989	.276

In this example, the ratio of rear impacts to other-type crash involvements is a statistically significant 6.6 percent lower in the model year 1986 cars than in the 1985 cars, indicating a benefit for CHMSL. However, this analysis understates the effectiveness of CHMSL because it does not adjust for vehicle age. Newer cars in general have relatively more rear impacts (and are less often the striking vehicle), inflating the "ratio of rear/other crashes." The evaluation developed a regression analysis that adjusted for vehicle age and yielded unbiased estimates of the effectiveness of CHMSL. The effect was positive and statistically significant in each of the eight states. When the results from the eight states are combined, the weighted-average effectiveness of CHMSL is statistically significant in each separate calendar year. However, the effect was highest in 1987, dropped in half by 1989, and then leveled off:

Calendar Year	Rear Impact Reduction by CHMSL (%)	Calendar Year	Rear Impact Reduction by CHMSL (%)
1986	5.1	1991	5.0
1987	8.5	1992	5.0
1988	7.2	1993	3.2
1989	4.3	1994	3.4
1990	5.3	1995	3.9

---

[19] Kahane and Hertz, pp. 7-41; Kahane, C.J., *The Effectiveness of Center High Mounted Stop Lamps: A Preliminary Evaluation*, NHTSA Technical Report No. DOT HS 807 076, Washington, 1987; Kahane, C.J., *An Evaluation of Center High Mounted Stop Lamps Based on 1987 Data*, NHTSA Technical Report No. DOT HS 807 442, Washington, 1989.
[20] Assumed to be equipped with CHMSL retrofit kits (not explicitly noted in the crash data).

Effectiveness dropped significantly in 1988 and 1989, but there was no statistically significant downward trend after 1989. The average effectiveness in 1989-95, 4.3 percent, may be considered the long-term effectiveness of CHMSL. The initial and long-term effectiveness estimates, and their 90% confidence bounds[21], are:

Calendar Year Group	Rear Impact Reduction by CHMSL (%)	90% Confidence Bounds
1986	5.1	2.9 to 7.3
1987	8.5	6.6 to 10.5
1988	7.2	5.2 to 9.1
1989-95 (long-term effect)	4.3	3.1 to 5.5

The decline in effectiveness to 4.3 percent long-term from 8.5 percent in 1987 (and from 35 percent in the test fleets of taxicabs and telephone-company cars) is consistent with the hypothesis that motorists would become acclimatized to the lamps. CHMSL are no longer startling, but their good location continues to make them effective.

More detailed analyses of data from five states (Florida, Maryland, Missouri, Pennsylvania and Utah) show CHMSL effective in the dark but even more so in daytime. They are effective regardless whether the following driver is young or old, male or female. In general, the simpler the crash scene, and the less a driver is distracted by other lights or traffic features, the more effective the CHMSL – e.g., at locations away from traffic signals, in rural areas, in simple two-vehicle collisions.[22]

Injury reduction – passenger cars Detailed analyses from five states show CHMSL are almost equally effective in preventing nonfatal-injury and property-damage crashes. In other words, the lamps' long-term effect will be to prevent 4.3 percent of the nonfatal injuries to occupants of any of the vehicles involved in front-to-rear collisions, including occupants of the striking, frontally-impacting vehicles.[23]

Fatality reduction – passenger cars When the basic crash-avoidance analysis was performed with 1986-95 FARS data, it did not show a statistically significant reduction of fatal rear-impact collisions with CHMSL in any specific calendar year, or in all of the calendar years combined. The lamps had little or no effect on fatal crashes at any time during 1986-95. A closer look at the data reveals that, in the majority of fatal rear impacts, the struck vehicle was not slowing or braking at all prior to the impact: the CHMSL was not activated and could not have had an effect.[24]

---

[21] Kahane and Hertz, pp. 39-40 and 81-83 computes 95% confidence bounds; they have been scaled back to 90% confidence bounds here (by taking ± 1.645 rather than ± 1.96 standard deviations), for consistency with most of the other NHTSA evaluations.

[22] Kahane and Hertz, pp. 43-53.

[23] *Ibid.*, pp. 53-54.

[24] *Ibid.*, pp. 54-60.

Effectiveness – light trucks Pickup trucks, vans and SUVs received CHMSL in 1991, 1992, 1993 or 1994, depending on the make-model. NHTSA analyzed 1994-96 crash data from six states that record the impact site and the VIN, allowing identification of make-model and model year: Florida, Maryland, Missouri, North Carolina, Pennsylvania and Utah. The crash-avoidance analysis, similar to the one for passenger cars, estimated that CHMSL in light trucks reduced rear impacts by a statistically significant 5.0 percent. The 90% confidence bounds are 1.0 to 8.7 percent. For practical and statistical purposes, this is more or less the same as the 4.3 percent long-term effectiveness estimate for passenger cars. The data did not show any significant variations in CHMSL effectiveness between pickup trucks, vans and SUVs.[25]

Summary of benefits Using 1994 as the "baseline" year, and based on a 4.3 percent long-term effectiveness in light trucks as well as cars, the evaluation estimated that CHMSL would prevent 92,000 to 137,000 police-reported rear-impact crashes per year, plus 102,000 unreported crashes, when all cars and light trucks on the road are equipped with the lamps. That will save 43,000 to 55,000 people from injury in the police-reported crashes, and 15,000 in the unreported crashes, for a total of 58,000 to 70,000 injuries avoided per year. At 1994 costs and prices, the value of the property damage avoided, even without counting anything for the injuries prevented, far exceeds the consumer cost of CHMSL, making them a highly cost-effective safety device.[26]

Retroreflective tape on heavy trailers

Regulatory history After extensive research, on-the-road testing and analysis, NHTSA amended FMVSS 108 in December 1992 to require retroreflective tape or sheeting, or reflex reflectors, on the sides and rear of heavy trailers (i.e., at least 80 inches wide and with GVWR over 10,000 pounds), effective December 1, 1993. Retroreflective tape has been used almost exclusively for meeting the standard, and it is the subject of NHTSA's evaluation. On new trailers manufactured on or after December 1, 1993, the tape outlines the bottom of the sides of the trailers, and the top corners, bottom and underride guard of the rear of the trailers. It is applied in a pattern of alternating red and white segments to the side, rear and underride guard, and in white to the upper rear corners. FMVSS 108, like other FMVSS, applies to new vehicles at the time they are manufactured or imported.[27]

In addition, the Federal Motor Carrier Safety Administration (FMCSA; and its predecessor, the Bureau of Motor Carrier Safety within the Federal Highway Administration) issued a regulation in March 1999 requiring all heavy trailers on the road on or after June 1, 2001, including older trailers manufactured before December 1, 1993, to have a conspicuity treatment. The FMCSA regulation allows older trailers some flexibility on the location and colors of the tape. Before 1992-93, quite a few manufacturers voluntarily equipped their trailers with tape, but not always according to the pattern subsequently chosen by NHTSA. In other words, FMCSA accepts those patterns for now and does not require an immediate retrofit to the NHTSA pattern. However, as of June 1, 2009, FMCSA will require all heavy trailers on the road to have the NHTSA pattern: many pre-1993 trailers would be retired by then, but if not, they would need a retrofit.

---

[25] *Ibid.*, pp. 61-70.

[26] *Ibid.*, pp. 71-75.

[27] Morgan, C., *The Effectiveness of Retroreflective Tape on Heavy Trailers*, NHTSA Technical Report No. DOT HS 809 222, Washington, 2001, pp. 1-6; *Federal Register* 57 (December 10, 1992): 58406.

Furthermore, the FMCSA regulation applies to trailers-in-use. Conspicuity treatments have to be maintained "good-as-new" and may be inspected as long as trailers are on the road. When tape deteriorates or peels over time, it must be repaired or replaced.[28]

Although these regulations do not apply directly to passenger cars and light trucks, they are included here because their benefits – lives saved, injuries and damages avoided – will accrue primarily to the occupants of the passenger vehicles that are at risk of hitting the trailers.

How retroreflective tape works Heavy trailers are quite difficult for drivers of other vehicles to see in the dark, and even harder to recognize for what they are. The surfaces of most trailers are not shiny and are barely seen until they suddenly loom up, too close to stop in time. Tail lamps, clearance lamps and side marker lamps theoretically outline a trailer, but they are isolated points of light on a large surface. Drivers might not see enough of them at the same time to understand, "There's a heavy trailer in front of me." Collisions of passenger vehicles into heavy trailers are intrinsically dangerous, day or night, because of the weight and height mismatch of the vehicles. In the dark, those collisions were resulting in over 500 fatalities per year during the 1980's.

Human-factors research suggested that continuous or semi-continuous retroreflective tape with alternating red and white segments would highlight and identify trailers. The tape reflects a car's headlights strongly and it is highly visible in the dark. The red and white pattern flags its bearer as a heavy trailer. Since the pattern extends around the sides and rear, a driver can identify a trailer as soon as the headlamps shine on any part of the tape. The standard-length segments in alternating colors additionally help drivers gauge their distance from the trailer and their rate of approach.[29]

A 1983-85 road test of 1,910 trailers equipped with tape and 1,910 control-group trailers showed statistically significant reductions of side and rear impacts with the tape, 21 percent in the dark and, interestingly, 16 percent in daylight. NHTSA proceeded with the rulemaking, predicted a 15-25 percent reduction of side and rear impacts in the dark but, conservatively, did not presume a reduction of daylight crashes.[30]

Expected benefits As stated above, retroreflective tape ought to reduce impacts by other vehicles into the side or rear of a heavy trailer, in the dark. The tape should not affect single-vehicle crashes of tractor-trailers, or crashes where a truck frontally runs into a car. Highest effectiveness is expected on unlighted roads, where a trailer is least visible without the tape, and

---

[28] *Federal Register* 64 (March 31, 1999): 15587.

[29] Burger, W.J., et al., *Improved Commercial Vehicle Conspicuity and Signalling Systems, Task I – Accident Analysis and Functional Requirements*, NHTSA Technical Report No. DOT HS 806 100, Washington, 1981; Ziedman, K., et al., *Improved Commercial Vehicle Conspicuity and Signalling Systems, Task II – Analyses, Experiments and Design Recommendations*, NHTSA Technical Report No. DOT HS 806 098, Washington, 1981. The research indicated that the tape did not have to be fully continuous on the sides of trailers – as little as 50 percent coverage accomplished safety objectives – and FMVSS 108 accordingly requires at least 50 percent coverage on the sides and continuous tape on the rear; however, many trailers have fully continuous coverage on the sides and rear

[30] Burger, W.J., et al., *Improved Commercial Vehicle Conspicuity and Signalling Systems, Task III – Field Test Evaluation of Vehicle Reflectorization Effectiveness*, NHTSA Technical Report No. DOT HS 806 923, Washington, 1985; *Preliminary Regulatory Evaluation – Proposed Amendment to FMVSS No. 108 to Require Retroreflective Material on the Side and Rears of Heavy Trailers*, Submission to NHTSA Docket No. 80-02-N09, NHTSA, 1991.

the reflection from the tape stands out the most strongly relative to the background. Less benefit is expected on lighted roads, or in twilight, because trailers are more visible, and the tape does not stand out as much. Even less benefit, if any, is expected in daylight, because trailers are easily visible without the tape, although the red and white pattern might help as an additional warning.

When the tape prevents a crash, it prevents the injuries to the occupants of any of the vehicles that would have been involved in the crash: the vehicle striking the trailer as well as the tractor pulling the trailer. Of course, most of those injuries would have been in the striking vehicle, and most of the striking vehicles would have been cars or LTVs. Unlike side marker lamps and CHMSL, there is every reason to hope that the tape will save lives, since impacts of passenger vehicles into trailers can easily be fatal at the speeds and sighting distances where the tape is likely to be effective.

The tape might be more effective in rear impacts than side impacts, if only because in most rear-impacts the passenger vehicle's headlights would point at the trailer, and the driver of the passenger vehicle is culpable for not slowing down or stopping in time – whereas in many side impacts the passenger vehicle's headlights might not point at the trailer (e.g., sideswipes), or they might not illuminate the trailer soon enough for the car/LTV driver to take action (e.g., high-speed intersection collisions), or trailer conspicuity is not an issue because the truck driver's action (e.g., sudden lane departure) precipitates the crash.

Crash avoidance NHTSA's evaluation was published in 2001. The analysis classifies crash involvements of heavy trailers (and the tractors that pull them) according to three binomial parameters: (1) tape-equipped vs. unequipped trailer; (2) dark vs. daytime crash; (3) rear- or side-impact vs. control group crash (where trailer visibility is not an issue – single-vehicle crashes and frontal impacts by tractor trailers). Since none of NHTSA's crash data at hand identified whether trailers had retroreflective tape, NHTSA worked out 1997-99 agreements with Florida and Pennsylvania state police to supplement their reports on crashes involving tractor trailers with a special form describing each trailer and its presence/condition of tape.[31]

The basic analysis[32] is a three-dimensional contingency table of the crash involvements of heavy trailers by the three binomial parameters. Here are the combined results for Florida and Pennsylvania:

---

[31] Morgan (2001 Retroreflective Tape), pp. 7-21.
[32] *Ibid.*, pp. 23-37.

	Side & Rear Impacts	Control Group Crashes	Ratio: S & R/ Control Gp	Excess S & R in the Dark
TRAILERS WITHOUT TAPE				
Crashes in the dark[33]	499	597	.836	1.199
Crashes in daylight	930	1,334	.697	
TRAILERS WITH TAPE				
Crashes in the dark	1,038	1,609	.645	0.851
Crashes in daylight	1,897	2,502	.758	

Among trailers without tape, in the dark, there are almost as many side and rear impacts (499) as control group involvements (597 single vehicle crashes and frontal impacts): a ratio of .836. In daylight, that ratio is only .697. Side and rear impacts are overrepresented in the dark by a factor of 1.199 = .836/.697, presumably, because other road users are unable to see the trailers in time to avoid a crash.

Among trailers with the tape, the statistics are reversed. The ratio of side and rear impacts to control group crashes is lower in the dark (.645) than by day (.758) – as it ought to be, since night is a time of lower traffic density, more single-vehicle crashes, and fewer multivehicle crashes resulting in side and rear impacts. Side and rear impacts are now, appropriately, underrepresented in the dark by a factor of 0.851. This is a 29 percent reduction of side and rear impacts in the dark, relative to trailers without the tape, and it is statistically significant.[34] The actual effectiveness of the tape has exceeded the predictions in NHTSA's regulatory analysis (15 to 25 percent).

As expected, the tape is by far the most effective on dark-not-lighted roads, reducing side and rear impacts by a statistically significant 41 percent. In dark-lighted, dawn and dusk conditions, the tape did not significantly reduce crashes. The tape also did not significantly reduce crashes in daylight. Also as expected, the tape appears to be somewhat more effective in preventing rear impacts than side impacts. The tape is more effective when the driver of the impacting vehicle is 15-50 years old than when he or she is over 50 years old. A possible explanation of the difference is that older drivers are less able to see, recognize and/or react to the tape in time to avoid hitting the trailer.[35]

The evaluation estimates that as of June 1, 2001, now that all heavy trailers on the road must have retroreflective tape, it is preventing approximately 7,800 police-reported crashes per year.[36]

---

[33] Includes dark not lighted, dark lighted, dawn and dusk.
[34] 1 – (0.851/1.199) = 29 percent; 90% confidence bounds: 19 to 39 percent; statistical significance was tested by the CATMOD procedure in SAS®, see *SAS/STAT® User's Guide*, Vol. 1, Version 6, 4th Ed., SAS Institute, Cary, NC, 1990.
[35] Morgan (2001 Retroreflective Tape), pp. 39-50.
[36] *Ibid.*, pp. 51-52.

Injury reduction Estimates of injury reduction were obtained by limiting the basic analysis to nonfatal, injury-producing crashes – i.e., crashes where at least one occupant in any of the involved vehicles (usually a passenger vehicle that hit the heavy trailer) was injured. The tape is especially effective, reducing injury-producing side and rear impacts in the dark by a statistically significant 44 percent. The evaluation estimates that retroreflective tape is preventing approximately 3,100 to 5,000 nonfatal injuries per year.[37]

Fatality reduction When these Florida and Pennsylvania analyses are limited to the fatal crashes, they show a statistically significant reduction of side and rear impacts in the dark, with tape. The evaluation estimates that retroreflective tape saves 191 to 350 lives per year.[38]

---

[37] *Ibid.*, pp. 48-49 and 52; 44% estimate includes dark not lighted, dark lighted, dawn and dusk conditions; 3,100 to 5,000 is not an interval estimate, but two point estimates, one based on 29% effectiveness (result for analysis of crashes of all severities), the other based on 44%.

[38] *Ibid.*, pp. 48-49 and 52-53; 191 and 350 are two point estimates, based on 29% and 44% effectiveness, respectively, as in the analysis of nonfatal injuries.

FMVSS 201: Occupant protection in interior impact

This standard is indirectly associated with one vehicle modification whose safety benefits have been evaluated by NHTSA and more directly associated with two recent innovations that the agency is just beginning to evaluate:

- *Redesign of mid and lower instrument panels with improved occupant protection*
- *Head injury protection upgrade (not yet evaluated)*
- *Head air bags (not yet evaluated)*

The first paragraph of FMVSS 201 says, quite simply, "This standard specifies requirements to afford impact protection for occupants."[1] Primarily, FMVSS 201 consists of performance requirements limiting the amount of force allowed when a headform is impacted into various sections of the vehicle interior that are typically contacted by occupants' heads during crashes. FMVSS 201 was one of NHTSA's initial safety standards, effective for passenger cars on January 1, 1968. It was extended to pickup trucks, vans and SUVs up to 10,000 pounds GVWR, effective September 1, 1981. More recently, performance requirements for cars and LTVs were upgraded by adding a 15 mph free-motion headform (FMH) impact test of the upper interior components, with a phase-in from September 1, 1998 to September 1, 2002.

NHTSA believes that most passenger cars met the original head-impact requirements of FMVSS 201 quite a few years before their effective date of January 1, 1968. The substantial vehicle modifications close to 1968 were in the mid and lower instrument panel, typically areas for contact by the occupant's torso and knees. These areas had been contemplated, but ultimately not included for regulation by FMVSS 201.

Redesign of mid and lower instrument panels with improved occupant protection

Regulatory history Instrument panel tops that were padded in some form were standard or optional on most passenger cars by 1963. The Society of Automotive Engineers issued Recommended Practice J921, defining a 15 mph headform impact test with the instrument panel, in June 1965. NHTSA's original, December 1966 NPRM for FMVSS 201 incorporated the SAE Recommended Practice and extended it to other likely head contact areas such as seatbacks, sun visors, armrest and projections in head impact areas. The August 16, 1967 final rule finalized specifications for instrument panels, seatbacks, sun visors, and armrests. The final rule set a peak deceleration of 80 g's over 3 milliseconds for impact tests of instrument panels and seatbacks and certain design criteria for the sun visor and armrest components. Most cars were apparently meeting the various head-impact requirements of FMVSS 201 well before 1968.[2]

---

[1] *Code of Federal Regulations*, Title 49, Government Printing Office, Washington, 2002, Part 571.201.

[2] Kahane, C.J., *An Evaluation of Occupant Protection in Interior Impact for Unrestrained Front Seat Occupants of Cars and Light Trucks*, NHTSA Technical Report No. DOT HS 807 203, Washington, 1988, pp. 2-3; Campbell, B.J., *A Study of Injuries Related to Padding on Instrument Panels*, HSL Publication No. 00427812, Report No. VJ-1823-R2, Cornell Aeronautical Laboratory, Buffalo, 1963; *Federal Register* 31 (December 3, 1966): 15212; *1967 SAE Handbook*, Society of Automotive Engineers, New York, 1967, pp. 881-884.

On the other hand, this NPRM also contemplated force limits for "knee" and "child" impact areas – i.e., the lower and mid instrument panel areas anterior to the right-front passenger – but they did not become part of FMVSS 201. In 1970, NHTSA again proposed a knee impact test, but did not issue a final rule. Thus, even though FMVSS 201 never actually regulated the mid and lower instrument panels, the repeated proposals to do so may have influenced the ultimately voluntary improvements to crashworthiness in those areas.[3]

NHTSA's 1970 NPRM also proposed to extend the original FMVSS 201 requirements to light trucks. That part of the proposal was adopted and took effect on September 1, 1981.[4]

Vehicle modifications, and their purpose In the 1960's, most right-front passengers were unrestrained. Safety researchers such as Daniel at Ford and Wilson at GM grasped that the instrument panel was the principal energy-absorbing "ride-down" mechanism for the unrestrained passenger in a frontal impact. The panel should decelerate the occupant as completely, but gradually as possible, at no time applying a dangerous level of force to the occupant. For the most part, instrument panels of the 1960's were too rigid and needed to be softened – but not so much that the occupant would break through to the even more rigid structures in front of the panel. Daniel believed panels with 1,200 pounds maximum force deflection for the chest in a 20 mph impact, and 1,400 pounds optimum for the knees in a 12 mph impact could protect even unrestrained occupants in 30 mph barrier impacts.[5]

A second objective was to "tune" the relative rigidity of the mid and lower instrument panels so that the occupant would remain upright during a frontal impact. For example, an excessively rigid lower panel might cause the occupant to pitch forward, head first. A third objective was to bring the panel as close to the occupant as possible, and to extend it downwards, consistent with the occupant's comfort. The sooner contact begins, and the more extensive the engagement with the occupant's knees, the more gradual the ride-down.

Happily, the new insight on the design of the panel coincided with the increasing feasibility and economy of plastics or thinner steel rather than heavy steel panels for assembling interior components.

NHTSA measured the geometry and force deflection characteristics of instrument panels in 21 cars ranging from MY 1965 to 1983. The analysis confirmed that manufacturers, especially in the late 1960's, and continuing in the early 1970's, significantly changed instrument panel design in the directions that their researchers believed to be safety-favorable. Panels became less rigid,

---

[3] Kahane (1988 Interior Impact), pp. 2-3; Williams, A.F., Wong, J. and O'Neill, B., "Occupant Protection in Interior Impacts – An Analysis of FMVSS No. 201," *Proceedings of the Twenty-Third Conference of the American Association for Automotive Medicine*, American Association for Automotive Medicine, Morton Grove, IL, 1979; *Federal Register* 31 (December 3, 1966): 15212, 35 (September 25, 1970): 14936.
[4] *Federal Register* 44 (November 29, 1979): 68470.
[5] Kahane (1988 Interior Impact), pp. 9-15; Daniel, R.P., "Vehicle Interior Safety Constraint Systems," *1970 International Automobile Safety Conference Compendium*, Paper No. 700423, Society of Automotive Engineers, New York, 1970; Wilson, R.A., "Evaluating Knee-to-Instrument Panel Impacts," *Thirteenth Stapp Car Crash Conference*, Paper No. 690801, Society of Automotive Engineers, New York, 1969.

extended downwards, and contoured in a manner that occupants would feel comfortable sitting closer to them.[6]

The implementation dates for these changes in LTVs are unknown. However, other safety improvements of that era, such as 3-point safety belts and energy-absorbing steering assemblies, typically appeared in LTVs about 5-10 years after cars. That suggests the panels may have been modified in LTVs more or less throughout the 1970's, which also were the decade between the NPRM to extend FMVSS 201 to light trucks (1970) and the final effective date (September 1, 1981).

Expected benefits Since the instrument panel was by far the most common source of life-threatening as well as moderate injury for unrestrained right-front passengers in frontal crashes, the improvements of the late 1960's and early 1970's could be expected to reduce injuries and fatalities.[7]

Injury reduction – passenger cars NHTSA's study, published in 1988, used data from the 1977-79 National Crash Severity Study[8] (NCSS). It compared non-minor (AIS $\geq$ 2) injury rates, due to instrument panel contact, per 100 front-seat passengers involved in frontal towaway crashes, for four model-year groups of cars: 1960-66, 1967-70, 1971-74 and 1975-78. After control for Delta V (an indicator of crash severity), the three later model-year groups each had significantly lower injury rates from instrument-panel contacts than the 1960-66 cars. The reductions were 23, 36 and 21 percent, respectively. The average reduction for all 1967-78 cars, relative to the 1960-66 cars, was a statistically significant 29 percent.[9] The differences among the three post-standard groups (1967-70, 1971-74 and 1975-78) are not statistically significant. The reduction of injuries due to contact with the instrument panel was not offset by any increase of injuries from other sources. Using 1982 as the "baseline" year, the evaluation estimated that when the pre-1967 cars are phased out and all cars on the road have the improved instrument panels, AIS $\geq$ 2 injuries to right-front passengers will decrease by an estimated 8,000 per year.[10]

Computer simulations of frontal crashes were consistent with the NCSS results. MVMA2D was a computer program to simulate an occupant's motion and compute the injury-criterion scores for contacts with the vehicle interior in a straight-ahead frontal crash. As stated above, NHTSA measured the geometry and force-deflection characteristics of instrument panels (and other structures) in 21 cars ranging from MY 1965 to 1983. These data were used to set up MVMA2D, which was run six times for each car to simulate the motion of an unbelted right-front passenger: for 50th percentile male, 5th percentile female and 95th percentile male dummies, each at 25 mph and 30 mph frontal barrier impacts. In each of these 126 simulations, MVMA2D estimated the Head Injury Criterion (HIC), the chest g's and deflection, the femur load, and six other injury criteria for the head, chest and neck.[11]

---

[6] Kahane (1988 Interior Impact), pp. 43-109.
[7] *Ibid.*, pp. xxvii and 15-20.
[8] Kahane, C.J., Smith, R.A. and Tharp, K.J., "The National Crash Severity Study," *Report on the Sixth International Technical Conference on Experimental Safety Vehicles*, NHTSA Report No. DOT HS 802 501, Washington, 1977, pp. 493-515.
[9] 90% confidence bounds: 10 to 48 percent.
[10] Kahane (1988 Interior Impact), pp. 21-41.
[11] *Ibid.*, pp. 43-109 and 205-220.

Injury scores were compared on a nonparametric basis[12] – i.e. the analysis shows if one group of cars is significantly safer than another, but does not estimate the percentage reduction of injury risk. The trends are the same as in the NCSS data. The 1969-71 cars and 1976-83 cars both had significantly lower overall injury risk than the 1965-66 cars. The average injury risk in 1976-83 was slightly, but not significantly lower than in 1969-71. Injury risk to the head/neck, chest and femur were all lower in 1969-71 and 1976-83 than in 1965-66.

Moreover, the MVMA2D simulations generated results that were consistent with researchers' intuition about what sort of panel modifications would benefit safety. Reductions in the rigidity of the lower and mid panels, and lengthening the panels by extending them downwards and towards the occupant, were all associated with significant injury reductions.

Injury reduction – light trucks NHTSA analyzed overall AIS $\geq$ 2 injury rates per 100 unrestrained right-front passengers in frontal towaway crashes in 1982-85 NASS and 1977-79 NCSS data. In those days, LTVs accounted for a smaller share of the market, and there weren't enough cases to adjust the rates for Delta V or other factors, or to single out injuries due to contact with the instrument panel. The analysis did not show a clear reduction in the injury rates. In NASS, the injury rate for model year 1971-74 LTVs was 43 percent lower than in 1966-70 LTVs. But the reduction relative to 1966-70 subsequently dwindled to 29 percent in 1975-78, 5 percent in 1979-81, and 16 percent in 1982-85. In NCSS, the 1971-74 LTVs had 33 percent lower injury rate than 1961-70 LTVs, but the 1975-78 LTVs had the highest rate, 24 percent higher than in 1961-70. While the reductions in 1971-74 are effects in the right direction, the subsequent trends toward higher injury rates are inconsistent with the idea that panel improvements continued throughout the 1970's. Small samples and inability to control for Delta V or other factors made it difficult to obtain statistically meaningful results.[13]

Fatality reduction – passenger cars The fatality "risk index" of right-front passengers in frontal crashes was computed by model year (1964-84), based on a two-step analysis of 1975-84 FARS data. The first step was the computation of an absolute risk index for drivers. The analysis compares the ratio of driver fatalities in head-on (front-to-front) collisions between cars of two different model years. For example, in head-on collisions between MY 1975 and 1979 cars, 37 drivers of the MY 1975 cars and 90 drivers of the MY 1979 cars were fatalities. In these unadjusted data, relative vehicle weight was the dominant factor: since many of the MY 1979 cars had been downsized, they had higher risk in the head-on collisions with the 1975's. Next, a logistic regression calibrated the expected fatality risk of each driver in a head-on collision, given only the curb weights of the two cars, and the age, gender and belt use of the two drivers. In each model year's cars, the actual number of fatalities in the head-on collisions is compared to the number that would have been expected, based on their own, and their collision partners' curb weight, driver age, gender and belt use. The fatality risk index is the ratio of actual to expected fatalities (normalized to 100).

The fatality risk index for drivers averaged 117.3 in model years 1964-66, had dropped sharply to 103 by 1968-70, and remained close to 100 throughout 1973-84. The drop from 117.3 to 103,

---

[12] Using a weighted average of the normalized rank order scores for the various criteria, for the six simulations on each car.

[13] *Ibid.*, pp. 165-173.

a 12 percent reduction, corresponds closely to the effectiveness estimate for energy-absorbing steering assemblies.[14] In other words, energy-absorbing steering assemblies significantly reduced the fatality risk of unrestrained drivers in frontal crashes in 1967-68, and no other vehicle modification in 1964-84 had any large effect.[15]

The second step was the computation of a risk index for right-front passengers relative to drivers. It is based on the ratio of right-front passenger to driver fatalities in frontal impacts (including, but not limited to head-on collisions) in cars of a given model year that were occupied by a right-front passenger as well as a driver, both unrestrained. For example, in frontal impacts of MY 1975 cars with both seats occupied, 1410 passengers and 1472 drivers were fatalities. Next, a logistic regression calibrated the expected fatality risk of each driver and each passenger in these impacts, given only the age and gender of the two occupants and the exact impact location (pure frontal, left front, right front). In each model year's cars, the actual ratio of right-front to driver fatalities is compared to the ratio that would have been expected, based on the two occupants' age/gender and the impact location. The fatality risk index is the actual ratio divided by the expected ratio (normalized to 100).

The relative risk index was 120-125 in model years 1960-65, dropped sharply in the late 1960's, and continued decreasing slowly, leveling out at close to 100 in 1975-84. In other words, the right-front seat was at one time substantially more dangerous than the driver's seat. But the safety gap was closed, despite the introduction of energy-absorbing steering assemblies that benefited only drivers. Something(s) that was even more effective than the energy-absorbing steering assemblies must have been introduced for passengers.[16]

The absolute fatality risk index for right-front passengers is the product of the driver and the passenger-relative-to-driver indices. This index averaged 135.7 in model years 1964-66, 109.1 in 1968-70, and 99.7 in 1973-84. The drop from 135.7 to 109.1 is a 20 percent reduction, equivalent to 665 lives saved per year. The more gradual change from 109.1 to 99.7 is an additional 9 percent reduction, equivalent to 235 lives saved per year.[17]

The redesign of instrument panels was the most important crashworthiness improvement for unrestrained right-front passengers in frontal crashes and must have accounted for many of these 665 + 235 = 900 lives saved per year, but it is difficult to estimate how many. The trend in the fatality risk index closely parallels the trends in the NCSS data and MVMA2D simulations: a strong improvement in the late 1960's, with some additional improvements in the early 1970's and leveling out after that. That is consistent with the hypothesis (but not a proof) that the fatality reduction largely reflects the instrument panel improvements. It is estimated from NCSS data that before FMVSS 201, 616 right-front passengers per year received fatal injuries from contact with the instrument panel, while Huelke's in-depth investigations suggested the number could be twice as high. If the fatality reduction for improved panels is the same as the AIS ≥ 2 injury reduction, 29 percent, that would amount to annually preventing 180-360 fatalities due to contact with the instrument panel. However, the MVMA2D simulations suggest that the benefits

[14] See the FMVSS 203/204 chapter.
[15] *Ibid.*, pp. 111-140.
[16] *Ibid.*, pp. 140-153.
[17] *Ibid.*, pp. 153-156 and 161.

of improved panels are not limited to preventing injuries directly caused by contact with the panel: by providing "ride-down" and keeping the occupant upright, they can reduce the severity of contacts with other interior surfaces. Other safety standards that benefited unrestrained passengers in frontal impacts, such as FMVSS 205 or 212, are unlikely to have saved more than 100 passengers per year. Taking all these considerations into account, NHTSA estimated that improved panels account for approximately 400 to 700 of the 900 passenger lives saved per year in frontal crashes. The point estimate is 550, and it is equivalent to a 16 percent reduction of unrestrained right-front passenger fatalities in frontal impacts.[18]

Fatality reduction – light trucks Using 1975-84 FARS data, NHTSA developed fatality risk indices for drivers and right-front passengers of light trucks, analogous to the indices for cars. The absolute risk index for right-front passengers dropped from an average of 127 in model years 1966-68 to an average of 98 in 1975-80. The drop was sharpest in 1969-75 and gradually leveled out afterwards. That is a 23 percent fatality reduction during the decade when instrument panels were modified in light trucks. That is similar to the long-term overall reduction in the risk index for car passengers, and it suggests that the effectiveness of improved panels may be about the same in light trucks and cars: a 16 percent reduction of unrestrained right-front passenger fatalities in frontal impacts.[19]

Head injury protection upgrade

On August 14, 1995, NHTSA issued a Final Rule extending the head injury protection requirements of FMVSS 201. The existing requirements of FMVSS 201 remained for the original target areas. However, a new list of target areas in the vehicle's upper interior includes the A-, B- and other pillars, the front and rear roof header, the roof side rails, and the upper roof, among others. These structures are sources of life-threatening head injuries in existing vehicles. The impact speed for the free-motion headform (FMH) impact test for the new areas is 15 mph, as in the original FMVSS 201, but for these targets, the Head Injury Criterion (HIC) may not exceed 1000 for any 36-millisecond period. Impacts may be directed from a range of vertical and horizontal angles, not just head-on. Manufacturers were offered a choice of several alternative phase-in schedules from September 1, 1998 to September 1, 2002. For example, they could certify the new requirements on at least 10 percent of cars and LTVs manufactured from September 1, 1998 through August 31, 1999; at least 25 percent of cars and LTVs manufactured from September 1, 1999 through August 31, 2000; at least 40 percent of cars and LTVs manufactured from September 1, 2000 through August 31, 2001; at least 70 percent of cars and LTVs manufactured from September 1, 2001 through August 31, 2002; and all cars and LTVs manufactured on or after September 1, 2002. This will be a high-priority evaluation for NHTSA.[20]

---

[18] *Ibid.*, pp. 15-18, 38-39, 92-106 and 161-163; Huelke, D.F. and Gikas, P.W., "Causes of Death in Automobile Accidents," *Police*, Vol. 13, November/December 1968, pp. 81-89.

[19] Kahane (1988 Interior Impact), pp. 174-197.

[20] *Federal Register* 60 (August 18, 1995): 43031; *National Highway Traffic Safety Administration Evaluation Program Plan, Calendar Years 2004-2007*, NHTSA Report No. DOT HS 809 699, Washington, 2004, p. 8.

Head air bags

During the 1990's, manufacturers and suppliers developed head air bags to provide additional protection, especially during lateral impacts, and possibly to act as a barrier to occupant ejection through side windows. NHTSA believes they are a promising safety technology. On July 29, 1998, NHTSA amended FMVSS 201 to facilitate the introduction of head air bags. Recognizing that the 15 mph headform test might be a problem in target areas where the undeployed air bag is stored (and, furthermore, an inappropriate test if the bag usually deploys at that speed), NHTSA offered an alternative compliance procedure. Manufacturers have the option to reduce the speed of the headform test to 12 mph on target areas where the bag is stored, provided they can meet an 18 mph lateral (90 degree) crash test for the full vehicle into a pole – with HIC < 1000. The pole test simulates a head impact with the deployed bag. In model year 1999, BMW and Volvo introduced head air bags in some of their cars. By 2001, many of the large manufacturers offered them on a variety of cars and/or LTVs. Designs of head air bags include "curtains" or "tubes" that drop down from the roof rail into the side-window area, and side air bags that deploy from the seat but extend upward far enough to protect the head impact zones around the side window. NHTSA's annual *Buying a Safer Car* brochures inform the public what make-models are equipped with head air bags. This, too, will be a high-priority evaluation for NHTSA.[21]

---

[21] *Federal Register* 63 (August 4, 1998): 41451; *Buying a Safer Car 2000*, NHTSA Publication No. DOT HS 809 046, Washington, 2000; *Buying a Safer Car 2001*, NHTSA Publication No. DOT HS 809 152, Washington, 2000; *Buying a Safer Car 2002*, NHTSA Publication No. DOT HS 809 409, Washington, 2002; *Buying a Safer Car 2003*, NHTSA Publication No. DOT HS 809 546, Washington, 2003; *NHTSA Evaluation Program Plan, CY 2004-2007*, p. 13.

FMVSS 202: Head restraints

This standard regulates one safety technology:

- ***Head restraints for front-outboard seats***

They are the one safety device aimed at mitigating injury risk in rear impacts, cushioning the heads of drivers and right-front passengers. They were introduced in passenger cars over model years 1964-69 and in light trucks over a longer period, extending until 1992. The first paragraph of FMVSS 202 states, "this standard specifies requirements for head restraints to reduce the frequency and severity of neck injury in rear-end and other collisions."[1]

Regulatory history FMVSS 202, applying at first only to passenger cars, was proposed in December 1967, became a Final Rule in February 1968 and had an effective date of January 1, 1969. It incorporates parts of General Services Administration Standard 515/22, effective October 1967 for federally purchased vehicles. The extension of FMVSS 202 to light trucks with GVWR up to 10,000 pounds was issued in September 1989 and took effect on September 1, 1991.[2]

How head restraints work "Whiplash" is by far the most frequent injury in rear-impact crashes. It is common even in low-speed, non-towaway crashes. It results from strains, tears or microscopic damage to muscles, tendons, ligaments, vertebra or nerves in or near the neck. The most familiar symptoms are pain or stiffness in the neck and upper back. Also common are pain or weakness in the arms or shoulders, and symptoms involving the central nervous system, such as headache, sight or hearing disturbances. Symptoms may appear immediately or not for several days. They may end in a day or two or they may go on, continually or intermittently, for a long time, even years.[3]

The first causal mechanism identified for whiplash was hyperextension of the neck. In a rear impact, a low, pre-FMVSS 202 seatback holds the occupant's back in place but the head jerks backward, over the top of the seatback, twisting the neck and injuring it. The remedy was to develop a head restraint that effectively makes the seatback higher, keeps the head upright and in line with the torso, and prevents twisting of the neck. However, subsequent research suggests that hyperextension is by no means the only mechanism of whiplash; other causes may include the longitudinal movement of the head relative to the torso, or rapid changes in spinal column pressure during the impact.

---

[1] *Code of Federal Regulations*, Title 49, Government Printing Office, Washington, 2002, Part 571.202.

[2] Kahane, C.J., *An Evaluation of Head Restraints - Federal Motor Vehicle Safety Standard 202*, NHTSA Technical Report No. DOT HS 806 108, Washington, 1982, pp. 104-106; Walz, M.C., *The Effectiveness of Head Restraints in Light Trucks*, Technical Report No. DOT HS 809 247, Washington, 2001, p. 1; *Federal Register* 31 (March 8, 1966): 4096, 31 (July 15, 1966): 9637, 32 (December 28, 1967): 20865, 33 (February 14, 1968): 2945, 54 (September 25, 1989): 39183.

[3] Kahane (1982 Head Restraints), pp. 80-83; Walz (2001), pp. 2-5; Mertz, H.J., Jr., and Patrick, L.M., "Investigation of the Kinematics and Kinetics of Whiplash," *Proceedings of Eleventh Stapp Car Crash Conference*, Society of Automotive Engineers, New York, 1967; States, J.D., Korn, M.W. and Massengill, J.B., "The Enigma of Whiplash Injuries," *Proceedings of the Thirteenth Annual Conference of the American Association of Automotive Medicine*, Morton Grove, IL, 1969; *Accident Analysis and Prevention*, Vol. 32, No. 2, March 2000.

Research on rear impacts and development of prototype head restraints began at the University of California, Los Angeles in the mid-1950's. The manufacturers have developed three main types of head restraints for production vehicles:

- Adjustable head restraints that may be lifted or lowered to suit the occupant. The head restraint should be lifted until the top of the restraint is at least level with the tops of the occupant's ears, or preferably as high as it can go without blocking the driver's vision. Some head restraints can also be adjusted forwards and backwards; if so, they should be moved forward as far as possible consistent with the occupant's comfort.

- Integral head restraints, such as "captain's chairs" or other seatbacks that extend upwards behind the driver's head.

- Fixed head restraints, such as padded tubes, shaped like an inverted "U," attached to the top of the seatback.

The test for FMVSS 202 requires the top of the head restraint to reach at least 27.5 inches above the "seating reference point" (the level of the seat cushion when the seat is occupied). That height provides adequate support for the head and neck of occupants up to 5'10" tall. If it is an adjustable head restraint, it only needs to achieve this height in the "up" position. The restraint must be able to withstand a static force of 200 pounds in the rearward direction, or meet a dynamic test.[4]

In actual pre-standard cars, the median seatback, without a head restraint, was 23 inches tall, well below the level that would be set by FMVSS 202. In actual cars of the 1970's, the median integral or fixed head restraint – and the median adjustable head restraint in the "up" position – reached 29 inches, exceeding the FMVSS 202 requirement. However, at that time occupants left 75 percent of adjustable head restraints in the "down" position, at a median height of 26 inches, and did not obtain full protection if they were 5'8" or taller. Integral or fixed head restraints predominated on small cars with bucket seats, but were becoming infrequent on bench or split-bench seats. In all, 30 percent of cars had integral or fixed head restraints, while 70 percent had adjustable head restraints in the 1970's. Although adjustable head restraints were often mispositioned, consumers seemed to prefer them. A 1974 NHTSA proposal to require integral or fixed head restraints for drivers, increase height requirements, and establish a minimum height for adjustable restraints in the "down" position was not adopted.[5]

By the mid-1990's, the situation had deteriorated in some ways and improved in others. Just 12 percent of new cars had integral or fixed head restraints. On the other hand, occupants were now leaving only 50 percent of the adjustable restraints "down," and the median height of adjustable

[4] Kahane (1982 Head Restraints), pp. 103-107; Severy, D.M., Brink, H.M. and Baird, J.D., *Backrest and Head Restraint Design for Rear-End Collision Protection*, Paper No. 680079, Society of Automotive Engineers, New York, 1968; *Code of Federal Regulations*, Title 49, Part 571.202.
[5] Kahane (1982 Head Restraints), pp. 113-123 and 257-259; *Federal Register* 39 (March 19, 1974): 10268.

head restraints in the "down" position had improved to 27 inches – close to the FMVSS 202 requirement for the "up" position. Also, integral head restraints predominated in light trucks.[6] Some questions about head restraints have been raised time and again during the past 30 years. (1) Are integral/fixed restraints more effective than adjustable head restraints, as used (often incorrectly adjusted), and if so, why don't more cars have integral or fixed restraints? (2) Should head restraints be increased in height (both in the "up" and the "down" positions) to protect taller-than-average occupants? How much? Would consumers accept them? (3) How important is it for the head restraint to be close to the head? Should head restraints be adjustable backwards and forwards, and/or should FMVSS 202 specify a maximum longitudinal distance from the head restraint to the head? (4) In combination with a head restraint, is a rigid seatback or a flexible seatback more desirable for mitigating whiplash? (5) Will head restraints be less effective in pickup trucks with conventional cabs than in other vehicles to the extent that the back wall of the cab, immediately behind the front seat, provides some degree of protection?[7]

Some passenger cars were equipped with head restraints as early as 1964, well before the January 1, 1969 effective date. However, only 3 percent of model year 1967 cars and 12 percent of model year 1968 cars had them. In model year 1969, as FMVSS 202 took effect in mid-year, 88 percent of cars were equipped with them. The extension of FMVSS 202 to light trucks was proposed in 1988 and took effect in September 1991. Most models of domestic pickup trucks were first equipped with head restraints between the proposal and the effective date (model years 1989-92). By contrast, the majority of vans and SUVs were already equipped with head restraints or high seatbacks potentially meeting FMVSS 202 as early as model year 1983.[8]

Expected benefits Drivers and right-front passengers can expect to receive benefits primarily in rear-impact crashes, since other impacts are far less likely to propel occupants backwards, directly toward the head restraint. "Whiplash" is usually a complaint-of-pain injury, seldom life-threatening, but it can manifest itself in many body regions. A significant fatality reduction, or a significant reduction in lacerations or related injuries is not anticipated. There is likely to be a reduction of minor/moderate non-bleeding injuries to the neck, the head and quite possibly the arms and back, even in crashes of low severity (non-towaways).

Injury reduction – passenger cars NHTSA's analysis, published in 1982, is based on 1972, 1974 and 1977 Texas data.[9] Texas files have records on every crash-involved driver, injured or not, and permit computation of the overall injury rate per 100 drivers in rear-impact crashes. Injury rates are compared for model year 1968, when only 6 percent of cars had adjustable and 6 percent had integral head restraints, and model year 1969, when 81 percent of cars had adjustable, and 7 percent had integral head restraints:

---

[6] *Head Restraints – Identification of Issues Relevant to Regulation, Design, and Effectiveness*, NHTSA Performance Standards, Washington, 1996 (www.nhtsa.dot.gov/cars/rules/crashworthy/headrest/status9/status9.html#28), pp. 19-23.
[7] Walz (2001), pp. 2-5.
[8] Kahane (1982 Head Restraints), p. 104; Walz (2001), p. 2.
[9] Kahane (1982 Head Restraints), pp. 181-197.

	N of Drivers	n Injured	Percent Injured
Model year 1968 (6% with adj. head restraints)	20,214	1,531	7.57
Model year 1969 (81% with adj. head restraints)	23,051	1,605	6.96

The injury rate is significantly lower in the 1969 cars. The arithmetic works out to a 10.4 percent injury reduction for adjustable head restraints relative to no head restraints. An identical estimate was obtained when the reduction in rear-impact injury rates was measured relative to a control group of side impacts, over a wider range of "before" and "after" model years.

Next, the analysis compared a group of 1970-77 make-models equipped almost exclusively (96%) with integral restraints to another group of make-models equipped with 97 percent adjustable head restraints.[10] In Texas, the injury rates in rear impacts were:

	N of Drivers	n Injured	Percent Injured
Models with 97% adjustable head restraints	17,758	1,394	7.85
Models with 96% integral head restraints	21,205	1,552	7.32

The injury rate is significantly lower in the make-models with primarily integral head restraints. It works out to a 7.3 percent injury reduction for integral relative to adjustable head restraints. The injury reduction for integral head restraints relative to no head restraints at all is 16.9 percent – i.e., $1 - [(1 - .104)(1 - .073)]$. Head restraints have significantly reduced the frequency of whiplash in crashes, but they have not nearly eliminated this type of injury.

Using 1979 as the "baseline" year, the evaluation estimated that an on-the-road fleet with 100 percent adjustable head restraints would prevent 52,000 injuries per year in passenger cars, but a fleet with 100 percent integral or fixed head restraints would prevent 85,000 injuries per year.[11]

Fatality reduction – passenger cars In-depth investigations of fatal rear-impact crashes showed that most of the fatalities involved fire, occupant ejection or catastrophic intrusion: mechanisms unlikely to be influenced much, one way or the other, by head restraints. NHTSA analyzed 1975-80 FARS data by various techniques: rear-impact fatalities, before and after FMVSS 202, relative to control groups of fatalities in other crash modes; regression analyses; rear-impact

---

[10] *Ibid.*, pp. 212-225.

[11] Ibid., pp. 237-245; 90% confidence bounds for effectiveness of adjustable head restraints: 4.0 to 16.8 percent, for integral head restraints: 9.1 to 24.7 percent; for benefits of adjustable head restraints: 17,000 to 87,000 injuries prevented, for integral head restraints: 40,000 to 130,000 injuries prevented per year.

fatality rates per million vehicle years. None of these analyses showed a statistically significant effect for head restraints.[12]

Injury reduction – light trucks NHTSA's 2001 evaluation of head restraints for light trucks used 1993-98 data from eight states: Florida, Indiana, Maryland, Missouri, North Carolina, Pennsylvania, Texas and Utah. Statistics were based on seven domestic make-models of pickup trucks that shifted during 1990-92 from few or no head restraints in one model year to 100 percent or near-100 percent equipped with head restraints in the next year. Injury rates were compared for the first two model years with head restraints and the last two years before head restraints. Vans and SUVs were mostly equipped with head restraints from the start of production, precluding computation of "before" injury rates, and could not be included in the analyses. Injury rates were computed for drivers and right-front passengers in rear-impact crashes and for a control group of frontal crash involvements.[13] Injury rates for the combined eight state date files were:

	N of Occupants	n Injured	Percent Injured
REAR IMPACTS			
Last 2 MY before head restraints	43,520	11,500	26.42
First 2 MY with head restraints	50,434	12,307	24.40
FRONTAL IMPACTS			
Last 2 MY before head restraints	98,541	23,581	23.93
First 2 MY with head restraints	106,417	24,631	23.15

Clearly, the injury reduction is greater in the rear impacts (where head restraints are effective) than in the frontal impacts (where head restraints ought to have no effect, and any difference between the "before" and "after" is due to various biases in the data). When the data are weighted by the number of registered pickup trucks in each state, rather than merely aggregated, and after corrections for the small percentages of trucks in the "before" group that already had head restraints and for those in the "after" group that still didn't, this works out to a statistically significant 6.1 percent injury reduction for head restraints in rear impacts (relative to the control group). Using 1999 as the "baseline" year, the evaluation estimated that when all light trucks on the road (pickups, vans and SUVs) are equipped with head restraints, they will prevent 14,900 injuries per year.[14]

It is unknown why head restraints are more effective in passenger cars (10.4 – 16.9 percent) than in light trucks (6.1%). Perhaps the large mass and extensive rear structure of light trucks, especially pickup trucks, offers a degree of protection in rear impacts that makes it less

---

[12] *Ibid.*, pp. 91-95 and 161-177.
[13] Walz (2001), pp. 9-27 and 41-46.
[14] *Ibid.*, pp. 65-68; 90% confidence bounds for effectiveness: 3.5 to 8.7 percent; for benefits: 8,300 to 21,700 injuries prevented, per year.

necessary to have head restraints. Also, this evaluation did not show a significant difference in the effectiveness of adjustable vs. integral head restraints in pickup trucks.[15]

Fatality reduction – light trucks NHTSA's 2001 evaluation does not analyze fatal crashes separately. However, it is reasonable to infer that, just as in passenger cars, head restraints have little or no effect on fatalities in light trucks.

Proposed FMVSS 202 upgrade On January 4, 2001, NHTSA published a proposed rulemaking to upgrade FMVSS 202. The new rule would require:

- Front-seat head restraints in cars and light trucks to reach a height of 31.5 inches in the "up" position (vs. 27.5 inches in the current FMVSS).
- Front-seat head restraints must reach at least 29.5 inches in the "down" position (vs. no requirement in the current FMVSS).
- The longitudinal distance ("backset") between the restraint and the head of the 50th percentile dummy shall not exceed 2 inches.
- Head restraints for rear-outboard seats, reaching at least 29.5 inches in the "down" position, and meeting the 2-inch limit for backset.

The upgrade would provide whiplash protection for tall occupants, safeguard people of average height even if they leave adjustable restraints in the "down" position, and protect rear-seat passengers. The proposal resembles, to some extent, an unsuccessful bid NHTSA made in 1974 to upgrade FMVSS 202.[16]

---

[15] *Ibid.*, pp. 58-63 and 68-69.

[16] *Federal Register* 66 (January 4, 2001): 967; integral or fixed head restraints would have to reach 31.5 inches in the front seat and 29.5 inches in the back seat; *National Highway Traffic Safety Administration Evaluation Program Plan, Calendar Years 2004-2007*, NHTSA Report No. DOT HS 809 699, Washington, 2004, p. 33.

FMVSS 203: Impact protection for the driver from the steering control system
FMVSS 204: Steering control rearward displacement

These two standards regulate one safety technology:

- ### Energy-absorbing and telescoping steering assembly

It is a basic protection for drivers in frontal crashes, designed to cushion their impact into the steering assembly. It was introduced in passenger cars in model years 1967-68 and in light trucks over a longer period, extending approximately from 1970 to 1982.

The first paragraph of FMVSS 203 states, "this standard specifies requirements for steering control systems that will minimize chest, neck and facial injuries to the driver as a result of impact." FMVSS 204 "specifies requirements limiting the rearward displacement of the steering control into the passenger compartment to reduce the likelihood of chest, neck or head injury."[1]

Regulatory history FMVSS 203 and 204, applying at first only to passenger cars, were among NHTSA's initial safety standards, with a Notice of Proposed Rulemaking in December 1966, a Final Rule in February 1967 and an effective date of January 1, 1968. They basically incorporate General Services Administration Standard 515/4a, effective October 1967, for federally purchased vehicles, and draw on SAE Recommended Practices J850 and J944. FMVSS 203 and 204 were extended to light trucks with GVWR up to 10,000 pounds, effective September 1, 1981.[2]

How it works Before FMVSS 203 and 204, the steering assembly was by far the most common source of fatal or serious injuries for drivers in frontal crashes. The pre-standard steering column was essentially a rigid pole attached to the vehicle's front structure. When drivers' forward momentum propelled them into the column, it did not "give." Even worse, when the front structure deformed rearwards in a frontal impact, it pushed the column upwards and backwards toward the driver.[3]

The safety technology associated with FMVSS 203 and 204 consists of devices that arrest, absorb or deflect the rearward motion of the column in crashes and that deform at a safe level of force upon impact by the driver. The single, rigid column was replaced by assemblies of concentric tubes. The inner tube slides within the outer tube (telescopes) to make the assembly shorter in case of impact. Usually there are two telescoping sections, each about eight inches

---

[1] *Code of Federal Regulations*, Title 49, Government Printing Office, Washington, 2002, Parts 571.203 and 571.204.

[2] Kahane, C.J., *An Evaluation of Federal Motor Vehicle Safety Standards for Passenger Car Steering Assemblies*, NHTSA Technical Report No. DOT HS 805 705, Washington, 1981, pp. 86-87; *Federal Register* 31 (March 8, 1966): 4091, 31 (July 15, 1966): 9631, 31 (December 3, 1966): 15212, 32 (February 3, 1967): 2414, 44 (November 29, 1979): 68470; *1968 SAE Handbook*, Society of Automotive Engineers, New York, 1968, pp. 915-916 and 923-925.

[3] Kahane (1981 Steering Assemblies), pp. 67-80; Huelke, D.F., and Gikas, P.W., "How Do They Die? Medical Engineering Data from On-Scene Investigation of Fatal Automobile Accidents", *Highway Vehicle Safety*, Society of Automotive Engineers, New York, 1968; Voight, G. and Wilfert, K., *Mechanisms of Injury to Unrestrained Drivers in Head-On Collisions*, Paper No. 690811, Society of Automotive Engineers, New York, 1969.

long: one in the engine compartment to prevent frontal damage to the vehicle from translating into rearward displacement of the steering wheel, and one in the passenger compartment to allow the driver to compress the steering assembly forward. The telescoping section in the passenger compartment, however, does not slide freely but contains retardant devices such as crushable mesh that collapse at a safe, controlled force level, absorbing the energy of the driver's impact. The manufacturers developed about five distinct types of energy-absorbing devices.[4]

The shear capsule, which brackets the column to the instrument panel, is a one-way gate that resists the rearward movement of the column due to vehicle damage but freely allows the forward movement of the column upon driver impact. The assembly may contain universal joints to give it a flexible Z-shape, further isolating the steering-wheel area from damages within the engine compartment. Some vehicles may have a deformable canister in the steering wheel instead of/in addition to a telescoping column in the passenger compartment. Steering wheel rims and spokes may also be deformable. Manufacturers padded steering-wheel hubs and removed hazardous appurtenances such as horn rings.

In the test for FMVSS 203, a body block simulating an adult torso (the "black tuffy"), moving straight ahead, impacts a steering assembly mounted at the manufacturer's installation angle. The steering assembly is intact and has been removed from the vehicle. At an impact speed of 22 feet per second, the force on the body block may not exceed 2,500 pounds. The design of most columns and the test approach have raised several issues over the years. (1) Columns are well-designed to compress under axial (straight-ahead) impacts, but there are questions about their performance in off-center impacts. (2) The driver might have an easier time compressing an intact column (as in the test) than one that has been damaged or displaced during the vehicle impact phase. (3) A force level of 2,500 pounds may be acceptable for the healthy adult torso but excessive for older adults or when the head or abdomen contacts the steering wheel; a 1970 proposal for a more stringent requirement was not adopted. (4) It was difficult for some full-sized vans to meet FMVSS 203 since their steering columns were nearly vertical, perpendicular to the motion of the body block; the lead-time allowed for extending FMVSS 203 to light trucks essentially permitted the phase-out of vans with near-vertical columns.[5]

FMVSS 204 is tested in the 30 mph, full-vehicle impact with a frontal barrier that simultaneously tests FMVSS 208, 219 and 301 as well. The steering assembly is not allowed to intrude more than 5 inches rearward, into the passenger compartment on this test. Vertical or sideways motion is permitted and has been substantial in a few test vehicles. That has also raised questions about the ability of these displaced or damaged columns to compress upon impact by the driver.

Vehicles with driver air bags – i.e., all new cars and light trucks up to 8,500 pounds GVWR – are exempt from the FMVSS 203 test. However, they continue to have energy-absorbing steering columns that, in combination with air bags, are still important for providing "crush space" and

---

[4] Kahane (1981 Steering Assemblies), pp. 87-93; Marquis, D.P., *The General Motors Energy Absorbing Column*, Paper No. 670039, Society of Automotive Engineers, New York, 1967; Huelke, D.F., *Accident Investigations of Performance Characteristics of Energy Absorbing Steering Columns*, Paper No. 690184, Society of Automotive Engineers, New York, 1969.
[5] Kahane (1981 Steering Assemblies), pp. 93-99; Hill, A., *Steering Wheel Oscillations and Vertical Movements in 30 mph Barrier Impacts*, NHTSA Technical Report No. DOT HS 803 606, Washington, 1978.

cushioning the driver's impact. The FMVSS 208 test, which measures the driver's injury risk in a 30 mph crash, implicitly tests the performance of the energy-absorbing steering assembly as part of the overall safety system.

The manufacturers were developing energy-absorbing and telescoping steering assemblies before NHTSA existed. All 1967 GM, Chrysler and American Motors cars built in the United States or Canada were equipped with them. All model-year 1968 passenger cars had them, even though the standards did not take effect until mid-model year (January 1, 1968). Implementation in light trucks took place over a longer period and is not as clearly documented: the Volkswagen "bus" apparently had them in 1970, all Jeep and GM trucks in 1973, and other trucks in later years, some not until the effective date, September 1, 1981.[6]

Expected benefits Only drivers can expect to receive benefits, since passengers rarely make significant contact with the steering assembly. Benefits are primarily in frontal crashes, since other impacts rarely propel the driver into the steering assembly directly and with enough force to compress the column. The principal benefit would be expected for injuries caused by driver contact with the steering assembly; it is theoretically possible, but not likely, that the behavior of the steering assembly would affect driver contacts with other parts of the vehicles. The largest reductions are expected for chest injuries, although head, neck and abdominal injuries could be mitigated as a result of less steering-column intrusion, more deformable steering wheels, and removal of hazardous appurtenances. The largest reductions are expected for unbelted drivers, since belts absorb much of the driver's momentum prior to contact with the steering assembly.

Fatality reduction – passenger cars NHTSA's analysis, published in 1981, was its first comprehensive evaluation of a FMVSS. It is based on 1975-79 FARS data.[7] Since FARS data do not indicate the source or body region of fatal injuries, or provide information on nonfatal crashes to allow computation of fatality rates, they are best suited for comparing overall fatality risk of drivers in frontal crashes, before and after FMVSS 203/204, relative to a control group. One possible control group is the passenger fatalities in these same frontal crashes. The 1975-79 FARS statistics were:

	Passenger Frontals	Driver Frontals
Model year 1966 ("before")	1,048	2,119
Model year 1968 ("after")	1,463	2,573

[6] Kahane (1981 Steering Assemblies), p. 92; *Multidisciplinary Accident Investigation Data File, Editing Manual and Reference Information*, Vol. 2, NHTSA Technical Report No. DOT HS 802 412, Washington, 1977; *1978 Model Year Passenger Car and Truck Accident Investigator's Manual*, Motor Vehicle Manufacturers Association, Detroit, 1978; Ludtke, N.F., "1980 and 1979 Ford F-150 Weight and Cost Analysis," *Third Automotive Fuel Economy Research Contractors' Coordination Meeting – Summary Report*, NHTSA Technical Report No. DOT HS 805 875, Washington, 1980; *Preliminary Evaluation of the Proposed Extension of Standards No. 201, 203 and 204 to Light Trucks, Buses, and Multipurpose Vehicles*, NHTSA Plans and Policy, Washington, 1978.
[7] Kahane (1981 Steering Assemblies), pp. 197-203.

This is a statistically significant 13 percent fatality reduction for the drivers, relative to the passengers, in model year 1968: $1 - [(2,573/2,119) / (1,463/1,048)] = .13$.

Another control group is driver fatalities in side or rear impacts. The 1975-79 FARS statistics were:

	Driver Side/Rear	Driver Frontals
Model year 1966 ("before")	1,103	2,119
Model year 1968 ("after")	1,508	2,573

This is a statistically significant 11.1 percent fatality reduction in frontals, relative to side and rear impacts, in model year 1968: $1 - [(2,573/2,119) / (1,508/1,103)] = .111$.

The average of these two estimates is a 12.1 percent fatality reduction for drivers in frontal crashes. Using 1978 as the "baseline" year, the evaluation estimated that energy-absorbing steering assemblies would eventually save 1,300 lives per year in passenger cars, when all cars on the road met FMVSS 203 and 204. Only safety belts and air bags save more lives. [8]

Injury reduction – passenger cars NHTSA's 1981 analysis uses data from the National Crash Severity Study (NCSS), the predecessor of NASS. These data specify, like NASS, the severity, body region and contact source of injuries. The injury rate – the number of fatal or hospitalizing injuries due to contact with the steering assembly per 100 drivers involved in frontal towaway crashes – was computed for cars with energy-absorbing steering assemblies and compared to the rate in cars without the assemblies. These injury rates were controlled for factors such as driver age and gender by multidimensional contingency table analysis, a predecessor of the logistic regression method widely used today. [9]

The analysis showed a statistically significant 38.4 percent reduction of fatal or hospitalizing injuries due to contact with the steering assembly. Using 1978 as the "baseline" year, the evaluation estimated that energy-absorbing steering assemblies would prevent 23,000 nonfatal hospitalizations per year when all cars on the road met FMVSS 203 and 204. [10]

The NCSS data did not show any one of the five types of energy-absorbing columns to be significantly more effective, or less effective than the other types, nor did they show a significant difference between energy-absorbing columns and energy-absorbing canisters in the steering

---

[8] 90% confidence bounds for effectiveness: 8.5 to 15.5 percent; for benefits: 900 to 1800 lives saved.
[9] *Ibid.*, pp. 138-197; Kahane, C.J., Smith, R.A. and Tharp, K.J., "The National Crash Severity Study," *Report on the Sixth International Technical Conference on Experimental Safety Vehicles*, NHTSA Report No. DOT HS 802 501, Washington, 1977, pp. 493-515.
[10] 90% confidence bounds for effectiveness: 28 to 48 percent; for benefits: 14,900 to 33,500 hospitalizations prevented.

wheel. In other words, the United States data did not duplicate the results of a British study finding much higher effectiveness for the canisters and much better performance under nonaxial loading.[11]

NHTSA's evaluation and other analyses showed many frontal crashes in which compression of the column was slight or minimal, especially in the presence of nonaxial forces. These results, however, tended to be inconclusive because of the difficulty of establishing cause-and-effect relationships. It is difficult to determine if columns were "binding" and causing injury to drivers, or if the compression of the column was slight because it was not strongly impacted by the driver. The effect, if any, of vehicle damage on the column's ability to compress in response to occupant loading is also unclear. In other words, the data did not point to an unequivocal need or method for strengthening the FMVSS 203 regulation, and the issue eventually became moot with the introduction of air bags and with higher use of safety belts.[12]

The NCSS data demonstrated the exceptional success of FMVSS 204. Intrusion of the steering column in any direction (rearward, upward or sideways) was reduced by 68 percent in frontal towaway crashes and rearward intrusion, in particular, was reduced by 81 percent.[13]

The introduction of energy-absorbing and telescoping steering assemblies, along with deformable steering wheels, padded hubs, and removal of hazardous appurtenances, reduced injuries to all body regions: chest, abdomen, and head/neck. It did not have any undesirable side effects, such as increased driver injuries due to contacts with other parts of the vehicle (windshield, instrument panel, etc.).[14]

Fatality reduction – light trucks NHTSA's 1988 evaluation of occupant protection in frontal interior impact addresses the effect of energy-absorbing steering assemblies in light trucks. Since there was uncertainty about exactly when these assemblies were first installed, a direct "before-after" comparison, as shown above for cars, seemed unsuitable. Instead, a more extended time-series analysis considered head-on collisions between early- and late-model vehicles. After controlling for vehicle weight, driver age and gender, who fared better: the driver of the late-model truck or the early-model truck? This analysis indicated a 16 percent reduction in fatality risk for unrestrained drivers in frontal crashes during the model years that energy-absorbing steering assemblies were primarily installed in light trucks (1973-80). The corresponding analysis for passenger cars showed a 12 percent reduction in fatality risk during the model years that those assemblies were installed in cars (1967-68). Since the results are quite similar (well within sampling-error bounds), NHTSA concludes that energy-absorbing

---

[11] Kahane (1981 Steering Assemblies), pp. 226-236; Gloyns, P.F. and Mackay, G.M., "Impact Performance of Some Designs of Steering Assembly in Real Accidents and under Test Conditions," *Proceedings of Eighteenth Stapp Car Crash Conference*, Society of Automotive Engineers, Warrendale, PA, 1974.

[12] Kahane (1981 Steering Assemblies), pp. 249-269; Garrett, J.W. and Hendricks, D.L., "Factors Influencing the Performance of the Energy Absorbing Steering Column in Accidents," *Report on the Fifth International Technical Conference on Experimental Safety Vehicles*, NHTSA, Washington, 1975; Horsch, J.D., Petersen, K.R. and Viano, D.C., *Laboratory Study of Factors Influencing the Performance of Energy Absorbing Steering Systems*, Paper No. 820475, Society of Automotive Engineers, Warrendale, PA, 1982.

[13] Kahane (1981 Steering Assemblies), pp. 220-226.

[14] *Ibid.*, pp. 239-248.

steering assemblies are as effective in light trucks as in passenger cars – i.e., an estimated 12.1 percent fatality reduction for drivers in frontal crashes.[15]

Injury reduction – light trucks NHTSA's crash files, at the time of the 1988 evaluation, did not have enough data for a statistically meaningful analysis of the injury-reducing effectiveness of energy-absorbing steering assemblies in light trucks.[16]

---

[15] Kahane, C.J., *An Evaluation of Occupant Protection in Frontal Interior Impact for Unrestrained Front Seat Occupants of Cars and Light Trucks*, NHTSA Technical Report No. DOT HS 807 203, Washington, 1988, pp. xxx-xxxii, xxxiv and 192-194.

[16] *Ibid.*, pp. 165-173.

FMVSS 205: Glazing materials

This standard regulates two innovative safety technologies that have been evaluated by NHTSA:

- *High-Penetration Resistant (HPR) windshields*
- *Glass-plastic windshields*

HPR windshields, an essential protection against facial lacerations, have been standard equipment in motor vehicles since model year 1966. Glass-plastic windshields, on the other hand, although envisioned as even better protection against lacerations, were only installed on a few make-models of passenger cars during 1985-87, and then discontinued.

The first paragraph of FMVSS 205 states, "The purpose of this standard is to reduce injuries resulting from impact to glazing surfaces, to ensure a necessary degree of transparency in motor vehicle windows for driver visibility, and to minimize the possibility of occupants being thrown through the vehicle windows in collisions."[1] FMVSS 205 regulates all automotive glazing, including windshields, side and rear windows, etc. However, only windshields experienced dramatic innovations during the 1965-99 time frame that were evaluated by NHTSA. The two evaluations will now be discussed, one-by-one.

High Penetration Resistant (HPR) windshields

Regulatory history FMVSS 205 applies to all motor vehicles. The first version of FMVSS 205 was one of NHTSA's initial safety standards, with a Notice of Proposed Rulemaking in December 1966, a Final Rule in February 1967 and an effective date of January 1, 1968. It primarily incorporates the American National Standards Institute (ANSI) Safety Code Z26.1 (July 1966) and SAE Standard J938, (October 1965).[2]

How HPR windshields work Laminated "safety" windshields, with a thin sheet of plastic sandwiched between two layers of glass, date back to 1927 in the United States. Nevertheless, the pre-HPR windshield caused large numbers of facial lacerations requiring surgical care. The glass layers readily broke and tore through the plastic layer in moderately severe frontal crashes with head-impact speeds around 13 mph. The HPR windshield incorporated two strategies to improve performance. (1) The plastic (polyvinyl butyral) layer was doubled in thickness to increase its tear resistance and energy absorption. (2) Even more important, the bond between the glass and the plastic layers was loosened by adding moisture to the plastic; as a result, the plastic flexed away from the glass pane rather than being torn by the glass. This allowed the windshield to bulge at a safe force level, without tearing, upon head-impacts ranging up to 22-29 mph, a quite severe crash. The test for FMVSS 205 specifies that a 5-pound steel ball must not penetrate the windshield at 19 mph – i.e., fast enough that pre-HPR windshields fail but slow

---

[1] *Code of Federal Regulations*, Title 49, Government Printing Office, Washington, 2002, Part 571.205.

[2] Kahane, C.J., *An Evaluation of Windshield Glazing and Installation Methods for Passenger Cars*, NHTSA Technical Report No. DOT HS 806 693, Washington, 1985, pp. 1-2; *Federal Register* 31 (March 8, 1966): 4091, 31 (July 15, 1966): 9631, 31 (December 3, 1966): 15219, 32 (February 3, 1967): 2414; *Safety Code for Safety Glazing Materials for Glazing Motor Vehicles Operating on Land Highways*, Standard Z26.1, American National Standards Institute, New York, 1966; *1968 SAE Handbook*, Society of Automotive Engineers, New York, 1968.

enough that HPR windshields succeed. The glass and auto manufacturers developed HPR windshields before NHTSA existed. All 1965 Ford Thunderbirds and all domestic passenger cars beginning in model year 1966 have HPR windshields.[3]

Expected benefits Drivers and front-seat passengers can expect to benefit in frontal crashes that propel them into the windshield. The zone of benefits is for head impacts between 13 and 22-29 mph, where the pre-HPR windshield tears but the HPR windshield remains intact while crumpling and bulging. The most common benefit ought to be a reduction of "nonminor" (AIS $\geq$ 2) facial lacerations from torn glass – injuries that cause disfigurement, extend into the subcutaneous tissue or require more than first aid or simple closure. There should be a large reduction in serious injuries to the eyes, lips or tongue that occur, in pre-HPR windshields, when the occupant's head rebounds through the break in the windshield. There may be a reduction in fractures to the nose or facial bones caused by edges of broken windshields. There would likely be a smaller reduction of minor lacerations, since they can occur even without tears in the windshield. Blunt impact trauma is less likely to be mitigated. Fatal head injuries are hardly ever due to windshield contact alone, and unlikely to be affected by the HPR windshield.

Injury reduction – passenger cars NHTSA's evaluation was published in 1985. It uses data from the National Crash Severity Study (NCSS), the predecessor of NASS. These data specify, like NASS, the severity (AIS), body region, lesion and contact source of injuries. The injury rate – the number of AIS $\geq$ 2 lacerations of the face or head due to contact with the windshield, per 1000 front-seat occupants involved in frontal towaway crashes – was computed for cars of the first model years with HPR windshields and compared to the rate in cars of the last model years with pre-HPR windshields.[4] For example, a comparison of cars of the first five years with HPR windshields and the last five years without them showed a dramatic, statistically significant 74 percent injury reduction with HPR:

	AIS $\geq$ 2 Lacerations	N of Occupants	Lacerations per 1000 Occupants
Model year 1961-65 ("before")	38	4,958	7.66
Model year 1966-70 ("after")	52	26,110	1.99

Nearly identical reductions were observed when the data were limited to four, three, two or just one model year before and after the shift to HPR. The 74 percent injury reduction is one of the

---

[3] Kahane (1985 Windshields), pp. 5-11; Patrick, L.M. and Daniel, B.P., "Comparison of Standard and Experimental Windshields," *Eighth Stapp Car Crash and Field Demonstration Conference*, Wayne State University Press, Detroit, 1966; Rieser, R.G. and Michaels, G.E., "Factors in the Development and Evaluation of Safer Glazing," *Ninth Stapp Car Crash Conference*, University of Minnesota, Minneapolis, 1966; Widman, J.C., *Recent Developments in Penetration Resistance of Windshield Glass*, Paper No. 650474, Society of Automotive Engineers, New York, 1965.

[4] Kahane (1985 Windshields), pp. 77-80; Kahane, C.J., Smith, R.A. and Tharp, K.J., "The National Crash Severity Study," *Report on the Sixth International Technical Conference on Experimental Safety Vehicles*, NHTSA Report No. DOT HS 802 501, Washington, 1977, pp. 493-515.

highest ever observed for any of the FMVSS. After 1966, there was an almost miraculous reduction of disfiguring facial injuries in motor vehicle crashes.[5] Using 1982 as the "baseline" year, the evaluation estimated that HPR windshields would prevent 39,000 AIS $\geq$ 2 lacerations per year when all cars on the road were equipped with them.[6]

HPR windshields reduce facial fractures involving windshield contact by 56 percent, preventing 8,000 fractures per year. In all, the windshields prevent 47,000 AIS 2-4 injuries per year[7] (39,000 lacerations plus 8,000 fractures). Many of these injuries involved the eyes, nose or mouth, and were especially distressing. HPR windshields reduced injuries to the eyes, nose or mouth involving contact with windshields by 72 percent, preventing 19,000 such injuries per year.

HPR windshields were also rather effective in preventing minor (AIS 1) facial lacerations due to windshield contact, reducing them by a statistically significant 25 percent. However, HPR windshields had little or no observed effect on concussions, contusions or complaints of pain.[8]

NHTSA also analyzed 1974 New York State crash data to corroborate the NCSS results. These data indicate the body region and lesion of an occupant's most severe injury, but not the contact source. Severity of nonfatal injuries is indicated by a three-letter scale (A, B, C) but not the AIS. Facial injuries with level A (severe) bleeding in frontal crashes [due to any contact source] were reduced by approximately 50 percent with HPR windshields; facial fractures by 25-30 percent. In other words, the New York results support the NCSS findings.[9]

Fatality reduction – passenger cars NHTSA's 1985 analysis is based on 1975-82 FARS data. Since FARS does not indicate the source or body region of fatal injuries, or provide information on nonfatal crashes to allow computation of fatality rates, it is best suited for comparing overall fatality risk of front-seat occupants in frontal crashes, before and after HPR windshields, relative to a control group of nonfrontal fatalities. Since energy-absorbing steering columns were introduced in model years 1967-68, 1-2 years after the HPR windshield, the analysis must either be limited to model year 1966 and earlier, or it must somehow control for the energy-absorbing column. NHTSA used a log-linear regression of the frontal/nonfrontal fatality ratio, by calendar year and model year, with independent variables: HPR windshield, energy-absorbing column, vehicle age, and calendar year. The effect attributed to the HPR windshield was negligible, and it was not statistically significant. The evaluation concluded that the HPR windshield, although highly effective in reducing nonfatal injuries, has little effect on fatalities.[10]

Injury and fatality reduction – light trucks NHTSA's 1985 evaluation only addressed the performance of windshields in passenger cars. However, since the mechanisms of injuries

---

[5] Huelke, D.F., Grabb, W.C., Dingman, R.O. and Oneal, R.M., "The New Automotive Windshield and its Effectiveness in Reducing Facial Lacerations," *Plastic and Reconstructive Surgery*, Vol. 41, No. 6, June 1968.
[6] 90% confidence bounds for effectiveness: 65 to 83 percent; for benefits: 25,000 to 53,000 lacerations prevented. See Kahane (1985 Windshields), pp. 127-128 and 225-235.
[7] 90% confidence bounds: 31,000 to 62,000 injuries.
[8] *Ibid.*, pp. 77-95 and 225-235.
[9] *Ibid.*, pp. 101-116.
[10] *Ibid.*, pp. 122-127.

involving windshield contact are presumably similar in cars and light trucks, the effectiveness estimates for cars are probably appropriate for light trucks as well.

Glass-plastic windshields

Since facial lacerations are so often due to contact with shattered, torn or crumpled glass, it would seem preferable to line the interior surface of the windshield with shatterproof, energy-absorbing plastic. However, glass is scratch-resistant and its long-term optical qualities are far superior to most plastics, making it essential for windshields and other windows that drivers must look through. Circa 1980, glass manufacturers developed a polyurethane sheet that was transparent and durable enough for use as a windshield's inner lining. In 1983, NHTSA amended FMVSS 205 to permit the use of glass-plastic glazing at the option of the motor vehicle manufacturer. The amendment includes numerous tests to assure that the plastic resists scratching, clouding and corrosion by chemicals. The inner lining is applied to the HPR laminated windshield and makes the glass-plastic windshield a four-ply material.[11]

It was hoped that glass-plastic windshields, by largely eliminating any exposure of occupants to glass surfaces in frontal crashes, would substantially reduce laceration injuries relative to the conventional HPR windshield, especially the minor (AIS 1) lacerations where the HPR windshield did not have that large an effect.

Glass-plastic windshields were standard equipment in the 1985-87 Cadillac Seville, 1986-87 Cadillac Eldorado, 1986-87 Buick Riviera and 1986-87 Oldsmobile Toronado as well as in substantial test fleets of several manufacturers, over 200,000 cars in all. However, the manufacturers discontinued installing glass-plastic windshields after 1987, largely because of durability problems in day-to-day use and high replacement costs. An additional consideration, perhaps, was that increased use of safety belts and installation of air bags would soon be reducing the frequency of occupant contacts with the windshield, anyway.[12]

Injury reduction NHTSA's evaluation was published in 1993. It is based on crash data from New York (1986-88), Pennsylvania (1985-88) and Indiana (1985-89). These state files distinguish between injuries with bleeding (primarily lacerations), injuries without bleeding (e.g., contusions or complaints of pain) and uninjured occupants. They identify the vehicle's make-model or VIN. However, not all of these files identify the impact location (frontal, side, etc.) or details about an injury's body region, severity or specific lesion. The rate of injuries with bleeding per 100 crash-involved front-seat occupants was computed for the four make-models that had glass-plastic windshields as standard equipment and compared to the rate in three "control-group" make-models that had conventional HPR windshields; 1985-87 Cadillac Deville, 1986-87 Buick Electra and 1986-87 Oldsmobile 98.[13] For example, the injury rates for drivers in the combined New York and Pennsylvania data were as follows:

---

[11] Parsons, G.G., *An Evaluation of the Effects of Glass-Plastic Windshield Glazing in Passenger Cars*, NHTSA Technical Report No. DOT HS 808 062, Washington, 1993, Chapter 1; *Federal Register* 48 (November 16, 1983): 52065; *Code of Federal Regulations*, Title 49, Part 571.205; *Final Regulatory Evaluation, Anti-Lacerative Glazing, FMVSS 205*, NHTSA Plans and Policy, Washington, 1983.

[12] Parsons (1993), Chapters 3 and 4.

[13] *Ibid.*, Chapter 2.

	Injuries with Bleeding	N of Drivers	Injuries with Bleeding per 100 Occupants
Control group	295	7,843	3.76
Glass-plastic windshields	85	1,942	4.38

Since the observed injury rate is actually slightly higher in the cars with glass-plastic windshields than in the control-group make-models, these data, as well as the other tables in the evaluation report, do not suggest that glass-plastic windshields reduced the overall rate of injuries with bleeding. (It is unknown whether or not they reduced the average severity of the injuries.)

More detailed (but less numerous) crash data from the operators of the test fleets are consistent with the statistical finding. They suggest that glass-plastic windshields were quite successful in keeping occupants away from contact with glass, but occupants nevertheless sustained lacerations due to skin tissue "splitting" or "tearing" from blunt impact with the plastic surface, or from sliding of the head or face against the plastic inner layer.[14]

Potential future applications Although the glass-plastic windshields actually installed during the 1980's were not successful in terms of durability or in reducing the overall frequency of lacerations, NHTSA is confident that glass-plastic glazing could have great potential benefits in other applications. Specifically, side windows of motor vehicles are currently made of tempered glass. The entire pane breaks into many small pieces upon impact. Although that has the advantage of preventing serious lacerations, it results in wide-open windows that become portals for occupant ejection. An inner layer of plastic with appropriate strength, durability and optical qualities could become one component in a system that would keep the occupant inside the vehicle.[15]

---

[14] *Ibid.*, p. 2-32.
[15] Willke, D., Summers, S., Wang, J., Lee, J., Partyka, S. and Duffy, S., *Ejection Mitigation Using Advanced Glazing: Status Report II*, NHTSA, Washington, 1999 (www-nrd.nhtsa.dot.gov/pdf/nrd-11/glazing0999.pdf)

FMVSS 206: Door locks and door retention components

This standard regulates one safety technology that has been evaluated by NHTSA:

- ***Stronger locks, latches and hinges for side doors***

The first paragraph of FMVSS 206 states unequivocally that its purpose is to prevent occupant ejections: "This standard specifies requirements for door locks and door retention components including latches, hinges and other supporting means, to minimize the likelihood of occupants being thrown from the vehicle as a result of impact."[1] FMVSS 206 took effect for doors on the side of passenger cars on January 1, 1968, multipurpose passenger vehicles on January 1, 1970, and other light trucks on January 1, 1972. Its 1997-98 extension to doors on the back ends of vehicles that access seating areas, such as station-wagon tailgates or minivan hatches (but excluding trunk lids), has not yet been evaluated by NHTSA.[2]

Regulatory history Self-regulation of side door locks, latches and hinges by the industry began in November 1962, well before NHTSA existed, with SAE Standard J839, "Passenger Car Side Door Latch Systems." In 1965, SAE substantially upgraded J839 to J839b, nearly doubling the strength requirements, and introduced another standard, J934, "Vehicle Passenger Door Hinge Systems." J839b and J934 were essentially incorporated into FMVSS 206, one of NHTSA's initial safety standards, with a Notice of Proposed Rulemaking in December 1966, a Final Rule in February 1967 and an effective date of January 1, 1968.[3]

What happened Safety researchers in the 1950's learned that car doors were forced open in many crashes, especially rollovers; that occupants were often ejected via the opened doors; and that being "thrown clear" in a crash was not an advantage but a terrible risk, especially in rollovers, where an occupant could be wedged between the car and the ground. Safety door latches were introduced in the 1955-56 domestic cars to assure that doors would not pop open when occupants on the inside, or objects/ground on the outside brushed against door handles during crashes. During about 1962-68, the manufacturers began an almost continuous process of strengthening latches and hinges in their various make-models. Doors can open in crashes if latches or hinges are pulled apart by impact damage, or if latches and hinges are pulled loose from their supporting structures. The net effect of these cumulative improvements was a dramatic reduction of door-opening, as seen in Automotive Crash Injury Research (ACIR), a crash data file of the 1950's and 60's:[4]

---

[1] *Code of Federal Regulations*, Title 49, Government Printing Office, Washington, 2002, Part 571.206.

[2] *Federal Register* 32 (February 3, 1967): 2414, 34 (January 24, 1969): 1150, 60 (September 28, 1995): 50124.

[3] Kahane, C.J., *An Evaluation of Door Locks and Roof Crush Resistance of Passenger Cars – Federal Motor Vehicle Safety Standards 206 and 216*, NHTSA Technical Report No. DOT HS 807 489, Washington, 1989, pp. 2-4; *1965 SAE Handbook*, Society of Automotive Engineers, New York, 1965, p. 893; *1967 SAE Handbook*, Society of Automotive Engineers, New York, 1967, pp. 904-906; *Federal Register* 31 (March 8, 1966): 4091, 31 (July 15, 1966): 9631, 31 (December 3, 1966): 15219, 32 (February 3, 1967): 2414.

[4] Kahane (1989 Door Locks/Roof Crush Resistance), pp. 27-30; Garrett, J.W., *The Safety Performance of 1962-63 Automobile Door Latches and Comparison with Earlier Latch Designs*, Report No. VJ-1823-R7, Cornell Aeronautical Laboratory, Buffalo, 1964; Garrett, J.W., *Comparison of Door Opening Frequency in 1967-68 Cars with Earlier Model U.S. Cars*, NHTSA Technical Report No. DOT HS 800 231, Washington, 1969.

Model Year	Percent of ACIR Crashes Where Door Opened
Pre-1956	43
1956-61	28
1962-63	23
1964-66	17
1967-68	12

Expected benefits The fewer doors open in crashes, the fewer ejections via that portal, especially in rollovers. Since it is safer to stay inside a vehicle than be ejected, there ought to be a significant fatality reduction, especially in rollovers. This should benefit all occupants who are at risk of being ejected via a side door, drivers and passengers, front-seat and second-seat.

Fatality reduction – passenger cars Frankly, it is difficult to compute the benefits of this technology. Since improvements entered gradually over several years (and NHTSA does not know exactly when on what models), a diffuse time-series approach must replace the customary "just before-just after" comparison. Moreover, door locks were just one influence on rollover fatalities in the 1960's: above all, car size changed from year to year, and when cars grow, rollover fatalities decrease. Nevertheless, there is no doubt that improved door locks have saved lives. The side door was the leading ejection portal in cars of the 1950's and early 1960's, but door ejections have decreased to the point that the side window became the leading portal in the 1970's. The problem is quantifying the benefit.

NHTSA's evaluation, published in 1989, computed several rollover-fatality risk indices that control for car size, using logistic regression.[5] For example, one analysis used 1975-86 FARS data to calibrate the ratio of rollover fatalities to frontal-fixed-object fatalities, as a function of model year, track width and curb weight. This ratio was graphed, by model year, for cars with 55 inch track width and 3000 pound curb weight (median values for cars on the road in 1975-86), and then indexed relative to the average value for model years 1975-80. The averages of the two indices defined in the evaluation[6] were:

Model Year	Rollover Fatality Risk Index
1963-64	123
1968-70	107
1975-80	100

The risk indices dropped substantially between 1963-64 and 1968-70, when door locks steadily improved. However, door locks were not the only factor that influenced rollover fatalities in those model years. The evaluation specifically estimated that the 1962-68 improvements to door

---

[5] Kahane (1989 Door Locks/Roof Crush Resistance), pp. 127-204.
[6] *Ibid.*, p. 202, average of the indices NEWROLL2 and NEWROLL3.

locks reduced fatalities involving occupant ejection in rollovers by 15 percent. Using 1982 as the "baseline" year, the evaluation estimated these improvements were saving approximately 400 lives per year.[7]

Injury reduction – passenger cars NHTSA's evaluation does not include any analysis or estimate of the effect of door locks on nonfatal injuries. However, other studies such as NHTSA's 1985 evaluation of windshields suggest that occupant ejections cause similar numbers of serious nonfatal injuries (AIS 3-5) and fatalities.[8] It is likely that the improvements to door locks, latches and hinges are about as effective in reducing serious injuries as fatalities, and that they are preventing comparable numbers of AIS 3-5 injuries and fatalities.

Fatality and injury reduction – light trucks NHTSA's 1989 evaluation only addressed the performance of door locks in passenger cars. Since FMVSS 206 was extended to light trucks a few years after it took effect for cars, and since the door components of trucks and cars are similar, it seems plausible that the gradual improvement of door locks in cars was matched by a comparable effort in light trucks, with a similar effectiveness in reducing fatalities associated with door ejections. However, National Crash Severity Study (NCSS) data show that, proportionately, the door is the ejection route in only 2/3 as often in LTV rollovers as in car rollovers. Thus, the overall reduction of ejection fatalities in rollovers, estimated to be 15 percent in cars, would be 10 percent in LTVs.[9]

Potential FMVSS 206 upgrade Whereas improvements to door locks of the 1960's greatly reduced the incidence of doors opening in crashes and occupant ejection via the open door, they did not eliminate these safety problems. NHTSA has been studying door latch integrity for some years, assessing the potential for benefits from strengthening the existing test procedures and/or adding new procedures to FMVSS 206.[10]

---

[7] *Ibid.*, pp. xxix and 222-225.

[8] Kahane (1985 Windshields), pp. 161-163.

[9] An analyis of MY 1990-2003 cars and LTVs in 1995-2003 NASS data confirms the dramatic, long-term reduction in ejections via the open door. In NCSS, the open door accounted for 32 percent of ejections in car rollovers, 20 percent in LTVs. In NASS, the open door accounted for just 8 percent of ejections in car rollovers, 7 percent in LTVs. Thus, our conclusions that FMVSS 206-related technology reduced overall rollover fatality risk by 15 percent in cars and 10 percent in LTVs may even be a little bit conservative if the long-term, continuing improvement of doors is taken into account.

[10] *B.01.04 Door Latch Integrity.* NHTSA Research & Development, www-nrd.nhtsa.dot.gov/departments/nrd-01/summaries/b0104.html , June 2000; *Federal Register* 60 (July 12, 1995): 35889.

FMVSS 207: Seating systems

"This standard establishes [a variety of strength] requirements for seats, their attachment assemblies, and their installation to minimize the possibility of their failure by forces acting on them as a result of vehicle impact."[1] It is, however, linked with just one tangible safety modification that has been evaluated by NHTSA:

- **Seatback locks for 2-door cars with folding front seatbacks**

The locks are designed to mitigate injury risk in frontal impacts by keeping seatbacks upright and in place. They are supposed to prevent seatbacks from folding over and pushing on front-seat occupants and to keep unrestrained back-seat occupants in the back-seat area. They were introduced in 2-door passenger cars primarily in model years 1967-68.

Regulatory history FMVSS 207, one of NHTSA's initial safety standards, took effect for passenger cars on January 1, 1968 and was extended to trucks, buses and multipurpose vehicles, effective on January 1, 1972. Except for the seatback lock requirement, FMVSS 207 largely incorporated SAE Recommended Practice J879, which had been in place since November 1963. Thus, the locks were the only substantial modification in domestic cars during the time frame of the FMVSS 207 rulemaking.[2]

How seatback locks are supposed to work Two-door cars generally need front seatbacks that fold down and out of the way to give passengers room to get in and out of the back seat. Without the locks, these seatbacks are likely to fold over in frontal crashes, even at low severities. That could potentially increase injury risk for front-seat occupants by adding some or all of the momentum of the 25-35 pound seatbacks to the momentum of the occupants as they slam into structures at the front of the passenger compartment. The locks could be even more beneficial when there are unrestrained back-seat occupants. In a car without the locks, the back-seat passenger could move forward with the seatback and this combined load could weigh upon the front-seat occupant. As the seatback folds over, it can act as a ramp for the back-seat passenger, who will vault head-first into the front compartment, contact the windshield or header, or make head-to-head contact with front-seat occupants. With the locks, a seatback could remain upright and be a relatively benign surface for the rear-seat passenger to impact.[3]

However, for seatback locks to be effective, they must endure the impact forces to which they are exposed. FMVSS 207 (which makes no explicit claim of effectiveness when back-seat passengers are present) only requires the locks to withstand a dynamic load of 20 times the weight of the seatback. When the 150-pound weight of an adult back-seat passenger is added to the 25-35 pound weight of the seatback, even a low-severity crash with acceleration levels well below 10 g's can produce loads exceeding the FMVSS 207 level.

---

[1] *Code of Federal Regulations*, Title 49, Government Printing Office, Washington, 2002, Part 571.207.
[2] Kahane, C.J., *A Preliminary Evaluation of Seat Back Locks*, NHTSA Technical Report No. DOT HS 807 067, Washington, 1987, pp. 2-3; *Federal Register* 31 (December 3, 1966): 15219, 32 (February 3, 1967): 2414, 36 (December 2, 1971): 22902, *1968 SAE Handbook*, Society of Automotive Engineers, New York, 1968, pp. 954-955.
[3] Kahane (1987 Seatback Locks), pp. 10-12; Severy, D.M., Brink, H.M., Baird, J.D. and Blaisdell, D.M., "Safer Seat Designs," *Thirteenth Stapp Car Crash Conference*, Society of Automotive Engineers, New York, 1969, pp. 323-325.

The manufacturers have developed two types of seatback locks for production vehicles:

- Manual seatback locks: occupants disengage the lock by operating a lever or pressing a button (not necessarily in a convenient location) every time they get in or out of the car.

- Automatic inertial seatback locks that allow free movement of the seatback during normal vehicle operation and only lock up during impacts, road bumps, or emergency braking. The technology is similar to, and probably inspired by inertial locking retractors for safety belts.

Volkswagen and Opel installed manual seatback locks in their model year 1966 cars; GM, Fiat, Renault and Datsun in 1967; and Ford and Chrysler in the full model year 1968 – all before the January 1, 1968 effective date. Automatic seatback locks are obviously more convenient for passengers, and they have completely superseded the manual type in domestic cars since about 1980. In addition to being more convenient, automatic locks preclude an imaginable (but unsubstantiated) safety problem: during a fire, immersion or other emergency egress situation, back-seat passengers could be delayed while they locate and operate manual locks, but can exit immediately from cars with automatic locks that allow free movement of the seatback.[4]

Potential benefits Drivers and right-front passengers might experience a reduction of fatalities and injuries in frontal crashes, when the back seat is unoccupied, provided that a locked seatback puts less load on them than a freely moving, pre-FMVSS 207 seatback. When the back seat is occupied, both front-outboard and rear-outboard occupants might experience a more substantial reduction of fatalities and injuries in frontal crashes, provided that the locks are able to withstand the impact force of the back-seat occupant. In addition, upright seatbacks might help contain back-seat occupants within their seating area and reduce their risk of ejection from the car.

Evaluation findings NHTSA's study, published in 1987, includes statistical analyses of crash data comparing injury and fatality rates in 2-door cars just before and after the introduction of seatback locks; sled tests simulating frontal crashes of seat systems with and without the locks, to study the effect of the locks on occupant force levels and kinematics; and a review of in-depth crash investigations to determine the ability of seatback locks to endure impact forces, as a function of Delta V.

Crash data analyses Injury rates per 100 crash-involved occupants are computed for domestic 2-door cars of the model year(s) just before and just after the introduction of manual seatback locks (GM in 1967, Ford and Chrysler in 1968). Corresponding rates are computed for a control group of 4-door cars of these same model years (i.e., just before and after seatback locks were introduced in the 2-door cars made by that manufacturer). Four-door cars are a useful control group because they received all of the safety improvements that 2-door cars got during that time frame – except seatback locks. In other words, did the injury reduction in 2-door cars (after vs. before seatback locks) surpass the corresponding reduction in 4-door cars? Injury rates were computed from 1973-77 Washington State data, 1972-74 Texas data and 1974 New York State

---

[4] Kahane (1987 Seatback Locks), pp. 4-5 and 11-12.

data.[5] Similarly, fatality rates per million vehicle registration years are computed from 1975-85 FARS data and R.L. Polk registration data (their "National Vehicle Population Profile").[6]

None of these analyses showed a statistically significant reduction of injury or fatality rates in 2-door cars above and beyond the corresponding reduction in 4-door cars. There was no observed benefit for front-seat occupants, regardless of whether the seat behind them was unoccupied, occupied by an adult, or occupied by a child. There was no reduction of injury or fatality rates, or of ejection, for back-seat occupants, adults or children.

Sled tests NHTSA sponsored 28 sled tests, involving 84 unrestrained 50th percentile Part 572 dummies.[7] Fourteen of these tests used a sled buck that simulated the passenger compartment of a 1981 domestic sedan with operational seatback locks: speeds ranged from 15 to 30 mph, impacts were full-frontal or oblique-frontal, dummies occupied the front-outboard seats in all tests and the rear-outboard seats in eight tests. The other 14 tests exactly replicated the above series, except that the seatback locks were removed to allow the seatbacks to pivot freely in the crash, simulating a pre-FMVSS 207 car. That provided 42 matched pairs of dummies tested under identical conditions, except one of them had the seatback lock and the other did not. The Head Injury Criterion (HIC), chest g's, femur loads, mean strain criterion, total laceration index and other measures were computed for each dummy.

The statistical analysis of the 42 matched pairs considers the average difference of the injury criterion in the test with seatback locks and in the corresponding test without seatback locks. These average differences were not significantly different from zero for any of the injury criteria. In other words, the dummies had the same average injury severity with and without the seatback locks. When the matched pairs were divided into three principal subgroups – front-seat occupants with nobody sitting behind them, front-seat occupants with somebody sitting behind them, and back-seat occupants – there was no significant benefit for seatback locks in any subgroup.

The films from the sled tests explain why seatback locks did not significantly reduce injury risk:

- Seatbacks (with or without the locks) only "follow" front-seat occupants gradually in a frontal crash, barely touch occupants at the crucial moment when they first slam into the steering assembly or instrument panel, and subsequently only apply force over an extended time period that hardly spikes up the injury criteria.
- At 26.5 mph the impact of a back-seat dummy sheared the seatback lock or tore the entire front seat out of its track every time.
- At 22.5 mph, although seatback locks remained intact, rear-seat dummies deflect the front seatback forward quite a bit and are not well contained in the back-seat area.

In-depth crash investigations NHTSA's Multidisciplinary Accident Investigation (MDAI) File provides detailed information on the performance of the seat and seatback lock in 1968-78 crashes, plus a description of the vehicle damage that allows a rough estimate of Delta V in

---

[5] *Ibid.*, pp. 55-96.
[6] *Ibid.*, pp. 97-135.
[7] *Ibid.*, pp. 21-54.

frontal crashes.[8] "Seat separation" is defined to include breakage of seatback locks and/or (much less frequently) seats tearing loose from tracks, seat tracks tearing loose from floors, seatbacks tearing loose from seat cushions. The MDAI data suggest front-seat separation occurred quite frequently in frontal impacts of 2-door cars equipped with seatback locks, when there was a back-seat occupant weighing 100 pounds or more:

Delta V (mph)	Percent of Front Seats Separated
10 – 14.9	30
15 – 19.9	37
20 – 29.9	51
30 +	85

This performance obviously limited any potential benefits for seatback locks when there were back-seat occupants.

Given the absence of statistically significant positive findings in the crash-data analyses, the negligible effect of seatback locks on dummy kinematics in sled tests, and the high rates of front-seat separation when there are back-seat occupants, NHTSA's evaluation concluded that seatback locks do not have a measurable overall effect on fatalities and injuries. It is conceivable that the locks are effective in certain narrowly defined crash types, but not to the extent of producing a statistically discernable overall effect.

---

[8] *Ibid.*, pp. 146-152.

FMVSS 208: Occupant crash protection
FMVSS 209: Seat belt assemblies
FMVSS 210: Seat belt assembly anchorages

These three standards regulate safety belts. FMVSS 208 also regulates frontal air bags. Safety belts and air bags are probably the two best-known systems to protect occupants from death and injury in crashes. However, they include ten somewhat distinct safety technologies that have been evaluated, or will be evaluated by NHTSA:

SAFETY BELTS
- *Lap belts for front-seat occupants*
- *Lap belts for back-seat occupants*
- *Manual 3-point lap-shoulder belts for front-seat occupants*
- *3-point lap-shoulder belts for back-seat occupants*
- *Automatic safety belts*
- *Pretensioners and load limiters for safety belts (not yet evaluated)*

AIR BAGS
- *Frontal air bags*
- *On-off switches for passenger air bags in pickup trucks*
- *1998-99 redesign of air bags (not yet evaluated)*
- *Advanced air bags (not yet evaluated)*

Safety belts are the basic protection system for all occupants in most types of crashes, designed to keep occupants within the vehicle and close to their original seating position, provide "ride-down" by gradually decelerating the occupant as the vehicle deforms and absorbs energy, and, if possible, prevent occupants from contacting harmful interior surfaces or one another. Lap belts for front outboard occupants were first offered as options in 1956 and have been standard on passenger cars since 1964. Shoulder harnesses were added in 1968-69 in passenger cars; modern 3-point lap-shoulder belts became standard in 1974. Automatic belts that require no action by the occupant were furnished on many cars starting in 1987, but had been phased out by 1997.

Air bags are an effective supplemental protection system for front-outboard occupants in frontal and partially frontal crashes. They significantly added to the energy absorption and ride-down offered by previously existing structures (safety belts, energy-absorbing steering columns and deformable instrument panels). They also help prevent head contacts with the windshield or windshield header.

FMVSS 208 "specifies performance requirements for the protection of vehicle occupants in crashes." FMVSS 209 "specifies requirements for seat belt assemblies." FMVSS 210 "establishes requirements for seat belt assembly anchorages to insure their proper location and to reduce the likelihood of their failure." The three standards work together as a single unit to protect occupants in crashes.[1]

---

[1] *Code of Federal Regulations*, Title 49, Government Printing Office, Washington, 2002, Parts 571.208, 571.209 and 571.210.

Overview of regulatory history Here are the highlights of the long regulatory history of FMVSS 208-209-210 – the amendments most associated with tangible changes in occupant protection technology. In many cases, manufacturers implemented the new technologies a few years, or even many years before the effective dates. Not included are a large number of primarily technical or interpretative amendments, or proposals and rules that were cancelled or changed before they took effect.

FMVSS 208, 209 and 210 were among NHTSA's initial safety standards, with a Notice of Proposed Rulemaking in December 1966, a Final Rule in February 1967, and an effective date of January 1, 1968. FMVSS 208 originally required lap belts at each designated seating position in passenger cars only, plus shoulder belts at the front-outboard seats if lap belts alone could not prevent dummies from contacting the windshield header in static tests. Most new passenger cars received shoulder belts at that time. FMVSS 209 specified performance requirements for the belts themselves, while FMVSS 210 specified requirements for the anchorages that secure the belt systems to the vehicle.[2]

NHTSA extended the original FMVSS 208 requirements to LTVs up to 10,000 pounds GVWR, effective July 1, 1971. The extension did not result in much immediate change, because most LTV make-models had been equipped with lap belts by 1968 or earlier and continued to have only lap belts up to the mid 1970's.[3]

Responding to (1) the inadequate restraint provided by lap belts alone for the head and torso, (2) the inadequate restraint provided by loosely worn belts and (3) low belt use, NHTSA amended FMVSS 208, effective January 1, 1972[4]:

- To require shoulder belts at the front outboard seats of all passenger cars (dropping the test of contact with the windshield header). Shoulder belts at that time could still be separate or integral with the lap belt.
- To require emergency-locking or automatic-locking retractors.
- To require a warning to sound when the front-outboard lap belts were not buckled.
- To permit an automatic restraint system, such as air bags or automatic belts, as an alternative to the manual shoulder belt and the buzzer.
- To include, for the first time, a 30 mph frontal barrier test (on which manual belts must remain intact, and optional automatic systems must meet injury criteria on the dummies).

Responding to (1) very low use of the separate shoulder belt and (2) continued low use of lap belts, NHTSA amended FMVSS 208, effective September 1, 1973 to require[5]:

- Integral, 3-point lap-shoulder belts at the front outboard seats of passenger cars (or, alternatively, automatic protection).
- Ignition interlocks whereby front-outboard belts must be buckled before a car can be started.

---

[2] *Federal Register* 31 (December 3, 1966): 15219, 32 (February 3, 1967): 2414.
[3] *Federal Register* 35 (September 30, 1970): 15222.
[4] *Federal Register* 36 (March 10, 1971): 4600; *Code of Federal Regulations*, Title 49, Part 571.208 S4.1.1.
[5] *Federal Register* 38 (June 20, 1973): 16072; *Code of Federal Regulations*, Title 49, Part 571.208 S4.1.2.

The ignition interlock was not a successful measure. NHTSA amended FMVSS 208, effective October 29, 1974, to delete this requirement. A 4-8 second visible and audible warning took its place.[6]

As part of its effort to bring safety requirements for LTVs up to the same level as cars, NHTSA amended FMVSS 208 to require integral, 3-point lap-shoulder belts (or, alternatively, automatic protection) at the front-outboard seats of most LTVs effective January 1, 1976 and all LTVs with GVWR up to 10,000 pounds effective September 1, 1981.[7]

Low use of manual belts continued into the early 1980's. On July 17, 1984, NHTSA amended FMVSS 208 to phase automatic protection, such as air bags or automatic belts, into the front-outboard seats of passenger cars between September 1, 1986 and September 1, 1989. To encourage the development of air bags, NHTSA exempted the right-front seat from the automatic protection requirement until August 31, 1993 in cars with driver air bags. NHTSA, the manufacturers and the safety community dedicated themselves to a successful effort to encourage buckle-up laws in the States. Comfort and convenience standards for belts were also added to FMVSS 208, effective September 1, 1986. During the implementation of automatic protection, automatic belts initially predominated, then driver air bags with manual 3-point belts, and, after September 1, 1993, dual air bags with manual 3-point belts.[8]

The superior protection of lap-shoulder belts, as compared to lap belts alone, was extended to the rear-outboard seats. Cars had to have 3-point belts, effective December 11, 1989, and LTVs, starting September 1, 1991.[9]

Automatic protection was to be phased into the front-outboard seats of LTVs with GVWR 8,500 pounds or less from September 1, 1994 to September 1, 1997. Manufacturers used air bags with manual belts.[10]

On-the-road experience and consumer reaction soon demonstrated that the combination of manual 3-point belts with air bags was the most effective and desirable system. All cars manufactured after September 1, 1997, and all LTVs after September 1, 1998 were required to have manual 3-point belts and air bags for the driver and the right-front passenger. Automatic belts were phased out.[11]

Air bags of the early 1990's presented risks to infants, children under the age of 12, and certain other individuals. NHTSA urged that high-risk individuals travel in the back seat when possible.

---

[6] *Federal Register* 39 (October 31, 1974): 38380, 39 (December 6, 1974): 42692.

[7] *Federal Register* 40 (July 9, 1975): 28805; *Code of Federal Regulations*, Title 49, Part 571.208 S4.2.1.

[8] *Federal Register* 46 (January 8, 1981): 2064, 49 (July 17, 1984): 28962, 50 (August 23, 1985): 34152; *Code of Federal Regulations*, Title 49, Parts 571.208 S4.1.3, 4.1.4 and S7.4

[9] From December 11, 1989 to August 31, 1990, cars were allowed separate lap and shoulder belts as an alternative to 3-point belts, but nobody exercised that option; *Federal Register* 54 (June 14, 1989): 25275, 54 (November 2, 1989): 46257; *Code of Federal Regulations*, Title 49, Parts 571.208 S4.1.4.2 and S4.2.4. The requirement does not apply to some types of seats/vehicles.

[10] *Federal Register* 56 (March 26, 1991): 12472; *Code of Federal Regulations*, Title 49, Part 571.208 S4.2.5.

[11] *Federal Register* 58 (September 2, 1993): 46551; *Code of Federal Regulations*, Title 49, Part 571.208 S4.1.5.3 and S4.2.6.2.

The agency also amended FMVSS 208 with measures to reduce risk when these people must travel in the front seat:

- Effective June 22, 1995, NHTSA permitted on-off switches for the passenger air bag in pickup trucks without back seats or other vehicles that cannot accommodate child safety seats in the back seat. This facilitated the implementation of passenger air bags in pickup trucks. Effective January 19, 1998, NHTSA also enabled people who must transport high-risk individuals in the front seats of any vehicle to obtain aftermarket on-off switches at their own expense.[12]
- Effective September 1, 1997, the agency relaxed some aspects of the frontal impact test in order to facilitate the introduction of "redesigned" air bags that deploy less forcefully.[13]
- From September 1, 2003 to September 1, 2005, "advanced" air bags will be phased in that do not deploy at all ("suppression") or deploy only at a low level of force ("low-risk deployment") if a small child is present, or if an older child/small adult is out-of-position, close to the air bag.[14]

Lap belts for front-seat occupants

History Ford and Chrysler began to offer front-outboard lap belts as an option on some of their 1956 models. Even before then, consumers were outfitting their vehicles with aftermarket lap belts; the American Medical Association and the National Safety Council, among others recommended belts in cars. In 1961, Wisconsin required front-outboard lap belts in all new cars sold there; New York State required anchorages to allow installation of aftermarket belts. By 1964, about half of the States had such laws. During that time, SAE issued Recommended Practices J4 for belts and J787 for anchorages; they were the starting points for FMVSS 209 and 210. The domestic manufacturers phased in lap belts at the front-outboard positions of all new cars by 1964-65, extending to the front-center seat by 1968. FMVSS 208 took effect in passenger cars on January 1, 1968 and in LTVs on July 1, 1971, requiring a shoulder belt on a front-outboard seat if a dummy restrained by just a lap belt could reach the windshield header. Lap-shoulder belts superseded lap belts at the front-outboard seats, immediately in passenger cars during MY 1968, more gradually in LTVs during 1969-81. However, front-center seats continue to be equipped with just lap belts.[15]

How lap belts work The principal benefit of a lap belt is to reduce greatly the likelihood of total ejection of an occupant from the vehicle during a crash. Ejection increases fatality risk in a crash by a factor of nearly four, all else being equal, because the ground is a less forgiving surface than the vehicle interior, and because the occupant may be caught between his or her own vehicle, the ground, and/or other vehicles. During 1977-99, 28 percent of unrestrained fatalities in cars, and

---

[12] *Federal Register* 60 (May 23, 1995): 27233, 62 (November 21, 1997): 62406 *Code of Federal Regulations*, Title 49, Part 571.208 S4.5.4 and Part 595.

[13] *Federal Register* 62 (March 19, 1997): 12960; *Code of Federal Regulations*, Title 49, Part 571.208 S13.

[14] *Federal Register* 65 (May 12, 2000): 30679; *Code of Federal Regulations*, Title 49, Part 571.208 S14.

[15] *Performance of Lap Belts in 26 Frontal Crashes*, National Transportation Safety Board Report No. NTSB/SS-86/03, Publication No. PB86-917006, Washington, 1986, pp. 225-230; *1965 SAE Handbook*, Society of Automotive Engineers, New York, 1965, pp. 868-872; *Federal Register* 32 (February 3, 1967): 2414, 35 (September 30, 1970): 15222.

48 percent in LTVs were ejectees; the percentages may have been higher before improved door locks (FMVSS 206).[16]

There was also hope, initially, that lap belts would be valuable for preventing, or reducing the severity of contacts with the vehicle interior, or with other occupants. These hopes were only partially realized – e.g., in farside or oblique frontal impacts, where the interior is furthest from the occupant. However, in direct frontal crashes, lap belts do not necessarily prevent head contacts with the interior. Furthermore, restraining the lower torso but not the thorax can have two effects that potentially increase rather than mitigate injury. Concentration of the occupant's force on the lap belt can result in abdominal injury, especially if the belt rides up from the pelvis to the waist ("submarining"). The head and chest can jackknife forwards and downwards at a high velocity.[17]

Expected benefits Substantial fatality reduction should be expected in rollovers, where the majority of unrestrained fatalities are ejectees. Fatality and injury reduction may be expected in farside and oblique frontal crashes, and in crashes with multiple harmful events, where occupants are ejected or tossed around within the vehicle. Possible adverse effects, especially in frontal crashes, could detract from overall effectiveness. Lap belts should be more effective in LTVs than in cars, since there are more rollovers, more ejections, and more space within the vehicle.

Early effectiveness studies Early analyses were usually based on fatalities or injuries per 100 crash-involved occupants. For example, if 10 percent of unrestrained occupants were injured, and 5 percent of lap-belted occupants, the effectiveness estimate was 50 percent. These studies estimated reductions of fatalities and/or serious injuries ranging from 35 percent upwards to 78 percent. Researchers grasped that several factors were definitely or possibly inflating the effectiveness attributed to belts:[18]

- People who buckle up also tend to drive more carefully than those who do not. They have less severe crashes, thus, lower injury rates per 100 crash involvements.
- Additional bias in favor of belts may occur if uninjured, unrestrained people report themselves, or are reported as belt users. That would lower the injury rate in the "belted" population, and raise it for the "unrestrained."
- Statistically meaningful results on fatality reduction are difficult to achieve because there are relatively few fatality cases on individual State crash files.

---

[16] Kahane, C.J., *Fatality Reduction by Safety Belts for Front-Seat Occupants of Cars and Light Trucks*, NHTSA Technical Report No. DOT HS 809 199, Washington, 2000, pp. 31-33; Evans, L., *Traffic Safety and the Driver*, Van Nostrand Reinhold, New York, 1991, pp. 52-54; Sikora, J.J., *Relative Risk of Death for Ejected Occupants in Fatal Traffic Accidents*, NHTSA Report No. DOT HS 807 096, Washington, 1986.
[17] *Performance of Lap Belts in 26 Frontal Crashes*; whereas that report focuses on back-seat occupants, the conclusions are also pertinent to front-seat occupants.
[18] *Safety Belt Usage – A Review of Effectiveness Studies – Suggestions for State Programs*, NHTSA Report No. DOT HS 801 988, Washington, 1976; Tourin, B. and Garrett, J.W., *Safety Belt Effectiveness in Rural California Automobile Accidents*, Cornell Aeronautical Laboratory, Buffalo, 1960; Levine, D.N. and Campbell, B.J., *Effectiveness of Lap Seat Belts and the Energy Absorbing Steering Column in the Reduction of Injuries*, HSL Publication No. 00221617, Highway Safety Research Center, University of North Carolina, Chapel Hill, 1971; Kahane, C.J., *Usage and Effectiveness of Seat and Shoulder Belts in Rural Pennsylvania Accidents*, NHTSA Technical Note No. DOT HS 801 398, Washington, 1974.

NHTSA's Restraint Systems Evaluation Project (RSEP) of 1974-75 addressed the first two problems. Multidisciplinary teams investigated a probability sample of towaway crashes involving 21,000 front-outboard occupants. Detailed information about vehicle damage made it possible to adjust the injury rates for differences in crash severity, using multidimensional contingency table analysis. Investigators were trained to report belt use as accurately as possible, based on an extensive checklist of evidence from the vehicle, driver interviews, the police report, and medical information. RSEP estimated that lap belts reduce moderate (AIS ≥ 2) injuries by 31 percent and severe (AIS ≥ 3) injuries by 46 percent.[19]

In 1984, in support of the rulemaking on automatic protection, NHTSA supplemented the RSEP data with NCSS plus all available NASS data. The additional data permitted a more extensive and realistic adjustment for crash severity, producing somewhat lower, but statistically significant injury reductions. They remain NHTSA's most reliable statistical results, and they will be discussed below. Based on these analyses and other studies available at that time, NHTSA's best estimates were that lap belts reduce fatalities by 30-40 percent, AIS 2-5 (moderate to critical) injuries by 25-35 percent, and AIS 1 (minor) injuries by 10 percent, for front-seat occupants of passenger cars.[20]

Fatality reduction – passenger cars Double-pair comparison, as originally applied by Evans to study lap-shoulder belts, greatly facilitated the estimation of fatality reduction.[21] Double-pair comparison allows the direct use of FARS data that have a much higher N of fatalities than NASS or State files. It implicitly adjusts or controls for the differences in the severity of crashes involving belted and unrestrained occupants. Until 1985, before buckle-up laws, FARS data were accurate or at least unbiased in the reporting of belt use.

A new double-pair analysis of lap-belt effectiveness was performed for this report, using the technique developed in NHTSA's evaluation of 3-point belts (and similar to Evans').[22] The analysis is based on MY 1956-73 cars (before 3-point belts) in CY 1975-85 FARS data. It is limited to cars with a driver and a right front (RF) passenger (and perhaps other passengers). The driver, or the RF passenger, or both were fatally injured. Only lap-belted and unrestrained drivers and RF passengers, age 14-97, are included. The 34,889 vehicle cases tabulate as follows:

---

[19] Reinfurt, D.W., Silva, C.Z. and Seila, A.F., *A Statistical Analysis of Seat Belt Effectiveness in 1973-75 Model Cars Involved in Towaway Crashes, Volume 1*, NHTSA Report No. DOT HS 802 035, Washington, 1976, pp. vii-viii.

[20] Kahane, C.J., *Addendum to "Seat Belt Effectiveness Estimates Using Data Adjusted for Damage Type"*, NHTSA Docket No. 74-14-N35-229-05, 1984; *Final Regulatory Impact Analysis, Amendment to Federal Motor Vehicle Safety Standard 208, Passenger Car Front Seat Occupant Protection*, NHTSA Report No. DOT HS 806 572, Washington, 1984, p. IV-2.

[21] Evans, L., "Double Pair Comparison - A New Method to Determine How Occupant Characteristics Affect Fatality Risk in Traffic Crashes," *Accident Analysis and Prevention*, Vol. 18, June 1986, pp. 217-227; Evans, L., "The Effectiveness of Safety Belts in Preventing Fatalities," *Accident Analysis and Prevention*, Vol. 18, June 1986, pp. 229-241.

[22] Kahane (2000), pp. 5-10.

Vehicles	Driver Died RF Survived	Driver Survived RF Died	Both Died
Both unrestrained	13,064	14,781	5,926
Driver unrestrained, RF lap-belted	111	93	32
Driver lap-belted, RF unrestrained	104	201	73
Both lap-belted	178	261	65

This can be tabulated as fatality rather than vehicle cases, by adding the "both died" column to each of the preceding columns:

Fatalities	Driver Fatalities	RF Fatalities	Driver/RF Risk Ratio
Both unrestrained	18,990	20,707	0.917
Driver unrestrained, RF lap-belted	143	125	1.144
Driver lap-belted, RF unrestrained	177	274	0.646
Both lap-belted	243	326	0.745

It is clear that (1) the overwhelming majority of people killed in CY 1975-85 crashes were unrestrained; (2) in these older cars, unrestrained drivers are at slightly lower risk than their RF passengers; and (3) whoever buckled up with lap belts reduced their risk. The four rows of data allow a total of four double-pair comparisons, two for computing the effectiveness of belts for drivers, and two for RF passengers. The first comparison for the driver is based on the first and third rows of data:

		Driver Fatalities	RF Fatalities	Driver/RF Risk Ratio
Driver unrestrained	RF unrestrained	18,990	20,707	0.917
Driver lap-belted	RF unrestrained	177	274	0.646

In both pairs, the driver's fatality risk is compared to the same control group: the unrestrained RF passenger. The unrestrained driver has slightly lower fatality risk than the unrestrained RF in the same crash, the lap-belted driver, substantially lower. The fatality reduction for lap belts is:
1 - (0.646/0.917) = 30 percent.

The other comparison for the driver is based on the second and fourth rows of data:

		Driver Fatalities	RF Fatalities	Driver/RF Risk Ratio
Driver unrestrained	RF lap-belted	143	125	1.144
Driver lap-belted	RF lap-belted	243	326	0.745

Here, the control group is the lap-belted RF passenger. The unrestrained driver has higher fatality risk than the lap-belted RF in the same crash, the lap-belted driver, lower. The fatality reduction is: 1 - (0.745/1.144) = 35 percent. The effectiveness estimates for drivers are similar with the two control groups.

The first double-pair comparison to estimate belt effectiveness for the RF passenger is obtained by using the first two rows of data, reversing the order of the columns and computing the RF/Driver rather than the Driver/RF risk ratio:

		RF Fatalities	Driver Fatalities	RF/Driver Risk Ratio
RF unrestrained	Driver unrestrained	20,707	18,990	1.090
RF lap-belted	Driver unrestrained	125	143	0.874

The control group is the unrestrained driver. The fatality reduction for the lap-belted RF passenger is: 1 - (0.874/1.090) = 20 percent. The second estimate uses the last two rows of data:

		RF Fatalities	Driver Fatalities	RF/Driver Risk Ratio
RF unrestrained	Driver lap-belted	274	177	1.548
RF lap-belted	Driver lap-belted	326	243	1.342

The control group is the lap-belted driver. The fatality reduction for the lap-belted RF passenger is: 1 - (1.342/1.548) = 13 percent. Again, the two control groups produce similar estimates. Also, lap-belt effectiveness is lower for the RF passenger than for the driver.

NHTSA's 2000 evaluation of 3-point belts develops a weighting procedure that combines the two driver estimates and the two RF estimates into a single number, the overall, average fatality reduction for front-seat lap belts in passenger cars: 27 percent.

By crash mode, the estimated fatality reduction for lap belts is: 43 percent in rollovers, 22 percent in frontals, and 21 percent in side impacts, rear impacts and other crashes.

Fatality reduction – light trucks A new double-pair comparison analysis was similarly performed for light trucks. The fatality reduction by lap belts for front-seat occupants averaged 48 percent. There were not enough FARS cases for statistically meaningful separate analyses by crash mode. That is very close to a previous NHTSA estimate of 50 percent fatality reduction by lap belts in the front-center seat of LTVs, not obtained directly from crash data, but inferred by examining the relationships between other estimates: lap-belt vs. 3-point-belt effectiveness in cars and 3-point belt effectiveness in LTVs vs. cars.[23]

Injury reduction – passenger cars As stated above, NHTSA analyzed a database in 1984 that combined the RSEP, the National Crash Severity Study (NCSS) plus all available NASS cases. (NCSS was the predecessor of NASS.) Injury rates were computed per 100 towaway-involved occupants: unrestrained, lap-belted, and lap-shoulder belted. The rates were adjusted for differences in impact location (front, side, rollover) and damage severity (CDC extent zone), using a Mantel-Haenszel model, a simpler approach than the logistic regression method widely used today. Lap belts reduced the adjusted overall (AIS $\geq$ 1) injury rate by 8 percent, moderate (AIS $\geq$ 2) injuries by 22 percent and severe (AIS $\geq$ 3) injuries by 33 percent.[24]

Injury reduction – light trucks NHTSA has not analyzed the injury reduction for lap belts for front-outboard occupants of LTVs. However, the agency estimated that lap belts in front-center seat of LTVs reduce AIS 2-5 injuries by 55 percent and AIS 1 injuries by 10 percent. The estimate was not obtained directly from crash data, but inferred by examining the relationships between other estimates: lap-belt vs. 3-point-belt effectiveness in cars, 3-point belt effectiveness in LTVs vs. cars, and fatality reduction vs. injury reduction.[25]

Lap belts for back-seat occupants

History The domestic manufacturers were providing lap belts at the rear-outboard positions by MY 1967. FMVSS 208 required safety belts at all rear seats, including the rear-center and the third row of seats in a station wagon, effective January 1, 1968 in passenger cars and July 1, 1971 in LTVs. As late as MY 1987, over 90 percent of new cars and LTVs had lap belts, not 3-point belts at the rear-outboard seats. In cars, over 90 percent of rear-center seats had lap belts (not 3-point belts) as late as 1998.[26]

How lap belts work Here, too, the principal benefit of a lap belt is to reduce greatly the likelihood of total ejection. Lap belts may prevent, or reduce the severity of some interior contacts – e.g., in farside or oblique frontal impacts. There was, perhaps, some optimism that, in direct frontal crashes, lap belts would have fewer shortcomings in the back seat than in the front seat. Back-seat occupants would hit the front seatback with their heads and chests. It is a well-padded surface, more benign than anything front-seat occupants might contact. However, the National Transportation Safety Board's 1986 study indicated that lap-belted back-seat occupants

---

[23] *Final Regulatory Impact Analysis, Extension of the Automatic Restraint Requirements of FMVSS 208 to Trucks, Buses and Multi-Purpose Passenger Vehicles*, NHTSA Docket No. 74-14-N70-001, 1990, p. 23.

[24] Kahane (1984 Seat Belts); 90% confidence bounds for effectiveness: AIS $\geq$ 1 injuries, 6 to 13 percent; AIS $\geq$ 2 injuries, 13 to 42 percent; AIS $\geq$ 3 injuries, 20 to 76 percent.

[25] *Final Regulatory Impact Analysis, Extension of the Automatic Restraint Requirements of FMVSS 208 to Trucks, Buses and Multi-Purpose Passenger Vehicles*, p. 23.

[26] *Performance of Lap Belts in 26 Frontal Crashes*, pp. 225-230.

had a high risk of abdominal injuries and head injuries in frontal crashes. These occupants experienced a concentrated force of the lap belt on their abdomens and their upper bodies jackknifed over the belt at a high velocity.[27]

Fatality reduction – passenger cars NHTSA's evaluation, published in 1999, is based on double-pair comparison analysis of 1988-97 FARS data.[28] Front-seat occupants (drivers and RF passengers) were the control groups used to compare the relative fatality risk of lap-belted vs. unrestrained rear-outboard occupants, as, for example, in the following table:

Back-Seat Belt Use	Front-Seat Belt Use	Back-Seat Fatalities	Front-Seat Fatalities	Back/Front Risk Ratio
Unrestrained	3-point belts	3,028	2,098	1.443
Lap belts	3-point belts	1,135	1,079	1.052

In this example, unrestrained back-seat occupants had considerably higher fatality risk than front-seat occupants protected by 3-point belts. However, lap-belted back-seat occupants had only slightly higher risk than 3-point-belted front-seat occupants of the same cars: the greater safety of the back seat nearly offsets the superior restraint system in the front seat.

The overall, average fatality reduction for lap belts at the rear-outboard seats of passenger cars was 32 percent.[29] However, the effectiveness of lap belts varies greatly by crash mode. The estimated fatality reduction is 76 percent in rollovers and 39 percent in side impacts, but it is zero in frontals, rear impacts and other crashes.

A supplement to FARS called the Multiple Cause of Death (MCOD) file gleans injury information from death certificates. These data show why lap belts have no net benefit in frontal crashes. The rate of life-threatening abdominal injuries is three times as high for a lap-belted occupant as for an unrestrained occupant. That offsets any benefits of lap belts in frontals, such as a reduction of ejections or head injuries. In frontal crashes, occupants age 55 or older actually have higher fatality risk with lap belts than unrestrained; they have exceptionally high risk of abdominal injury with lap belts.[30]

Fatality reduction – light trucks NHTSA's 1999 evaluation similarly analyzes vans and SUVs. The fatality reduction by lap belts for back-seat occupants averaged 63 percent.[31] By crash mode, the estimated fatality reduction for lap belts is: 80 percent in rollovers, 44 percent in frontals, and 64 percent in side impacts, rear impacts and other crashes. A high percentage of rear-seat fatalities in LTVs are ejectees, even in frontals.

---

[27] *Ibid.,* pp. 1-11.
[28] Morgan, C., *Effectiveness of Lap/Shoulder Belts in the Back Outboard Seating Positions,* NHTSA Technical Report No. DOT HS 808 945, Washington, 1999.
[29] *Ibid.,* pp. x and 13-30; 90% confidence bounds, 23 to 40 percent.
[30] *Ibid.,* pp. xi-xiii, 47-52 and 99-111.
[31] *Ibid.,* pp. xii and 83-97; 90% confidence bounds, 52 to 71 percent.

Injury reduction – passenger cars A 1987 NHTSA report estimated the effectiveness of lap belts in reducing nonfatal injuries, based on double-pair comparison analyses of 1982-85 Pennsylvania crash data. Compared to unrestrained back-seat occupants, lap belts reduced fatal-and-serious (K + A) injury risk by 37 percent, moderate-to-fatal (K + A + B) injury risk by 33 percent, and overall (K + A + B + C) injury risk by 11 percent.[32]

Injury reduction – light trucks The agency has not analyzed crash data to estimate the injury reduction for lap belts in LTVs. However, a NHTSA Regulatory Impact Analysis in 1990 projected that lap belts in back seats of LTVs could reduce AIS 2-5 injuries by 35.5 percent and AIS 1 injuries by 5.5 percent. Here, the agency assumed the same effectiveness in cars and LTVs; in retrospect, the AIS 2-5 number would appear to be an underestimate for the LTVs.[33]

Manual 3-point lap-shoulder belts for front-seat occupants

History Volvo introduced 3-point lap-shoulder belts as standard equipment in Sweden in 1959 and in the United States in 1963. In 1968, Volkswagen and, apparently, Audi, BMW and Mercedes made 3-point belts standard. At that time, the domestic and, apparently, the Japanese manufacturers installed separate lap and shoulder belts in all new cars. Retractor systems were required effective January 1, 1972, and had been phased in by MY 1972. Effective January 1, 1968, FMVSS 208 originally required shoulder belts only if lap belts alone could not prevent contact with the windshield header. Effective January 1, 1972, FMVSS 208 required shoulder belts on all passenger cars, and effective September 1, 1973, shoulder belts had to be integral with the lap belt. Thus, from model year 1974 onwards, all passenger cars, except those with automatic belts, have had the modern system of integral 3-point belts with automatic locking retractors.[34]

LTVs were equipped with 3-point belts later than cars, but they went directly from lap belts to 3-point belts, skipping separate lap and shoulder belts. FMVSS 208 required 3-point belts effective January 1, 1976 for all LTVs with GVWR up to 10,000 pounds except forward-control vehicles and some other types, and effective September 1, 1981 on the forward-control vehicles. However, Volkswagen had introduced 3-point belts by 1969. NCSS and NASS data indicate that other manufacturers began to phase them in during 1972, 1973 or 1974. The phase-in was completed in pickup trucks and SUVs by January 1, 1976, and in vans by MY 1980.[35]

How 3-point belts work The combination of a lap and shoulder belt furnishes at least as much protection against ejection as a lap belt alone, and possibly more, because the shoulder strap

---

[32] Kahane, C.J., "Fatality and Injury Reducing Effectiveness of Lap Belts for Back Seat Occupants," *Restraint Technologies: Rear Seat Occupant* Protection, Paper No. 870486, Society of Automotive Engineers Publication No. SP-691, Warrendale, PA, 1987, pp. 45-52; 90% confidence bounds for effectiveness: K + A injuries, 23 to 51 percent; K + A + B injuries, 26 to 40 percent; all injuries, 7 to 15 percent.

[33] *Final Regulatory Impact Analysis, Extension of the Automatic Restraint Requirements of FMVSS 208 to Trucks, Buses and Multi-Purpose Passenger Vehicles*, p. 23; *Final Regulatory Evaluation, Rear Seat Lap Shoulder Belts in Passenger Cars*, NHTSA Docket No. 87-08-N03-001, 1989, p. IV-1.

[34] *Performance of Lap Belts in 26 Frontal Crashes*, pp. 225-230; *Federal Register* 32 (February 3, 1967): 2414, 36 (March 10, 1971): 4600, 38 (June 20, 1973): 16072; *Code of Federal Regulations*, Title 49, Part 571.208 S4.1.1 and S4.1.2.

[35] *Federal Register* 40 (July 9, 1975): 28805; *Code of Federal Regulations*, Title 49, Part 571.208 S4.2.1.

might sometimes prevent partial ejection through the side window. But lap-shoulder belts also significantly reduce the injury severity of occupants who remain within the vehicle, and they avoid most of the additional injury risks associated with lap belts. The lap-shoulder belt is the basic protection system for all occupants in most types of crashes. Ideally, belts provide "ride-down" by linking the occupant to the vehicle structure. They gradually decelerate the occupant as the vehicle deforms and absorbs energy and as the belts themselves "give" a little – but not so much as to allow forceful contacts with interior surfaces. They keep the occupant in an upright, seated position and reduce the risk of flexion/tension injuries. In general, lap-shoulder belts accomplish these missions, although they cannot always prevent head contacts with interior surfaces, may sometimes place excessive force on the occupant in severe crashes, and are less effective if the passenger compartment is damaged, as in a nearside impact.

Separate shoulder belts were uncomfortable and inconvenient, and were only used by about 15-25 percent of the people who buckled the lap belt. The most important feature of the integral 3-point belt is that everybody who buckled up received the protection of both the shoulder and the lap restraint, except for a small percentage of occupants who misused the shoulder belt by running it behind their backs or under their arms. Furthermore, the public did not consider the double belt excessively burdensome, because use of 3-point belts was as high or higher than use of the lap belt alone.[36] Technological advances made the 3-point belt acceptable to the public. Most important was the locking retractor. Whereas "airline style" lap belts are merely inconvenient, a shoulder belt that is not self-adjusting would be unacceptable because drivers must lean forward or sideways in the course of normal driving (and passengers also desire some flexibility). The locking retractor is a system that spools out more webbing when occupants wish to move around during normal driving, locks the spool when it senses a crash or other sudden decelerations (e.g., driving over potholes), and picks up excessive slack in the belt. Slack is comfortable, up to a point, in normal driving operations but most undesirable in a crash because it defeats the early engagement of the occupant with the belts. The system had to be fine-tuned. Belts should never feel too tight or too loose, lock too seldom or too often, or allow excessive slack. NHTSA's comfort and convenience standards for belts addressed these issues.[37]

Two problems with 3-point belts were excessive slack in some crashes, and excessive force on some occupants in other crashes. Pretensioners and load-limiters, discussed later, address those shortcomings.

Belt use trends Safety belt use in the United States has increased from about 11 percent in 1979-82 to 80 percent in 2004. This has saved more lives than any other vehicle safety program. Finally, the vast majority of people are using the basic occupant protection that has been in their vehicles since the 1970's.

The National Occupant Protection Use Surveys (NOPUS) of 1994, 1996, and every year since 1998, give unbiased estimates of belt use by front-outboard occupants of cars and LTVs on the nation's streets, roads and highways during daylight hours, based on direct observation of a

---

[36] Reinfurt, D.W., Silva, C.Z. and Seila, A.F, p. 18; Phillips, B.M., *Safety Belt Usage Among Drivers*, NHTSA Report No. DOT HS 805 398, Washington, 1980, pp. 4-5.
[37] *Federal Register* 46 (January 8, 1981): 2064, 50 (August 23, 1985): 34152; *Code of Federal Regulations*, Title 49, Part 571.208 S7.4

probability sample of vehicles and roadways. While unbiased, NOPUS results can have sampling error; 95% confidence bounds were initially $\pm$ 4 percent but in recent years have shrunk to $\pm$ 2 percent or less.[38]

From 1976 to 1991, NHTSA estimated national safety belt use by drivers in the "19 City Survey." It was not a probability sample of the nation's roadways. Contractors observed belt use at randomly selected controlled intersections in 19 metropolitan regions purposively chosen as a cross-section of the nation's urban areas. Surveyors returned to the same intersections and cities year after year.[39] The point estimates generated by the 19 City Survey are not directly comparable to NOPUS and do not have estimates of sampling error.[40]

The 19 City estimates of belt use in the United States from 1976 to 1991 and the NOPUS estimates from 1994 to 2004 are the following:

---

[38] *Observed Safety Belt Use in 1996*, NHTSA Research Note, Washington, 1997; *Observed Safety Belt Use, Fall 2000 National Occupant Protection Use Survey*, NHTSA Research Note, Washington, 2001; Glassbrenner, D., *Safety Belt and Helmet Use in 2002 – Overall Results*, NHTSA Technical Report No. DOT HS 809 500, Washington, 2002; Glassbrenner, D., *Safety Belt Use in 2003*, NHTSA Technical Report No. DOT HS 809 646, Washington, 2003; Glassbrenner, D., *Safety Belt Use in 2004 – Overall Results*, NHTSA Research Note No. DOT HS 809 783, Washington, 2004.

[39] Stowell, C. and Bryant, J., *Safety Belt Usage: Survey of the Traffic Population*, NHTSA Report No. DOT HS 803 354, Washington, 1978 (this initial survey encompassed only 15 of the 19 cities); Phillips, B.M., *Safety Belt Usage Among Drivers*, NHTSA Report No. DOT HS 805 398, Washington, 1980; Phillips, B.M., *Restraint System Usage in the Traffic Population*, NHTSA Report No. DOT HS 806 424, Washington, 1983; Perkins, D.D., Cynecki, M.J. and Goryl, M.E., *Restraint System Usage in the Traffic Population*, NHTSA Report No. DOT HS 806 582, Washington, 1984; Goryl, M.E. and Cynecki, M.J., *Restraint System Usage in the Traffic Population*, NHTSA Report No. DOT HS 806 714, Washington, 1985; Goryl, M.E., *Restraint System Usage in the Traffic Population*, NHTSA Report No. DOT HS 806 987, Washington, 1986; Goryl, M.E. and Bowman, B.L., *Restraint System Usage in the Traffic Population*, NHTSA Report No. DOT HS 807 080, Washington, 1987; Bowman, B.L. and Rounds, D.A., *Restraint System Usage in the Traffic Population 1987 Annual Report*, NHTSA Report No. DOT HS 807 342, Washington, 1988; Bowman, B.L. and Rounds, D.A., *Restraint System Usage in the Traffic Population 1988 Annual Report*, NHTSA Report No. DOT HS 807 447, Washington, 1989; Datta, T.K. and Guzek, P., *Restraint System Use in 19 U.S. Cities 1989 Annual Report*, NHTSA Report No. DOT HS 807 595, Washington, 1990; Datta, T.K. and Guzek, P., *Restraint System Use in 19 U.S. Cities 1990 Annual Report*, NHTSA Report No. DOT HS 808 147, Washington, 1991; Datta, T.K. and Guzek, P., *Restraint System Use in 19 U.S. Cities 1991 Annual Report*, NHTSA Report No. DOT HS 808 148, Washington, 1992.

[40] NOPUS and the 19 City Survey never overlapped. However, both studies overlapped with a NHTSA computation of national averages based on individual State surveys during the 1990's. These averages cannot be considered unbiased estimates of the nation's belt use for various reasons – e.g., the exclusion of pickup trucks (that have lower belt use) in up to half the State surveys before 1997. In 1991, belt use was 51 percent in the 19 City Survey and averaged 58 percent in the State surveys; in 1994, belt use was 58 percent in NOPUS and averaged 66 percent in the State surveys. Thus, the State survey averages exceed the 19 City and NOPUS estimates by approximately the same amounts (7 or 8 percentage points) in 1991 and 1994, respectively. See also Kahane (2000), pp. 63 and 75-76.

SAFETY BELT USE IN THE UNITED STATES, 1976-2004
(19 City Survey, 1976-1991; NOPUS, 1994-2004)

Year	Percent
1976	19 %
1977	
1978	13
1979	11
1980	
1981	11
1982	11
1983	14
1984	14
1985	21
1986	37
1987	42
1988	46
1989	46
1990	49
1991	51
1992	
1993	
1994	58
1995	
1996	61
1997	
1998	69
1999	
2000	71
2001	73
2002	75
2003	79
2004	80

Safety belt use actually deteriorated from 19 percent in 1976 to 11 percent in 1979-82, in the wake of the unsuccessful ignition interlock and prior to national buckle-up campaigns. NHTSA responded in the early 1980's with "Get It Together," its first national campaign for State, community and private organizations to promote safety belts. Belt use has improved ever since 1982.[41]

The greatest jump in belt use, however, took place in 1985-1986-1987 as buckle-up laws took effect in 29 States, including all ten most populous States. New York's law was the first, effective December 1, 1984, followed by New Jersey, Illinois and Michigan. Simultaneous with the July 1984 decision on automatic crash protection, NHTSA, the manufacturers – through their Automotive Coalition for Traffic Safety (ACTS) – and the safety community dedicated themselves to a successful effort to encourage buckle-up laws in the States. By 1995, 49 States, the District of Columbia and Puerto Rico had buckle-up laws.[42]

Even though most of the States had buckle-up laws by 1990, belt use hardly leveled off, but rather continued to increase steadily, from 49 percent in 1990 to 80 percent in 2004. Sustained national campaigns, spearheaded by DOT/NHTSA and supported by States, communities and the private sector, have played an important role. They combine public information and education about the life-saving benefits of belts, law enforcement by States and communities, and additional information to make the public aware of the enforcement programs. Campaigns include "70% By 92" and "Avoid The Summertime Blues" in the early 1990's, "Buckle-Up America" starting in the late 1990's, and, in this decade, "Click It Or Ticket." There has also been a push for "primary" buckle-up laws that allow police to make a traffic stop solely upon observation of a belt law violation; there are 22 primary laws in 2004, up from 9 in 1993. Finally, after 20 years of cumulative and sustained education about safety belts, a generation of Americans has been buckling up since childhood, and views safety belts as a basic part of riding in a vehicle.[43]

---

[41] *Get It Together*, NHTSA Publication No. DOT HS 806 254, Washington, 1982; Stowell and Bryant; Phillips (1980); Phillips (1983).

[42] *Third Report to Congress – Effectiveness of Occupant Protection Systems and Their Use*, NHTSA Report No. DOT HS 808 537, Washington, 1996; ACTS, Automotive Coalition for Traffic Safety, Inc., State Laws Database www.actsinc.org .

[43] *Buckle Up: Avoid the Summertime Blues*, HSL Publication No. DOT HS 041 653, NHTSA, Washington, 1991; "70 by 92: Increased Safety-Belt Use," *Journal of the American Medical Association*, Vol. 268, July 15, 1992, p. 318; *Buckle Up America*, NHTSA Publication No. DOT HS 808 628, Washington, 1997; "'Click It Or Ticket' Expands beyond North Carolina," *Status Report*, Vol. 36, November 15, 2001, p. 4, Insurance Institute for Highway Safety, Arlington, VA; Eby, D.W., Vivoda, J.M. and Fordyce, T.A., *Direct Observation of Safety Belt Use in Michigan: Fall 1999*, Report No. UMTRI-99-33, University of Michigan Transportation Research Institute, Ann Arbor, 1999; *Safety Belt Use Laws*, Insurance Institute for Highway Safety, Arlington, VA, 2004 (http://www.hwysafety.org/safety%5Ffacts/state%5Flaws/restrain3.htm).

SAFETY BELT USE (%) IN THE UNITED STATES, BY CALENDAR YEAR

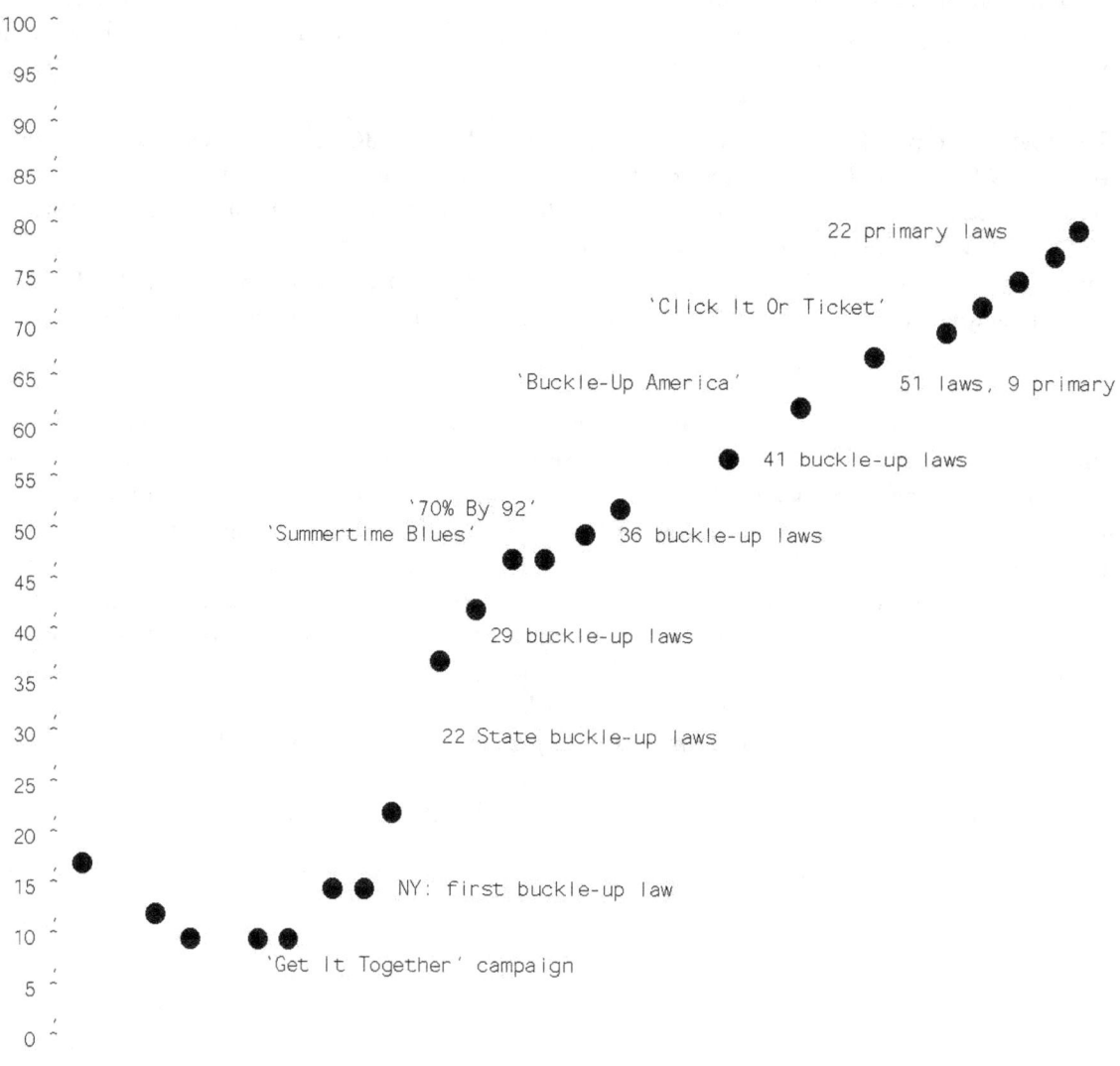

CALENDAR YEAR

Furthermore, belt use "where it counts" – i.e., by typical drivers and passengers in potentially fatal crashes – is quite close to the belt use observed on the road by surveys such as NOPUS. Drivers who drink or engage in other antisocial behaviors[44] have lower belt use. But among other drivers, even those held culpable for the crash, the proportion of fatalities that wore safety belts is consistent with the belt use in NOPUS and the fatality-reducing effectiveness of belts.[45]

Early effectiveness studies NHTSA's Restraint Systems Evaluation Project (RSEP) of 1974-75 was the first statistical analysis in the United States with a large sample of crash-involved occupants who wore lap-shoulder belts. After controlling for damage location and severity, occupant age and vehicle weight, RSEP estimated that lap-shoulder belts reduce moderate (AIS ≥ 2) and severe (AIS ≥ 3) injuries by 57 percent. RSEP hinted that fatality reduction might be close to 60 percent.[46]

By 1984, during the rulemaking on automatic protection, NHTSA realized that fatality reduction had to be somewhat lower than 60 percent. Buckle-up laws outside the United States were saving lives, but not as many as would be predicted based on 60 percent effectiveness. A review of 781 unrestrained fatalities in NCSS suggested that belts would have had little effect in nearly half of the cases. The agency performed two new analyses of belt effectiveness. RSEP data were supplemented with NCSS data plus all available NASS cases. The additional data permitted a more extensive adjustment for crash severity, and showed a statistically significant 49 percent reduction of fatalities and 45 percent reduction of severe (AIS ≥ 3) injuries. NHTSA compared the increase in belt use with the fatality reduction in 11 nations or Canadian Provinces that enacted buckle-up laws. The fatality reduction, on the average, suggested belts were 47 percent effective when used. In 1984, based on these analyses, NHTSA lowered its "best estimates" to 40-50 percent fatality reduction, 45-55 percent AIS 2-5 reduction, and 10 percent AIS 1 reduction.[47]

---

[44] Drugs, speeding, reckless driving, escaping police, racing, driving without a license.

[45] Partyka, S.C., *Belt Use in Serious Impacts Estimated from Fatality Data*, NHTSA Technical Report No. 807 519, 1989; Kahane (2000), pp. 43-51 and 64-70.

[46] Reinfurt, D.W., Silva, C.Z. and Seila, A.F., pp. iv-viii; *Safety Belt Usage – A Review of Effectiveness Studies – Suggestions for State Programs*.

[47] *Final Regulatory Impact Analysis, Amendment to Federal Motor Vehicle Safety Standard 208, Passenger Car Front Seat Occupant Protection*, pp. IV-2 – IV-16; Kahane, C.J., *Estimates of Fatality Reduction for Air Bags and Lap/Shoulder Belts Based on Case by Case Analysis of Unrestrained Fatalities*, NHTSA Docket No. 74-14-N35-012, 1984; Kahane (1984 Seat Belts), 90% confidence bounds for effectiveness: fatalities, 37 to 68 percent; AIS ≥ 3 injuries, 38 to 55 percent.

Expected benefits Large fatality reductions should be expected in rollovers, where the majority of unrestrained fatalities are ejectees. Substantial fatality and injury reduction may be expected in direct-frontal, oblique-frontal and farside crashes, except in impacts too severe for belts to effectively protect against contacts with interior surfaces. Effectiveness will be lowest in nearside impacts or other catastrophic crashes that reduce the space available to the occupant within the vehicle. Lap-shoulder belts should be more effective in LTVs than in cars, overall and specifically in side impacts, since there are more rollovers, more ejections, and fewer catastrophic damages.

Fatality reduction – passenger cars Evans' double-pair comparison analyses of 1985-86 were an important advance in the estimation of fatality reduction. They confirmed NHTSA's estimate that effectiveness was in the 40's.[48] Double-pair comparison allows the direct use of FARS data. It implicitly controls for crash severity. Until 1985, before buckle-up laws, FARS data were accurate or at least unbiased in the reporting of belt use. NHTSA's current estimate of overall belt effectiveness, in its 2000 evaluation, is based on a double-pair analysis of MY 1975-86 cars in CY 1977-85 FARS data. It is limited to cars with a driver and an RF passenger (and perhaps other passengers). The driver, or the RF passenger, or both were fatally injured. Only 3-point-belted and unrestrained drivers and RF passengers, age 14-97, are included. This basic table demonstrates the life-saving benefit of 3-point belts.[49]

	Driver Fatalities	RF Fatalities	Driver/RF Risk Ratio
Both unrestrained	16,503	16,786	0.983
Driver unrestrained, RF belted	374	226	1.655
Driver belted, RF unrestrained	288	589	0.489
Both belted	739	895	0.826

When the driver and RF passenger were both unrestrained, they had nearly equal risk. Unrestrained drivers had nearly double the risk of the belted RF passengers riding with them. Belted drivers had less than half the risk of unrestrained RF passengers. The evaluation report developed a procedure that combined these statistics into a single estimate of the overall fatality reduction for 3-point belts in passenger cars, 47.8 percent. This is very close to 45 percent, the midpoint of the agency's 1984 "best estimate" range of 40-50 percent and it essentially confirms that earlier estimate.

Immediately after 1985, double-pair comparison began producing inflated results. Whereas FARS apparently continued to report accurately the belt use of fatally injured front-outboard occupants, buckle-up laws gave survivors a motive to declare they had been belted in some of the cases where they were actually unrestrained. In CY 1986, belt "effectiveness" jumped up over 60 percent, and it stayed there. However, numerous FARS analyses revealed that the degree of exaggeration in the effectiveness estimate varied little from year to year, State to State, by crash type, vehicle type or occupant characteristics. NHTSA developed an empirical tool

---

[48] Evans (1986 Double Pair Comparison); Evans (1986 Belt Effectiveness).
[49] Kahane (2000), pp. 5-10.

called the "Universal Exaggeration Factor" to adjust the biases out of double-pair comparison estimates based on 1986-99 FARS data. That opened up the 1986-99 FARS database, with its incomparably greater number of belted cases than the earlier FARS, permitting point-estimation of belt effectiveness by crash type, occupant age and gender, vehicle type, etc:[50]

FATALITY REDUCTION (%) BY MANUAL 3-POINT BELTS IN PASSENGER CARS
For Drivers and Right-Front Passengers

		Fatality Reduction (%)	Sampling Error Range[51]
IN ALL CRASHES		45	x
BY	Frontal impacts	50	x
IMPACT	Side impacts	21	xx
TYPE:	Nearside	10	xx
	Farside	39	x
	Rollovers (primary)	74	x
	Rear impacts & other crashes	56	x
BY	Single vehicle	58	x
CRASH	Multivehicle	32	x
TYPE:	With a car	41	x
	With a light truck	31	x
	With a heavy truck	25	xx
BY SEAT	Driver	48	x
POSITION:	Right-front passenger	37	x
BY	5- 9 (RF passengers)	34	xxx
OCCUPANT	10-14 (RF passengers)	35	xx
AGE:	15-29	50	x
	30-54	49	x
	55-69	43	x
	70-79	38	x
	80 and older	27	xx
BY	Male	45	x
GENDER:	Female	45	x

---

[50] *Ibid.*, pp. 10-43.

[51] Minimum sampling error range: "x" denotes 2-sigma sampling error at least in the $\pm$ 4-10 percentage point range (plus possible, additional, unknown non-sampling error); "xx" = at least $\pm$ 10-20 percentage points; "xxx" = at least $\pm$ 20-50 percentage points.

		Fatality Reduction (%)	Sampling Error Range[52]
BY	No air bags.	45	x
AIR BAG	MY 1975-79	47	x
TYPE:	MY 1980-85	46	x
	Dual air bags.	48	x
BY	2499 pounds or lighter	48	x
CURB	2500-3149 pounds	44	x
WEIGHT:	3150 pounds or heavier	41	x

Three-point belts are most effective in rollovers (74 percent), but they are also highly effective in frontals (50 percent), farside impacts (39 percent) and all crash types except nearside impacts (10 percent). They are effective in single- and multivehicle crashes (but especially the single-vehicle crashes, where most of the rollovers are); slightly more effective for drivers (48 percent) than RF passengers (37 percent); beneficial for all age groups, but especially at age 15-54; equally effective for males and females; consistently beneficial from MY 1975 onwards, including in cars without air bags and with air bags; and effective in cars of all sizes.

Fatality reduction – light trucks In 1989 the agency concluded, based on double-pair comparison analyses by Partyka, that the overall fatality-reducing effectiveness of 3-point belts in LTVs was greater than in cars, and that it was close to 60 percent.[53] NHTSA's 2000 evaluation reconfirms that estimate and provides more detailed estimates of belt effectiveness in LTVs by crash type, occupant age and gender, vehicle type, etc:[54]

---

[52] Minimum sampling error range: "x" denotes 2-sigma sampling error at least in the $\pm$ 4-10 percentage point range (plus possible, additional, unknown non-sampling error); "xx" = at least $\pm$ 10-20 percentage points; "xxx" = at least $\pm$ 20-50 percentage points.

[53] *Final Regulatory Impact Analysis, Extension of the Automatic Restraint Requirements of FMVSS 208 to Trucks, Buses and Multi-Purpose Passenger Vehicles*, p. 23; Partyka, S.C., "Belt Effectiveness in Pickup Trucks and Passenger Cars by Crash Direction and Accident Year (May 1988)," *Papers on Adult Seat Belts - Effectiveness and Use*, NHTSA Technical Report No. DOT HS 807 285, Washington, 1988, pp. 99-102.

[54] *Ibid.*, pp. 10-43.

FATALITY REDUCTION (%) BY MANUAL 3-POINT BELTS IN LTVs
For Drivers and Right-Front Passengers

		Fatality Reduction (%)	Sampling Error Range[55]
IN ALL CRASHES		60	x
BY	Frontal impacts	53	x
IMPACT	Side impacts	48	x
TYPE:	Nearside	41	xx
	Farside	58	x
	Rollovers (primary)	80	x
	Rear impacts & other crashes	81	x
BY	Single vehicle	70	x
CRASH	Multivehicle	43	x
TYPE:	With a car	57	x
	With a light truck	45	xx
	With a heavy truck	28	xx
BY SEAT	Driver	61	x
POSITION:	Right-front passenger	58	x
BY	5- 9 (RF passengers)	59	xxx
OCCUPANT	10-14 (RF passengers)	63	xx
AGE:	15-29	63	x
	30-54	64	x
	55-69	53	xx
	70-79	42	xx
	80 and older	30	xxx
BY	Male	60	x
GENDER:	Female	62	x
BY	No air bags.	60	x
AIR BAG	MY 1975-79	49	xx
TYPE:	MY 1980-85	63	x
	Dual air bags.	63	xx
BY TRUCK	Pickup truck	58	x
TYPE:	Van or SUV	63	x

---

[55] Minimum sampling error range: "x" denotes 2-sigma sampling error at least in the $\pm$ 4-10 percentage point range (plus possible, additional, unknown non-sampling error); "xx" = at least $\pm$ 10-20 percentage points; "xxx" = at least $\pm$ 20-50 percentage points.

Belts are exceedingly effective in rollovers (80 percent), but they are also very effective in all other types of crashes, including even nearside impacts (41 percent), because LTVs are better able than cars to resist catastrophic damage in side impacts. They are effective in single-vehicle crashes and multivehicle crashes (but least so in impacts with heavy trucks); beneficial for all age groups; consistently effective from MY 1980 onwards, including in LTVs without air bags and with air bags; and beneficial in pickup trucks, vans and SUVs.

Injury reduction – passenger cars As stated above, NHTSA analyzed a database in 1984 that combined the RSEP, NCSS, plus all available NASS cases. Injury rates were computed per 100 towaway-involved occupants: unrestrained, lap-belted, and lap-shoulder belted. The rates were adjusted for differences in impact location (front, side, rollover) and damage severity (CDC extent zone), using a Mantel-Haenszel model, a simpler approach than the logistic regression method widely used today. Lap-shoulder belts reduced the adjusted overall (AIS ≥ 1) injury rate by 13 percent, moderate (AIS ≥ 2) injuries by 46 percent and severe (AIS ≥ 3) injuries by 45 percent.[56]

Injury reduction – light trucks NHTSA has not statistically analyzed the injury reduction by 3-point belts for front-outboard occupants of LTVs. However, the agency estimated that they reduce AIS 2-5 injuries by 65 percent and AIS 1 injuries by 10 percent. The estimate was inferred by examining the relationships between other estimates: 3-point belt fatality reduction in LTVs vs. cars, and fatality reduction vs. injury reduction in cars.[57]

Three-point lap-shoulder belts for back-seat occupants

History Volvo made 3-point belts standard at the rear-outboard positions in MY 1971, and Mercedes in MY 1974. The 1982 Honda Accord was the first high-sales make-model equipped with 3-point belts. The majority of new cars had them at least one or two years before the December 11, 1989 effective date specified in FMVSS 208. The first LTVs with 3-point belts were some Isuzu Troopers in 1987, all Troopers and all full-sized GM vans in 1988. FMVSS 208 required them in all LTVs starting September 1, 1991. Retrofit kits could be purchased for many earlier vehicles equipped only with lap belts at the rear-outboard positions.[58]

"Anton's Law," signed on December 4, 2002, directs NHTSA to issue by December 2004, a regulation requiring lap/shoulder belts for each rear seating position – i.e., including rear-center seats – in passenger vehicles with GVWR ≤ 10,000. Phase-in is scheduled from September 1, 2005 to September 1, 2007. In fact, manufacturers have already been installing 3-point belts at rear-center seats of many vehicles, especially from 1999 onward, well in advance of the new

---

[56] Kahane (1984 Seat Belts); 90% confidence bounds for effectiveness: AIS ≥ 1 injuries, 9 to 17 percent; AIS ≥ 2 injuries, 40 to 54 percent; AIS ≥ 3 injuries, 38 to 55 percent.

[57] *Final Regulatory Impact Analysis, Extension of the Automatic Restraint Requirements of FMVSS 208 to Trucks, Buses and Multi-Purpose Passenger Vehicles*, p. 23.

[58] *Federal Register* 54 (June 14, 1989): 25275, 54 (November 2, 1989): 46257; *Code of Federal Regulations*, Title 49, Parts 571.208 S4.1.4.2 and S4.2.4; from December 11, 1989 to August 31, 1990, cars were allowed separate lap and shoulder belts as an alternative to 3-point belts, but nobody exercised that option; Document Nos. 001, 002, 003, 005, 016, 018-H and 035 in Docket No. 89-01-N01 and Nos. 006, 007, 008, 009, 011, 012, 014, and 015 in NHTSA Docket No. 89-01-N03, 1992 (Lap-Shoulder Belts in the Back Seat); *Lap/Shoulder Belt Kits for Rear Seats*, NHTSA Publication No. DOT HS 807 881, Washington, 1992.

requirement. By MY 2001, the majority of new passenger cars had 3-point belts at the rear-center seats.[59]

How 3-point belts work – expected benefits As in the front seat, a lap and shoulder belt ought to provide excellent protection against ejection, provide "ride-down," keep occupants in an upright, seated position and reduce the severity of interior contacts. They should be effective in most crash modes, including frontals (but less so in severe nearside impacts or other catastrophic crashes). They should greatly alleviate the problem of abdominal injuries experienced with lap belts alone.

Fatality reduction – passenger cars NHTSA's 1999 evaluation is based on double-pair comparison analysis of 1988-97 FARS data. Front-seat occupants (drivers and RF passengers) were the control groups used to compare the relative fatality risk of belted vs. unrestrained rear-outboard occupants, in the following table:[60]

Back-Seat Belt Use	Front-Seat Belt Use	Back-Seat Fatalities	Front-Seat Fatalities	Back/Front Risk Ratio
Unrestrained	Unrestrained	4,953	7,248	0.683
3-point belts	Unrestrained	119	344	0.346
Unrestrained	Air bag alone	650	820	0.793
3-point belts	Air bag alone	49	148	0.331
Unrestrained	3-point belts	3,028	2,098	1.443
3-point belts	3-point belts	807	880	0.917
Unrestrained	3-pt. + air bag	670	371	1.806
3-point belts	3-pt. + air bag	431	490	0.880

In each of these four comparisons with a control group, the back-seat occupants protected with 3-point belts had substantially lower risk than the unrestrained back-seat occupants. For example, relative to unrestrained front-seat occupants, unrestrained back-seat occupants had a risk ratio of 0.683, belted back-seat occupants just 0.346. The overall, average fatality reduction for 3-point belts at the rear-outboard seats of passenger cars was 44 percent.[61]

The preceding table also shows that the back seat is a safer place to ride than the front seat. Unrestrained back-seat occupants had 32 percent lower fatality risk than unrestrained front-seat occupants of the same car, and they had lower fatality risk than unbelted front-seat occupants in cars with air bags. Back-seat occupants using 3-point belts had lower fatality risk than front-seat

---

[59] *Anton's Law*, Public Law 107-318, December 4, 2002.
[60] Morgan (1999), p. 19.
[61] *Ibid.*, pp. x and 13-30; 90% confidence bounds, 38 to 50 percent.

occupants using 3-point belts, even if the car was also equipped with an air bag in the front seat. (For additional discussion, see the FMVSS 213 chapter.)

Three-point belts are effective in all crash modes, including frontals. The estimated fatality reduction is: 77 percent in rollovers, 42 percent in side impacts, 29 percent in frontals, and 31 percent in rear impacts and other crashes.

Analyses of the FARS Multiple Cause of Death (MCOD) file showed that 3-point belts reduce the risk of life-threatening abdominal injuries by 52 percent and head injuries by 47 percent relative to lap belts only, in frontal crashes. (Occupants wearing 3-point belts, however, still have higher risk of abdominal injury than unrestrained people, but their risk of head injury is 63 percent lower.) Three-point belts are beneficial for occupants of all age groups, including those age 55 or older.[62]

Fatality reduction – light trucks NHTSA's 1999 evaluation analyzes vans and SUVs. Three-point belts are highly effective: the fatality reduction for back-seat occupants averaged 73 percent.[63] By crash mode, the estimated fatality reduction is: 84 percent in rollovers, 50 percent in frontals, and 70 percent in side impacts, rear impacts and other crashes.

Injury reduction – passenger cars NHTSA's 1999 study did not attempt to obtain point estimates of the effectiveness of 3-point belts in reducing nonfatal injuries, because belt use appeared to be substantially overreported in available State crash files. However, an "as-used" analysis of 1990-96 Pennsylvania data showed rear-outboard occupants of cars equipped with 3-point belts (regardless of whether or not these occupants wore the belts) had a statistically significant 12 percent fewer moderate-to-fatal (K + A + B) injury rate in frontal crashes than occupants of cars equipped with lap belts only. In 1990-96 Florida data, the reduction was 6 percent, also statistically significant.[64] In 1989, the agency had projected that 3-point belts in back seats of cars could reduce AIS 2-5 injuries by 45-55 percent and AIS 1 injuries by 10 percent. The estimate was not obtained directly from crash data, but inferred by examining the relationships between other estimates: lap-belt vs. 3-point-belt effectiveness in the front seat of cars, lap belt effectiveness in the rear seat vs. the front, and fatality reduction vs. injury reduction.[65]

Injury reduction – light trucks The agency has not analyzed crash data to estimate the injury reduction for 3-point belts in LTVs. NHTSA's Regulatory Impact Analysis in 1990 projected that 3-point belts in back seats of LTVs could reduce AIS 2-5 injuries by 50 percent and AIS 1 injuries by 5.5 percent. Here, the agency assumed the same effectiveness in cars and LTVs; in retrospect, the AIS 2-5 number would appear to be an underestimate for the LTVs.[66]

---

[62] *Ibid.*, pp. xi-xiii, 47-52 and 99-111.
[63] *Ibid.*, pp. xii and 83-97; 90% confidence bounds, 64 to 79 percent.
[64] *Ibid.*, pp. 75-82.
[65] *Final Regulatory Evaluation, Rear Seat Lap Shoulder Belts in Passenger Cars*, pp. IV-1 – IV-26.
[66] *Final Regulatory Impact Analysis, Extension of the Automatic Restraint Requirements of FMVSS 208 to Trucks, Buses and Multi-Purpose Passenger Vehicles*, p. 23; *Final Regulatory Evaluation, Rear Seat Lap Shoulder Belts in Passenger Cars*, p. IV-1.

Automatic safety belts

Types of automatic belts – how they worked There have been two fundamentally different types of automatic safety belts for front-outboard occupants of passenger cars: 2-point and 3-point. A 2-point automatic belt was a diagonal torso belt, running from the roof rail or the top of the door down to a floor anchor near the center of the car. An entirely separate manual lap belt and/or knee bolster had to be added for lower-torso restraint. Motorized 2-point belts were anchored to a track on the roof rail. A motor automatically rolled the anchor backward, moving the torso belt into place, when the ignition was turned on, and rolled it forward, away from the occupant, when he or she prepared to leave the car. All motorized belt systems came with a manual lap belt. Non-motorized 2-point belts were anchored to the door. They were designed to move into place when the door shut and away when it opened. Before 1990, some of them did not come with a lap belt. All 2-point belts could be loosened for emergency egress, by a device that spooled the belt out temporarily and/or by a buckle that could be permanently disconnected or reconnected as the occupant wished.

Automatic 3-point belts, when in use, closely resemble manual 3-point belts, but they are anchored to the door rather than the roof rail/B-pillar and fashioned so as to move into place around the occupant when the door shuts and move away when the door opens. They have a buckle close to the anchor point at the center of the car. If the occupant so chooses, he or she may unbuckle it and let the belts retract toward the door; thereafter, the belts may be manually buckled and unbuckled just like conventional, manual 3-point belts, each time he or she rides in the car.

To qualify as automatic crash protection under FMVSS 208, cars with automatic belts had to meet that standard's 30 mph frontal test requirement. Manual lap belts were not compulsory since the automatic belt, by itself, satisfied the FMVSS 208 requirement for a belt system.[67]

History Twelve model years before FMVSS 208 required automatic protection on any vehicle, Volkswagen introduced non-motorized 2-point belts, with knee bolsters but no lap belt, as part of a popular option package on their MY 1975 Rabbit. From 1975 to 1984, approximately 30 percent of Rabbits were sold with automatic belts, comprising nearly 400,000 cars. Toyota made the motorized 2-point belt standard on all Cressidas, starting in 1981, accounting for over 200,000 cars in 1981-86.[68]

The phase-in of automatic protection began on September 1, 1986. During model year 1987, the first year of the phase-in, 91 percent of the cars with automatic protection had automatic belts (only 9 percent had air bags), and in MY 1988, 93 percent. Starting in 1989, sales of cars with air bags steadily gained relative to automatic belts. Nevertheless, in MY 1990, the first year after FMVSS 208 was fully phased in, over 70 percent of the cars had automatic belts, comprising over 6 million cars. Sales of cars with automatic belts decreased each year after 1990. By 1997, one year before FMVSS 208 required it, all new cars had dual air bags and manual 3-point belts – but not before 31 million cars with automatic belts were sold. Close to 20 million of them

---

[67] *Code of Federal Regulations*, Title 49, Parts 571.208 S4.1.2.1 and S4.5.3.
[68] *Final Regulatory Impact Analysis, Amendment to Federal Motor Vehicle Safety Standard 208, Passenger Car Front Seat Occupant Protection*, pp. II-3 and IV-23.

were still on the road as of 2003. Over the years, sales of automatic 2-point belts (17 million cars) slightly outnumbered automatic 3-point belts (14 million cars). LTVs were not equipped with automatic belts.

Automatic belt use In 1983-84, when NHTSA developed the automatic protection requirement of FMVSS 208, automatic belts were a promising alternative. Manual belt use was 14 percent, whereas automatic belt use was observed to be 96 percent in the Toyota Cressida (motorized) and 75 percent in the Volkswagen Rabbit (non-motorized 2-point) in the 19-City Study.[69] State buckle-up laws quickly narrowed the gap between manual and automatic belt use. In the 19-city survey of 1991, manual belt use in MY 1987-90 cars had risen to 56 percent, while automatic belt use was 91-97 percent in cars with motorized 2-point belts, 74 percent with non-motorized 2-point belts, and 64 percent with 3-point belts.[70] In the 1994 NOPUS, use of manual belts in late-model (MY 1989-94) cars was up to 69 percent, exactly the same as 3-point automatic belts; use of 2-point automatic belts was not much higher: 78 percent for motorized, 75 percent for non-motorized.[71] NHTSA believes that, during most of the 1990's, use rates of manual and automatic 3-point belts were almost equal, increasing from year to year, while use of 2-point belts was close to 84 percent, changing little from year to year.[72]

In 1987, the Insurance Institute for Highway Safety observed 253 parked, unoccupied cars with 3-point automatic belts. Already at this early date, over 99 percent of them had been unbuckled and allowed to retract toward the door; functionally rendering them a manual belt system. It is little wonder, then, that automatic 3-point belts had the same use rates as manual within a few years after their introduction. By contrast, over 95 percent of the 753 2-point belts in their survey were still connected in the "automatic" mode (although that percentage must have decreased in later years).[73]

All of the preceding statistics for 2-point belts are use rates for the automatic shoulder belt. However, most of those cars also had a manual lap belt. Lap belt use has been lower than shoulder belt use, but it is unclear by how much. (Lap belt use is, among other things, more difficult to observe than shoulder belts.) Lap belt use was only 29 percent in a 1989-90 North Carolina observational survey, 43 percent in a 1987 IIHS study, but 69 percent in 1992 Michigan observations. Shoulder belt use was over 90 percent in all three of these surveys.[74]

---

[69] Perkins, Cynecki and Goryl, p. 16.

[70] *Evaluation of the Effectiveness of Occupant Protection, Interim Report*, NHTSA Report No. DOT HS 807 843, 1992, pp. 11-17.

[71] The 1994 NOPUS controlled-intersection survey was the only one to record license plates and obtain VINs from DMVs; NHTSA decoded the VINs to identify the model year and type of occupant protection.

[72] Kahane (2000), pp. 41 and 52-55.

[73] Williams, A.F., Wells, J.K., Lund, A.K. and Teed, N., "Observed Use of Automatic Seat Belts in 1987 Cars," *Accident Analysis and Prevention*, Vol. 21, October 1989, pp. 427-433.

[74] *Ibid.*; Hunter, W.W. et al., *Analysis of Occupant Restraint Issues from State Accident Data*, HSL Publication No. 00578847, Report No. HSRC-TR75, Highway Safety Research Center, University of North Carolina, Chapel Hill, 1990, Chapter 6; Streff, F.M., Molnar, L.J. and Christoff, C., "Automatic Safety Belt Use in Michigan: A Two-Year Follow-Up," *Journal of Safety Research*, Vol. 25, Winter 1994, pp. 215-219; Schmidt, R.A., Young, D.E. and Ayres, T.J., "Automobile Seat Belts: Usage Patterns in Automatic Belt Systems," *Human Factors*, Vol. 40, March 1998, pp. 126-135.

Expected benefits Automatic 3-point belts closely resemble manual 3-point belts and may provide equally good protection. One caveat is that, being door-mounted, they might release in a crash that damages and opens the door, and allow more ejections than manual 3-point belts. Similarly, an automatic 2-point belt in combination with the manual lap belt may be functionally equivalent to a manual 3-point belt and equally effective. Without the lap belt, it stands to reason that 2-point belts would be less effective overall, but it is not clear where benefits would be reduced. Since the vehicle must meet the FMVSS 208 frontal barrier test without the lap belt, effectiveness could still be high in frontal crashes. However, without the lap belt, occupants might be more prone to slide beneath the shoulder belt (submarining) and they might not be as well protected from ejection.

Fatality reduction, automatic 3-point belts NHTSA's 2000 evaluation estimated the fatality reduction for automatic 3-point belts, based on the same double-pair comparison analysis of 1986-99 FARS data used for manual belts. Automatic and manual belts are equally effective: they reduce fatalities by 45 percent.[75] Furthermore, an analysis of 1986-91 FARS data did not show any significant increases in occupant ejections as make-models shifted from manual belts to automatic 3-point belts during 1987-89.[76] Automatic and manual 3-point belts appear to have essentially the same effectiveness and use rates.

Fatality reduction, automatic 2-point belts The 2000 study estimated a 32 percent overall fatality reduction for automatic 2-point belts. That is significantly less than the 45 percent effectiveness of manual 3-point belts. However, this 32 percent effect is an average, comprising people who did not buckle the manual lap belt as well as those who did. (FARS data cannot be relied on to identify use of the lap belt.) If lap belt use has been as low as suggested by the North Carolina survey – less than one-third of shoulder belt use – it is easy to see how net effectiveness could drop to 32 percent. But if lap belt use were as high as the Michigan survey – over two-thirds of shoulder belt use – it would imply these systems are not very effective without the lap belt.[77] In any case, given recent use rates for manual belts, such as 80 percent in 2004, the combination of manual belts with air bags is clearly more beneficial than any automatic belt system.

Pretensioners and load limiters for safety belts

Safety belt pretensioners retract the safety belt almost instantly in a crash to remove excess slack. They are mechanical or pyrotechnic devices located within the belt's retractor assembly. By pulling in slack, they reduce occupants' impacts with interior surfaces and help the belt engage with the occupant to expedite "ride-down." Load limiters and other energy management systems allow safety belts to yield in a crash, preventing the shoulder belt from exerting too much force on the chest of an occupant. In NHTSA's frontal New Car Assessment Program (NCAP) test, a 35 mph impact into a rigid barrier with belted dummies, the combination of pretensioners and

---

[75] Kahane (2000), pp. 40-41; the point estimate for the 3-point automatic belts was actually 48 percent; since it was not significantly different from the 45 percent estimate for manual belts, the two systems may be considered equally effective, since they essentially work the same way.

[76] *Evaluation of the Effectiveness of Occupant Protection, Interim Report*, pp. 43-46.

[77] Kahane (2000), pp. 40-41; the point estimate for the 3-point automatic belts was actually 48 percent; since it was not significantly different from the 45 percent estimate for manual belts, the two systems may be considered equally effective, since they essentially work the same way.

load limiters reduced (i.e., improved) Head Injury Criterion (HIC) by 232, chest acceleration by an average of 6.6 g's, and chest deflection by 10.6 mm, for drivers and right front passengers. Each of these reductions was statistically significant. This is a promising indication that pretensioners and load limiters make belts more effective in at least one type of crash. Approximately 63 percent of MY 2002 cars and LTVs were equipped with pretensioners, and 84 percent with load limiters or other energy management systems. When these vehicles have been on the road long enough to accumulate sufficient crash experience, NHTSA will analyze FARS data, based on double-pair comparison, to see if pretensioners and load-limiters enhance the overall fatality-reducing effectiveness of safety belts.[78]

Frontal air bags

How air bags work A dual air bag system includes sensors at various locations in the vehicle that send an electrical signal if they experience a substantial deceleration and velocity change from a frontal direction. A control module commands the air bags to deploy if these signals imply that the vehicle has been in a relatively severe frontal impact. Two air bag assemblies include a charge of Sodium Azide propellant that will generate Nitrogen gas upon firing and/or a cylinder of compressed Argon gas and an inflatable bag made of fabric. All bags have vents to release the gas gradually after a deployment. Many have tethers (internal straps) that limit how far the bag can deploy toward the occupant, and make it spread outward, over a larger area. These assemblies are located in the steering wheel hub for the driver, in the instrument panel for the passenger. A driver-only air bag system has all of these except the passenger bag assembly.

In a severe frontal crash such as a 30 mph barrier impact, even while the front bumper comes to an immediate stop against the barrier and the sheet metal deforms, the occupants remain in their seats for about the first 50 milliseconds or more, and the compartment interior continues forward at close to 30 mph as if nothing had happened yet. In the next 50-75 milliseconds, the compartment is slowed to a stop while unrestrained occupants continue to move forward at close to 30 mph, fly out of their seats, and strike the steering assembly, instrument panel and other structures at a high relative speed.

The technological marvel of the air bag is that it fully deploys in less than 50 milliseconds, before a correctly positioned occupant has even begun to move out of the seat. In that time, the sensors detect the crash, the control module sends signals to fire the propellant or open the cylinders, and gas is generated or released at high enough pressure for the bags to burst through "doors" in the steering wheel hub and instrument panel, and to fill up entirely. As the occupant begins to move forward, he or she will almost immediately contact an energy-absorbing surface that is ideally "tuned": soft enough to cushion the head and neck without serious injury, yet rigid enough to absorb much of the torso's kinetic energy. This energy is absorbed as the occupant compresses the bag and pushes the gas out through the vents.

The air bag is one key component in a chain of deformable devices that allow the occupant to "ride down" from 30 mph to 0 mph as gradually as possible, remaining in an upright position.

---

[78] Walz, M.C., *NCAP Test Improvements with Pretensioners and Load Limiters*, NHTSA Technical Report No. DOT HS 809 562, Washington, 2003; *National Highway Traffic Safety Administration Evaluation Program Plan, Calendar Years 2004-2007*, NHTSA Report No. DOT HS 809 699, Washington, 2004, p. 4.

The other components are: the continuing crumple of the sheet metal in the front of the vehicle even after the occupant moves out of the seat; the compression of the energy-absorbing steering assembly (FMVSS 203/204), facilitated by transmission of force through the air bag; deformation of the instrument panel by the knees (FMVSS 201); and the "give" in the seat belt system (especially with load limiters) and the seat structure.

An air bag must deploy rapidly and with great force if it is to get its job done in less than 50 milliseconds. For adequate gas pressure (energy-absorbing capability) in the fully deployed bag, there must be much higher gas pressure and temperature as the bag begins to deploy. Initial bag velocities of 200 mph or more were not unusual in the 1985-97 generation of air bags. Whereas air bags are of great value to the correctly positioned occupant, they can increase risk for an out-of-position occupant who is touching a deployment "door," or is within about 4 inches of it, at the moment of deployment.

How do people get close to the air bag? Infants in rear-facing child safety seats would be adjacent to the deployment door at all times, and must *never* be placed in the front seat of a vehicle with a working passenger air bag. Other people may be jostled forward from their seats during hard braking, bumps on or off the road, and/or minor impacts before the one that deploys the bag. A correctly worn safety belt or forward-facing safety seat and/or the inertial weight of an adult reduce the tendency to slide forwards. Thus, unrestrained children 1-12 years old are especially vulnerable. Drivers of short stature (usually less than 5 feet tall) who must sit within 10 inches of the steering wheel in order to operate the pedals are also more prone than others to be thrown within 4 inches of the deployment door. Temporary proximity to the air bag can result from leaning forward to adjust the radio or other controls. People who lean their arms across the steering wheel can sustain arm injuries from the air bag that, of course, are rarely life-threatening.

The only two guaranteed methods to stay away from an air bag are to ride in the back seat or to turn the on-off switch "off," if one is available. NHTSA recommends that children up to age 12 ride in the back seat if possible; it is the safest place in the vehicle. Risk can be greatly reduced, although not totally eliminated, by using safety belts/forward-facing safety seats, moving the passenger seat backward, moving adjustable pedals backward if this feature is available, and not leaning one's arms across the steering wheel. For the overwhelming majority of occupants age 13 or older who wear safety belts, the risk of serious injury from air bags is so negligible relative to their potential benefits that they should not even consider obtaining an aftermarket on-off switch.

Safety belts remain the primary occupant protection system and need to be worn in vehicles with air bags. In side impacts, rear impacts and primary rollovers, air bags generally do not deploy, and even if they do, they have limited benefits. Air bag deployments are, of course, beneficial in relatively severe frontal and oblique-frontal crashes, but there is no guarantee an air bag will deploy in a particular crash. While brochures indicate that air bags begin to deploy in frontals equivalent to 8-14 mph barrier crashes, this is at best a rule of thumb. Deployments occasionally happen at lower speeds and sometimes do not take place at substantially higher Delta V. This is not an evidence of malfunction; rather, the sensors take into account various aspects of the crash pulse and don't just measure Delta V. Frontal air bags do not prevent ejection and they may have limited benefits if a frontal impact is followed by a rollover, side impact, fire or immersion. They deflate rapidly after deploying and may not be as helpful in multiple impacts. Notwithstanding these caveats, air bag systems usually deploy when they are needed, don't deploy when they aren't needed, and seldom malfunction. The manufacturers and NHTSA constantly review the performance of the systems and promptly correct problems that may develop over time.[79]

History Air bags to cushion occupants in frontal crashes were conceptualized in the 1950's. Prototypes were built and tested in the early 1960's. NHTSA announced in 1969 it was considering air bags or other automatic protection, especially in view of low belt use, and issued its first NPRM for automatic protection in 1970. The original January 1, 1972 effective date was postponed or cancelled several times for reasons such as revision of the test requirements, alternative proposals for occupant protection, risk to children/out-of-position occupants and lack of industry/public support. However, NHTSA's July 17, 1984 rule to phase automatic protection into the front-outboard seats passenger cars was successfully implemented during 1986-89.[80]

On-the-road experience began in 1972-73: 831 MY 1972 Mercury Montereys with passenger air bags and 1000 MY 1973 Chevrolet Impalas with dual air bags. These cars were driven by corporate personnel or leased to corporate fleets. In MY 1974-75, the general public bought 10,281 full-size Buicks, Cadillacs and Oldsmobiles with optional dual air bags. No air bags were offered on new MY 1976-83 cars. During MY 1985 (or possibly in late MY 1984), driver air bags of the next design generation (1984-97) began to appear in some Mercedes models for sale to the public and in 5,000 Ford Tempos purchased by the Federal government. Driver air bags were standard on all 1986 Mercedes.[81]

The FMVSS 208 phase-in of automatic protection began on September 1, 1986. To help get air bags into cars, recognizing that the development of passenger air bags was taking longer than driver air bags, NHTSA exempted the right-front seat from the automatic protection requirement until August 31, 1993 in cars with driver air bags. During MY 1987-88, fewer than 10 percent of cars with automatic protection had air bags, while over 90 percent had automatic belts. The public, however, soon expressed a preference for air bags. In mid-1988, Chrysler committed itself to shift from automatic belts to air bags within the next few years. By 1990-91 all domestic manufacturers made driver air bags standard on some models with high sales; in 1992-93 driver

[79] *Air Bags & On-Off Switches*, NHTSA Publication No. DOT HS 808 629, Washington, 1997; *Fifth/Sixth Report to Congress – Effectiveness of Occupant Protection Systems and Their Use*, NHTSA Report No. DOT HS 809 442, Washington, 2001, pp. 2-3.

[80] Clark, C., Blechschmidt, C. and Gordon, F., "Impact Protection with the 'Airstop' Restraint System," *Eighth Stapp Car Crash and Field Demonstration Conference*, Wayne State University Press, Detroit, 1966; *Federal Register* 34 (July 2, 1969): 11148, 35 (May 7, 1970): 7187, 42 (July 5, 1977): 34289, 46 (October 29, 1981): 53419, 49 (July 17, 1984): 28962; "Air Bag Chronology," *USA Today*, July 13, 1999.

[81] *Final Regulatory Impact Analysis, Amendment to Federal Motor Vehicle Safety Standard 208, Passenger Car Front Seat Occupant Protection*, p. IV-40; Smith, R.A. and Kahane, C.J., *1974 Accident Experience with Air Cushion Restraint Systems*, NHTSA Technical Note No. DOT HS 801 565, Paper No. 750190, Society of Automotive Engineers, Warrendale, PA, 1975; Bloch, B., "Advanced Designs for Side Impact and Rollover Protection," *Sixteenth International Technical Conference on the Enhanced Safety of Vehicles*, NHTSA Report No. DOT HS 808 759, Washington, 1998, p. 1780; *Recent Air Bag Field Performance*, Research Note, NHTSA National Center for Statistics and Analysis, 1986; Kahane, C.J., *Fatality Reduction by Air Bags*, NHTSA Technical Report No. DOT HS 808 470, Washington, 1996, p. 89.

air bags peaked at 49 percent of new car sales. Starting September 1, 1993, cars could not have just a driver air bag and manual belts, but they could have a driver air bag and an automatic belt for the RF passenger; in fact 25 percent of MY 1994 cars had that configuration.[82]

Dual air bags were available on Porsche in 1987-89 and standard in 1990. They became available on a few other models in 1990-91, were in 5 percent of MY 1992 cars and 14 percent of MY 1993, including, for example, all the Chrysler LH cars. By MY 1994, the manufacturers could place dual air bags in 58 percent of their cars, and in 1995, 90 percent. All new cars had dual air bags with manual 3-point belts from MY 1997 onward, one year in advance of the FMVSS 208 requirement.[83]

Chrysler equipped some MY 1991 minivans and all 1992's with driver air bags. Chrysler and Toyota minivans were the first with dual air bags, in MY 1994. The FMVSS 208 phase-in of automatic protection into light trucks started on September 1, 1994. By then, the majority of LTVs at least had driver air bags and 16 percent had dual air bags. Most SUVs and vans had dual air bags by MY 1996-97, two to three years in advance of the FMVSS 208 requirement. In pickup trucks without back seats, and in other vehicles that could not accommodate a child restraint in the back seat, however, passenger air bags would have posed a risk to children obligated to ride in the front seat. After NHTSA permitted on-off switches for the passenger air bag in such vehicles, effective June 22, 1995, dual air bags with the switches began to appear in 1996 Ford Rangers, and were standard in all pickup trucks with GVWR < 8,500, excluding crew-cab models, by MY 1998. LTVs with GVWR over 8,500 pounds are exempt from the automatic protection requirement and often do not have air bags.[84]

Expected benefits Air bags ought to be quite effective in frontal crashes, the closer to directly frontal, the more effective. They should very substantially reduce life-threatening head and chest injuries in these crashes. They should be effective, but to a lesser extent, in oblique frontal crashes, because occupants are often injured by components off to the side, not in a trajectory to the air bag. Air bags ought to save lives of belted as well as unrestrained occupants, because safety belts alone cannot prevent all contacts with the compartment interior in frontal crashes.

Air bags will have no effect in crashes where they do not deploy, including most 90-degree side impacts, rear impacts and rollovers. Frontal air bags generally will not block occupant ejection from the vehicle, except possibly through the windshield portal. When frontal crashes are followed by other harmful events such as secondary impacts, rollovers or fires, the air bags might not prevent injuries associated with the subsequent events. Intuitively, air bags would be less effective in catastrophic crashes where the integrity of the occupant compartment is lost.

Passenger air bags would clearly be dangerous to infants in rear-facing safety seats in the front seat of a vehicle, and they could also increase the net fatality risk for some other types of child

---

[82] *Federal Register* 49 (July 17, 1984): 28962, 55 (January 17, 1990): 1586; *Code of Federal Regulations*, Title 49, Parts 571.208 S4.1.4; Kahane (1996), pp. 77-95.

[83] *Federal Register* 58 (September 2, 1993): 46551; *Code of Federal Regulations*, Title 49, Part 571.208 S4.1.5.3; Kahane (1996), pp. 77-95.

[84] *Federal Register* 56 (March 26, 1991): 12472, 58 (September 2, 1993): 46551, 60 (May 23, 1995): 27233; *Code of Federal Regulations*, Title 49, Part 571.208 S4.2.5, 4.2.6.2 and S4.5.4; Kahane (1996), pp. 97-101.

passengers. As of October 1, 2004, NHTSA reported 23 confirmed cases of infants in rear-facing safety seats and 132 other children who received fatal injuries from air bag deployments in relatively low-speed crashes.[85] However, the latter is not necessarily the "net" increase, since air bags may have saved some children in other crashes.

Early predictions of fatality reduction In 1974, NHTSA predicted that air bags would reduce overall fatality risk by 32 percent relative to an unrestrained occupant, and that air bags would save 9,000 lives per year. In 1977, after the energy crisis, 55 mph speed limit and other factors had dramatically reduced overall traffic fatalities, NHTSA upped the effectiveness estimate for air bags to 40 percent and still predicted they would save 9,000 lives per year. Because air bags protected dummies in 35 mph frontal barrier crashes, the analyses assumed that air bags would essentially eliminate fatalities in frontal crashes on the highway with Delta V $\leq$ 35 mph (57 to 65 percent of all unrestrained frontal fatalities) – regardless of what contact points or injury mechanisms were actually causing the fatalities. The analyses also assumed air bags would have some benefits in side impacts, rear impacts and rollovers if these crashes included a secondary frontal impact or a frontal-force component in the main impact.[86]

During the 1984 rulemaking on automatic protection, NHTSA took a hard second look at air bag effectiveness, based on the agency's data systems created during 1977-84. Two NHTSA analysts independently reviewed approximately 800 unrestrained front-outboard fatality cases on the NCSS file. Both studies concluded that 72-77 percent of crash fatalities were unlikely to be mitigated by air bags because they were attributed to contacts with the vehicle's side interior or other non-frontal areas, ejection through side doors or windows, catastrophic damage to the vehicle interior, fire or immersion. Thus, the overall fatality reduction for air bags could not exceed 23-28 percent and presumably would be some fraction of that. NHTSA at that time essentially halved its predictions, ranging them down to 20 percent fatality reduction and 4,570 lives saved per year.[87]

Fatality reduction – passenger cars – drivers The four NHTSA analyses of the actual crash experience of cars equipped with air bags, in 1992, 1996, 1999 and 2001, have had remarkably consistent results. Air bags reduced overall fatality risk by 12 percent, 11 percent, 11 percent and 12 percent, respectively. The last study is based on the most data and it is NHTSA's best estimate.[88]

---

[85] NCSA Special Crash Investigations, Downloadable Files. NHTSA, Washington (www-nrd.nhtsa.dot.gov/departments/nrd-30/SCI.html).

[86] *Final Regulatory Impact Analysis, Amendment to Federal Motor Vehicle Safety Standard 208, Passenger Car Front Seat Occupant Protection*, pp. IV-37 – IV-39; *Analysis of Effects of Proposed Changes to Passenger Car Requirements of FMVSS 208*, NHTSA Docket No. 74-14-N01-104, Washington, 1974; *Standard No. 208 – Passive Restraint Amendment, Explanation of Rulemaking Action*, NHTSA Docket No. 74-14-N10-011, NHTSA Report No. DOT HS 802 523, Washington, 1977.

[87] Partyka, S.C., "Assessment of the Potential of Air Bags to Prevent Car Occupant Fatalities Using NCSS Data," *Final Regulatory Impact Analysis, Amendment to Federal Motor Vehicle Safety Standard 208, Passenger Car Front Seat Occupant Protection*, pp. 3 and IV-47 – IV-67; Kahane (1984 Air Bags).

[88] *Evaluation of the Effectiveness of Occupant Protection, Interim Report*, pp. 21-25; Kahane (1996), pp. 7-12; *Fourth Report to Congress – Effectiveness of Occupant Protection Systems and Their Use*, NHTSA Report No. DOT HS 808 919, Washington, 1999, pp. 8-11; *Fifth/Sixth Report to Congress*, pp. 6-11.

Each of these studies estimated fatality reduction based on two double-pair comparison analyses, and averaged the two results. The first analysis is feasible because a large number of make-models were initially equipped with only a driver air bag at some point before 1994. The driver's seats changed from no air bags to air bag-equipped, while the RF passenger seat stayed the same: no air bag. That allows double-pair comparison, with the RF passenger as the control group. NHTSA's 2001 study was based on CY 1986-2000 FARS data involving MY 1985-2000 cars equipped with 3-point belts and no air bags, or with 3-point belts and a driver-only air bag. It is limited to cases where the driver's and right-front seats were both occupied; the RF passenger was at least 5 years old; and the driver, or the RF passenger, or possibly both died:

Driver's Seat	RF Passenger Seat	Driver Fatalities	RF Passenger Fatalities	Driver/RF Risk Ratio
No air bag	No air bag	22,535	24,168	0.932
Air-bag equipped	No air bag	3,654	4,420	0.827

In cars with no air bags, drivers are nearly as much at risk as RF passengers. In cars with driver air bags, there are substantially fewer driver fatalities than passenger fatalities. Drivers experienced a statistically significant 11 percent fatality reduction with air bags:

$$1 - [(3,654/4,420) / (22,535/24,168)] = .113$$

This is called an "as used" analysis, because it includes all types of crashes, and it is not limited to crashes in which the air bag deployed. It estimates the overall effectiveness of air bags as a system.

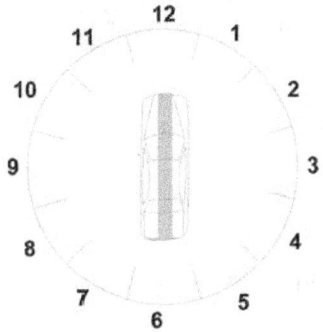

The second analysis is based on another distinctive characteristic of air bags: they are primarily designed for action in frontal crashes. With an inclusive definition of "frontal and partially frontal" crashes (initial or principal impact location between 10:00 and 2:00), it can be assumed that air bags have little effect in the remaining "non-frontal" crashes. These non-frontal fatalities are a control group. NHTSA's 2001 study was based on CY 1986-2000 FARS data and limited to selected MY 1985-2000 make-models equipped with 3-point belts that shifted from no air bags to driver (or dual) air bags. Driver fatalities in cars of the last 3 model years before air bags were compared to fatalities in cars of the first 3 model years with driver (or dual) air bags. (In this analysis, it does not matter if the RF seat is occupied or unoccupied, or what sort of occupant protection it has.):

	Frontal (10-2:00) Fatalities	Non-Frontal Fatalities	Frontal/Non-Frontal Risk Ratio
Driver seat without air bag	5,434	2,860	1.900
Driver seat equipped with air bag	3,360	2,223	1.511

Drivers of cars equipped with air bags experienced a statistically significant 20 percent reduction of frontal fatalities:

$$1 - [(3,360/2,223) / (5,434/2,860)] = .204$$

Because 65 percent of fatalities were frontal (by this rather inclusive definition of "frontal"), that amounts to a 13 percent overall reduction of fatalities:

$$.204 \ [5,434 / (5,434 + 2,860)] = .134$$

The two statistically significant estimates of overall fatality reduction (11 and 13 percent) average out to 12 percent.

Air bags are most effective in "purely frontal" crashes where the principal impact is at a 12:00 location and it is also the most harmful event – i.e., the most harmful event is not a subsequent rollover, fire, or immersion. The first analysis method demonstrates the high effectiveness of air bags:

Purely Frontal Crashes

Driver's Seat	RF Passenger Seat	Driver Fatalities	RF Passenger Fatalities	Driver/RF Risk Ratio
No air bag	No air bag	7,597	8,241	0.922
Air-bag equipped	No air bag	950	1,438	0.661

The average for the two analysis methods is a 29 percent fatality reduction in purely frontal crashes. Effectiveness drops off steeply as the impact moves away from purely frontal. When the principal impact is at an 11:00 or 1:00 location and is also the most harmful event, air bags reduce fatality risk by 15 percent; when the principal impact/most harmful event is at a 10:00 or 2:00 location, fatality reduction for air bags is 6 percent. When the principal impact is 3-9:00 or to the top of the car, or when the most harmful event is a rollover, fire, immersion or other non-collision, effectiveness is zero.

Enough data were available to obtain statistically significant estimates of fatality reduction for subpopulations of drivers:

110

FATALITY REDUCTION (%) BY AIR BAGS IN PURELY FRONTAL CRASHES
Drivers of Passenger Cars

		Fatality Reduction (%)
IN ALL PURELY FRONTAL CRASHES		29
BY	14-29	29
OCCUPANT	30-55	30
AGE:	56-69	30
	70 and older	20
BY	Male	29
GENDER:	Female	29
BY	Belted	21
BELT USE:	Unbelted	34
BY	2778 pounds or lighter	30
CURB	2779-3119 pounds	28
WEIGHT:	3120 pounds or heavier	26

Air bags are about equally effective for young and middle-aged drivers (29-30 percent), and perhaps slightly less effective for drivers age 70 or older (20 percent). As of October 1, 2004, NHTSA reported 84 confirmed cases of adult drivers, including 28 age 70 or older, who had received fatal injuries from air bag deployments in relatively low-speed crashes – but, clearly, the lives saved by air bags in more severe crashes far exceed these losses, in all age groups.[1] Air bags are equally effective for male and female drivers, and about equally effective in cars of various sizes. They are quite effective for unrestrained drivers (34 percent), but also provide significant supplemental protection for belted drivers (21 percent). Since 3-point belts, by themselves, reduce fatality risk by 50 percent in frontal crashes (see above), the combination of 3-point belts with air bags reduces fatality risk by an estimated 61 percent in purely frontal crashes:

$$1 - [(1 - .50) / (1 - .21)] = .605$$

NHTSA was also able to estimate the fatality reduction in all crashes (not just purely frontal) separately for belted and unbelted drivers. Air bags reduced overall fatality risk by 11 percent for belted drivers and 14 percent for unbelted drivers. Since 3-point belts, by themselves, reduce fatality risk by 45 percent, the combination of 3-point belts with air bags reduces fatality risk by an estimated 51 percent:

$$1 - [(1 - .45) / (1 - .11)] = .5105$$

---

[1] NCSA Special Crash Investigations, Downloadable Files, NHTSA, Washington (www-nrd.nhtsa.dot.gov /departments/nrd-30/SCI.html).

Fatality reduction – passenger cars – right-front passengers age 13 and older Essentially the same two double-pair comparison analyses can be used to compute fatality reduction for RF passengers. Separate estimates are obtained for passengers age 13 or older, who ought to benefit significantly from air bags, and children age 12 or younger, who might experience increased fatality risk.

The first analysis method works because quite a few make-models had only driver air bags up to MY 1993 and were equipped with dual air bags starting in MY 1994. The RF seats changed from no air bags to air bag-equipped, while the driver's seat stayed the same: air bag-equipped. That allows double-pair comparison, with the driver as the control group. The following example from NHTSA's 2001 study was based on purely frontal crashes where the driver's and right-front seats were both occupied; the RF passenger was at least 13 years old; and the driver, or the RF passenger, or possibly both died:[2]

Purely Frontal Crashes

RF Passenger Seat	Driver's Seat	RF Passenger Fatalities	Driver Fatalities	RF/Driver Risk Ratio
No air bag	Air-bag equipped	1,427	914	1.561
Air-bag equipped	Air-bag equipped	1,282	1,179	1.087

In these purely frontal crashes, when the driver had an air bag and the passenger did not, the fatality risk was substantially higher for the passenger. But with dual air bags, the fatality risk was about the same for drivers and passengers. The average for this analysis method and the other method (based on purely frontal vs. non-frontal fatalities) is a statistically significant 32 percent fatality reduction in purely frontal crashes. In other words, air bags are just as effective, and perhaps a bit more effective for RF passengers age 13 and older than they are for drivers (29 percent fatality reduction).

Effectiveness drops off to 24 percent in 11:00 or 1:00 principal impact/most harmful events, 4 percent in 10:00 or 2:00 impacts, and near zero in non-frontal crashes or when most harmful event is a rollover, fire, immersion or other non-collision. In all types of crashes, air bags reduce the fatality risk of RF passengers age 13 and older by an average of 14 percent.

Air bags are about equally effective for passengers age 13-29, 30-55, 56-69 and 70+, and in cars of various sizes.

Fatality reduction – light trucks – drivers NHTSA's 2001 study had enough FARS data for statistically significant estimates of the effect of air bags in pickup trucks, SUVs and vans, based on the same analysis methods as for cars. Driver air bags reduced fatality risk in purely frontal crashes by 29 percent, and in all crashes by 12 percent. These estimates are exactly the same as

---

[2] *Fifth/Sixth Report to Congress*, pp. 10-11.

the corresponding results for passenger cars, and they suggest that air bags are about equally effective for LTV and car drivers.[3]

Fatality reduction – light trucks – right-front passengers age 13 and older NHTSA's 2001 report had fewer data on LTV passengers than on drivers or car passengers, because most LTVs were not equipped with dual air bags until 1996-98. Nevertheless, there are enough data for a statistically significant estimate that passenger air bags reduced fatality risk in purely frontal crashes by 32 percent. It is exactly the same as the corresponding result for car passengers, and it suggests that air bags are about equally effective for LTV and car passengers age 13 and older.[4]

Effect on fatalities of right-front passengers age 12 and younger NHTSA has repeatedly advised the public about the hazards of air bags to infants and children, and it periodically updates a list of Special Crash Investigations (SCI) of children who received fatal injuries from air bags in relatively low-speed crashes. NHTSA's 1996 evaluation recommends a double-pair comparison method for estimating the effect of air bags on child passenger fatalities. When that analysis is updated with the FARS database of the 2001 report, it shows that, in all crashes, child passengers age 0-12 in the RF seat have 22 percent higher fatality risk with air bags than without them. In crashes where the principal and/or initial impact is at 12:00, risk is 42 percent higher with air bags. Both increases are statistically significant. However, the added risk is not at all uniform, but depends a lot on the age of the child and restraint use. For infants in rear-facing safety seats, air bags increase fatality risk in 12:00 impacts by an estimated factor of 5½. For unrestrained infants and children up to age 5, air bags approximately double fatality risk in 12:00 impacts; for children age 1-5 in safety seats or belts (not necessarily used correctly) and unrestrained children age 6-10, air bags increase risk by 70 percent. Each of these three increases is statistically significant. Air bags have smaller effects for these groups in oblique-frontal crashes (10-11:00 or 1-2:00). They do not significantly increase or decrease the net fatality risk of belted children age 6-10, or of any children age 11-12, and, of course, they have little or no effect in non-frontal crashes.[5]

Based on these statistically derived effectiveness estimates, NHTSA calculates that air bags may have resulted in a *net* increase of 136 child passenger fatalities in FARS-reportable traffic crashes on public roads during 1993-2002.[6] That number is somewhat lower than the 159 SCI cases (152 "confirmed" and 7 "unconfirmed") of infants and children with fatal injuries from air bags in 1993-2002 crashes, reported as of October 1, 2004. In fact, the two numbers are quite compatible: SCI only counts cases where air bags were harmful, not the net increase after deducting whatever lives were saved by air bags; furthermore, SCI includes a fair number of incidents in private parking lots and other situations that are not reportable to FARS.

---

[3] *Ibid.*, p. 11.

[4] *Ibid.*, p. 11.

[5] NCSA Special Crash Investigations, Downloadable Files, NHTSA, Washington (www-nrd.nhtsa.dot.gov /departments/nrd-30/SCI.html); Kahane (1996), pp. 48-49; these estimates are based on all FARS up to MY and CY 2000 – i.e., a mix of vehicles consisting of many 1st generation air bags and relatively few redesigned (1998-2000) air bags; in the future, when FARS includes more cases of redesigned air bags, it will be possible to obtain separate estimates of the effects of 1st generation and redesigned air bags.

[6] 136 is the sum of 130 in Table 2-39 of this report (estimated net fatality increase in vehicles without on-off switches) and 6 in Table 2-40 (estimated net increase in pickup trucks with switches that were left on).

Injury reduction NHTSA's reports to the Congress estimated injury reduction based on injuries per 100-crash involved occupants in NASS data, with logistic regression to control for Delta V and other variables. The estimates are biased upwards because:

- Belt use of uninjured occupants appears to be overreported.
- Vehicles equipped with air bags are substantially newer than vehicles without air bags.

Neither bias is fully addressed by the logistic regression method. The effectiveness estimates increased with each successive report. Furthermore, the use of weighted NASS data increased sampling error. For example, the last report to Congress estimated that air bags plus safety belts reduced AIS $\geq$ 3 injury risk by 67 percent relative to an unrestrained occupant, while safety belts alone reduced it by 69 percent. Both estimates are implausibly high, and the latter ought not have been higher than the former, since it implies that air bags increased the injury risk of a belted occupant.[7]

Less biased and more precise estimates can be obtained from unweighted 1989-2003 NASS casualty cases by using double-pair comparison analysis of frontal vs. non-frontal casualties, almost exactly as with FARS. To maximize sample size, the analysis comprises fatality cases of drivers and RF passengers of cars as well as LTVs, belted as well as unrestrained. First, as a "calibration run," fatality reduction is estimated from NASS data; the overall fatality reduction is 12 percent, statistically significant and exactly the same as the fatality reduction in FARS. The same analysis estimates that air bags reduce AIS $\geq$ 3 injury risk by 15 percent and AIS $\geq$ 2 injury risk by 11 percent, in all crashes. Both reductions are statistically significant.[8]

Effectiveness can be estimated separately for unrestrained and belted occupants.[9] For unrestrained occupants, air bags reduce both AIS $\geq$ 3 and AIS $\geq$ 2 injury risk by 21 percent, in all crashes. For occupants who buckle up, air bags reduce AIS $\geq$ 3 injuries by 12 percent and AIS $\geq$ 2 injuries by 4 percent. Since 3-point belts, by themselves, reduce AIS $\geq$ 3 injuries by 45 percent and AIS $\geq$ 2 by 46 percent in passenger cars (RSEP-NCSS-NASS estimates, see above), the injury reduction for the combination of 3-point belts with air bags can be estimated:

PASSENGER CARS ALL CRASHES	Fatality Reduction (%)	AIS $\geq$ 3 Reduction (%)	AIS $\geq$ 2 Reduction (%)
3-point belt plus air bag	51	52	48
Air bag alone	14	21	21
3-point belt alone	45	45	46

---

[7] *Third Report to Congress*, p. 20; *Fourth Report to Congress*, p. 17; *Fifth/Sixth Report to Congress*, p. 13.

[8] Analysis of unweighted NASS data is unacceptable for estimating national totals or rates (the customary use of NASS), but may be empirically defensible for effectiveness analyses employing double-pair comparison, based entirely on fatal or non-minor injury cases. In this analysis, "frontal" crashes are those with GAD1 = F and/or DOF1 = 11, 12 or 1.

[9] Belt use is overreported by uninjured occupants in NASS, but can be assumed accurately reported for people with AIS $\geq$ 2 injuries; the double-pair comparison analysis is based entirely on the injury cases.

In purely frontal crashes,[10] air bags reduced AIS $\geq$ 3 injuries by 32 percent and AIS $\geq$ 2 injuries by 23 percent. For unbelted occupants alone, these reductions are 39 percent and 38 percent, respectively. For occupants who buckle up, they are 27 percent and 10 percent, respectively. In other words, the AIS $\geq$ 3 and AIS $\geq$ 2 injury reductions for air bags are generally similar (and sometimes even a little higher) than the fatality reductions, with the one exception that air bags apparently do not have that large an effect on the AIS $\geq$ 2 injuries of belted occupants.

On-off switches for passenger air bags in pickup trucks

History NHTSA recognized that passenger air bags of the early 1990's were absolutely incompatible with rear-facing child safety seats and also presented risks to child passengers and certain other individuals. NHTSA urged that children and other high-risk passengers travel in the back seat when possible. The agency also amended FMVSS 208 with measures to reduce risk when these people must travel in the front seat. Effective June 22, 1995, more than three years before the September 1, 1998 effective date for dual air bags in all LTVs, NHTSA permitted (but did not require) on-off switches for the passenger air bag in pickup trucks without back seats or other vehicles that could not accommodate rear-facing child safety seats in the back seat. This facilitated the implementation of passenger air bags in pickup trucks, beginning with some Ford Rangers in MY 1996 and extending to all models with passenger air bags by 1998. Effective January 19, 1998, NHTSA also enabled people who must transport high-risk individuals in the front seats of any vehicle to obtain aftermarket on-off switches at their own expense.[11]

As of July 1, 2001, there were 11,700,000 pickup trucks factory-equipped with the switches, and 171,000 other vehicles without back seats, such as some cargo vans and sporty cars. In August 2002, only 12,513 vehicles were known to have aftermarket switches. Thus, over 97 percent of the switches were in pickup trucks. Since MY 2000, an increasing number of pickup trucks have full crew cabs whose back seats can accommodate rear-facing infant seats; they do not have the switches. Needless to say, pickup trucks over 8,500 pounds GVWR that do not have air bags do not have the switches, either. All other pickup trucks in MY 1998-2003 had the switches: regular cabs with no back seat at all, but also extended cabs with back seats not designed to accommodate a rear-facing infant seat. Although NHTSA's regulation permits (but does not require) switches in new pickup trucks until MY 2012, they may be phased out earlier – e.g., during or after the MY 2004-2006 phase-in of advanced systems that automatically suppress the deployment of air bags when there is a child passenger.[12]

How it works An on-off switch is usually located on the driver's part of the instrument panel, but well over to the right where both the driver and the passenger can see it. The ignition key operates it. The switch has two settings, "air bag on" and "air bag off." It remains at its current setting until somebody uses the key to change it. When the switch is "off," a light on the dashboard advises "air bag off," and that light stays on as long as the ignition is on.

---

[10] GAD1 = F, DOF1 = 12, and this event is a collision with another vehicle or a fixed object.
[11] *Federal Register* 60 (May 23, 1995): 27233, 62 (November 21, 1997): 62406 *Code of Federal Regulations*, Title 49, Part 571.208 S4.5.4 and Part 595.
[12] Morgan, C., *Results of the Survey on the Use of Passenger Air Bag On-Off Switches*, NHTSA Technical Report No. DOT HS 809 689, Washington, 2003, pp v and 4-5.

Public use and understanding of the switches During July-November 2000, NHTSA sponsored a survey of 3,182 pickup trucks (including 617 with child passengers) at 79 sites in 4 States. "Since the recommended switch setting depends on who is in the front seat at the moment, the survey was performed while the vehicles were occupied." Investigators observed the actual setting of the switch and recorded the passenger's age. Then they interviewed the driver to find out his or her awareness of the switch and its current setting, and the driver's own criteria for turning the switch on or off.[13]

NHTSA recommends turning the switch "off" if there is a child passenger age 12 or younger in the front seat. With very rare exceptions, it should be "on" if the passenger(s) are age 13 or older. The principal finding of the survey was that many switches were "on" for child passengers in the RF seat, especially age 7-12, while quite a few were "off" for passengers age 13 or older (the percentage of switches at the recommended setting is shown in bold type)[14]:

Age of the RF Passenger	Switch "On" (%)	Switch "Off" (%)
Less than 1 year[15]	14	**86**
1-6 years	26	**74**
7-8	41	**59**
9-10	53	**47**
11-12	70	**30**
13-15	**78**	22
16-19	**83**	17
20-59	**85**	15
60-79	**81**	19
70 years and older	**44**	56

While 86 percent of switches are turned "off" for infants in the RF seat, that percentage decreases to 74 at age 1-6 and to 30 by age 12. The percentage of switches turned "on" for adults does not exceed 85 in any age group, and it drops to 44 at age 70 or older.

Interviews showed that these drivers were well informed in 2000 about some aspects of the switch: 98 percent knew their truck had a switch that could turn the air bag on or off, and 97 percent correctly identified its setting; 65 percent who had used the switch and even 35 percent of those who had never used it volunteered (i.e., without being given a list of choices) that the primary purpose of the switch was to prevent "children" from being exposed to a deployment. But they were not as well informed about what age children should not be exposed. Some thought that air bags were safe for 3- or 4-year-olds, while others worried they were unsafe for

---

[13] *Ibid.*, pp. 4-8.

[14] *Ibid.*, pp. 13-15 and 22; the report's statistics for infants in rear-facing seats, forward-facing seats and unrestrained are combined here, but cases of infants in the center-front seat are not included in the table.

[15] 29 RF passengers less than 1 year old and/or seated in a rear-facing child safety seat, 4 switches "on." In these four cases, the driver was not the owner of the truck, or mistakenly believed the switch was "off." Not a single driver intentionally left the switch "on" in his or her own truck.

their 6'1" teen-aged children. Thus, the proportion of switches turned off declines gradually as the passengers get older, rather than dropping abruptly at age 13.[16]

Other factors that can result in switches being "off" for adults or "on" for children:[17]

- People who sometimes transport children may prefer to leave the switch "off" all the time, rather than try to change the setting each time an adult or child rides in the truck. They do not want to risk forgetting to turn it off for the child.
- Quite a few people turn the air bag off for older passengers. In fact, older passengers derive a significant net benefit from air bags (see above). The older occupant/air bag problem is short drivers who must sit close to the steering wheel, not passengers.
- Some adult passengers who "don't like air bags" turn off the factory-equipped switches in pickup trucks, even though they would not incur the effort and expense to obtain aftermarket switches in other vehicles.
- With an adult in the RF seat and a child in the center-front seat, or vice-versa, the driver cannot optimally protect both.
- People who borrow somebody else's truck may be unfamiliar with the switch or reluctant to change its setting.

Effect on fatalities On-off switches have been a necessary and fairly successful interim measure that made it possible to offer life-saving passenger air bags in pickup trucks as early as 1996-98, while allowing the opportunity to protect infants and children from those air bags. Without the switches, passenger air bags might not have been feasible until the development of automatic suppression technology (scheduled for 2004-2006 phase-in). Table 2-40 of this report estimates that the combination of passenger air bags and on-off switches in MY 1996-2002 pickup trucks had already saved 117 adults by the end of CY 2002. NHTSA's earlier evaluation report had estimated that they would eventually save 329 adults over the service life of MY 1996-2001 trucks.[18]

On the other hand, the combination of air bags and switches has not fully achieved the potential benefits it could accrue if they were always at the recommended setting. Table 2-40 estimates that 30 adults had died as of December 2002 because the switches were turned off in MY 1996-2002 pickup trucks. There could be 68 additional fatalities over the service life of MY 1996-2001 trucks if switches continue to be turned off for 17 percent of adult passengers.[19]

Of course, the purpose of the switches is to protect infants and children from air bag deployments. On October 1, 2004, SCI had 5 actual, "confirmed" cases of infants and children with fatal injuries from air bags in pickup trucks where the switch was left "on." Their ages ranged from 2 months to 8 years. The statistical analysis of FARS data in Table 2-40, based on the estimated effects of air bags and the percentages of switches left "on," produces a similar estimate: it concludes that the combination of air bags and on-off switches may have resulted in a net increase of 6 child passenger fatalities in 1996-2002. Without the switches, air bags in these

---

[16] *Ibid.*, pp. vii, 10-13, 15-16 and 22-23.
[17] *Ibid.*, pp. vi-vii, 10-13, and 20-21.
[18] *Ibid.*, pp. vi, ix and 31-36.
[19] *Ibid.*, pp. vi, ix and 31-36.

pickup trucks would have increased child passenger fatalities by a net 24; the public saved 18 of these potential fatalities by turning the switches "off."[20]

Thus, in 1996-2002, the combination of air bags and switches saved an estimated 117 adults in pickup trucks, with a net increase of 6 child passenger fatalities. That record can be improved in the future with information and education campaigns urging owners of pickup trucks to turn switches "off" for child passengers and leave them "on" for adults.

1998-99 redesign of air bags

The design of air bags involves a trade-off between supplying enough gas to absorb a substantial portion of the kinetic energy of a large, unrestrained person, and minimizing risk to out-of-position occupants by reducing the force of deployments. In the early 1980's, when belt use was less than 20 percent, the unrestrained occupant was a priority. Air bags had to pass a 30 mph crash test with unrestrained dummies. By the mid 1990's, the majority of occupants buckled up, whereas injuries to out-of-position occupants had become an issue. Effective September 1, 1997, the agency relaxed some aspects of the frontal impact test with unrestrained dummies in order to facilitate the introduction of "redesigned" air bags that deploy less forcefully: instead of a crash test with an actual vehicle, manufacturers could use a sled test with a deceleration pulse characteristic of a large passenger car – i.e., relatively gradual. In some cases, suppliers achieved less forceful deployments by literally "depowering" air bags: removing some of the gas-generating propellant. Others replaced or supplemented the propellant with a cylinder of stored argon gas. The gas can be vented at different rates, depending on the occupant's belt use and the crash severity. Changing the volume of the air bag or its arrangement of tethers were other options. This redesign took place in time for the 1998, or at the latest, the 1999 model year. However, some air bags had little change from 1997 to 1999. NHTSA will analyze FARS, NASS and SCI data to investigate the effects of redesigned air bags for child passengers, older adults, out-of-position occupants, and unrestrained occupants.[21]

Advanced air bags

In 2000, NHTSA amended FMVSS 208 to make future air bags substantially less hazardous to out-of-position occupants, but also more effective for correctly positioned occupants. These "advanced" air bags will be implemented step-by-step. From September 1, 2003 to September 1, 2005, "advanced" air bags will be phased in that do not deploy at all ("suppression") or deploy only at a low level of force ("low-risk deployment") if a small child is present, or if an older child/small adult is out-of-position, close to the air bag. The technology for suppression and low-risk deployment includes sensors that detect the weight and/or position of an occupant. Furthermore, these air bags will need to pass a barrier and offset test with 5th percentile female dummies in addition to the current barrier test with a 50th percentile male dummy. MY 2008-2010 will phase in a 35 mph barrier test with the belted 50th percentile male dummy, an increase

---

[20] NCSA Special Crash Investigations, Downloadable Files, NHTSA, Washington (www-nrd.nhtsa.dot.gov /departments/nrd-30/SCI.html).

[21] *Federal Register* 62 (March 19, 1997): 12960; *Code of Federal Regulations*, Title 49, Part 571.208 S13; *NHTSA Evaluation Program Plan, CY 2004-2007*, pp. 8-9.

from the current 30 mph. The evaluation of advanced air bags will be a high priority for NHTSA.[22]

---

[22] *Federal Register* 65 (May 12, 2000): 30679; *Code of Federal Regulations*, Title 49, Part 571.208 S14; *NHTSA Evaluation Program Plan, CY 2004-2007*, pp. 21-22 and 23-24.

FMVSS 212: Windshield mounting

This standard is associated in the public mind with a specific safety technology that has been evaluated by NHTSA:

- *Adhesive windshield bonding*

In fact, FMVSS 212 is simply a performance standard requiring vehicles to retain not less than 75 percent of the windshield periphery (50 percent in vehicles with automatic occupant protection) after the 30 mph frontal barrier test that NHTSA uses for FMVSS 204, 208, 212, 219 and 301.[1] The purpose of windshield retention (and tighter bonding) is to prevent occupant ejection. FMVSS 212 took effect for passenger cars on January 1, 1970, two years after NHTSA's initial group of safety standards, and it was extended to most light trucks, vans and multipurpose vehicles up to 10,000 pounds GVWR, effective September 1, 1978.[2]

Bonding of the windshield directly to its frame with adhesives gradually (1963-83) superseded the earlier method of first enclosing the windshield in a rubber gasket and then attaching the gasket to the frame. Thus, the shift to adhesive bonding began in some vehicles well before anybody anticipated FMVSS 212, but rubber gaskets persisted in other make-models for quite a few years after the standard. Although rubber gaskets are generally a looser installation than adhesive bonding, they can readily be designed to meet FMVSS 212. Each installation method has advantages, and the gradual shift from one to the other was motivated by various factors, sometimes including FMVSS 212. However, NHTSA's evaluation estimates the effects and benefits of adhesive bonding in all make-models of passenger cars, regardless of FMVSS 212's role in motivating its installation.

How adhesive bonding works Before 1963, windshields were sealed inside a rubber gasket or molding that, in turn, was attached and sealed to the frame. It was a relatively loose attachment. In low-speed impacts, the rubber gasket had some energy-absorbing "give." At higher speeds, the gasket could partly or completely tear away from the frame, beginning during the initial vehicle collision and deformation, and continuing as occupants impact the windshield. The advantages of gaskets are their durability, water- and sound-proofing, and perhaps, with pre-HPR windshields[3] (pre-1966), hope that windshields that "give" or "pop out" would be less prone to lacerate occupants. The important disadvantage of too loose an installation is that occupants are more likely to be ejected via the windshield portal. Such ejections almost always follow partial or complete separation of the windshield bond. Hardly ever are occupants ejected through the glazing while the bond remains intact.[4]

In the early 1960's, General Motors found that butyl rubber tape could be used to attach windshields directly to the frame, making a firmer bond than rubber gaskets, apparently at lower

---

[1] *Code of Federal Regulations*, Title 49, Government Printing Office, Washington, 2002, Part 571.212.
[2] Kahane, C.J., *An Evaluation of Windshield Glazing and Installation Methods for Passenger Cars*, NHTSA Technical Report No. DOT HS 806 693, Washington, 1985, p. 3; *Federal Register* 33 (August 16, 1968): 11652, 41 (August 30, 1976): 36493, 42 (July 5, 1977): 34288.
[3] Please see the chapter on FMVSS 205, earlier in this report.
[4] *Ibid.*, pp. 8 and 11-17.

cost, with acceptable durability, water- and sound-proofing. GM began using the tape on some 1963 Buick Specials and Oldsmobile F-85's, and on all their domestic passenger cars by 1968, well before the FMVSS 212 effective date. In 1973, GM shifted from butyl tape to a polyurethane sealant, perhaps an even stronger adhesive bond. Ford began adhesive bonding in 1965 on full-sized cars but continued using rubber gaskets on some car models through 1977; Chrysler started in 1969 but continued gaskets on some models through 1976. Obviously, the cars with rubber gaskets from 1970 onwards could and did meet FMVSS 212: the American rubber-gasket installation was hardly a "pop-out" windshield.

Overseas manufacturers mostly continued to use rubber gaskets until the late 1970's. However, Volkswagen (and perhaps other German manufacturers) had loose enough bonds that specific actions were needed to meet FMVSS 212. VW began to install continuous plastic clips between the gasket and the frame's pinchweld flange in 1970. It appears that Japanese manufacturers had shifted to adhesive bonding, or to a rubber gasket installation as tight as adhesive bonding by MY 1980, and European manufacturers at most a few years after that.[5]

Pickup trucks, vans and SUVs also kept rubber gaskets during most of the 1970's, and in many cases after FMVSS 212 was extended to light trucks (September 1, 1978). Manufacturers may have been especially concerned that operation on rough roads could accelerate deterioration of adhesive bonds, as compared to rubber gaskets. Adhesive bonding was gradually phased in during approximately 1978-85.[6] The final transition to adhesive bonding may have been spurred by anticipation of safety benefits, cost advantages with the second-generation sealants, and a 1976 rule allowing NHTSA to conduct the FMVSS 212 test in a wider range of temperatures, from 15 to 110 degrees Fahrenheit: the performance of rubber gaskets can vary widely with temperature, adhesive bonding, less so.[7]

Expected benefits; concerns about a possible increase in lacerations Windshields installed in domestic cars with adhesive bonding ought to have a lower percentage of bond separation, in frontal crashes, at any speed, than windshields mounted with rubber gaskets in those domestic make-models. Similarly, German cars meeting FMVSS 212, with plastic clips between the rubber gasket and the frame, ought to have less bond separation than pre-FMVSS 212 German cars also with rubber gaskets, but without the clips.

As a consequence, drivers and front-seat passengers should have a significantly lower risk of ejection via the windshield portal – in domestic cars after they switched to adhesive bonding, and in German cars after they got plastic clips in the rubber gaskets. Since an ejected occupant has substantially higher risk of fatality and serious injury than a non-ejected occupant under the same crash conditions, that will result in a reduction of fatalities and serious injuries.

---

[5] Mazda and Volvo used adhesive bonding from 1970 (*Ibid.*, p. 14); Ebay.com offers extensive "windshield seals," "windshield gaskets," "windshield rubber" or "windshield weatherstrips" for domestic cars in model years when they had rubber gaskets, and for imported cars in the 1970's, but few after 1980; a recent analysis of NCSS data suggests that Japanese cars of the 1970's with rubber gaskets had about the same probability of ≥ 25 % bond separation in frontal crashes as domestic cars with adhesive bonding.

[6] Richard Humphrey of General Motors (retired) advised the author that GM phased adhesive bonding into their trucks specifically during MY 1978-85; Ebay.com offers rubber gaskets for a wide variety of LTVs up to approximately MY 1985, but after that only for SUVs with removable windshields (such as Jeep Wrangler).

[7] *Federal Register* 41 (August 30, 1976): 36493.

Conversely, if adhesive bonding makes windshields less prone to "give" or "pop out" upon head impacts by occupants, it is conceivable that facial lacerations could increase. Specifically, a 1968 statistical study by Fargo of cars with pre-HPR windshields showed significantly higher laceration rates with adhesive bonding than with rubber gaskets.[8] At that time, there were insufficient crash cases of cars with HPR windshields to compare laceration rates with rubber gaskets vs. adhesive bonding.

Effect on windshield bond separation in crashes NHTSA's 1985 analysis uses data from the National Crash Severity Study (NCSS), a predecessor of NASS.[9] Investigators measured the percentage of bond separation and computed the Delta V in actual towaway crashes. Linear regression calibrated the average percentage of bond separation, as a function of Delta V, in frontal crashes, by windshield installation method. For example, in frontal crashes with Delta V = 30 mph, the average percentage of bond separation was:

Country of Manufacture	Windshield Installation Method	Pre or Post FMVSS 212?	Average Bond Separation at ΔV 30 mph (%)
United States	rubber gasket	pre	22
		post	23
	butyl tape	pre	16
		post	15
	polyurethane	post	14
Germany	rubber gasket	pre	59
		post	37

In domestic cars, both types of adhesive bonding substantially reduced bond separation at 30 mph, relative to rubber gaskets. Within each method, separation rates were about the same before and after FMVSS 212. In German cars, the plastic clips added to the rubber gaskets in response to FMVSS 212 greatly reduced bond separation, although to a level that was still higher than pre-standard domestic cars. Similarly, at all other speeds, adhesive bonding reduced the percentage of bond separation in domestic cars, and the plastic clips reduced it in German cars.

Effect on occupant ejection via the windshield portal Combining the 1977-79 NCSS with 1968-78 Multidisciplinary Accident Investigations (MDAI) generates a large data file with information on the windshield installation method of each case vehicle. It specifies whether

---

[8] Fargo, R.B., *Windshield Glazing as an Injury Factor in Automobile Accidents*, Cornell Aeronautical Laboratory, Buffalo, 1968.
[9] Kahane (1985 Windshields), pp. 143-154; Kahane, C.J., Smith, R.A. and Tharp, K.J., "The National Crash Severity Study," *Report on the Sixth International Technical Conference on Experimental Safety Vehicles*, NHTSA Report No. DOT HS 802 501, Washington, 1977, pp. 493-515.

occupants were ejected and, if so, via what portal. The rate of ejection via the windshield portal per 100 crash-involved front-seat occupants was 68 percent lower in domestic cars with adhesive bonding than in domestic cars with rubber gaskets. However, after examining the rates of ejection via other portals, which also decreased to some extent in the later-model cars for reasons obviously unrelated to the windshield installation method, the evaluation concluded that, more realistically, a 50 percent reduction of ejection via the windshield portal could be attributed to adhesive bonding. Similarly, in German cars, the installation of plastic clips within the rubber gaskets reduced ejection via the windshield portal by an estimated 50 percent.[10]

Fatality reduction – passenger cars The great advantage of preventing ejection is that, all else being equal, an occupant is safer remaining inside the vehicle than being ejected from it. Yet, not every ejection prevented is a life saved. A portion of ejectees would have sustained fatal injuries from interior components even if they had stayed within the car. This is especially true of the windshield portal, where an estimated 70 percent of the fatally injured ejectees in the NCSS-MDAI data had already received mortal wounds from the steering assembly, instrument panel, windshield header and/or other components before the ejection, and would have died even if they had not been ejected. In other words, reducing ejection-fatality cases by 100 means saving 30 lives (the other 70 would still not survive). The 50 percent reduction of ejection attributed to adhesive bonding in domestic cars, and to plastic clips in German cars corresponds to saving 15 percent of the fatalities who were ejected via the windshield portal before these safety improvements. Using 1982 as the "baseline" year, the evaluation estimated that adhesive bonding would be saving 105 lives per year in domestic cars, and plastic clips, 7 lives per year in German cars by the time all cars built before these safety improvements had been phased out.[11]

Serious injury reduction – passenger cars Likewise, 30 percent of the people ejected via the windshield portal who survived, but sustained one or more serious, nonfatal AIS 3-5 injuries received all of these injuries after exiting the vehicle, while 70 percent suffered at least one such injury prior to the ejection. Thus, as above, a 50 percent reduction of ejection corresponds to a 15 percent injury reduction. Again using 1982 as the "baseline" year, the evaluation estimated that adhesive bonding would eventually save 160 people from AIS 3-5 injury per year in domestic cars, and plastic clips, 11 AIS 3-5 casualties per year in German cars.

Effect on lacerations – passenger cars As stated above, if adhesive-bonded windshields "give" less upon head impacts by occupants, facial lacerations could increase. Fargo's 1968 study of pre-HPR windshields showed significantly higher laceration rates with adhesive bonding than with rubber gaskets and raised a worry this could persist with HPR windshields (1966 and later). NHTSA extensively analyzed crash data from NCSS, New York (1974) and Texas (1972-74) on cars equipped with HPR windshields. Cars with adhesive bonding and rubber gaskets had equally low rates of facial lacerations and other type of minor-moderate injuries involving windshield contact. Windshield penetration by occupant impact was no more common with adhesive bonding than with rubber gaskets. NHTSA concluded that adhesive bonding does not have any side effect of increasing minor-moderate injuries in cars with HPR windshields. Incidentally, the NHTSA analysis confirmed Fargo's finding that adhesive bonding was

---

[10] Kahane (1985 Windshields), pp. 154-161.
[11] *Ibid.*, pp. 161-163, 179-182 and 236-243; 90% confidence bounds for life-saving effectiveness: 7 to 22 percent; for lives saved in domestic cars: 35 to 175.

associated with higher injury rates in cars with pre-HPR windshields, now long phased out except for occasional "historic" vehicles.[12]

Effectiveness – light trucks NHTSA's 1985 evaluation only addressed the performance of windshield bonding in passenger cars. However, since the kinematics of occupant ejection via the windshield portal are presumably similar in cars and light trucks, the estimated reduction (50 percent) for this type of ejection from cars is probably appropriate for light trucks as well.

---

[12] *Ibid.*, pp. 185-220.

FMVSS 213: Child restraint systems
FMVSS 225: Child restraint anchorage systems

These two standards work together as a single unit to protect child passengers. FMVSS 213 regulates child safety seats, safety equipment that is not part of the vehicle. FMVSS 225 regulates vehicles, specifically the anchors whereby safety seats are secured in the vehicle. FMVSS 213 took effect on April 1, 1971, but FMVSS 225 did not begin to phase in until September 1, 1999. Three technologies to advance child passenger safety have been evaluated, or will be evaluated by NHTSA. A fourth "technology" that has always been available – riding in the back seat – is not related to any particular FMVSS but will be discussed here because it has important benefits for child passengers:

- *Child safety seats*
- *Upper tethers and anchorages (not yet evaluated)*
- *LATCH (lower anchors and tethers for children – not yet evaluated)*
- *Riding in the back seat*

Child safety seats are the basic protection system for passengers who are too small to obtain full benefits from safety belts. Newborns start with rear-facing infant seats, graduate to forward-facing toddler seats, booster seats and, finally, safety belts. Child seats have typically been secured in a vehicle at one place, by a lap belt; upper tethers secure them at a second place, significantly reducing the tendency of seats to move or rotate in crashes. LATCH supersedes the vehicle's lap belts as a method of securing the seat in the vehicle; the new method will be more effective, more convenient and harder to misuse. The back seat is intrinsically safer than the front seat, given equal safety equipment, because there are a lot more severe frontals than rear impacts, and it is safer to sit further away from the point of impact.

FMVSS 213 "specifies requirements for child restraint systems ... to reduce the number of children killed or injured." FMVSS 225 "establishes requirements for child restraint anchorage systems to ensure their proper location and strength for the effective securing of child restraints, to reduce the likelihood of the anchorage systems' failure, and to increase the likelihood that child restraints are properly secured and thus more fully achieve their potential effectiveness."[1]

Child safety seats

History Early child seats and car beds, dating back to 1933, were designed for comfortable transportation, not crash protection. The first safety seats include the Ford Astro-Guard (1965) and Tot-Guard (1967) and the GM "Love" infant and child seats (1970). NHTSA issued FMVSS 213, effective April 1, 1971, to establish performance criteria for any device marketed as a "safety car seat," and to discourage the marketing of non-safety devices for use in vehicles. This version of FMVSS 213 required that safety seats be designed for attachment within a car by car's safety belts, and it set limits for forward motion under a static load, but it did not include a dynamic test simulating a crash. By the mid-1970's, frontal 30 mph sled testing by the Consumers Union and others demonstrated that some seats offered substantially less protection

---

[1] *Code of Federal Regulations*, Title 49, Government Printing Office, Washington, 2002, Parts 571.213 and 571.225.

than others. NHTSA revised FMVSS 213, effective January 1, 1981, incorporating the 30 mph test. The seat must remain structurally intact during the test, limit the forward motion of the dummy, and also limit the Head Injury Criterion (HIC) to 1000 and chest acceleration to 60 g's on the 3-year-old dummy. Ever since the mid-1970's, several manufacturers of juvenile equipment have offered a selection of crashworthy seats for children of various ages.[2]

How safety seats work The goal of safety seats is to provide small children with the same protection, or better, that adults obtain from safety belts and the other safety equipment built into a vehicle. They are designed to keep children within the vehicle and close to their original seating position, prevent contacts with harmful interior surfaces or other occupants and provide "ride-down" by gradually decelerating the child as the vehicle deforms and absorbs energy. Key terminology is the "excursion" of the head and knee, the greatest distance that those parts of a dummy extend forward from the seatback at any time during the 30 mph test. Excursion is allowed by the looseness of the devices that secure the seat to the vehicle and the dummy within the seat, and by the tendency of seats to tilt or rotate during the test. Obviously, the less the excursion, the lower the probability of contacting any part of the vehicle interior.

Rear-facing infant seats are recommended from birth until an infant reaches a manufacturer-specified weight, often 20 pounds (typically before the first birthday), sometimes higher. These infants are not yet sitting up. The safety seat is like a cushioned box that holds the infant in a semi-recumbent position, with the head higher than the torso and legs. Since the seat faces rearwards, the infant's entire body presses into the cushion during a frontal crash, and rides down with the vehicle. That is the safest way to absorb the infant's kinetic energy, since force is not concentrated on one part of the body. It minimizes the risk of neck injury, to which infants are especially vulnerable. In side impacts, the infant presses against cushions on the side of the seat. A small harness keeps the infant within the seat in rollovers, rear impacts or multiple impacts. The vehicle's lap belt secures the safety seat to the vehicle. A rear-facing seat must never be exposed to a deploying air bag, because the front of the seat is directly in the path of the deploying air bag. Infants must ride in the back seat, or if that is impossible but the vehicle has an on-off switch for the air bag, that switch must be turned off.

Forward-facing child seats are recommended for a manufacturer-specified range of weights, often 20-40 pounds (typically ages 1-4). The safety seat is like an armchair. The child sits up in it; the child's head and torso are protected in back and, to some extent, on each side by the seat's shell and cushion; the child's legs hang down over the edge of the seat. Typically a 5-point harness, fitting over both shoulders, hips and abdomen, restrains the child in rollovers, frontals and side impacts, without concentrating force on one part of the body. Some place a cushioned

---

[2] Radovich, V.G., "Development of Infant and Child Restraint Regulations and Their Application," *SAE Child Injury and Restraint Conference Proceedings*, Paper No. 831655, Society of Automotive Engineers Publication No. P-135, Warrendale, PA, 1983, pp. 101-111; Kahane, C.J., *An Evaluation of Child Passenger Safety: The Effectiveness and Benefits of Safety Seats*, NHTSA Technical Report No. DOT HS 806 890, Washington, 1986, pp. 2-4; *Federal Register* 35 (September 23, 1970): 14778, 44 (December 13, 1979): 72131; *Consumer Reports*, Vol. 37, August 1972, p. 484.

shield in front of the child in addition to the harness, or sometimes instead of the harness. The back of the seat and its cushion protect the child from whiplash in rear impacts.[3]

Until August 31, 2002, the vehicle's lap belt secured the safety seat to the vehicle. Belts were routed through the frame of the child seat, as specified by the manufacturer, or less frequently around the seat and the child. Beginning September 1, 2002, seats must have lower tethers for use in new vehicles equipped with LATCH anchorages, but must also be securable by lap belts in older vehicles. Starting September 1, 1999, all seats have to be equipped with upper tethers for an additional attachment to the vehicle. Before that, some seats were equipped with upper tethers on a voluntary basis dating back to 1970. The advantage of the upper tether is that it reduces the tilting or rotation of the seat during a crash, because it is firmly attached to the vehicle at the top as well as the bottom.

Effective January 1, 1981, FMVSS 213 limited forward head excursion to 32 inches and knee excursion to 36 inches on the 30 mph frontal test, measured from the seatback pivot point. Those levels are sufficient to prevent contact with interior surfaces in an average-sized passenger car. Upper tethers improve performance: as of September 1, 1999, FMVSS 213 has further limited head excursion to 720 mm (28.35 inches), providing an additional margin of safety. FMVSS 213 also sets numerous other requirements for seats, including a limit of 1000 HIC and 60 chest g's on the 3-year-old dummy and maintenance of structural integrity in the 30 mph test, clear labeling of instructions it their use, convenient seat registration, and harness buckles that are convenient for adults but cannot be opened by small children.[4]

Convertible seats can be used in a rear-facing mode for an infant, and then tilted 90 degrees to become a forward-facing seat when the infant "graduates" to that type of seat. The floor of the infant seat becomes the back of the child seat, while the front of the infant seat becomes the bench of the child seat.

Booster seats are recommended for a manufacturer-specified range of weights, often 40-80 pounds (typically ages 4-8), or until a child is 4'9'' tall. Booster seats essentially raise the child up high enough that the vehicle's manual 3-point belt fits over one shoulder and diagonally across the thorax, as it would for an adult. The lap portion of the 3-point belt should fit snugly around the hips. When the 3-point belt is thus correctly positioned, it should provide similar benefits for these children as for adults. Some booster seats also have their own shields or 5-point harnesses for use by children at the lower end of the weight range. Some booster seats have their own seatbacks, while others use the vehicle's seatback.[5]

Safety seat use trends In 1979, NHTSA's first national survey indicated that 45 percent of infants under age 1 and 9 percent of toddlers age 1-4 rode in some kind of child "seat." Half of these "seats" were not secured by the car's belt, hardly protecting the child (many of them were

---

[3] *NHTSA Child Seat Safety Information*, NHTSA, Washington, 2004 (http://www.nhtsa.dot.gov/people/injury/childps/csr2001/csrhtml/safetyFeatures.html#forward); Kahane (1986), pp. 19-26.
[4] *Code of Federal Regulations*, Part 571.213 S5.1.3. and Figure 1B; the seatback pivot point is at the lower-rear edge of the seatback; Kahane (1986), p. 3.
[5] *A Parent's Guide to Booster Seats*, NHTSA Publication No. DOT HS 808 671, Washington, 1998; *Buying a Safer Car for Child Passengers 2003*, NHTSA Publication No. DOT HS 809 545, Washington, 2003.

probably not safety seats designed to FMVSS 213). However, Tennessee had recently become the first State with a child passenger protection law, effective January 1, 1978. Rhode Island followed in 1980, West Virginia in 1981 and ten States in 1982. By June 1, 1985, every State and the District of Columbia had a law to protect child passengers. In NHTSA's 19-City Study, safety seat use rose quickly during the 1980's, exceeding 80 percent for infants and toddlers by the end of the decade. By 2002, NOPUS showed that 99 percent of infants and 94 percent of toddlers age 1-4 were in some kind of restraint (not necessarily age-appropriate or correctly used). Many of the laws initially allowed children to "graduate" too early from safety seats to the vehicle's belts, but as of 2003, 44 States require children to be in a safety seat at least up to their fourth birthday.[6]

Safety seat misuse For full benefits, safety seats need to be installed correctly in the vehicle and children properly harnessed within the seat. Misuse is a potential problem because there are many styles of seats, and the mode of use can also vary by the age of the child, the type of vehicle, or even by seat position within the same vehicle. NHTSA's evaluation of child passenger safety reviewed 957 observations during 1984 of children age 0-4 riding in safety seats:[7]

- 40 percent of the seats were correctly used, exactly as recommended by the manufacturer or very close to that.
- 18 percent were "grossly" misused: the seat was not secured to the vehicle by a belt and/or the seat's harness or shield was not used. In a crash, the child would immediately become a projectile and derive little or no protection from the seat.

---

[6] Lawless, E.W. and Siani, T.A., "The State of the Art of Child Passenger Safety Legislation in North America," *SAE Child Injury and Restraint Conference Proceedings*, Paper No. 831650, Society of Automotive Engineers Publication No. P-135, Warrendale, PA, 1983, pp. 229-242; Kahane (1986), pp. 4-5; Phillips, B.M., *Safety Belt Usage Among Drivers*, NHTSA Report No. DOT HS 805 398, Washington, 1980; Phillips, B.M., *Restraint System Usage in the Traffic Population*, NHTSA Report No. DOT HS 806 424, Washington, 1983; Perkins, D.D., Cynecki, M.J. and Goryl, M.E., *Restraint System Usage in the Traffic Population*, NHTSA Report No. DOT HS 806 582, Washington, 1984; Goryl, M.E. and Cynecki, M.J., *Restraint System Usage in the Traffic Population*, NHTSA Report No. DOT HS 806 714, Washington, 1985; Goryl, M.E., *Restraint System Usage in the Traffic Population*, NHTSA Report No. DOT HS 806 987, Washington, 1986; Goryl, M.E. and Bowman, B.L., *Restraint System Usage in the Traffic Population*, NHTSA Report No. DOT HS 807 080, Washington, 1987; Bowman, B.L. and Rounds, D.A., *Restraint System Usage in the Traffic Population 1987 Annual Report*, NHTSA Report No. DOT HS 807 342, Washington, 1988; Bowman, B.L. and Rounds, D.A., *Restraint System Usage in the Traffic Population 1988 Annual Report*, NHTSA Report No. DOT HS 807 447, Washington, 1989; Datta, T.K. and Guzek, P., *Restraint System Use in 19 U.S. Cities 1989 Annual Report*, NHTSA Report No. DOT HS 807 595, Washington, 1990; Datta, T.K. and Guzek, P., *Restraint System Use in 19 U.S. Cities 1990 Annual Report*, NHTSA Report No. DOT HS 808 147, Washington, 1991; Datta, T.K. and Guzek, P., *Restraint System Use in 19 U.S. Cities 1991 Annual Report*, NHTSA Report No. DOT HS 808 148, Washington, 1992; *National Occupant Protection Use Survey: Controlled Intersection Study*, NHTSA NCSA Research Note, Washington, 1995; *National Occupant Protection Use Survey – 1996: Controlled Intersection Study*, NHTSA NCSA Research Note, Washington, 1997; *Traffic Safety Facts 1999 – Children*, NHTSA Publication No. DOT HS 809 087, Washington, 2000, p. 5; Glassbrenner, D., *The Use of Child Restraints in 2002*, NHTSA Research Note No. DOT HS 809 555, Washington, 2003; *Child Restraint Laws*, Insurance Institute for Highway Safety, Arlington, VA, 2004 (http://www.hwysafety.org/safety%5Ffacts/state%5Flaws/restrain2.htm).
[7] Kahane (1986), pp. 30-47, especially p. 45.

- 42 percent were "partially" misused, substantively differing from the method recommended by the manufacturer. Depending on the type of misuse, effectiveness could be slightly or seriously degraded. Typical misuse modes include:
  - Incorrectly routing the lap belt through/around the frame, where the frame is weaker or where it will allow more excursion.
  - Using a seat in the forward facing mode for an infant who should be in a rear-facing seat, or vice-versa.
  - Not attaching the upper tether, if one is available.

NHTSA, other government and private organizations and the safety-seat manufacturers have made an effort to design seats that are easy to install and use correctly, to provide clear instructions that are prominently displayed on the seat, and to establish a national network of instructors who can advise people about the type of seat they need for their child and their vehicle, and show them how to install it correctly. The public may find an instruction station close to their home at http://www.nhtsa.dot.gov/CPS/CPSFitting/Index.cfm.

A 1997 survey of 2,965 safety seats in actual use showed that gross misuse of seats had greatly decreased since 1984, but misuse of safety seats remained a problem. Approximately 2 percent were not secured by the vehicle's belt and 3.3 percent did not have the safety seat's harness connected (a big reduction from 18 percent gross misuse in the earlier survey). Misrouting of the vehicle's belt, once quite common, had decreased to about 3 percent. On the other hand, 24 percent of infant seats were used in a forward facing mode. Locking clips, a separate device employed in some vehicles to better secure lap-shoulder belts, were not used in over 60 percent of the vehicles where they were needed. Also common were improper use or fit of the vehicle's belts or the seat's harness.[8]

Expected benefits Correctly used safety seats ought to have the same types of benefits that safety belts have for adults, and probably more so, because they can do a better job preventing contacts with the vehicle interior, and because their typically 5-point harnesses will more evenly spread forces than a lap-shoulder belt. Persistent misuse of safety seats, however, is likely to diminish "as used" effectiveness – i.e., if the restrained population includes the actual mix of correctly used and misused seats, effectiveness will be lower than for the population consisting exclusively of children who are correctly restrained. "As used" effectiveness can increase over time if the proportion of correctly used seats increases.

Fatality reduction – passenger cars In 1984, at about the same time that Evans developed a method to estimate belt effectiveness and called it "double-pair comparison analysis," Partyka at NHTSA independently developed essentially the same method to estimate safety seat effectiveness. NHTSA expanded and updated the analysis in its 1986 evaluation, in 1988, and finally in 1996. The last study provides NHTSA's current estimates of the "as used" fatality reduction by child safety seats. It is based on 1988-94 FARS data. The analysis of safety seats resembles the procedures used to estimate the effectiveness of safety belts, especially back-seat safety belts (see the FMVSS 208 chapter). For example, in cars with unrestrained drivers, the fatality ratio of unrestrained children to their drivers is compared to the fatality ratio of children

---

[8] Decina, L.E. and Knoebel, K.Y., "Child Safety Seat Misuse Patterns in Four States," *Accident Analysis and Prevention*, Vol. 29, January 1997, pp. 125-132.

in safety seats to *their* drivers. "Safety seat users" include those that FARS reported in "child safety seats" or "child safety seats used improperly." Since the percentage of seats "used improperly" is far lower in FARS than in observational surveys, a substantial but unknown percentage of the children that FARS merely reported "in safety seats" were also in improperly used seats. Thus, the two groups have been combined to provide an overall, "as used" effectiveness estimate. For infants less than 1 year old, safety seats reduced overall fatality risk by a statistically significant 71 percent. For toddlers age 1-4, the seats reduced fatality risk by a statistically significant 54 percent.[9]

The preceding "as used" estimates include misused seats. In 1983, the Tennessee Highway Patrol estimated that correctly used safety seats could reduce fatality risk by 71 percent, an estimate widely quoted in the literature. It is not based on statistical analysis of fatalities to restrained vs. unrestrained children, but on a detailed case-by-case analysis of 51 unrestrained fatalities, judging what would have happened if the children had been protected by safety seats. However, it continues to appear plausible, especially given that the preceding "as used" estimates including misused seats are not much lower.[10]

Over the years, a moderate proportion of toddlers age 1-4, especially at the higher ages have only been restrained by the vehicle's belts, not by a safety seat. The restraint is essentially a lap belt only, for even if the seat is equipped with a 3-point belt, the shoulder harness will not fit across the torso of a child that small, and it will offer little protection. NHTSA urges that children under 4'9'', and especially in the 1-4 age group not be restrained with belts alone, but with a FMVSS 213 seat appropriate to their weight. Nevertheless, belts have some benefits for children age 1-4. NHTSA's 1986 evaluation, based on double-pair comparison analysis of 1975-84 FARS data, estimates that lap belts reduce fatality risk by 33 percent. That is substantially lower than the 54 percent "as used" effectiveness of the actual mix of correctly used and misused safety seats, let alone the effectiveness of correctly used safety seats. It is about the same as the effectiveness of lap belts for adults (27 percent in the front seat and 32 percent in the back seat – see the FMVSS 208 chapter).[11]

Fatality reduction – light trucks NHTSA's 1996 study also showed statistically significant fatality reductions for child safety seats in LTVs. For infants less than 1 year old, safety seats reduced overall fatality risk by 58 percent. For toddlers age 1-4, the reduction was 59 percent. The study showed a 48 percent fatality reduction by lap belts for toddlers age 1-4. This is the same as the estimated effectiveness of lap belts for adults in the front seat.[12]

---

[9] Partyka, S.C., *Restraint Use and Fatality Risk for Infants and Toddlers*, NHTSA National Center for Statistics and Analysis, Washington, 1984; Kahane (1986), pp. 140-163; Partyka, S.C., *Lives Saved by Child Restraints from 1982 through 1987*, NHTSA Technical Report No. DOT HS 807 371, Washington, 1988; Hertz, E.C., *Revised Estimates of Child Restraint Effectiveness*, NHTSA NCSA Research Note, Washington, 1996 (this is a summary report that does not show the detailed analyses, but they are set up the same way as in Partyka's 1988 report).
[10] *51 Young Lives*, HSL Report No. DOT HS 034 896, Tennessee Highway Patrol, Nashville, 1983; Kahane (1986), pp. 96-99; *National Child Passenger Safety Week: One-Minute Child Safety Seat Facts*, http://www.nhtsa.dot.gov/people/outreach/safesobr/OPlanner/ncpsw/kidseat3.html.
[11] Kahane (1986), pp. 161-164; these data are entirely before 1986, when State buckle-up laws began to bias belt use reporting on FARS.
[12] Hertz (1996).

Injury reduction – passenger cars NHTSA's 1986 evaluation furnished three sets of estimates of injury reduction by safety seats in passenger cars. In a file that combined all available detailed investigations of crashes selected by probability sampling – the 1974-75 RSEP, the 1977-79 NCSS, the 1979 pilot test of NASS at the NCSS sites, and the 1979-83 NASS – the rate of fatal or hospitalizing injuries per 100 children age 0-5 involved in towaway crashes was 56 percent lower for children in safety seats than for unrestrained children. The reduction was statistically significant. However, it is an overestimate of the overall "as used" effectiveness, because these investigators classified children in grossly misused seats as "unrestrained." If they had been classified as safety seat users, injury reduction would have been an estimated 46 percent in 1984.[13]

An analysis of 1981-83 Pennsylvania crashes computed injury rates per 100 children, adjusting for the child's age and seat position. Compared to unrestrained children, safety seats reduced fatal-and-serious (K + A) injury risk by 43 percent, moderate-to-fatal (K + A + B) injury risk by 45 percent, and overall (K + A + B + C) injury risk by 31 percent. Here, too, many children in grossly misused seats may have been classified as "unrestrained." The 1986 report concludes that K + A + B injury reduction would have been closer to 37 percent if they had all been classified as safety seat users.[14]

As part of the 1986 evaluation, NHTSA conducted 36 frontal or oblique-frontal sled tests, each with 3 or 4 3-year-old dummies, at speeds ranging from 11 to 35 mph. Dummies were unrestrained, or in safety seats of four different types – correctly used, grossly misused, or in various modes of partial misuse. NHTSA calibrated the probability of hospitalizing head or torso injury as a function of speed for the unrestrained dummy – and various seat types/use modes. A mathematical model computed overall effectiveness from the calibrated relationships. Correctly used, the various types of safety seats reduced the risk of hospitalizing injuries in frontal crashes by an average of 61 percent, relative to the unrestrained dummy. The average effectiveness of the partially misused seats was 38 percent. Not surprisingly, the dummies in grossly misused seats had the same injury risk as unrestrained dummies. When correct use, partial misuse and gross misuse were weighted by their frequency of occurrence in the observational survey, the average "as used" effectiveness of safety seats in frontal crashes was 40 percent.[15]

Injury reduction – light trucks NHTSA has not estimated the injury reduction by safety seats for children who are occupants of LTVs.

Upper tethers and anchorages
LATCH (lower anchors and tethers for children)

On March 5, 1999, NHTSA published a Final Rule amending FMVSS 213 to require safety seats to have upper and lower tethers, and establishing FMVSS 225 to require motor vehicles to have anchorages for attaching the tethers. On child safety seats (except certain types), upper tethers

---

[13] Kahane (1986), pp. xxxiv and 201-214.
[14] *Ibid.*, pp. xxxiv and 169-184; rates were adjusted for age and seat position by multidimensional contingency table analysis.
[15] *Ibid.*, pp. 235-305, especially p. 301.

were required on September 1, 1999 and lower tethers on September 1, 2002. In motor vehicles, the phase-in of upper tether anchorages extended from September 1, 1999 to September 1, 2000, and the lower anchorages from September 1, 2000 to September 1, 2002. The upper anchorages are required at all positions on the rear-most forward-facing row of seats (i.e., the back seat of a car, but the third seat of most minivans), and the lower anchorages at two of those positions. The objectives of the regulation are to secure seats more tightly in the vehicle, make their installation easier and more uniform, and reduce the probability of misuse.[16]

Safety seats with upper tethers have been available dating back to 1970. Upper tethers reduce the tilting or rotation of the seat during a crash, because they provide a second attachment of the seat to the vehicle (in addition to the lap belt or lower tethers). Sled testing shows unequivocally that tethered seats, when correctly used, have less head excursion than other types. However, observational surveys showed that most seats did not have tethers, and among those that did, only 15 percent of tethers were attached in 1984, in part because many vehicles did not have tether anchorages. The new regulation guarantees that every seat will have this performance-enhancing feature, and every vehicle will have a method of attaching it.[17]

The method of securing a seat by the vehicle's safety belts can vary considerably from seat to seat and from vehicle to vehicle, creating opportunities for misuse. The fit of the belts is often loose, especially when, more often than not, a locking clip needed to prevent spooling of the belt is not used. LATCH will establish a more uniform attachment system that does not rely on the safety belt.

NHTSA will conduct an observational survey to find out if consumers are using upper tethers and LATCH system to install child safety seats, if they are easy to install, and the percentage that are being installed correctly. Subsequently, the agency will analyze crash data to see if upper tethers and LATCH have made safety seats more effective.[18]

Safety benefits of riding in the back seat

Why the back seat is safer The closer the occupant sits to the point of impact, the greater the risk of fatal injury. That is because fatal crashes often result in severe damage, but rarely to the entire vehicle. They can demolish the portion of the passenger compartment closest to the impact, while the furthest portion remains nearly intact. A frontal impact is twice as dangerous for front-seat occupants as back-seat occupants, whereas a rear impact is twice as dangerous for

---

[16] *Federal Register* 64 (March 5, 1999): 10786; *Code of Federal Regulations*, Title 49, Parts 571.213 S5.9 and 571.225 S4.

[17] Bayer, A.R., Jr. and Peterson, B.S., *Child Restraint Systems Testing*, NHTSA Report No. DOT HS 803 408, Washington, 1978; Kelleher, B.J. and Walsh, M.J., "Sled Test Comparisons of Child Restraint Performance," *Twenty-Second Stapp Car Crash Conference*, Paper No. 780903, Society of Automotive Engineers, Warrendale, PA, 1978; Kelleher, B.J., Walsh, M.J , Dance, D.M. and Gardner, W.T., "An Experimental Study of the Effects of Child Restraint Improper Installation and Crash Protection for Larger Size Children," *SAE Child Injury and Restraint Conference Proceedings*, Paper No. 831602, Society of Automotive Engineers Publication No. P-135, Warrendale, PA, 1983, pp. 31-51; Radovich, *op. cit.*; Kahane (1986), pp. xliv, 30-47 and 106-134.

[18] *National Highway Traffic Safety Administration Evaluation Program Plan, Calendar Years 2004-2007*, NHTSA Report No. DOT HS 809 699, Washington, 2004, pp. 9-10.

the back-seat occupants.[19] But this is not a trade-off. High-speed frontal impacts are far more common than high-speed rear impacts, resulting in greater overall fatality risk to front-seat occupants:

- Collisions with fixed objects are frontal, unless the vehicle has yawed out of control, and even then they are rarely rear impacts.
- When two vehicles collide, usually at least one of them, and sometimes both have frontal damage, but rear-to-rear collisions are rare.
- Head-on collisions often involve a dangerous closing speed because the two vehicles are moving in opposite directions; rear impacts are often of low severity because both vehicles are moving in the same direction.

A second advantage for the unrestrained back-seat occupant in a frontal crash is that he or she will contact the back of the front seat, a more benign surface than the steering assembly, instrument panel, or windshield header contacted by the unrestrained front-seat occupant. This advantage may be lost if both occupants are correctly restrained. Nevertheless, a back-seat occupant, restrained or unrestrained, ought to have lower fatality risk than a front-seat occupant with the same safety equipment. For many years, however, the front-seat has been the first place to get new safety equipment (3-point belts), and sometimes the only place (air bags).

Fatality reduction Evans demonstrated that unrestrained drivers and adult right-front passengers of passenger cars have almost exactly the same fatality risk in a crash. In other words, in crashes where both seats were occupied by people of the same age and gender, the ratio of driver to RF fatalities was 1:1. But back-seat outboard occupants have 26 percent lower fatality risk than drivers and RF passengers of the same age and gender. Back-seat center occupants have 37 percent lower fatality risk than drivers and RF passengers: the rear-center seat is the safest place in a car, because it is far away from the sides as well as the front of the car. These are the results for unrestrained occupants of passenger cars without air bags, age 16 and older.[20]

Children, too, are substantially safer in the back seat. In NHTSA's 1986 evaluation, double-pair comparison analyses of 1975-84 FARS data showed that an unrestrained child age 0-4 had 27 percent lower fatality risk in the back seat than in the front seat. For a child in a safety seat, the fatality risk was 20 percent lower in the back seat than in the front seat.[21]

These benefits have persisted in vehicles without air bags. Three NHTSA publications in 1996-97 cite an analysis of 1988-94 FARS data showing that both restrained children and unrestrained children experienced a 26 [or 27] percent reduction in fatality risk in the back seat relative to the front seat.[22]

---

[19] Evans, L., *Traffic Safety and the Driver*, Van Nostrand Reinhold, New York, 1991, pp. 47-51.

[20] *Ibid.*

[21] Kahane (1986), pp. xxxiv and 164-167; in the double-pair comparison, the fatality ratio of children to their drivers is compared for children in the front seat and children in the back seat.

[22] Hertz (1996); *Federal Register* 62 (November 21, 1997): 62406, Footnote 23; *Air Bags & On-Off Switches*, NHTSA Publication No. DOT HS 808 629, Washington, 1997; the supporting analysis is not available, and it is unknown if it included children age 0-4 or 0-12, or if it was limited to passenger cars or included LTVs.

The following analysis of MY 1981-95 passenger cars without air bags in 1991-99 FARS data supplies more detailed results. Children of all ages and adolescents, restrained as well as unrestrained, obtained a substantial benefit for riding in a rear-outboard seat rather than the right-front seat, and an even larger fatality reduction if they traveled in the rear-center seat. Specifically, children age 0-12, both restrained and unrestrained, experienced statistically significant fatality reductions close to 30 percent for moving from the front-right seat to a rear outboard seat, and close to 45 percent for moving from the front-right seat to the rear-center seat:

CARS WITHOUT AIR BAGS	Fatality Reduction Relative to Right-Front Seat (%)			
	Rear-Outboard Seat		Rear-Center Seat	
Age Group	Restrained	Unrestrained	Restrained	Unrestrained
0 – 4	36	26	46	29
5 – 12	29	36	47	59
0 – 12	*33*	*31*	*46*	*47*
13 – 15	11	30	49	44

Air bags have made it much more important for children to ride in the back seat, less so for adolescents and adults. The benefits of air bags for right front passengers age 13 and older have narrowed the difference between the front and back seats for those groups. But children are much safer in the back seat, without air bags, than in the front seat of a car with air bags. For example, air bags with 1990's technology were estimated to increase the fatality risk of an infant RF passenger in a rear-facing safety seat by a factor of approximately 5½ (see the FMVSS 208 chapter). Thus, the fatality reduction for moving that infant to the back seat is not merely 36-46 percent, as shown in the above table, but close to 90 percent. Similarly, these air bags may have increased the fatality risk of a toddler in a forward-facing safety seat by approximately 70 percent; that toddler will be 62 percent safer in a rear-outboard seat, and 68 percent safer in the rear-center seat than in the right-front seat.[23]

More children are riding in the back seat Since the mid-1990's, NHTSA, the manufacturers and the safety community have campaigned to inform the public about the hazard of air bags to children and the need for children to ride in the back seat of vehicles with air bags. Moreover, the campaigns make it clear that the back seat is the safest place for a child in all vehicles, including those without air bags.[24]

---

[23] For example, if the fatality risk of an infant in the RF seat of a car without air bags is 1, the risk with air bags is 5.5; the risk in the rear-center seat is 1 - .46 = .54; the reduction is 1 - .54/5.5 = 90 percent.

[24] *Air Bag & Seat Belt Safety Campaign, Air Bag & Seat Belt Safety Tips*, National Safety Council, Chicago, November 17-30, 2003, http://www.nsc.org/partners/safetips.htm; *Air Bags & On-Off Switches, op. cit.*; *Buckle-Up America Child Passenger Safety Week, February 10-16, 2002, Talking Points*, NHTSA, http://www.nhtsa.dot.gov/people/injury/airbags/buckleplan/CPS%20Week%20Planner_files/talking1.html.

The public has responded. Substantially fewer children age 0-7 rode in the front seat in 2001 than in 1995. NHTSA analyzed State crash data from Florida, Maryland and Utah for 1995-2001. The distribution of seat positions of the 363,579 child passengers in these mostly low-severity crashes is probably about the same as in the general, non-crash-involved traffic stream. In 2001, only 8 percent of 0-3 year old infants and toddlers still rode in the front seat, down from 26 percent in 1995, and the proportion of 4-7 year old children in the front seat decreased to 19 percent from 33 percent. However, the proportion of 8-12 year old children in the front seat only declined to 35 percent from 39 percent.[25]

When a child is the only passenger in the vehicle, the driver may be reluctant to have the child ride alone in the back seat. But even here, there were dramatic changes between 1995 and 2001 for children age 7 or younger (but not for 8-12 year olds). Fewer children are riding in the front seat, not only in vehicles with passenger air bags, but even in vehicles without air bags:

Vehicles with Only One Passenger, a Child Age 0-12

Percent in Front Seat

		Age 0-3	Age 4-7	Age 8-12
All vehicles				
	1995	52	78	88
	2001	17	43	79
Vehicles with				
dual air bags	1995	49	75	88
	2001	14	36	76
Vehicles without				
air bags	1995	52	78	88
	2001	22	54	84

---

[25] Kindelberger, J. and Starnes, M., *Moving Children from the Front Seat to the Back Seat: The Influence of Child Safety Campaigns*, NHTSA Research Note No. DOT HS 809 698, Washington, 2003.

FMVSS 214: Side impact protection

This standard is associated with two vehicle modifications whose safety benefits have been evaluated by NHTSA and two recent innovations that the agency is just beginning to evaluate:

- *Side door beams*
- *Voluntary (pre-1994) TTI(d) improvements in 2-door cars*
- *Side impact protection, dynamic test standard, passenger cars (not yet evaluated)*
- *Side air bags (not yet evaluated)*

Before 1969, the side doors of passenger cars were nearly empty shells of sheet metal, and offered little protection to occupants in side impacts. Side door beams, running longitudinally inside the door, were a first step to provide some crush resistance and structural strength, but they were not enough to resist a high-speed impact by another vehicle. NHTSA, the manufacturers and the safety community researched additional structural improvements to resist crush and padding to cushion the occupant. They developed a dynamic crash test, a dummy, and an injury criterion TTI(d) to measure how well a vehicle protects its occupants. Side air bags provide additional cushioning for the occupant and they improve TTI(d) performance.

The second paragraph of FMVSS 214 says, "The purpose of this standard is to reduce the risk of serious and fatal injury ... in side impact crashes by specifying vehicle crashworthiness requirements in terms of accelerations measured on anthropomorphic dummies in test crashes [and] by specifying strength requirements for side doors."[1] That's exactly what it has done, but in reverse chronological order. First, FMVSS 214 set strength requirements for doors, effective January 1, 1973 in passenger cars and extended to pickup trucks, vans, buses and SUVs up to 10,000 pounds GVWR, effective September 1, 1993. Subsequently, it added a dynamic crash test requirement, phased in for passenger cars from September 1, 1993 to September 1, 1996, and effective for LTVs up to 6,000 pounds GVWR on September 1, 1998.

Side door beams

History During the 1960's, the manufacturers conducted research and tests to improve side impact protection. Hedeen and Campbell at GM developed side door beams, and the company also developed a static test for measuring a door's crush resistance. The beams were installed in MY 1969 full-size GM cars. The test procedure became Society of Automotive Engineers' Recommended Practice J367 in March 1970, but it did not specify pass-fail criteria. NHTSA announced its intention to regulate side door strength with an ANPRM in October 1968, followed by several NPRMs. The Final Rule issued in October 1970, with an effective date of January 1, 1973, built upon the SAE practice and added pass-fail criteria. A rigid steel cylinder is gradually forced into the middle of the door, its top nearly level with the bottom of the window. The cylinder must encounter an average crush resistance of at least 2,250 pounds during the first 6 inches of crush, an average of 3,500 pounds over the first 12 inches, and a peak of at least 7,000 pounds or twice the vehicle's curb weight (whichever is less) somewhere in the

---

[1] *Code of Federal Regulations*, Title 49, Government Printing Office, Washington, 2002, Part 571.214.

first 18 inches. Most domestic cars were equipped with the beams at some point during MY 1969-73 and, of course, all cars had them after January 1, 1973 (mid-MY 1973).[2]

LTVs, with their higher sills and more rigid construction, are more robust than cars when they are hit in the side by another vehicle, but they share cars' vulnerability in side impacts with fixed objects. As part of its effort to bring safety requirements for LTVs up to the same level as cars, NHTSA extended the static strength test of FMVSS 214 to LTVs, effective September 1, 1993. Most LTVs received side door beams in MY 1994, but a few make-models in MY 1991, 1992 or 1993.[3]

What they are and how they work The side door beam "is a metal bar of channel design, typically 8 inches wide and with channels 2 inches deep. It is located inside [each side] door, close to the outside surface, about 10 inches above the sill. It runs the length of the door, being attached to the door frame vertical members at the hinge and latch ends of the door. ... In some vehicles, the beam is accompanied by a beam cover, stiffener, ... mounting flanges ... [and/or] a local reinforcement of the B pillar at the floor level."[4] "Channel design" resembles fluting or corduroy, running from the front to the rear of the beam, and it is a feature that adds strength to metal structures. In MY 1979-81, the weight of the beams ranged from 5 to 7 pounds per door in 4-door cars, and 10 to 21 pounds per door in 2-door cars.[5]

In a direct side impact by another vehicle, the front of that vehicle will push the door structure inwards until it hits the occupant's torso at close to that vehicle's original impact speed. Injury risk is high. Therefore, any structure that engages the striking vehicle is welcome. By putting some crush on the front of the striking vehicle and/or transmitting force to the remainder of the struck vehicle and accelerating it sideways, the structure can reduce the amount of intrusion toward the occupant and slow down the rate of that intrusion.

Whereas any added strength is welcome, it is obvious that a 5-21 pound beam, stretching from one end of the door to the other without much support in the middle, has limited power to resist a severe and perpendicular impact into the middle of the door by a 2,000-5,000 pound vehicle. The 10-inch gap between the beam and the sill is an additional weak point. The beams might be somewhat effective in lower-speed impacts. NHTSA's 1982 evaluation found that side door beams reduced the depth of maximum crush from an average of 10 inches to 8 in compartment-centered towaway impacts by another vehicle.[6]

---

[2] Kahane, C.J., *An Evaluation of Side Structure Improvements in Response to Federal Motor Vehicle Safety Standard 214*, NHTSA Technical Report No. DOT HS 806 314, Washington, 1982, pp. 100-108; Hedeen, C.E. and Campbell, D.D., *Side Impact Structures*, Paper No. 690003, Society of Automotive Engineers, New York, 1969; *Federal Register* 33 (October 5, 1968): 14971, 33 (December 11, 1968): 18386, 35 (January 21, 1970): 813, 35 (April 23, 1970): 6513, 35 (October 30, 1970): 16801; *1971 SAE Handbook*, Society of Automotive Engineers, New York, 1971, pp. 962-963.

[3] Walz, M.C., *Evaluation of FMVSS 214 Side Impact Protection for Light Trucks: Crush Resistance Requirements for Side Doors*, NHTSA Technical Report No. DOT HS 809 719, Washington, 2004; *Federal Register* 56 (June 14, 1991): 27427.

[4] Kahane (1982 FMVSS 214), p. 106.

[5] *Ibid.*, p. 380.

[6] *Ibid.*, p. 301.

The beam ought to be considerably more effective when the impact has a strongly frontal component of force, especially in typical side impacts with fixed objects where a vehicle is traveling forward and skidding off the road at the same time. The side of the vehicle contacts the pole or tree, but the direction of force is oblique frontal. Rather than merely absorbing energy, the beam acts like an internal "guard rail" to deflect the vehicle and allow it to slide past the pole or tree, with a longer, shallower crush pattern and, ultimately, a lower Delta V. Furthermore, the simultaneous contact of the fixed object with the high beam and the low sill, rather than just the sill, keeps the vehicle upright and reduces the risk of damage to the "greenhouse" area of the occupant compartment. Indeed, NHTSA's 1982 evaluation showed that side door beams significantly changed damage patterns in side impacts with fixed objects. The ratio of the maximum depth of crush to the width of crush decreased from .24 to .15. There was less damage to the occupant compartment and more to the fenders. There was less greenhouse damage. Integrity of the side doors was better preserved.[7]

Expected benefits Given the effects of side door beams on damage patterns and severity, it will not be a surprise if beams are found to reduce fatalities in single-vehicle crashes, but not in multivehicle crashes. They may reduce nonfatal injuries in the less severe multivehicle crashes. Since the beams affect the overall damage pattern to the vehicle in single-vehicle crashes, and help reduce the severity of the crash, they may benefit the farside as well as the nearside occupants. Because there is a beam in every side door on a vehicle, drivers, front-seat passengers and back-seat passengers can benefit.

Fatality reduction – passenger cars NHTSA's analysis, published in 1982, was based on 1975-81 FARS data.[8] These data are best suited for comparing overall fatality risk of occupants of outboard seats (i.e., next to a door: drivers; right-front, left-rear and right-rear passengers) in side impacts, before and after the installation of side door beams, relative to a control group of fatalities in frontal impacts. In one analysis, the data encompassed cars of the first 2 model years with beams (e.g., 1969-70 Chevrolet Impala, 1973-74 Chevrolet Nova) vs. cars of the same or equivalent make-models in the last two model years before beams (e.g., 1967-68 Chevrolet Impala, 1971-72 Chevrolet Nova). Frontal crashes are an acceptable control group because there were few innovations that affect fatality risk in frontal during those model years; specifically, every car in the analysis was equipped with energy-absorbing steering assemblies. The 1975-81 FARS statistics for fatalities in single-vehicle crashes were[9]:

SINGLE-VEHICLE CRASHES	Frontal Fatalities	Side Impact Fatalities
Last 2 MY before beams	4,325	2,505
First 2 MY with beams	4,303	2,137

---

[7] *Ibid.*, pp. 33-43 and 281-329.
[8] *Ibid.*, pp. 22-25 and 143-182.
[9] *Ibid.*, p. 157.

This is a statistically significant 14 percent fatality reduction in single-vehicle side impacts with side door beams: $1 - [(2,137/2,505) / (4,303/4,325)] = .143$.[10]

Analyses by other techniques confirmed the 14 percent fatality reduction.[11] Effectiveness was nearly identical in nearside impacts (14 percent) and farside impacts (15 percent). When "single-vehicle side impacts" were limited to collisions with a single fixed object (and impacts by trains, rollovers with principally side damage and complex off-road excursions were excluded), effectiveness rose to 23 percent.[12]

Using 1980 as the "baseline" year, the evaluation estimated that side door beams would eventually save 480 lives per year in single-vehicle crashes of passenger cars, when all cars on the road were equipped with them. When NHTSA added the dynamic crash test for FMVSS 214, the agency retained the static test requirement, in part, to assure that the benefits of side door beams in single-vehicle crashes would be preserved.[13]

By contrast, none of the analyses showed a statistically significant effect for side door beams on fatalities in multivehicle crashes. Point estimates were close to zero.

Fatality reduction – light trucks Similar analyses were performed with 1989-2001 FARS data to estimate the effect of side door beams in light trucks, where they were introduced in MY 1994 or shortly before that. In single-vehicle side impacts, side door beams reduced fatality risk by a statistically significant 19 percent. The reduction was larger in nearside impacts (25 percent) than in farside impacts (11 percent). Using 1999-2002 as the "baseline" years, the evaluation estimated that side door beams would eventually save 151 lives per year in single-vehicle side impacts of LTVs, when all LTVs on the road have the beams.[14]

This evaluation, too, did not show a statistically significant effect for side door beams on fatalities in multivehicle crashes. Point estimates by all analysis methods were close to zero.

Injury reduction – passenger cars NHTSA's 1982 evaluation used data from the National Crash Severity Study (NCSS), the predecessor of NASS. The injury rate – the number of fatal or hospitalizing injuries per 100 occupants involved in side-impact towaway crashes – was computed for cars with side door beams and compared to the rate in cars without the beams. These injury rates were controlled for belt use, rural/urban, speed limit, frame or unibody construction, or other factors by multidimensional contingency table analysis, a predecessor of the logistic regression method widely used today. Also, in some analyses, the data were limited to cars produced 2 (or 5) model years just before and after the make-model was equipped with the beams.[15]

---

[10] 90% confidence bounds: 7 to 21 percent.

[11] Other techniques included: limiting the data to the first year with beams and the last year without them; regression of the ratio of side to frontal fatalities by beam status, vehicle age and other factors.

[12] *Ibid.*, pp. 157 and 354.

[13] *Ibid.*, pp. 385-388; 90% confidence bounds: 300-660.

[14] Walz (2004); 95% confidence bounds for effectiveness: 4 to 32 percent; for lives saved per year: 22-285.

[15] Kahane (1982 FMVSS 214), pp. 183-266; Kahane, C.J., Smith, R.A. and Tharp, K.J., "The National Crash Severity Study," *Report on the Sixth International Technical Conference on Experimental Safety Vehicles*, NHTSA Report No. DOT HS 802 501, Washington, 1977, pp. 493-515.

In single-vehicle side impact crashes, the analyses showed a statistically significant 25 percent reduction of fatal or hospitalizing injuries for side door beams.[16]

At the nonfatal level, beams were also effective in some multivehicle crashes. When a vehicle was struck in the side by another vehicle and the damage was centered in the occupant compartment area, side door beams reduced nearside occupants' rate of hospitalizing injuries by a statistically significant 25 percent. But for farside occupants in all multivehicle crashes, and for nearside occupants when the damage was not centered on the compartment, the effect of side door beams was close to zero.[17]

Using 1980 as the "baseline" year, the evaluation estimated that side door beams would prevent 4,550 nonfatal hospitalizations per year in single-vehicle crashes and 4,920 in multivehicle crashes when all cars on the road are equipped with the beams. That adds up to 9,470 nonfatal hospitalizations prevented per year.[18]

Injury reduction – light trucks NHTSA has not analyzed the effect of side door beams on nonfatal injuries in LTVs.

Voluntary (pre-1994) TTI(d) improvements in 2-door cars

History of the FMVSS 214 upgrade By the late 1970's, if not earlier, researchers suspected that side door beams alone were insufficient to significantly attenuate intrusion in a severe side impact by another vehicle and reduce fatality risk to the nearside occupant of the struck car. At a public Side Impact Conference on January 31, 1980, NHTSA outlined its plans to upgrade FMVSS 214 with a dynamic test.[19]

Unlike the early FMVSS that largely incorporated other organizations' test procedures and existing safety technologies, the FMVSS 214 upgrade necessitated many years of research, analysis and testing by NHTSA and the safety community to develop comprehensive and self-sufficient science:[20]

- A review of crash data indicated that a typical side impact fatality involved a fast-moving vehicle striking a slow-moving car in the door, at a right angle – e.g., at an intersection.
- Injury data indicated that nearside occupants' life-threatening injuries often involved contact between the sides of their torsos and the interior side surface (most frequently the door) of the car.
- The Thoracic Trauma Index (TTI) was found to be an excellent predictor of thoracic injury severity in experimental impacts to cadavers. $TTI = \frac{1}{2} (G_R + G_{LS})$, where G_R is the

---

[16] Kahane (1982 FMVSS 214), pp. 12 and 25-33; 90% confidence bounds: 11 to 35 percent.

[17] *Ibid.*, pp. 10, 12, 25-33 and 186; 90% confidence bounds: 6 to 38 percent; damage to the side of a car is "centered on the compartment" if the centerpoint of the damage (CRASH program parameter D) is in a range from 15 inches behind to 45 inches ahead of the longitudinal centerpoint of the car.

[18] *Ibid.*, pp. 15, and 388-397; 90% confidence bounds: 4,300 to 14,700 hospitalizations prevented.

[19] *Side Impact Conference*, NHTSA Report No. DOT HS 805 614, Washington, 1980.

[20] Kahane, C.J., *Evaluation of FMVSS 214 - Side Impact Protection: Dynamic Performance Requirement; Phase 1: Correlation of TTI(d) with Fatality Risk in Actual Side Impact Collisions of Model Year 1981-1993 Passenger Cars*, NHTSA Technical Report No. DOT HS 809 004, Washington, 1999, pp. 1-4.

greater of the peak accelerations of either the upper or the lower rib, expressed in g's and G_{LS} is the lower spine (T12 vertebra) peak acceleration. Pelvic g's are an additional injury criterion.

- A Side Impact Dummy (SID) was developed that reliably registered TTI and pelvic g's in side impact tests. The injury score measured on the dummy is called TTI(d).
- A Moving Deformable Barrier (MDB) was developed to represent a generic 3000-pound passenger vehicle. The test procedure simulates an MDB moving 30 mph hitting, at a right angle, the door area of a subject vehicle, traveling 15 mph. (It is accomplished by having the MDB travel at 33.5 mph at an angle of 63 degrees with the longitudinal centerline of a stationary test vehicle. The wheels of the MDB are "crabbed" 27 degrees toward the rear of the test vehicle to obtain a right-angle contact.)
- The baseline distribution of TTI(d) was learned by testing of a variety of production MY 1980-88 passenger cars, and monitored by testing other cars up to MY 1993, just before the phase-in period for the new FMVSS 214.
- Two technologies were demonstrated that, singly or in combination could significantly reduce TTI(d) from its baseline levels in production vehicles:
  - *Structure* modifications such as substantially strengthening pillars, sills, roof rails, seats or cross-members of a car, and stronger overlap between doors and pillars, sills, etc., to slow down and reduce the extent of door intrusion into the passenger compartment.
  - *Padding* capable of absorbing significant energy at a force-deflection rate safe for occupants. It is thick plastic foam - not a soft pad.
- Regulatory analysis estimated the lives saved by reducing TTI(d) to various levels and the extent of vehicle modifications needed to secure those levels – and, finally –
- NHTSA promulgated the new FMVSS 214 in 1990, allowing TTI(d) up to 90 in 2-door cars and up to 85 in 4-door cars, with phase-in scheduled for MY 1994-97.[21]

Voluntary improvements before model year 1994 The 1980-93 development of FMVSS 214 was an iterative process with feedback between NHTSA and the manufacturers, and extensive public participation. Manufacturers could test TTI(d) in their existing vehicles and compare it to levels proposed for FMVSS 214 or achieved by competitors in tests by public organizations.

TTI(d) in 2-door cars averaged 110 for those MY 1981-90 cars that were tested and the results made public, and it ranged as high as 131. Manufacturers had voluntarily improved that to an average 97 by MY 1993, the year before the phase-in for FMVSS 214 – much closer to the 90 allowed by the future standard. The best score among the pre-standard 2-door cars was 82. There was relatively less improvement in 4-door cars: from 80 in 1981-90 to 74 in 1993; furthermore, even the earlier average is better than the 85 eventually permitted by the future standard. Two-door cars are intrinsically more vulnerable than 4-door cars in side impacts, because the door of a 2-door car is usually much longer than the front door of a 4-door car. Impacting vehicles are less likely to strike pillar(s), more likely to hit the long, weakly supported door area between pillars. In any case, manufacturers took steps to redesign some of the worst performers, especially among 2-door cars, before FMVSS 214 took effect. For example, one

---

[21] *Federal Register* 55 (October 30, 1990): 45752.

manufacturer ran a cross member across the A-pillars through the dash, reinforced the B-pillar at the sill level and added some floor stiffeners.[22]

Fatality reduction NHTSA's 1999 evaluation used a cross-sectional approach, rather than "before-after" analysis, to estimate the benefit of TTI(d) improvements. It compared the side-impact fatality risk of cars with good TTI(d) scores to cars with poor scores. NHTSA has TTI(d) scores, from the test configuration subsequently used in FMVSS 214, for 43 pre-standard (MY 1980-92) make-model-model year combinations, 17 2-door and 26 4-door. The evaluation gathered 1980-98 FARS cases of front-outboard occupant fatalities in each of these 43 make-model-MY combinations, including cars from earlier/later model years and corporate "twins" of essentially identical design. "Side-impact fatality risk" is the ratio of occupant fatalities in side impact crashes (where the principal impact is at a 2-4:00 or 8-10:00 location) to fatalities in purely frontal crashes (where the principal impact is at a 12:00 location and the most harmful event is not a subsequent rollover). Purely frontal crashes are an acceptable control group because cars of the 1980's did not substantially change their technologies for occupant protection in frontal crashes (cars with air bags or automatic belts were excluded from the analysis). The data file comprised 10,983 cases of occupant fatalities in side impacts and 12,019 in frontals.[23]

Separate analyses were performed for the 2-door cars and the 4-door cars. Analysis techniques included:

- Correlation of TTI(d) with side-impact fatality risk across the various make-model-MY groups.[24]
- Comparing the fatality risk of the cars with the best TTI(d) scores to the cars with the poorest scores.[25]
- Logistic regression of the probability that a fatality is in a side impact (and not in a frontal), as a function of TTI(d), curb weight, driver age and other variables.[26]

In the 2-door cars, all three analysis methods showed significantly lower fatality risk in the cars with lower TTI(d) scores. For example, cars with TTI(d) > 115 had more fatalities in side impacts than in pure frontals, whereas cars with TTI(d) < 102 had substantially fewer[27]:

---

[22] Kahane (1999), pp. vii, 6 and 19-23.
[23] *Ibid.*, pp. 11-18 and 24-31.

[24] *Ibid.*, pp. 31-45.
[25] *Ibid.*, pp. 47-71.
[26] *Ibid.*, pp. 73-103.
[27] *Ibid.*, p. 51.

2-DOOR CARS	Purely Frontal Fatalities	Side Impact Fatalities
Models with TTI > 115 (average = 123)	760	812
Models with TTI < 102 (average = 94)	654	528

This is a statistically significant 24 percent reduction of side impact fatalities with lower TTI(d): $1 - [(528/654) / (812/760)] = .244$. More precisely, a logistic regression estimated that fatality risk was reduced by a statistically significant 0.927 percent per unit reduction of TTI(d). Thus, the reduction from the 1981-90 "baseline average" TTI(d) of 110 to the "pre-standard best-practices" score of 82 corresponds to a 23 percent reduction in the risk of side-impact fatalities. Importantly, unlike the preceding results on side door beams, this reduction is not limited to single-vehicle crashes, but is about equally large in multivehicle and single-vehicle crashes. In fact, the highest effect, a 39 percent reduction, was observed for nearside occupants of 2-door cars that were hit in the side by another passenger car.[28]

Corresponding analyses of 4-door cars, however, did not show statistically significant relationships between TTI(d) and side-impact fatality risk. Unlike the 2-door cars, a large portion of the crash data involved cars with TTI(d) scores clustered in a relatively narrow range from 78 to 94, making it statistically more difficult to find correlations between TTI(d) and risk. Direct comparisons of the fatality risk in 4-door cars with the best and worst TTI(d) scores (analogous to the preceding table for 2-door cars) showed 5-15 percent fatality reductions for the cars with the better scores, but these reductions were not statistically significant.[29]

Side impact protection, dynamic test standard, passenger cars

On October 39, 1990, NHTSA issued the Final Rule amending FMVSS 214 to phase in a dynamic test of side impact protection. It required at least 10 percent of passenger cars produced between September 1, 1993 and August 31, 1994 to meet the standard; at least 25 percent of cars produced between September 1, 1994 and August 31, 1995; at least 40 percent of cars between September 1, 1995 and August 31, 1996; and all cars after September 1, 1996. Effective September 1, 1998, LTVs up to 6,000 pounds GVWR were also required to meet the dynamic test. The upgraded FMVSS 214 aims especially to reduce fatal thoracic injuries when a car is struck in the side by another vehicle. The preceding section describes the rationale, development and requirements of the dynamic test.[30]

Recognizing the intrinsically greater vulnerability of 2-door cars, FMVSS 214 allows TTI(d) up to 90 in 2-door cars, but limits 4-door cars and LTVs to 85. In compliance or research tests of actual cars, TTI(d) improved from an average of 97 in 2-door cars of MY 1993, the last year before the phase-in, to 74 in MY 1997, the first year that all cars had to meet FMVSS 214. In

---

[28] *Ibid.*, pp. viii, 84 and 91-100; $1 - (1 - .00927)^{110 - 82} = 23$ percent.

[29] *Ibid.*, pp. ix, 37-45, 63-71 and 85-87.

[30] *Federal Register* 55 (October 30, 1990): 45752, 60 (July 28, 1995): 38749; manufacturers also had the option of 100 percent of their cars meeting the standard beginning on September 1, 1994 and no requirement before that date.

other words, average performance improved from a score that would not have been permitted under FMVSS 214 to a score that was well within the limits of FMVSS 214. In 4-door cars, TTI(d) improved by a smaller amount, from an average of 74 in 1993 to 65 in 1997; both averages are within the limit of 85 permitted by the standard.[31]

As described above, structural modifications and padding are the principal technologies used to meet FMVSS 214. The manufacturers have provided NHTSA with detailed lists and diagrams showing what changes were made to achieve compliance during the phase-in period. This information suggests that make-models accounting for approximately 56 percent of new cars received substantial structural modifications, usually accompanied with padding, during MY 1994-97; 27 percent of the fleet received padding only, or padding with minor structural modifications; whereas 17 percent of cars remained essentially unchanged from previous years, implying that even the pre-1994 models of these cars could have met FMVSS 214. In addition, side air bags began to appear on some cars in MY 1996.[32]

NHTSA's 1999 evaluation report includes an analysis plan for FMVSS 214. The agency will estimate the change in side-impact fatality risk after FMVSS 214 vs. just before the standard: for all cars, by car type (2-door vs. 4-door), by type of vehicle modification (structure, padding, side air bags, combinations of these), and as a function of how much the test criterion TTI(d) was reduced when the standard was implemented in a make-model. This will be a high-priority evaluation for NHTSA.[33]

Side air bags

During the 1990's, manufacturers and suppliers developed side air bags that deploy from the seat or the door to provide an energy-absorbing cushion between the occupant's torso and the vehicle's side structure during lateral impacts. Volvo made them standard on all their MY 1996 cars, while Audi, BMW and Cadillac began to offer them in 1997. By MY 2001, 22 percent of new cars and LTVs were equipped with side air bags. NHTSA believes they are a promising safety technology, as evidenced by substantial reductions in TTI(d). NHTSA's annual *Buying a Safer Car* brochures inform the public what make-models are equipped with side air bags. This, too, will be a high-priority evaluation for NHTSA.[34]

---

[31] Kahane (1999), pp. vii, 19-23 and 139; the peak lateral acceleration of the pelvis shall not exceed 130 g's in any vehicle.

[32] *Ibid.*, pp. vii and 139-143.

[33] *Ibid.*, pp. 143-155; *National Highway Traffic Safety Administration Evaluation Program Plan, Calendar Years 2004-2007*, NHTSA Report No. DOT HS 809 699, Washington, 2004, p. 5.

[34] *Ibid.*, p. 223 shows TTI(d) reduced from 72 to 52 in one make-model; *Buying a Safer Car 2000*, NHTSA Publication No. DOT HS 809 046, Washington, 2000; *Buying a Safer Car 2001*, NHTSA Publication No. DOT HS 809 152, Washington, 2000; *Buying a Safer Car 2002*, NHTSA Publication No. DOT HS 809 409, Washington, 2002; *Buying a Safer Car 2003*, NHTSA Publication No. DOT HS 809 546, Washington, 2003; *NHTSA Evaluation Program Plan, CY 2004-2007*, p. 13.

FMVSS 216: Roof crush resistance

This standard is associated in the public mind with one vehicle modification whose safety benefits have been evaluated by NHTSA:

- *Redesign of true hardtops as pillared hardtops or sedans*

In fact, FMVSS 216 is a performance standard limiting the amount of crush allowed when a static load is applied to the roof of a vehicle. The first paragraph of FMVSS 216 says, "The purpose of this standard is to reduce deaths and injuries due to the crushing of the roof into the occupant compartment in rollover crashes."[1] FMVSS 216 took effect for passenger cars on September 1, 1973. It was extended to most pickup trucks, vans and SUVs up to 6,000 pounds GVWR, effective September 1, 1993.[2]

Most passenger cars built since September 1, 1973 have easily complied with FMVSS 216, and it is also believed most cars built before that date could have met the standard. It was primarily full-sized true hardtops of the late 1960's and early 1970's that had typically borderline or worse performance This body style was phased out – redesigned as a pillared hardtop or sedan – a few years before or after FMVSS 216 took effect. Although some true hardtops were built after 1973 and did meet FMVSS 216, they were pretty much gone by the late 1970's. NHTSA evaluated the safety benefits of the redesign, regardless of FMVSS 216's role in motivating it.

As for light trucks, NHTSA believes they met FMVSS 216 well before 1993, and that there were no substantial changes in roof design around that time. For that reason, NHTSA does not plan to evaluate FMVSS 216 for light trucks.

Regulatory history In October 1967, almost six years before the eventual effective date of FMVSS 216, NHTSA issued an ANPRM broaching possible limits on the intrusion of a vehicle's roof, front, side and rear structures into the passenger compartment during crashes. The roof intrusion portion of the ANPRM was a starting point for FMVSS 216. The industry developed a procedure for measuring static roof crush resistance, SAE Recommended Practice J374, dated December 1968, without defining specific pass-fail levels. Soon after, the new generation of hardtops with more vulnerable roof structures spurred NHTSA to issue FMVSS 216, with a Notice of Proposed Rulemaking in January 1971, a Final Rule in December 1971 and an effective date of September 1, 1973. The FMVSS largely incorporates the J374 procedure and sets a 5-inch limit on crush given a load of 1½ times the curb weight, or 5,000 pounds, whatever is less, applied to one of the sides of the roof, at the forward edge.[3]

A rather simple static test was preferred to a staged rollover crash because there was no repeatable, standardized rollover test that would have worked for all make-models. However,

---

[1] *Code of Federal Regulations*, Title 49, Government Printing Office, Washington, 2002, Part 571.216.
[2] *Federal Register* 36 (December 8, 1971): 23299, 56 (April 17, 1991): 15510.
[3] Kahane, C.J., *An Evaluation of Door Locks and Roof Crush Resistance of Passenger Cars – Federal Motor Vehicle Safety Standards 206 and 216*, NHTSA Technical Report No. DOT HS 807 489, Washington, 1989, pp. 4-8; *Federal Register* 32 (October 13, 1967): 14278, 36 (January 6, 1971): 166, 36 (December 8, 1971): 23299; *1973 SAE Handbook*, Society of Automotive Engineers, New York, 1973, p. 1172.

the application of the load in the static test to the front and side of the roof resembles many rollover crashes. A car is required to support 1½ times its weight, rather than just its own weight, because rollovers involve an additional dynamic load when the car flips onto its top. However, cars weighing over 3,333 pounds need only support a 5,000 pound load, less than 1½ times their weight, and are essentially held to a less strict FMVSS 216 requirement, since they are much less rollover-prone than light cars.

How it works One obvious danger in rollover crashes is that when a vehicle flips or bounces onto its roof, that impact could push the roof into the compartment far enough to contact and harm an occupant. However, it is uncertain what proportion of the fatalities in rollover crashes involved occupant contact with the roof and could have been avoided if the roof had not been crushed. For the unrestrained occupant, injuries from being ejected or tossed against interior surfaces are far more common, especially if the rollover is just one event in a sequence of impacts. In fact, some of the early research doubted that roof crush strength had any practical correlation with injury risk. Nevertheless, at least intuitively, a weak enough roof in a severe enough rollover is likely to collapse and harm occupants. In more recent years, when higher belt use has reduced the risk of ejection in rollovers, roof crush surely has become more important.[4]

The principal structures that resist roof crush are the roof itself, its side, front and back rails, and the two, three or four pillars on each side that hold it up. The A-pillars are on both sides of the windshield. In a sedan, the B-pillars are prominently visible between the front and back side-windows; in a true hardtop, the B-pillar extends only from the floor to where the windows begin, and does not support the roof; in a "pillared hardtop," the B-pillar extends fully to the roof, somewhat hidden in the metal between the front side-window and the "opera" window. The C-pillars are behind the rear side-windows. Station wagons also have D-pillars on both sides of the tailgate.

Four design factors increase roof crush resistance. One, obviously, is thicker or stronger materials in the roof, rails or pillars. A second is to have as many pillars as possible, specifically a full B-pillar. A third is to build a smaller car, reducing the unsupported surface area of the roof and the distance between pillars. A fourth is to arch the roof and angle the pillars inward, an intrinsically better weight-bearing design. The last three factors are the exact opposite of the popular "look" in the late 1960's and early 1970's, featuring the largest cars ever, true hardtops with continuous glass along the side, and a "lower, longer, wider" profile with a wide, flat roof. It is not surprising that large hardtops of the 1968-75 era were, on the average, the worst performers on roof crush tests. The manufacturers phased out true hardtops, not all at once but year-by-year as they restyled their various model lines, either by redesigning them as pillared hardtops (e.g., Chevrolet Monte Carlo in 1973, Ford Mustang in 1974) or by completely dropping the hardtop option from their catalogs (e.g., Dodge Aspen in 1976, GM full-sized cars in 1977). By the later 1970's, downsizing was in and the "lower, longer, wider" look was passé.

---

[4] Kahane (1989 Door Locks/Roof Crush Resistance), pp. 32-36; Hight, P.V., Siegel, A.W. and Nahum, A.M., "Injury Mechanisms in Rollover Collisions," *Proceedings of Sixteenth Stapp Car Crash Conference*, Society of Automotive Engineers, New York, 1972; Huelke, D.F., Marsh, J.C. IV and Sherman, H.W., "Analysis of Rollover Accident Factors and Injury Causation," *Proceedings of the Sixteenth Conference of the American Association for Automotive Medicine*, American Association for Automotive Medicine, Morton Grove, IL, 1972; Huelke, D.F. and Compton, C., "Injury Frequency and Severity in Rollover Car Crashes as Related to Occupant Ejection, Contacts and Roof Damage," *Accident Analysis and Prevention*, Vol. 15, October 1983.

Expected benefits Improved roof crush resistance ought to reduce fatality risk in rollover crashes for occupants who were not ejected from the vehicle. It could benefit occupants at any seat position.

Evaluation findings NHTSA's study, published in 1989, includes roof crush testing by the FMVSS 216 procedure on cars built before and after the standard took effect; statistical analyses of average roof-crush depth in actual crashes; and statistical analyses of fatality risk for non-ejected occupants in rollover crashes. The analyses compare true hardtops to cars with full B-pillars, taking into account vehicle size and before/after FMVSS 216.

Roof-crush tests Compliance test results for 108 cars of model years 1974-85 were supplemented by running the FMVSS 216 test procedure on 14 pre-standard cars of model years 1964-72, selected from make-models that matched 1974-75 compliance tests. Most cars passed easily: 80 percent of them had less than half the crush allowed by FMVSS 216. Larger cars had significantly greater roof crush than smaller cars. Full-sized and intermediate hardtops had significantly more crush than sedans of those sizes. The worst performers were full-sized hardtops of the 1968-75 era: 10 were tested, only 2 had less than half the crush allowed by FMVSS 216, and they accounted for 4 of the 6 worst scores among all the cars.[5]

Roof-crush depth in rollover crashes The 1982-86 NASS, the 1977-79 National Crash Severity Study[6] (NCSS), and 1968-78 Multidisciplinary Accident Investigations (MDAI) were combined with appropriate case weighting. They generated a file of roof-damage rollover crashes with information on roof crush (as measured by the extent zone in the Collision Deformation Classification[7]) and the make-model and body style of each case vehicle. After adjusting for car size, the average extent of roof crush in cars with full B-pillars (sedans, coupes, pillared hardtops, hatchbacks, station wagons) changed very little between model years 1964 and 1982. Up until 1967, true hardtops had about the same roof crush as cars with full B-pillars. However, in 7 of the 8 model years between 1968 and 1975, the hardtops had more average crush than pillared cars of the same size – a statistically significant difference. After 1975, there are only a few true hardtops in the data.[8]

Fatality reduction NHTSA evaluated the risk of a non-ejection fatality in a rollover crash, based on logistic regression using 1975-86 FARS data. The ratio of non-ejection rollover fatalities to non-ejection frontal-fixed-object fatalities in model-year 1966-81 cars was calibrated by B-pillar status (true hardtop, full B-pillar), FMVSS 216 status (pre-1971, 1971-74 transition, post-1974), track width and curb weight. This ratio was a statistically significant 15 percent higher in true hardtops than in pillared cars of the same size, before 1976. In other words, phasing out the true hardtops has saved lives. The analysis showed little or no effect for FMVSS 216 in cars with full B-pillars – i.e., the safety benefit derived from phasing out true hardtops, not from any changes within the pillared cars. After controlling for car size, the average non-ejection rollover fatality

---

[5] Kahane (1989 Door Locks/Roof Crush Resistance), pp. 39-69.

[6] Kahane, C.J., Smith, R.A. and Tharp, K.J., "The National Crash Severity Study," *Report on the Sixth International Technical Conference on Experimental Safety Vehicles*, NHTSA Report No. DOT HS 802 501, Washington, 1977, pp. 493-515.

[7] *1985 SAE Handbook*, Vol. 4, Society of Automotive Engineers, Warrendale, PA, 1985, Recommended Practice No. J224 MAR 80.

[8] Kahane (1989 Door Locks/Roof Crush Resistance), pp. 71-86.

risk in cars of the late 1970's, nearly all equipped with full B-pillars, was 7.4 percent lower than in cars of the early 1970's, a near 50-50 mix of hardtops and sedans.[9] Using 1982 as the "baseline" year, the evaluation estimated the phasing out of true hardtops saved approximately 110 lives per year.[10]

---

[9] *Ibid.*, pp. 159-167.
[10] *Ibid.*, pp. 214-217.

FMVSS 301: Fuel system integrity

FMVSS 301 is a performance standard that limits the amount of fuel allowed to spill from a vehicle after a crash test impact from any one of several directions, or during a static rollover test. The standard was substantially upgraded in 1975-77; that upgrade was accompanied by substantive vehicle modifications and it was evaluated by NHTSA:

- *1975-77 upgrade: rollover, rear-impact and lateral-impact tests*

The first paragraph of FMVSS 301 says, "The purpose of this standard is to reduce deaths and injuries occurring from fires that result from fuel spillage during and after motor vehicle crashes."[1] Approximately 29,000 cars and LTVs per year catch fire after being involved in a crash.[2] The 2001 FARS indicates that 1,447 occupants of these 29,000 cars and LTVs were fatally injured (not necessarily from burns, but possibly due to interior impacts, ejection, or other reasons).

Regulatory history FMVSS 301 is one of NHTSA's initial safety standards, with an effective date of January 1, 1968 for passenger cars. Originally, cars only had to pass a frontal impact test, into a rigid barrier at 30 mph. Fuel spillage after the impact is not allowed to exceed 1 ounce while the car is still in motion and 5 ounces during the first 5 minutes after the car comes to a stop. During the next 25 minutes, fuel spillage may not exceed 1 ounce during any 1-minute interval.[3]

Cost analyses by the agency did not show any substantive changes from model year 1967 to 1968 needed to meet the original FMVSS 301.[4] Statistical analyses of post-crash fire rates did not show a significant difference between pre-1968 and MY 1968+ cars.[5] NHTSA believes that cars sold in the United States generally would have been capable of meeting the FMVSS 301 frontal impact test for some years before the regulation was proposed or issued.[6]

During the 1970's, FMVSS 301 was significantly upgraded over a three-year phase-in period. Effective September 1, 1975, passenger cars had to meet a static rollover test. Immediately after the frontal test, the damaged vehicle is slowly rotated 90 degrees, 180 degrees (upside-down) and 270 degrees, holding at each of these positions for 5 minutes. Fuel spillage may not exceed 1 ounce during any 1-minute interval in this process.

---

[1] *Code of Federal Regulations*, Title 49, Government Printing Office, Washington, 2002, Part 571.301.

[2] Parsons, G.G., *Motor Vehicle Fires in Traffic Crashes and the Effects of the Fuel System Integrity Standard*, NHTSA Technical Report No. DOT HS 807 675, Washington, 1990, pp. xvii and 3-25 – 3-28.

[3] *Ibid.*; *Federal Register* 32 (February 3, 1967): 2414.

[4] McLean, R.F., Eckel, C. and Cowan, D., *Cost Evaluation for Four Federal Motor Vehicle Standards, Volume I*, NHTSA Report No. DOT HS 803 871, Washington, 1978, pp. 63-77.

[5] Reinfurt, D.W., *A Statistical Evaluation of the Effectiveness of FMVSS 301: Fuel System Integrity*, NHTSA Report No. DOT HS 805 969, Washington, 1981; Flora, J.D., Beitler, P., Bromberg, J., Goldstein, N., and O'Day, J., *An Evaluation of FMVSS 301 – Fuel System Integrity*, Report No. UM-HSRI-79-42, UMTRI, University of Michigan, Ann Arbor, 1979.

[6] Parsons, G.G., *Evaluation of Federal Motor Vehicle Standard 301-75, Fuel System Integrity: Passenger Cars*, NHTSA Technical Report No. DOT HS 806 335, Washington, 1983, p. 3.

Effective September 1, 1976, cars had to meet 30 mph frontal, oblique-frontal and rear-impact tests, plus a 20 mph lateral test, each followed by a static rollover test, with the same limits on fuel spillage as in the original frontal test. The rear and lateral impacts involve a 4,000-pound moving barrier striking a stationary vehicle. Rear and lateral impacts are, in a sense, more stringent tests than the frontal impact, because they come closer to the location of the fuel tank in most vehicles. LTVs with GVWR less than or equal to 6,000 pounds had to meet 30 mph frontal and rear-impact tests followed by static rollover. LTVs with GVWR 6,000-10,000 pounds (including small LTV-based school buses) had to meet the frontal test without static rollover.

Effective September 1, 1977, all LTVs up to 10,000 pounds (including small school buses) had to meet the same requirements as passenger cars: frontal, oblique-frontal, rear and lateral tests with subsequent rollover. (Also, on April 1, 1977, the requirements were extended to large school buses with GVWR over 10,000 pounds, but excluding the subsequent rollover tests.)[7]

Vehicle modifications in the mid-1970's The type and extent of modifications near the time of the 1975-77 upgrade varied greatly by make-model. Strategies used by the manufacturers include:

- Strengthening the fuel tank or other components of the fuel delivery system.
- Strengthening the structures that hold the fuel tank in place.
- Shielding the fuel tank and delivery system from other parts of the vehicle.
- Relocating parts of the fuel system further away from other parts of the vehicle or areas likely to be damaged during impacts.
- Relocating other parts of the vehicle further away from the fuel system, or reshaping them to make them less likely to damage the fuel system.[8]

Expected benefits Improved fuel system integrity ought to reduce the number of post-crash fires that are primarily sustained by fuel spillage. The measures used to improve fuel system integrity might not be effective in certain crashes that are far more severe than the levels tested in FMVSS 301. Not all fatalities in vehicles with post-crash fires are due to the fire; occupants primarily injured by interior components, or due to ejection, could have had the same injury severity even without the fire.

Post-crash fire reduction – passenger cars NHTSA's evaluation, published in 1990, is based on 1982-87 State crash data from Maryland, Michigan and Ohio. Crash-involved cars are grouped by model year and calendar year. The proportion of crashes accompanied by post-crash fires is analyzed by a linear regression, with FMVSS 301 (MY 1976+ = post-standard) and vehicle age as the dependent variables. (Fire rates increased strongly as cars aged, but the rates were uncorrelated with curb weight.) The analysis indicated that post-standard cars had 14 percent fewer fires than pre-standard cars, after controlling for vehicle age.[9] The reduction is statistically

---

[7] Parsons (1990), p. 1-3; *Federal Register* 39 (March 21, 1974): 10588, 39 (November 21, 1974): 40857, 40 (August 6, 1975): 33036, 40 (October 10, 1975): 47790.

[8] *Ibid.*, pp. 4-11 – 4-22.

[9] 95% confidence bounds: 7 to 21 percent

significant. Using 1982-87 as the "baseline" years, the evaluation estimated that there would be 3,900 fewer fires annually when the entire car fleet meets FMVSS 301.[10]

Post-crash fire reduction – light trucks A similar analysis was performed on the proportion of LTV crash involvements with fires (except that here MY 1977+ = post-standard). However, this analysis did not show a statistically significant reduction of post-crash fires with FMVSS 301.[11]

Fatality reduction – passenger cars A similar analysis was also performed on 1975-88 FARS data: a regression on the proportion of fatal crash involvements followed by fires. After controlling for vehicle age, the observed effect of FMVSS 301 was close to zero and not statistically significant. NHTSA was unable to conclude that FMVSS 301 has had any effect on fatal fires. Another analysis, based on direct comparison of the fire rates N model years before vs. after FMVSS 301, also showed little or no effect for FMVSS 301 after controlling for vehicle age.[12] "No reduction in fire-related fatalities was found; the force levels encountered in fatal fire crashes may generally exceed the levels set by the Standard."[13] That suggests the 14 percent overall reduction of fires, discussed above, happened primarily in the less severe, nonfatal crashes.

The probability of fire increases with vehicle age. During 1975-88 people began to hold onto their vehicles longer and scrap them later. As the median age of vehicles increased, the number of fatal crash involvements followed by fires grew from 1,300 in 1975 to 1,800 in 1988.[14]

Fatality reduction – light trucks Similar analyses were performed on the proportion of fatal LTV crash involvements with fires (except that here MY 1977+ = post-standard). They did not show a statistically significant effect for FMVSS 301.[15]

Future FMVSS 301 upgrade On April 12, 1995 NHTSA issued an ANPRM to consider upgrading FMVSS 301, possibly including changes in the system performance tests, new tests of individual components, and criteria to address environmental and vehicle-aging effects on the integrity of the fuel system.[16] After reviewing the research and the public comments to the ANPRM, NHTSA issued, on November 6, 2000, an NPRM limiting, for now, the upgrade to more stringent rear- and lateral-impact test procedures for vehicles up to 10,000 pounds GVWR. The NPRM proposes to replace the current rear-impact test (30 mph by a 4,000-pound rigid barrier and 100 percent overlap) with a 50 mph impact by a 3,000-pound deformable barrier at 70 percent overlap with the test vehicle. The lateral impact (currently 20 mph by a 4,000-pound rigid barrier) would be replaced by the FMVSS 214 test configuration: a 33.5 mph impact by a 3,000-pound deformable barrier. For most vehicles, the new tests would be substantially more stringent. NHTSA concluded these changes would prevent fire-related fatalities and injuries.[17] The Final Rule was issued on December 1, 2003. It phases in the new standard for 40 percent of

---

[10] *Ibid.*, pp. xix, 2-1 – 2-33, and 4-2.
[11] *Ibid.*, pp. xx and 2-63 – 2-82.
[12] *Ibid.*, pp. xx and 2-53 – 2-62.
[13] *Ibid.*, p. xxiv.
[14] *Ibid.*, p. xviii.
[15] *Ibid.*, pp. xx and 2-91 – 2-98.
[16] *Federal Register* 60 (April 12, 1995): 18566.
[17] *Federal Register* 65 (November 13, 2000): 67693.

cars and LTVs produced from September 1, 2006 to August 31, 2007; 70 percent of cars and LTVs produced from September 1, 2007 to August 31, 2008; and all cars and LTVs produced on or after September 1, 2008.[18]

---

[18] *Federal Register* 68 (December 1, 2003): 67068

<u>*NCAP: New Car Assessment Program*</u>

NHTSA's New Car Assessment Program (NCAP) is not a regulation or a standard, but a program of testing the crashworthiness and crash avoidance capabilities of new cars and LTVs and publishing the results for the public. NCAP began in 1978 with frontal crash testing of MY 1979 passenger cars. LTVs were first tested in MY 1983. NCAP expanded to side impact testing in MY 1997, and it began to provide information on the rollover resistance of vehicles in MY 2001. NCAP information, and any vehicle modifications in response to it, may be divided into four categories. So far, only the first has been evaluated by NHTSA:

- *Frontal NCAP-related improvements in cars without air bags*
- *Frontal NCAP in vehicles with air bags (not yet evaluated)*
- *Side NCAP (not yet evaluated)*
- *Rollover-resistance NCAP (not yet evaluated)*

NCAP's frontal impact test is conducted at 35 mph. That is 5 mph faster than the test speed in FMVSS 204, 208, 212, 219 and 301, the NHTSA regulations that set basic performance requirements in frontal impacts. Similarly, NCAP's side-impact test is 5 mph faster than the dynamic test in FMVSS 214. One purpose of NCAP, of course, is to enable consumers to compare the performance of different vehicles they might want to buy. In addition, NCAP specifies tests above and beyond the minimum requirements of the FMVSS. Manufacturers know what they are, and they have the option to design their vehicles for optimum performance in these tests. Even though NCAP has no "pass-fail" criteria, manufacturers have improved performance over the years by introducing, modifying or fine-tuning safety equipment. For example, voluntary introductions of technologies such as pretensioners and load-limiters for safety belts or side/head air bags (not required by any FMVSS as of October 2004) have improved performance on NCAP tests. When consumers and manufacturers respond to the information, the end result is a fleet of vehicles with better test performance.

Frontal NCAP-related improvements in cars without air bags

NCAP history and procedure   Title II of the Motor Vehicle Information and Cost Savings Act of 1973 authorized NHTSA to develop consumer information on the crashworthiness of passenger vehicles. The agency developed a 35 mph frontal impact test into a rigid barrier with belted dummies at the driver's and right-front seats. The test setup was essentially identical to the one used in NHTSA's basic frontal crashworthiness standards, except 5 mph faster. The same injury criteria – originally the Head Injury Criterion (HIC), chest g's and femur load – were measured on the dummies.[1]

In MY 1979, the first year of testing, fewer than 25 percent of the cars were able to meet the FMVSS 208 criteria (HIC < 1,000, chest g's < 60 and femur load < 2,500 pounds on each leg) in the 35 mph test. (Of course, the only requirement of FMVSS 208 at that time was for vehicles with automatic occupant protection to meet those criteria at 30 mph.) By MY 1986-91, over 60 percent of passenger cars met the FMVSS 208 criteria at 35 mph. Manufacturers had begun

---

[1] *The Motor Vehicle Information and Cost Savings Act of 1973*, Public Law 92-513, as amended, 15 *United States Code* 1912-2012.

"designing to 35 mph" a large proportion of their vehicles. Specifically, during the first four years of NCAP, 1979-82, HIC averaged 1,052 and chest g's 54.9. By MY 1983, modifications that improved NCAP performance had worked their way through the production cycle. In 1983-86, average HIC had dropped to 915 and chest g's even more so, to 46.8. In 1987-91 HIC continued to improve to 827, while chest g's remained about the same, at 46.5.[2]

Vehicle modifications, and their purpose NHTSA's films and other data from NCAP tests are public information. Manufacturers, if they chose, had the opportunity to study results of the initial NCAP tests of their MY 1979-81 cars. They could identify the contacts that were driving up HIC or chest g's on the dummies and they could test out a variety of remedies by running their own 35 mph impacts. Subsequent modifications were not limited to a single, specific technology, but included adjusting the belt system, the steering assembly, the instrument panel and the seat structure, taking into account how belted dummies interacted with those systems in 35 mph tests. For example, manufacturers:

- Reduced spool-out of the safety belts in crashes. Excessive spool-out allowed head impacts with frontal components. Strategies to reduce spool-out included adding a web-locking mechanism, relocating the D-ring, modifying the retractor or shortening the belts.
- Found ways to reduce upward movement of the steering assembly in crashes and keep it away from the driver's head.
- Reduced dashboard intrusion into the occupant compartment.
- Stiffened the front seat cushion.
- Realigned the seats relative to the steering assembly and instrument panel to provide additional ride-down space for occupants.[3]

Expected benefits Strategies to optimize the performance of belted dummies in 35 mph frontal barrier impacts are likely to have the largest benefits for belted occupants in frontal crashes, generally, and especially in frontal crashes that have velocity change somewhere near 35 mph (i.e., severe, but not extreme) and resemble an impact with a flat barrier (hitting a wide object or another vehicle, with plenty of overlap).

Fatality reduction – passenger cars In 1994, NHTSA published a study of the *Correlation of NCAP Performance with Fatality Risk in Actual Head-On Collisions*. It was based on 1979-92 FARS data, specifically head-on collisions of two MY 1979-91 passenger cars, each with a belted driver, resulting in a fatality to one or to both of the drivers. A head-on collision is a special type of highway crash ideally suited for studying frontal crashworthiness differences between two cars. Both cars are in essentially the same frontal collision. It doesn't matter if one of them had a "safe" driver and the other, an "unsafe" driver; at the moment they collide head-on,

---

[2] Kahane, C.J., *Correlation of NCAP Performance with Fatality Risk in Actual Head-On Collisions*, NHTSA Technical Report No. DOT HS 808 061, Washington, 1994, pp. 1-2.

[3] Hackney, J. and Quarles, V., "The New Car Assessment Program – Status and Effect," *Ninth International Technical Conference on Experimental Safety Vehicles*, NHTSA Report No. DOT HS 806 341, Washington, 1983, pp. 809-824; Hackney, J. and Ellison, C., "A Review of the Effects of Belts Systems, Steering Assemblies, and Structural Design on the Safety Performance of Vehicles in the New Car Assessment Program," *Tenth International Technical Conference on Experimental Safety Vehicles*, NHTSA Report No. DOT HS 806 916, Washington, 1983, pp. 380-413.

how safely they were driving before the crash is nearly irrelevant to what happens in the crash. Which driver dies and which survives depends primarily on the intrinsic relative crashworthiness of the two cars, their relative weights, and the age and gender of the two drivers (vulnerability to injury).

If car 1 and car 2 weigh exactly the same, and both drivers are the same age and gender, the likelihood of a driver fatality in a head-on collision would be expected to be equal in car 1 and car 2. When vehicle weight or driver age/gender are not the same, the expected fatality risk for each driver was calibrated from the FARS data by a logistic regression. The expected fatality risk for driver 1 was

$$\frac{\exp[.616 - 5.427(\log W_1 - \log W_2) + .0531(A_1 - A_2) + .34(F_1 - F_2)]}{1 + \exp[.616 - 5.427(\log W_1 - \log W_2) + .0531(A_1 - A_2) + .34(F_1 - F_2)]}$$

where W_1 is the curb weight of car 1, A_1 is the age of driver 1 and F_1 is 1 if driver 1 was female, 0 if male. The expected fatality risk for driver 2 was

$$\frac{\exp[.616 + 5.427(\log W_1 - \log W_2) - .0531(A_1 - A_2) - .34(F_1 - F_2)]}{1 + \exp[.616 + 5.427(\log W_1 - \log W_2) - .0531(A_1 - A_2) - .34(F_1 - F_2)]}$$

For example, given 100 fatal head-on collisions between 3000 pound cars driven by belted, 20-year-old males and 2500 pound cars driven by belted, 50-year-old females, the formulas predict 9 deaths among the young males in the heavier cars and 97 deaths among the older females in the lighter cars (for a total of 106 fatalities in the 100 collisions, since some of them resulted in fatalities to both drivers).

Cars with average crashworthiness will experience an actual number of fatalities close to what is predicted by the formulas. If a group of cars consistently experiences fewer driver fatalities than expected in their head-on collisions, then this group of cars is more crashworthy than the average car of similar mass. More generally, given a set of head-on collisions between one group of cars A and another group of cars B, it is possible to compare the crashworthiness of the two groups. For example, if the actual fatalities in the crashes and the sums of the expected probabilities of fatality (based on the formulas) are as follows:

Head-On Collisions between Groups A and B

	Car Group A	Car Group B
Actual fatalities	60	100
Expected fatalities	68	92

Then the fatality reduction for Group A relative to Group B is

$$1 - [(60/68) / (100/92)] = 18.8 \text{ percent}$$

Specifically, NHTSA's evaluation compares actual and expected fatalities when Group A consists of cars with good scores on one or more NCAP parameters and Group B consists of cars with poor scores. For example, in 125 head-on collisions between cars with chest g's ≤ 56 on the NCAP test (a group averaging 44.6 chest g's on the test) and cars with chest g's > 56 (averaging 63.4 chest g's), with both drivers belted, the actual number of driver fatalities relative to the expected was 18.7 percent lower in the cars with the good scores than in the cars with the poor scores. The evaluation also considered many other sets of head-on collisions between groups of make-models with good and poor NCAP scores, where "good" and "poor" were defined on the basis of a single injury criterion – HIC, chest g's or femur load – or on various combinations of those criteria. Consistently, the actual fatality risk (relative to expected – i.e., after controlling for vehicle weight, driver age and gender) was significantly lower in the make-models the better NCAP scores.[4]

As stated above, the average NCAP performance of new cars improved substantially between MY 1979 and MY 1991, with the largest improvement coming around 1983-86. Therefore, we would expect a corresponding reduction over time of fatality risk in head-on collisions. Indeed, NHTSA's evaluation showed that MY 1983-86 cars had a statistically significant 20 percent reduction of fatality risk, relative to MY 1979-82 cars, for belted drivers in car-to-car head-on collisions.[5]

NHTSA's evaluation is based entirely on analyses of belted drivers in head-on crashes between two cars. The data were limited almost entirely to cars without air bags. It is unknown if the 20 percent fatality reduction in these crashes would extend to other types of crashes. Quite probably, there would be a similar effect for belted drivers in frontal impacts with the side or rear of another passenger car (i.e., car-to-car, but not head-on). More tenuously, in frontal impacts with LTVs, heavy trucks or fixed objects, all of which are usually more rigid than cars. The results are least likely to extend to unbelted drivers, or any kind of non-frontal impacts. Also, these results for drivers of cars without air bags need not apply to passengers, or to drivers of cars with air bags.

Frontal NCAP in vehicles with air bags

The proportion of new vehicles equipped with air bags increased greatly from 1990 onwards. The relationship between NCAP scores and fatality risk in actual head-on collisions in vehicles with air bags may be different from the result found in NHTSA's 1994 evaluation of MY 1979-91 cars, mostly without air bags, described above. Other developments in frontal NCAP not considered in the 1994 evaluation include the extensive test results and crash data now available on LTVs, and the five-star rating system initiated in 1994 to make NCAP results more useful to the public. The agency is considering an updated evaluation of frontal NCAP, based on crash data involving head-on collisions of cars and LTVs equipped with driver or dual air bags, using the five-star rating system, among other criteria, to distinguish between "good" and "poor" NCAP performance.[6]

---

[4] Kahane (1994 NCAP), pp. 79-118.

[5] *Ibid.*, 1994, pp. xvi and 129-142.

[6] Hackney, J.R. and Kahane, C.J., *The New Car Assessment Program: Five Star Rating System and Vehicle Safety Performance Characteristics*, Technical Paper No. 950888, Society of Automotive Engineers, Warrendale, PA, 1995; *New Car Assessment Program: Response to the NCAP FY 1992 Congressinal Requirements*, Report to the Congress, NHTSA, Washington, 1993; *National Highway Traffic Safety Administration Evaluation Program Plan, Calendar Years 2004-2007*, NHTSA Report No. DOT HS 809 699, Washington, 2004, p. 26.

Side NCAP

After September 1, 1996, when all cars were required to meet a dynamic side impact test as part of FMVSS 214, NHTSA also expanded NCAP to provide consumer information on vehicle performance in side impacts. The agency uses a five-star rating system for front-outboard and rear-outboard occupants. In the NCAP tests, the moving deformable barrier (MDB) strikes the side of the target vehicle at 38.5 mph, 5 mph faster than in the FMVSS 214 test. As part of its ongoing evaluation of side impact protection, NHTSA plans to study the correlation between test results on Side NCAP and fatality risk in actual side-impact crashes.[7]

Rollover-resistance NCAP

In MY 2001, NHTSA expanded NCAP with ratings of one to five stars for rollover resistance, based on the Static Stability Factor. The SSF equals one half of the track width divided by the height of the center of gravity above the road. This measurement identified the location of the center of gravity of the vehicle and correlated this to the risk of a tripped rollover crash. Linear and logistic regressions were run in 2000, using MY 1994 to 1998 vehicles, to determine and verify the relationship between SSF and rollover rate. SSF has a remarkably strong relationship to the rate rollovers per 100 single vehicle crashes. NHTSA has planned an evaluation of the long-term trend of the SSF for various types of cars, pickup trucks, SUVs and vans. Starting with 2004 model year vehicles, NHTSA has issued rollover resistance star ratings that combine the SSF (based on measurements of stationary vehicles) and the tip or no-tip results of a new, dynamic maneuvering rollover test.[8]

---

[7] *NHTSA Evaluation Program Plan, Calendar Years 2004-2007*, p. 5.
[8] *Federal Register* 66 (January 12, 2001): 3388; *NHTSA Evaluation Program Plan, Calendar Years 2004-2007*, pp. 14-15.

SUMMARY TABLES FOR PART 1

Tables 1-2, 1-3 and 1-4 summarize the effectiveness estimates for safety technologies obtained in past evaluation reports, as discussed in the preceding chapters of Part 1. Table 1-2 summarizes the estimated percentage reductions in fatalities; Table 1-3, nonfatal injuries; and Table 1-4, crashes (always specifying to what group of crashes/injuries/fatalities these percentages apply – but for more detail, refer back to the preceding chapters). The effectiveness estimates in Table 1-2 are also used in this report's model to assess lives saved by the FMVSS in 1960-2002. When NHTSA has evaluated a safety technology more than once (e.g., Center High Mounted Stop Lamps), the latest estimates are the ones shown in Tables 1-2, 1-3 and 1-4.

Tables 1-3 and 1-4 also summarize the estimated annual benefits of nonfatal injury reduction and crash avoidance, as stated in NHTSA's evaluation reports. However, those estimates of benefits, in general, are not directly comparable and do not add up to the overall crash avoidance and injury reduction by all the FMVSS in, say, 2002. NHTSA does not have enough "building blocks" to develop models for overall crash avoidance and injury reduction comparable to the analysis of fatal crashes in Part 2 of this report.

TABLE 1-2: ESTIMATES OF FATALITY REDUCTION
NHTSA EVALUATIONS OF SAFETY TECHNOLOGIES

FMVSS

    SAFETY TECHNOLOGY

Vehicle Type	Fatality-Reducing Effectiveness Estimate(s)
105: Hydraulic brake systems	
Dual master cylinders	
Cars and LTVs	0.7 % reduction of all fatal crash involvements
Front disc brakes	
Cars and LTVs	0.17 % reduction of all fatal crash involvements
Rear-wheel ABS	
LTVs	No significant overall effect
4-wheel ABS	
Cars and LTVs	No significant overall effect
	24 % reduction of multivehicle involvements on wet roads
	27 % reduction of collisions with non-occupants (in 1989-93 only?)
	28 % increase in run-off-road crashes (in 1989-93 only?)
Electronic stability control	
Cars	30 % reduction of single-vehicle run-off-road crashes (preliminary estimate)
LTVs	63 % reduction of single-vehicle run-off-road crashes (preliminary estimate)
108: Lamps, reflective devices	
Side marker lamps	
Cars and LTVs	No significant effect
Center High Mounted Stop Lamps	
Cars and LTVs	No significant effect
Retroreflective tape	
Heavy trailers	29 % reduction of cars and LTVs hitting the side or rear of a heavy trailer in the dark

TABLE 1-2 (Continued): ESTIMATES OF FATALITY REDUCTION
NHTSA EVALUATIONS OF SAFETY TECHNOLOGIES

FMVSS

SAFETY TECHNOLOGY

Vehicle Type	Fatality-Reducing Effectiveness Estimate(s)

201: Occupant protection in interior impact

Voluntary mid and lower instrument panel improvements

Cars and LTVs	15.9 % in frontal crashes, for unrestrained RF passengers

202: Head restraints

Head restraints for front seat

Cars and LTVs	No significant effect

203: Impact protection from the steering control; 204: Steering control rearward displacement

Energy-absorbing steering assemblies

Cars and LTVs	12.1 % in frontal crashes, for drivers

205: Glazing materials

HPR windshields

Cars and LTVs	No significant effect

Glass-plastic windshields

Cars	No significant effect

206: Door locks and retention components

Improved locks for side doors

Cars	15.38 % reduction of ejection fatalities in rollovers (excluding occupants riding outside the passenger compartment)
LTVs	9.8 % reduction of ejection fatalities in rollovers (excluding occupants riding outside the passenger compartment)

207: Seating systems

Seatback locks

2-door cars	No significant effect

FMVSS

SAFETY TECHNOLOGY

Vehicle Type	Fatality-Reducing Effectiveness Estimate(s)

208: Occupant crash protection; 209: Seat belt assemblies; 210: Seat belt assembly anchorages

Lap belts for front seat (age 5+)

Cars	27 % overall fatality reduction
LTVs	48 % overall fatality reduction

Lap belts for back seat (age 5+)

Cars	32 % overall fatality reduction
LTVs	63 % overall fatality reduction

Lap belts for child passengers age 1-4 (front or back seat)

Cars	33 % overall fatality reduction
LTVs	48 % overall fatality reduction

3-point belts for front seat (age 5+)

Cars	45 % overall fatality reduction
LTVs	60 % overall fatality reduction

3-point belts for back seat (age 5+)

Cars	44 % overall fatality reduction
LTVs	73 % overall fatality reduction

Automatic 2-point belts (age 5+)

Cars	32 % overall fatality reduction

Frontal air bags

Cars and LTVs	10.8 % overall fatality reduction for belted drivers
	14 % overall fatality reduction for unbelted drivers
	11.9 % overall fatality red. for belted RF age 13+
	15.4 % overall fatality red. for unbelted RF age 13+
	22 % fatality increase for child passengers age 0-12 (varies considerably depending on age/restraint use)

TABLE 1-2 (Continued): ESTIMATES OF FATALITY REDUCTION
NHTSA EVALUATIONS OF SAFETY TECHNOLOGIES

FMVSS

SAFETY TECHNOLOGY

Vehicle Type   Fatality-Reducing Effectiveness Estimate(s)

208: Occupant crash protection (continued)

On-off switches for air bags

Pickup trucks   86 % of switches were turned off for infants in a rear-facing child seat in the RF seat of the truck

74 % of switches turned off for children age 1-6

83 % of switches turned on for adults (age 13+)

212: Windshield mounting

Adhesive windshield bonding

Cars and LTVs   15 % fatality reduction for people ejected through the windshield portal

213: Child restraint systems; 225: Child restraint anchorage systems

Child safety seats

Cars   71 % overall fatality reduction for infants (age 0)

54 % overall fatality reduction for toddlers (age 1-4)

LTVs   58 % overall fatality reduction for infants (age 0)

59 % overall fatality reduction for toddlers (age 1-4)

Riding in the back seat, children age 0-12

Cars w/o air bags   33 % reduction for restrained in the rear-outboard seat relative to restrained in the right-front seat

31 % reduction for unrestrained in the rear-outboard seat relative to unrestrained in the right-front seat

46 % reduction for restrained in the rear-center seat relative to restrained in the right-front seat

47 % reduction for unrestrained in the rear-center seat relative to unrestrained in the right-front seat

TABLE 1-2 (Concluded): ESTIMATES OF FATALITY REDUCTION
NHTSA EVALUATIONS OF SAFETY TECHNOLOGIES

FMVSS

    SAFETY TECHNOLOGY

Vehicle Type	Fatality-Reducing Effectiveness Estimate(s)

214: Side impact protection

    Side door beams

Cars	14 % fatality reduction for occupants of outboard seats in single-vehicle side impacts
LTVs	19 % fatality reduction for occupants of outboard seats in single-vehicle side impacts

    Voluntary (pre-1994) TTI(d) improvements

2-door cars	23 % fatality reduction for front-outboard occupants in all side impacts

216: Roof crush resistance

    Eliminate true hardtops

Cars	7.4 % reduction of non-ejection fatalities in rollover crashes (excluding occupants riding outside the passenger compartment)

301: Fuel system integrity

    1975-77 upgrade: rollover, rear- and lateral-impact tests

Cars and LTVs	No significant effect

NCAP: New Car Assessment Program

    Frontal NCAP-related improvements

Cars w/o air bags	20 % fatality reduction for MY 1986 and later cars, relative to MY 1979 cars, for belted drivers in frontal collisions with other passenger cars

TABLE 1-3: ESTIMATES OF INJURY REDUCTION
NHTSA EVALUATIONS OF SAFETY TECHNOLOGIES

FMVSS

SAFETY TECHNOLOGY	Injury-Reducing Effectiveness Estimate(s)
Vehicle Type	Injuries Prevented Per Year When All Vehicles on the Road Have the Technology

105: Hydraulic brake systems

Dual master cylinders

| Cars and LTVs | 0.7 % reduction of all injury crash involvements |
| | 24,000 injuries of any severity prevented per year (computation used 1979-80 as the baseline year) |

Front disc brakes

| Cars and LTVs | 0.17 % reduction of all injury crash involvements |
| | 5,700 injuries of any severity prevented per year (computation used 1979-80 as the baseline year) |

Rear-wheel ABS

LTVs	30 % reduction of injuries in nonfatal rollovers
	20 % reduction of injuries in nonfatal side impacts with fixed objects
	No significant effect in multivehicle crashes

4-wheel ABS

| Cars | 28 % reduction of injuries in multivehicle involvements on wet roads |
| | 19 % increase of injuries in run-off-road crashes (in 1989-93 only?) |

FMVSS

SAFETY TECHNOLOGY	Injury-Reducing Effectiveness Estimate(s)
Vehicle Type	Injuries Prevented Per Year When All Vehicles on the Road Have the Technology

108: Lamps, reflective devices

Side marker lamps

Cars and LTVs	21 % reduction of injury-producing nighttime angle collisions when both vehicles in a crash have the lamps
	93,000 injuries of any severity prevented per year (computation used 1979-80 as the baseline year)

Center High Mounted Stop Lamps

Cars and LTVs	4.3 % reduction of injuries to people involved in front-to-rear collisions
	58,000-70,000 injuries of any severity prevented per year (computation used 1994 as the baseline year)

Retroreflective tape

Heavy trailers	29 % reduction of cars and LTVs hitting the side or rear of a heavy trailer in the dark
	3,100-5,000 injuries of any severity prevented per year (computation used 2001 as the baseline year)

201: Occupant protection in interior impact

Voluntary mid and lower instrument panel improvements

Cars	29 % reduction of AIS $\geq$ 2 injuries from instrument panel contact in frontal crashes, for RF passengers
	8,000 AIS $\geq$ 2 injuries prevented per year (computation used 1982 as the baseline year)

FMVSS

SAFETY TECHNOLOGY	Injury-Reducing Effectiveness Estimate(s)
Vehicle Type	Injuries Prevented Per Year When All Vehicles on the Road Have the Technology

202: Head restraints

Head restraints for front seat

Cars	16.9 % injury reduction in rear-impact crashes with fixed or integral head restraints
	10.4 % injury reduction in rear-impact crashes with adjustable head restraints
	85,000 injuries of any severity prevented per year with a fleet of fixed or integral head restraints (computation used 1979 as the baseline year)
	52,000 injuries of any severity prevented per year with a fleet of adjustable head restraints (computation used 1979 as the baseline year)
LTVs	6.1 % injury reduction in rear-impact crashes with any type of head restraint
	14,900 injuries of any severity prevented per year (computation used 1999 as the baseline year)

203: Impact protection from the steering control; 204: Steering control rearward displacement

Energy-absorbing steering assemblies

Cars	38.4 % reduction of fatal or hospitalizing injuries due to contact with the steering assembly in frontal crashes
	23,000 nonfatal hospitalizations prevented per year (computation used 1978 as the baseline year)

FMVSS

SAFETY TECHNOLOGY	Injury-Reducing Effectiveness Estimate(s)
Vehicle Type	Injuries Prevented Per Year When All Vehicles on the Road Have the Technology

205: Glazing materials

HPR windshields

Cars

74 % reduction of AIS $\geq$ 2 lacerations due to windshield contact

56 % reduction of AIS 2-4 facial fractures due to windshield contact

72 % reduction of AIS 2-4 injuries of the eyes, nose and mouth due to windshield contact

25 % reduction of AIS 1 lacerations due to windshield contact

39,000 AIS $\geq$ 2 lacerations prevented per year (computation used 1982 as the baseline year)

8,000 AIS 2-4 facial fractures

The preceding 47,000 AIS 2-4 injuries include 19,000 injuries to the eyes, nose or mouth

Glass-plastic windshields

Cars — No significant effect

207: Seating systems

Seatback locks

2-door cars — No significant effect

FMVSS

SAFETY TECHNOLOGY	Injury-Reducing Effectiveness Estimate(s)
Vehicle Type	Injuries Prevented Per Year When All Vehicles on the Road Have the Technology

208: Occupant crash protection; 209: Seat belt assemblies; 210: Seat belt assembly anchorages

Lap belts for front seat

    Cars   33 % overall AIS ≥ 3 injury reduction

                 22 % overall AIS ≥ 2 injury reduction

                 8 % overall AIS ≥ 1 injury reduction

Lap belts for back seat

    Cars   37 % overall fatal-and-serious (K + A) injury reduction

                 33 % overall moderate-to-fatal (K + A + B) injury reduction

                 11 % overall (K + A + B + C) injury reduction

3-point belts for front seat

    Cars   45 % overall AIS ≥ 3 injury reduction

                 46 % overall AIS ≥ 2 injury reduction

                 13 % overall AIS ≥ 1 injury reduction

Frontal air bags

    Cars and LTVs   12 % overall AIS ≥ 3 injury reduction for belted occupants age 13+

                 4 % overall AIS ≥ 2 injury reduction for belted occupants age 13+

                 21 % overall AIS ≥ 3 injury reduction for unbelted occupants age 13+

                 21 % overall AIS ≥ 2 injury reduction for unbelted occupants age 13+

TABLE 1-3 (Concluded): ESTIMATES OF INJURY REDUCTION
NHTSA EVALUATIONS OF SAFETY TECHNOLOGIES

FMVSS

    SAFETY TECHNOLOGY   Injury-Reducing Effectiveness Estimate(s)

        Vehicle Type   Injuries Prevented Per Year When All Vehicles on the Road Have the Technology

212: Windshield mounting

    Adhesive windshield bonding

        Cars and LTVs   15 % AIS 3-5 injury reduction for people ejected through the windshield portal

                                 171 AIS 3-5 injuries prevented per year (computation used 1982 as the baseline year)

213: Child restraint systems; 225: Child restraint anchorage systems

    Child safety seats

        Cars   46 % overall reduction of fatal or hospitalizing injuries for children age 0-5

                                 31 % reduction of overall (K + A + B + C) injuries

214: Side impact protection

    Side door beams

        Cars   25 % reduction of fatal or hospitalizing injuries for occupants of outboard seats in single-vehicle side impacts

                               25 % reduction of fatal or hospitalizing injuries for nearside occupants in side impacts by another vehicle, centered in the occupant compartment area

                               9,470 nonfatal hospitalizations prevented per year (computation used 1980 as the baseline year)

301: Fuel system integrity

    1975-77 upgrade: rollover, rear- and lateral-impact tests

        Cars   14 % reduction in all post-crash fires

                               3,900 fires prevented per year (unknown number of injuries; computation used 1982-87 as the baseline years)

        LTVs   No significant effect

TABLE 1-4: ESTIMATES OF CRASH AVOIDANCE
NHTSA EVALUATIONS OF SAFETY TECHNOLOGIES

FMVSS

SAFETY TECHNOLOGY	Crash-Reducing Effectiveness Estimate(s)
Vehicle Type	Crashes Avoided Per Year When All Vehicles on the Road Have the Technology
103: Windshield defrosting and defogging	
Rear-window defoggers	
Cars	No significant effect
105: Hydraulic brake systems	
Dual master cylinders	
Cars and LTVs	0.7 % reduction of all crash involvements
	40,000 police-reported crashes avoided per year (computation used 1979-80 as the baseline year)
Front disc brakes	
Cars and LTVs	0.17 % reduction of all injury crash involvements
	9,800 police-reported crashes avoided per year (computation used 1979-80 as the baseline year)
Rear-wheel ABS	
LTVs	30 % reduction of nonfatal rollovers
	20 % reduction of nonfatal side impacts with fixed objects
	No significant effect in multivehicle crashes
4-wheel ABS	
Cars	28 % reduction of multivehicle involvements on wet roads
	19 % increase of run-off-road crashes (in 1989-93 only?)
Electronic stability control	
Cars	35 % reduction of single-vehicle run-off-road crashes (preliminary estimate)
LTVs	67 % reduction of single-vehicle run-off-road crashes (preliminary estimate)

TABLE 1-4 (Concluded): ESTIMATES OF CRASH AVOIDANCE
NHTSA EVALUATIONS OF SAFETY TECHNOLOGIES

FMVSS

SAFETY TECHNOLOGY	Crash-Reducing Effectiveness Estimate(s)
Vehicle Type	Crashes Avoided Per Year When All Vehicles on the Road Have the Technology

108: Lamps, reflective devices

Side marker lamps

Cars and LTVs	16.8 % reduction of nighttime angle collisions when both vehicles in a crash have the lamps
	106,000 police-reported crashes avoided per year (computation used 1979-80 as the baseline year)

Center High Mounted Stop Lamps

Cars and LTVs	4.3 % reduction of rear impacts
	92,000-137,000 police-reported crashes plus 102,000 unreported crashes avoided per year (computation used 1994 as the baseline year)

Retroreflective tape

Heavy trailers	29 % reduction of cars and LTVs hitting the side or rear of a heavy trailer in the dark
	7,800 police-reported crashes avoided per year (computation used 2001 as the baseline year)

PART 2

LIVES SAVED BY THE FEDERAL MOTOR VEHICLE SAFETY STANDARDS AND OTHER VEHICLE SAFETY TECHNOLOGIES, 1960-2002

SUMMARY OF THE ESTIMATION METHOD

During 1960-2002, nearly 2,000,000 people died in motor vehicle crashes in the United States. Starting in 1975, the Fatality Analysis Reporting System (FARS) has furnished detailed information on over 1,200,000 of these fatality cases. Using the FARS data, we have built a model from which we can estimate how many additional people would have died had it not been for the Federal Motor Vehicle Safety Standards (FMVSS) and other vehicle safety technologies affecting passenger cars and light trucks. We can also estimate how many lives have been saved by each technology, year-by-year, from 1960 through 2002.

Each 100 actual fatality cases on FARS represent a potentially even greater number of fatalities that could have happened if the vehicles had not met any of the FMVSS. The actual fatality cases are the starting point for a model to estimate the lives saved by the FMVSS. The model is similar to the technique that NHTSA has been using since 1993 to estimate the number of lives saved each year by safety belts, air bags and child safety seats.[1] That technique is essentially expanded to the other safety standards.

For example, FARS might have records of 100 driver fatalities, wearing safety belts, in model year 1999 cars equipped with air bags, in frontal multivehicle crashes with primary impact point 12:00 (i.e., directly frontal). Air bags reduce fatality risk for belted drivers by 25.3 percent in 12:00 frontal crashes. Thus, if the cars had not been equipped with air bags, there would have been not 100 but

$$100/(1 - .253) = 134 \text{ potential fatalities.}$$

In other words, the existence of these 100 actual fatality cases with air bags on FARS implies that the MY 1999 vehicles were involved in 134 crashes potentially fatal to their drivers. However, 34 of those potentially fatal crashes did not become FARS fatality cases because the air bag saved the driver's life.

Three-point safety belts reduce drivers' fatality risk by 42 percent in multivehicle frontal impacts. If safety belts as well as air bags had been removed from those cars, the number of potential fatalities would have increased from 134 to

$$134/(1 - .42) = 231 \text{ potential fatalities.}$$

---

[1] *Traffic Safety Facts 2002 – Occupant Protection*, NHTSA Publication No. DOT HS 809 610, Washington, 2003; Kahane, C.J., *Fatality Reduction by Air Bags*, NHTSA Technical Report No. DOT HS 808 470, Washington, 1996, pp. 19-20.

That implies the use of safety belts saved 231 – 134 = 97 lives. But the benefits of the FMVSS do not stop here. Even after stripping the air bags and 3-point belts, these MY 1999 cars would still have had energy-absorbing steering assemblies to protect their drivers. Those assemblies reduce drivers' fatality risk in frontal collisions by 12.1 percent. If the energy-absorbing steering assemblies in these cars had been replaced by pre-1968 rigid columns, there would have been not 231 but

$$231/(1 - .121) = 263 \text{ potential fatalities.}$$

The energy-absorbing columns saved an implicit 263 – 231 = 32 lives. The 100 actual fatality cases on FARS imply the existence of 263 potentially fatal crash cases. They imply that the three safety technologies (air bags, safety belts, energy-absorbing columns) saved a combined total of 163 lives: 34 by air bags, 97 by safety belts and 32 by energy-absorbing columns.

The process works the same in reverse: consider a population of MY 1966 cars, not equipped with energy-absorbing columns, 3-point belts or air bags, involved in 12:00 frontal multivehicle crashes, resulting in fatalities to 263 drivers. If these cars had been equipped with energy-absorbing columns (introduced in 1967-68) but not yet 3-point belts (introduced primarily in 1974) or air bags (introduced in 1985-97), the columns would have reduced the fatalities by 12.1 percent, from 263 to 231. If, in addition, the cars had been equipped with 3-point belts and the drivers had buckled up, the belts would have reduced those fatalities by 42 percent, from 231 to 134. Finally, the addition of air bags would have reduced the remaining fatalities by 25.3 percent, from 134 to 100.

In other words, the model begins with the actual FARS fatality cases and inflates them step-by-step, using the effectiveness estimates from the NHTSA evaluations documented in Part 1 of this report[2], until all FMVSS and safety technologies applicable to this type of crash have been "removed" from the vehicle – until the vehicle has been degraded to a level of safety performance characteristic of the 1950's rather than its actual model year. The safety technologies are "removed" in the reverse chronological order that they were historically introduced into vehicles. Starting, say, with an MY 2002 vehicle, we first "remove" the safety device most recently introduced, in 2001, and downgrade the vehicle to the MY 2000 level of safety; next, the device introduced in 1991, diminishing the vehicle to the MY 1990 level, and so on until the vehicle is degraded to 1950's technologies without any of the modern safety devices. At each step into the past, the model tallies the lives saved by the latest safety technology – i.e., the additional fatalities that would have occurred if that technology had been removed.

The model produces unbiased estimates of the lives saved by the various technologies, and it is not an exercise in double counting, because each of the effectiveness estimates in Part 1 is based on analyses of vehicles produced *just after* vs. *just before* the FMVSS in question.[3] For

---

[2] Except electronic stability control (ESC) systems; they did not begin to appear in the late 1990's, and NHTSA has so far only conducted a preliminary evaluation.

[3] Even though the analyses are based on vehicles produced just after vs. just before the FMVSS, they usually include crashes involving these vehicles over a number of calendar years after the FMVSS. Specifically, in the case of crash-avoidance standards, vehicles have been on the road long enough for drivers to adapt to the technologies, if that is an issue.

example, the evaluation of air bags was limited to data on MY 1985 and later cars, some without air bags and some with air bags, but all equipped with 3-point belts and energy-absorbing columns. Thus, the above effectiveness estimate is no more than the incremental effect of air bags, given that a car already has energy-absorbing columns and that the driver buckled up. Similarly, the effectiveness estimate for 3-point belts was based on MY 1975-86 cars always equipped with energy-absorbing columns but almost never with air bags. It measured the incremental effect of belts, given that cars already had the columns, but not air bags. The evaluation of energy-absorbing columns was limited to data on MY 1966-68 cars, without air bags or 3-point belts, and measured the effect of the column for an (almost always) unrestrained driver of a pre-standard car. In other words, these effectiveness estimates are each incremental, and may be applied in sequence, as shown above, to estimate the total fatality reduction for the combination of the three technologies.[4]

One factor that simplifies the model is that rarely more than a few FMVSS apply to any specific crash situation. For example, the energy-absorbing column only benefits drivers in frontal crashes, and its effect will never need to be combined with, say, the benefit of back-seat belts. In addition, the safety technologies that have little or no effect on fatalities, such as head restraints or Center High Mounted Stop Lamps, do not need to be addressed in this model.

The model requires enough information about each FARS case of an occupant fatality in a passenger car or light truck to determine which safety technologies were present in the vehicle, and what effect, if any, each of these technologies would have had, given the type of crash and the type of occupant. In other words, the following data elements are needed to apply the model to an individual fatality case:

Vehicle
- Vehicle type: car or light truck (pickup, SUV, minivan or full-size van)
- Model year
- Make-model, if possible
- Impact locations – principal and initial
- Most harmful event

Occupant
- Seat position
- Age
- Restraint system use
- Ejected?

Crash
- Calendar year
- Single- or multivehicle

---

[4] Kahane (1996), pp. 7-9; Kahane, C.J., *Fatality Reduction by Safety Belts for Front-Seat Occupants of Cars and Light Trucks*, NHTSA Technical Report No. DOT HS 809 199, Washington, 2000, pp. 5-10; Kahane, C.J., *An Evaluation of Federal Motor Vehicle Safety Standards for Passenger Car Steering Assemblies*, NHTSA Technical Report No. DOT HS 805 705, Washington, 1981, pp. 197-203.

- First harmful event/manner of collision

Crash-avoidance technologies are more complicated because their life-saving benefits are not limited to the occupants of the vehicles equipped with them. Dual master cylinders and front disc brakes in passenger cars and light trucks will, in addition, benefit the pedestrians, bicyclists and motorcyclists who, as a result, are not struck by those vehicles. Conspicuity tape has only been installed on heavy truck trailers, but it saves occupants of cars and LTVs (light trucks and vans, including pickup trucks, SUVs, minivans and full-size vans) who, as a consequence, did not hit the trailers. This "full" accounting of the benefits of the FMVSS and other safety technologies for cars and LTVs includes not only the lives saved in cars and LTVs by car/LTV technologies or child safety seats (which turn out to be 99 percent of the total), but also the non-occupants and motorcyclists saved by car/LTV technologies and the car/LTV occupants saved by technologies in other vehicles (heavy trailers) – any technology where cars/LTVs produce and/or receive benefits. For that purpose it is necessary to know, for each non-occupant or motorcyclist fatality, the model year/type of the vehicle that hit it, and for each car/LTV occupant fatality in impacts with heavy trucks, the model year and impact location on the heavy truck, and the light conditions at the time of the crash.

The model does not address the benefits of safety technologies in vehicles other than cars and LTVs, except for the benefits to occupants of the cars/LTVs struck by those other vehicles. For example, it does not include motorcycle occupant lives saved by motorcycle helmets.

The model also does not estimate fatality reductions (or increases) due to shifts between existing vehicle types or seat positions such as: between cars and LTVs, large cars and small cars, 2-door and 4-door cars, or between the front seat and the back seat. They are not "new safety technologies" but shifts between existing vehicle designs. However, the effects of these shifts on safety may account for some of the year-to-year changes in the "base" of actual and potential fatalities.

The model consists of three principal sections:

Preprocessor to adjust for missing data elements The model needs complete information on a core group of data elements, listed above, to determine what safety equipment was in the case vehicle and what effect, if any, this equipment had on fatality risk, given the type of crash, the occupant's seat position, and other factors. But FARS cases may have "unknown" codes for one or more of these data elements. The original FARS data need to be transformed to a file that has complete information on the core data elements, and whose case weights add up to the same number of fatalities as on the original file. Two strategies adjust for missing data:

Vehicle type and vehicle model year are rarely missing (well under 1 percent in occupant fatality cases). But when they are missing, the remaining data elements rarely provide clues for an educated guess at the vehicle type or model year. The strategy here is to omit the cases with missing data and to give the leftover cases a weight factor slightly higher than 1, in order that their weights add up to the original fatality total.

176

More frequently unknown are the vehicle impact location/crash type (2 percent of car/LTV occupant fatalities), occupant seat position (2 percent) and restraint use (14 percent). Here, however, other variables can help you determine, case-by-case, the likely distributions of the unknowns. For example, if the "manner of collision" is front-to-rear (multivehicle), the impact location/crash type is most likely a frontal or a rear impact, rarely a side impact or rollover; but if the most harmful event is a collision with a tree, the impact location is very likely frontal, possibly a side impact, rarely a rear impact. A passenger with unknown seat location was more likely to have been in the back seat if he or she was a child, than if an adult; more likely to have occupied an "unenclosed part of the vehicle" if the vehicle was a pickup truck than if it was a minivan. An adult with unknown restraint use was quite possibly belted if the crash happened in 2002, much less so if it happened in 1980.

Several regression analyses estimate the distribution of impact type, seat position and belt use as a function of other core variables, based on the calibration dataset of cases with known values on all the core variables. For example, when the fatality is a non-ejected driver age 50 or older, the vehicle is a passenger car, and the first harmful event is an impact with a tree or highway sign, the distribution of these fatalities by impact type is 76 percent frontal, 19 percent side, 3 percent rollover and 2 percent rear/other. The strategy here is to replace the FARS fatality case of the 50+ year-old car driver who hit a tree with unknown vehicle-impact location by four new cases, each of which retain the values of the other core variables: one with a frontal impact, one with a side impact, one rollover and one rear/other impact. If the original case had a weight factor of 1, the new cases will have weight factors of .76, .19, .03 and .01, respectively.

The two strategies generate a file of fatality cases with non-missing values on all the core variables. Data inconsistencies are also corrected – e.g., "lap and shoulder belts used" is changed to "lap belt used" if a seat was only equipped with lap belts.

Calculation of lives saved, 1975-2002 The model runs through each of the actual fatality cases, removes the safety technologies from the vehicle in the reverse chronological order that they were implemented in the fleet, inflating the case weight factor as each technology is removed. In the example at the beginning of this discussion, if the case of the belted driver fatality in the MY 1999 car in a directly frontal, multivehicle crash has a case weight of 1, that case weight would have been inflated to 1.34 after removing the air bag, 2.31 after removing the safety belts, and 2.63 after removing the energy-absorbing column: because the combination of energy-absorbing columns, safety belts and air bags are estimated to reduce 2.63 fatalities to 1 fatality in this type of crash. Thus, this actual fatality case contributes evidence of .34 lives saved by air bags, .97 by safety belts and .32 by energy-absorbing columns, totaling up to 1.63 lives saved by all the FMVSS. Benefits are summed and tabulated by safety standard/technology, by vehicle type, by calendar year, and by whether the technology was mandated by a standard already in effect or was voluntarily furnished by the manufacturer at the time the vehicle was built.

The chronological order of the safety technologies is based on their *median implementation dates*, generally the first model year in which 50 percent or more of new vehicles were equipped with the technology. It is a sort of "average" introduction year for the technology. Table 2-1 shows the chronological order of the 22 safety improvements for passenger cars, evaluated to date, that significantly reduced fatalities.

TABLE 2-1

INSTALLATION DATES OF SAFETY TECHNOLOGIES IN PASSENGER CARS

		First Offered In	*Median Installation Year*[5]	FMVSS Effective Date
208A	Lap belts for front-outboard occupants	1956	*1962*	1/1/68
208B	Lap belt use by children age 1-4	1956	*1964*	1/1/68
206	Improved door locks	1962 (?)	*1965*	1/1/68
208C	Lap belts for back-outboard occupants	1961 (?)	*1965 late*	1/1/68
212	Adhesive windshield bonding	1963	*1966 early*	1/1/70[6]
208D	Lap belts for front-center occupants	1961 (?)	*1966*	1/1/68
105A	Dual master cylinders	1962	*1966 late*	1/1/68
208E	Lap belts for back-center occupants	1961 (?)	*1967 early*	1/1/68
203	Energy-absorbing steering assemblies	1967	*1967*	1/1/68
201	Voluntary instrument panel improvements	1967 (?)	*1968*	1/1/68[7]
105B	Front disc brakes	1965	*1971*	1/1/76
214A	Side door beams	1969	*1973 early*	1/1/73
216	Roof crush strength, eliminate hardtops	1970	*1973*	9/1/73
208F	3-point belts for front-outboard occupants	1963	*1974*	9/1/73
NCAP	NCAP-related improvements (w/o air bags)	1982	*1984 early*	N.A.
213	Child safety seats	Pre-1970[8]	*1985*[9]	4/1/71
208G	3-point belts for back-outboard occupants	1971	*1989*	12/11/89
208H	Automatic 2-point belts	1975[10]	*1991*[11]	9/1/86-9/1/89
214B	Voluntary[12] TTI(d) reductions in 2-door cars	1986 (?)	*1993*	N.A.
208I	Frontal air bags	1985	*1994*	9/1/86-9/1/89
108	Conspicuity tape – heavy trailers	1991 (?)	*1996*[13]	12/1/93[14]
208J	3-point belts for back-center occupants	Pre-1995	*2001*	9/2005-2007

---

[5] Unless otherwise noted, it is the first model year that over 50 percent of new vehicles were equipped with this safety technology. "Early" = close to 50 percent were already equipped in the preceding year and/or much more than 50 percent were equipped in this year or the following year. "Late" = far fewer than 50 percent were equipped in the preceding year and/or barely over 50 percent were equipped in this year and the following year.

[6] FMVSS 212 did not specifically require adhesive bonding; some cars continued to have rubber gaskets (meeting FMVSS 212) after the effective date.

[7] FMVSS 201 did not require these improvements to mid- and lower instrument panels; these improvements were not necessarily introduced before the effective date.

[8] Effective child safety seats were offered before 1970; child passenger protection laws took effect in all States, 1978-85.

[9] Safety seat use first exceeded 50 percent in 1985.

[10] Phased out of new cars after 1996.

[11] Never reached 50 percent of new vehicles; 1991 was the peak installation year.

[12] Pre-1994, prior to the phase-in of the dynamic test requirement of FMVSS 214.

[13] First calendar year when over 50 percent of heavy trailers on the road were equipped with the tape.

[14] FMCSA retrofit requirement effective 6/1/2001.

For example, lap belts for front-outboard occupants were offered on some new cars as early as 1956, and were required on all new cars after FMVSS 208 took effect on January 1, 1968. However, NCSS and NASS data show that 1962 is the first model year in which over 50 percent of new cars were equipped with lap belts. A description, regulatory history and effectiveness estimate for every safety improvement in Table 2-1 may be found in Part 1 of this report, within the chapter whose FMVSS number matches the code number for that technology (or in the "NCAP" chapter).

NHTSA's initial safety standards took effect on January 1, 1968 but, in many cases, the manufacturers had already implemented the safety technologies several years earlier. Three-point belts and side door beams were the major innovations of the 1970's. Automatic occupant protection and several other standards date to the late 1980's and the 1990's. The most recent technologies, including the dynamic side impact standard, and the head injury protection upgrade are not included in the model, because NHTSA has not yet evaluated their effectiveness.

Two of the 22 technologies are not applied to the cars themselves. Child safety seats are add-on equipment, while conspicuity tape is applied to heavy trailers and helps car drivers see the trailers at night and not hit them. For these two, the analog to "median installation year" is: the first calendar year when over 50 percent of children age 0-4 rode in safety seats (1985), and the first calendar year when over 50 percent of heavy trailers on the road were treated with tape (1996).

"Reverse chronological order" is just one of several general methods that could have been used to allocate lives saved among the safety technologies in those crashes where more than one technology is beneficial (e.g., air bags, safety belts and energy-absorbing columns for drivers in frontal crashes). Furthermore, "median installation year" is just one of several specific methods that could have been used to order the standards chronologically. However, in considering other allocation methods or orderings, it is important to note that all methods should produce the *same estimate of overall lives saved* (the primary objective of this report), only changing the allocation of the lives saved among the individual standards. In the example at the beginning of this section, all methods would agree that the combination of air bags, safety belts, and energy-absorbing columns saved 163 lives in the crashes where drivers were protected by all three. They would only differ on how to allocate these 163 lives among air bags, belts and energy-absorbing columns – a more theoretical issue because these drivers are being saved by the combination of all three devices, and there is no obvious way to pinpoint who was saved by what.

The reverse-chronological method is intuitively straightforward. It views each (at that time) new safety technology as building upon the baseline of previously existing technologies, and it allocates to the new technology only the benefits relative to that baseline: it computes how many lives are saved if technology X is added to cars that already have Y and Z. Another advantage is that the effectiveness estimates in NHTSA evaluations are computed on that basis, and enter directly into the model this report uses to estimate total lives saved.

A less attractive feature of the reverse-chronological method is that the allocation between technologies depends on their chronological order. In general, moving a standard earlier in the chronological order will increase the benefits allocated to that standard, and take some away from the standards that follow (leaving unchanged, however, the total lives saved by the combination of these standards). Two technologies might, in some intuitive way, be equally effective, but the model could allocate more lives saved to the earlier one.

An alternative approach that could allocate independently of chronological order is based on the effectiveness of each technology acting alone. For example, X alone might reduce fatality risk by 30 percent and Y alone by 10 percent. If the combination of X and Y saves 1,000 lives, 750 of them are allocated to X and 250 to Y (based on the ratio of their free-standing effectiveness estimates). That approach is feasible if there are data on vehicles that have only X and other vehicles that have only Y (e.g., safety belts without air bags, and air bags with unrestrained drivers). But it is essentially impossible for most of the non-belt technologies in this report, because they were introduced in nearly all vehicles in the same sequence: first X, then Y. For example, we know the effectiveness of energy-absorbing columns without air bags, but it is impossible to estimate, from crash data, the effectiveness of air bags without energy-absorbing columns, because there are no vehicles with air bags and rigid columns.

Even within the reverse-chronological method, there are alternatives to using the median installation year. The median year is fine if technologies X and Y were each introduced over relatively short, non-overlapping timeframes – e.g., energy-absorbing columns (1967-68) and air bags (1985-96). It works for the non-belt technologies in Tables 2-1 and 2-2 (because even though some technologies were introduced at about the same time, those usually do not apply to the same occupant or crash type – e.g., energy-absorbing columns for drivers and improved instrument panels for passengers).

A problem arises with safety belts, whose use has gradually increased over the years. Some people began to buckle up long before cars had air bags, others only started a few years ago. Theoretically, at least, we should ascertain when each individual started buckling up to decide the "order" of safety belts and air bags for that individual. FARS, however, only tells us if a fatality was belted, but it does not say in what year that person first started buckling up. We only know that 3-point belts became available in new cars in model year 1974, that most States had buckle-up laws by 1987, and that on-the-road belt use exceeded 50 percent by 1991. Whatever year we use as the "median" implementation year for safety belts, it is <u>before</u> the median implementation year for air bags (1994) and <u>after</u> the median implementation years for all the early "built-in" safety technologies up to side door beams, preserving the sequence in Table 2-1.

Table 2-2 shows the chronological order of the 18 life-saving improvements for LTVs. They are the same improvements as for cars, but excluding automatic belts (never installed in LTVs), roof crush strength, NCAP improvements and TTI(d) reductions (not evaluated for LTVs). Many FMVSS were not extended to LTVs until several years after they had been in effect for cars. As a result, the chronological order of the technologies in LTVs is slightly different from cars. In both vehicles, though, 3-point belts were installed before air bags but after energy-absorbing columns and voluntary instrument panel improvements.

180

TABLE 2-2

INSTALLATION DATES OF SAFETY TECHNOLOGIES IN LTVs

		First Offered In	*Median Installation Year*[15]	FMVSS Effective Date
208A	Lap belts for front-outboard occupants	1956 (?)	*1964*	7/1/71
208B	Lap belt use by children age 1-4	1956 (?)	*1966*	7/1/71
105A	Dual master cylinders	1962 (?)	*1966 late*	9/1/83
206	Improved door locks	1962 (?)	*1967*	1/1/72
208C	Lap belts for back-outboard occupants	?	*1968*	7/1/71
208D	Lap belts for front-center occupants	1965 (?)	*1968*	7/1/71
208E	Lap belts for back-center occupants	?	*1968*	7/1/71
105B	Front disc brakes	1965 (?)	*1971*	9/1/83
201	Voluntary instrument panel improvements	1969 (?)	*1972*	9/1/81[16]
203	Energy-absorbing steering assemblies	1970	*1976 early*	9/1/81
208F	3-point belts for front-outboard occupants	1969	*1976*	9/1/76
212	Adhesive windshield bonding	1978 (?)	*1980*	9/1/78[17]
213	Child safety seats	Pre-1970[18]	*1985*[19]	4/1/71
208G	3-point belts for back-outboard occupants	1987	*1990*	9/1/91
214A	Side door beams	1991	*1994*	9/1/93
208I	Frontal air bags	1991-92	*1995*	9/1/94-9/1/97
108	Conspicuity tape – heavy trailers	1991 (?)	*1996*[20]	12/1/93[21]
208J	3-point belts for back-center occupants	Pre-1995	*> 2001*	9/2005-2007

---

[15] Unless otherwise noted, it is the first model year that over 50 percent of new vehicles were equipped with this safety technology. "Early" = close to 50 percent were already equipped in the preceding year and/or much more than 50 percent were equipped in this year or the following year. "Late" = far fewer than 50 percent were equipped in the preceding year and/or barely over 50 percent were equipped in this year and the following year.

[16] FMVSS 201 did not require these improvements to mid- and lower instrument panels.

[17] FMVSS 212 did not specifically require adhesive bonding; some LTVs continued to have rubber gaskets (meeting FMVSS 212) after the effective date.

[18] Effective child safety seats were offered before 1970; child passenger protection laws took effect in all States, 1978-85.

[19] Safety seat use first exceeded 50 percent in 1985.

[20] First calendar year when over 50 percent of heavy trailers on the road were equipped with the tape.

[21] FMCSA retrofit requirement effective 6/1/2001.

If two technologies were implemented at the same time, and benefited the same people in the same types of crashes, there is a danger of double-counting the benefits – i.e., attributing the joint benefit to each of the individual technologies. However, the history of the safety measures, as shown in Tables 2-1 and 2-2 and the analysis methods in most NHTSA evaluations (with control groups) minimize that danger. Major technologies usually came in several years apart, but even when two were introduced at once, they did not benefit the same people because:

- They applied at different seat positions (e.g., front seat belts and rear seat belts).
- They applied in different crash types (e.g., side impacts and frontal impacts).
- Belt technologies only benefit belted occupants, while other technologies can benefit unrestrained occupants.
- 100-series FMVSS can be evaluated for their crash avoidance benefits, while 200-series FMVSS do not have such benefits.

Postprocessor: estimate lives saved in 1960-74 FARS data only exist from 1975. However, the FMVSS date back to 1968 and some of them incorporate technologies available well before that. The postprocessor extends the estimation model back to the years before FARS existed, specifically 1960-74.

The National Safety Council's *Accident Facts* tally the number of fatalities in motor vehicle crashes each year in the United States and apportion the fatalities among car occupants, "truck" occupants, motorcyclists, pedestrians/bicyclists and several other categories.[22] For CY 1975-80, when FARS and *Accident Facts* are both available, it is possible to compare the total fatalities for the two data sources (which use slightly different definitions of "motor vehicle crash fatality") and the number in each category, and establish ratios of FARS to *Accident Facts* case counts. These ratios are applied to the *Accident Facts* counts for each earlier year to estimate the "number of FARS fatality cases if FARS had existed that year."

The postprocessor essentially recreates a FARS file for each year from 1960-74, with just the core variables needed to operate the model. It runs these files through the basic model to estimate how many additional people would have died had it not been for the FMVSS and other vehicle safety technologies affecting passenger cars and light trucks.

The core variables are the number of vehicles in the crash (1, 2+); the vehicle's impact type (front, side, rollover, rear/other) and model year; and the occupant's age group (0, 1-4, 5+), seat position, belt/safety seat use and ejection status. The joint distribution of the core variables is inferred from 1975-80 FARS. Specifically, all the variables except model year, belt use and safety seat use are assumed to have the same distribution in earlier years as on 1975-80 FARS. The distribution of vehicle age is also assumed not to change, and the distribution of vehicles by model year is obtained by noting that

$$MY = CY - \text{vehicle age}$$

---

[22] For example, *Accident Facts 1978 Edition*, National Safety Council, Chicago, 1978, pp. 56 and 58.

Belt use depends on the vehicle's model year and the occupant's seat position, based on trends seen in 1975-80 FARS data; safety seat use depends on the child's age and the calendar year. The various assumptions add uncertainty to this part of the model, but it does not matter so much because, as we shall see, less than 5 percent of the lives saved in 1960-2002 were saved during 1960-1974. A file is created with one cell for each possible combination of values of the core variables, and with cell fatality counts that add up to the "number of FARS cases if FARS had existed that year."

For example, suppose this process had created a cell count of 100 "would-have-been" FARS cases of unbelted drivers in frontal crashes of MY 1968 cars in CY 1972. All MY 1968 cars were equipped with energy-absorbing steering assemblies that reduce drivers' fatality risk in frontal collisions by 12.1 percent. If the energy-absorbing steering assemblies in these cars had been replaced by pre-1968 rigid columns, there would have been not 100 but

$$100/(1 - .121) = 114 \text{ potential fatalities.}$$

The energy-absorbing columns saved an implicit $114 - 100 = 14$ lives in this cell.

Appendix A of this report contains step-by-step descriptions of the preprocessor, the main model and the postprocessor, and it discusses the auxiliary programs needed to run the model.

FINDINGS

Estimate of lives saved

Table 2-3 shows that safety technologies regulated by the Federal Motor Vehicle Safety Standards or introduced on a voluntary basis saved an estimated 328,551 lives in 1960-2002. The annual number of lives saved grew from 115 in 1960, when a small number of people used lap belts, to 24,561 in 2002, when most cars and LTVs were equipped with numerous modern safety technologies and belt use on the road achieved 75 percent. The number of lives saved grew every year except 1974 and 1981-82, when, among other things, external events (an energy crisis and a recession) substantially reduced baseline crashes from the previous year.

The total of 328,551 includes 44,483 lives saved by voluntary improvements and 284,069 saved by FMVSS that were in effect at the time. "Voluntary" improvements include those introduced before the actual effective date of a FMVSS (e.g., energy-absorbing steering assemblies before January 1, 1968) and distinct technologies that were never required for meeting a FMVSS (such as the improvements to the mid- and lower instrument panels). On the other hand, improvements directly pursuant to a specific FMVSS in effect at the time are credited to the "FMVSS in effect," even if the manufacturer exceeded the minimum requirements of the FMVSS. For example, lives saved by a frontal air bag in a 1998 car are credited to the air bag FMVSS, regardless of whether that car scored a HIC of 1000 (the minimum requirement) or 200 (exceptionally good performance) on the FMVSS 208 test.

Of course, before January 1, 1968, every life saved was due to a "voluntary" improvement – or at least one not required by the Federal government, although some States required safety belts in new cars well before 1968. After 1968, lives saved by voluntary improvements have remained close to 1,000 per year, comprising modifications to instrument panels, TTI(d) reductions before the FMVSS 214 upgrade, NCAP-related changes, and numerous improvements made a few years before a FMVSS took effect. For example, the energy-absorbing columns in MY 1967 cars, produced before the January 1, 1968 effective date, started generating "voluntary" benefits in late 1966, when those cars first began to sell, and continued to generate them as long as any of those cars were still on the road, even in 2002.

By contrast, the number of lives saved by FMVSS in effect has grown dramatically from a modest start of 107 in 1968, and they accounted for 22,999 of the 24,561 total lives saved in 2002.

Any non-belt technology "built into" new cars or LTVs takes years to build up substantial benefits. At first, only the newest vehicles on the road have the technology. Only after 6-8 years are they in the majority of vehicles on the road. By contrast, any program to increase belt use, such as buckle-up laws, can have an immediate large effect on lives saved. The belts are already in the vehicles; the program immediately increases the number of people who use them.

TABLE 2-3: LIVES SAVED BY VEHICLE SAFETY TECHNOLOGIES, 1960-2002

(Car and LTV occupants saved,
plus non-occupants and motorcyclists saved by car/LTV brake improvements)

L I V E S   S A V E D

CY	TOTAL	BY VOLUNTARY IMPROVEMENTS	BY FMVSS IN EFFECT
1960	115	115	0
1961	117	117	0
1962	135	135	0
1963	160	160	0
1964	203	203	0
1965	251	251	0
1966	339	339	0
1967	509	509	0
1968	816	709	107
1969	1,179	831	348
1970	1,447	866	581
1971	1,774	957	817
1972	2,226	1,096	1,130
1973	2,576	1,168	1,408
1974	2,518	1,005	1,513
1975	3,058	1,073	1,986
1976	3,240	1,089	2,151
1977	3,671	1,156	2,515
1978	4,040	1,185	2,855
1979	4,299	1,198	3,101
1980	4,539	1,218	3,322
1981	4,455	1,167	3,288
1982	4,057	1,044	3,013
1983	4,248	986	3,262
1984	4,835	956	3,879
1985	6,389	1,073	5,316
1986	8,523	1,207	7,316
1987	9,973	1,224	8,748
1988	11,265	1,291	9,974
1989	11,487	1,291	10,197
1990	11,711	1,272	10,439
1991	12,194	1,252	10,942
1992	12,483	1,266	11,217
1993	13,796	1,319	12,477
1994	15,154	1,411	13,744
1995	16,117	1,469	14,648
1996	17,813	1,553	16,260
1997	18,560	1,591	16,969
1998	19,380	1,550	17,830
1999	19,942	1,521	18,420
2000	21,789	1,569	20,220
2001	22,605	1,527	21,078
2002	24,561	1,562	22,999
	===========	===========	===========
	328,551	44,483	284,069

Figure 2-1 graphs the year-by-year trend in lives saved (the first column in Table 2-3). It took until 1969 to reach 1,000. Throughout the 1970's, new vehicles meeting the early FMVSS gradually superseded the older on-road fleet, contributing to steady growth in lives saved, but that was offset by declining belt use in the late 1970's, in the wake of the unsuccessful ignition interlock and prior to national buckle-up campaigns. The largest gains in both absolute and relative terms came with the buckle-up laws: lives saved rose from 4,835 in 1984 to 11,265 in 1988. Since then, continued increases in belt use, air bags and other recent FMVSS, and a steadily escalating "base" of more vehicles and more VMT (vehicle miles of travel) have helped the fatality reduction grow steadily, reaching 24,561 in 2002.

Table 2-4 compares the lives saved, year-by-year, by vehicle or person type. Car occupants are the largest group, comprising 232,255 of the 328,551 lives saved. However, commensurate with the growth in LTVs (light trucks and vans, including pickup trucks, SUVs, minivans and full-size vans), the number of lives saved in LTVs has accelerated in recent years, reaching 10,331 in 2002, compared to 14,175 car occupants. By contrast, in 1980, safety technologies saved only 608 LTV occupants vs. 3,848 car occupants.

The effect of car/LTV braking improvements on pedestrian, bicyclist and motorcyclist fatalities is small relative to the effect of all the FMVSS on car/LTV occupants. Non-occupants accounted for 1,782 and motorcyclists, 398 of the 328,551 lives saved. Once dual master cylinders and front disc brakes were in most of the cars and LTVs on the road, in the mid-1970's, non-occupant and motorcyclist lives saved stopped increasing and have, in fact, declined as the "base" of pedestrian and motorcyclist exposure decreased. A revival of motorcycle popularity reversed the decreasing trend after 1999.

Net effectiveness for car/LTV occupants

Table 2-5 concentrates on the occupant fatalities in cars and LTVs. The first column is the actual number of occupant fatalities. There were an estimated 1,443,030 actual occupant fatalities in cars and LTVs in 1960-2002.[1] The next column estimates how many fatalities would have happened in 1960-2002 if none of the safety technologies had existed: 1,769,401.[2] The third column, equal to column 2 minus column 1, is the lives saved by the safety technologies: 326,371 in 1960-2002. The fourth column is the overall effectiveness of the existing safety technologies in each calendar year: the lives saved divided by the number of fatalities there would have been without any safety improvements. The effectiveness grew in *every* year from 1960 to 2002, from a humble 0.40 percent in 1960 to a very substantial 42.81 percent fatality reduction in 2002. There were 32,737 actual car/LTV occupant fatalities in 2002, but there would have been 57,242 without the safety technologies. The 24,506 lives saved are 42.81 percent of the potential fatalities.

---

[1] This is the sum of the ORIGWT values in the programs LS2004 and OLDFA24.
[2] This is the sum of WEIGHTFA in LS2004 and OLDFA24, the inflated values of ORIGWT after all the safety technologies are "removed," one-by-one, from the vehicles.

FIGURE 2-1: LIVES SAVED BY VEHICLE SAFETY TECHNOLOGIES, 1960-2002

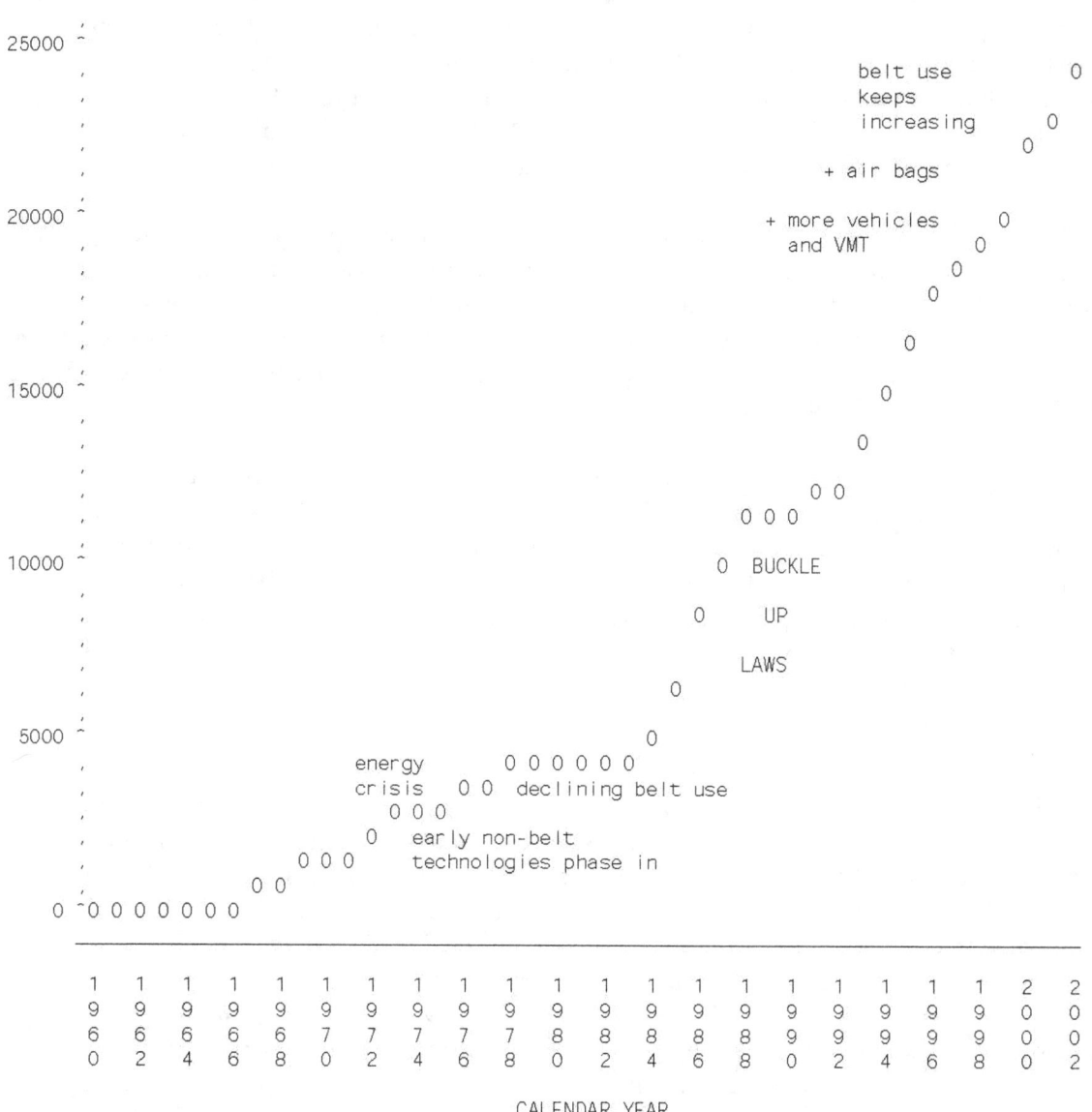

TABLE 2-4: LIVES SAVED BY VEHICLE SAFETY TECHNOLOGIES, 1960-2002
BY VEHICLE/PERSON TYPE

CY	TOTAL	CAR OCCUPANTS	LTV OCCUPANTS	PEDESTRIANS BICYCLISTS	MOTOR- CYCLISTS
1960	115	92	22	0	0
1961	117	95	22	0	0
1962	135	110	25	0	0
1963	160	131	28	1	0
1964	203	168	34	1	0
1965	251	211	38	2	0
1966	339	288	46	4	1
1967	509	444	55	9	1
1968	816	719	80	15	2
1969	1,179	1,042	114	21	2
1970	1,447	1,279	136	28	4
1971	1,774	1,574	161	35	4
1972	2,226	1,984	194	42	6
1973	2,576	2,282	238	48	8
1974	2,518	2,220	244	45	9
1975	3,058	2,723	272	53	10
1976	3,240	2,852	324	54	11
1977	3,671	3,190	409	58	14
1978	4,040	3,501	462	60	16
1979	4,299	3,657	561	64	18
1980	4,539	3,848	608	65	18
1981	4,455	3,758	614	64	18
1982	4,057	3,394	585	62	16
1983	4,248	3,534	641	57	15
1984	4,835	3,943	815	60	17
1985	6,389	5,196	1,118	59	17
1986	8,523	6,827	1,620	59	17
1987	9,973	7,783	2,115	60	15
1988	11,265	8,733	2,457	61	14
1989	11,487	8,677	2,741	58	12
1990	11,711	8,684	2,957	58	12
1991	12,194	8,943	3,188	53	11
1992	12,483	9,007	3,417	50	9
1993	13,796	9,916	3,818	52	10
1994	15,154	10,626	4,469	50	9
1995	16,117	11,115	4,942	51	9
1996	17,813	12,076	5,679	50	9
1997	18,560	12,146	6,356	49	9
1998	19,380	12,325	6,998	48	9
1999	19,942	12,401	7,486	45	9
2000	21,789	13,052	8,682	43	11
2001	22,605	13,532	9,016	45	12
2002	24,561	14,175	10,331	44	12
===========	===========	===========	===========	===========
	328,551	232,255	94,117	1,782	398

TABLE 2-5: ACTUAL OCCUPANT FATALITIES, POTENTIAL FATALITIES WITHOUT THE VEHICLE SAFETY TECHNOLOGIES, AND LIVES SAVED IN CARS/LTVs

CAR+LTV OCCUPANT FATALITIES

CY	ACTUAL	W/O SAFETY TECHS.	LIVES SAVED	PERCENT SAVED
1960	28,183	28,298	115	0.40
1961	28,087	28,204	117	0.41
1962	30,544	30,679	135	0.44
1963	32,664	32,823	159	0.49
1964	35,603	35,805	202	0.56
1965	36,518	36,767	249	0.68
1966	39,130	39,465	334	0.85
1967	39,327	39,826	499	1.25
1968	41,019	41,818	799	1.91
1969	42,117	43,273	1,156	2.67
1970	39,556	40,972	1,415	3.45
1971	38,916	40,651	1,735	4.27
1972	40,103	42,281	2,178	5.15
1973	38,739	41,258	2,520	6.11
1974	31,145	33,608	2,463	7.33
1975	31,361	34,355	2,995	8.72
1976	32,222	35,398	3,176	8.97
1977	33,173	36,772	3,599	9.79
1978	34,988	38,951	3,964	10.18
1979	35,108	39,325	4,217	10.72
1980	35,097	39,554	4,456	11.27
1981	33,911	38,284	4,373	11.42
1982	29,855	33,834	3,979	11.76
1983	29,209	33,384	4,176	12.51
1984	30,177	34,935	4,758	13.62
1985	30,044	36,357	6,314	17.37
1986	32,380	40,827	8,447	20.69
1987	33,306	43,203	9,898	22.91
1988	34,217	45,407	11,190	24.64
1989	33,709	45,127	11,418	25.30
1990	32,830	44,470	11,640	26.18
1991	30,928	43,060	12,131	28.17
1992	29,542	41,966	12,424	29.60
1993	30,182	43,917	13,735	31.27
1994	30,979	46,075	15,096	32.76
1995	32,057	48,113	16,056	33.37
1996	32,534	50,289	17,755	35.31
1997	32,501	51,003	18,502	36.28
1998	31,940	51,263	19,323	37.69
1999	32,151	52,038	19,887	38.22
2000	32,234	53,968	21,734	40.27
2001	32,009	54,558	22,548	41.33
2002	32,737	57,242	24,506	42.81
	===========	===========	===========	
	1,443,030	1,769,401	326,371	

190

Two graphs help to illustrate the trends in actual fatalities, potential fatalities and effectiveness. Figure 2-2 compares the actual car/LTV occupant fatalities (A) to the potential fatalities that could have happened without the safety technologies (P). Actual fatalities remained fairly level, usually in the low 30,000's, throughout 1960-2002. Only in 1968-72, when large numbers of baby-boomers were young drivers, did they crest just over 40,000. By contrast, potential fatalities have doubled from 28,298 in 1960 to 57,242 in 2002. They have experienced a long-term rising trend, with some exceptions such as the energy crisis in the mid-1970's, the recession in the early 1980's, and a smaller dip in 1990-92 perhaps associated with a high public emphasis on health and safety as the baby-boomers reached their 40's. Periods of prosperity such as the later 1980's and most of the 1990's saw large rises in potential fatalities, commensurate with rising VMT, vehicle sales, and trips for both work and recreation.

Until the mid-1970's, there is little distance between the A's and the P's on Figure 2-2. Then a gap opens between them, and it really widens after buckle-up laws were enacted in the mid-1980's. From that time onward, the potential fatalities continue to grow, but the FMVSS and other safety technologies have held the line on actual fatalities, working a little harder each year, keeping them in the low 30,000's year after year.

Figure 2-3 graphs the *percentage* of potential fatalities saved by the safety technologies. It looks a lot like Figure 2-1, a graph of the absolute number of lives saved. The difference is that Figure 2-3 essentially controls for VMT and other factors that make the "base" grow or shrink. The trend of the effectiveness of the vehicle safety technologies becomes even clearer:

- Not much effect until the late 1960's.
- Steady growth in the early-to-mid 1970's as new vehicles meeting the early FMVSS gradually phased in.
- Little progress in 1978-82, when belt use declined, in the wake of the unsuccessful ignition interlock and prior to national buckle-up campaigns.
- The largest absolute and relative gains came with the buckle-up laws in the mid-to-late 1980's.
- Steady progress since the late 1980's thanks to continued increases in belt use, air bags and other recent FMVSS.

FIGURE 2-2: ACTUAL VS. POTENTIAL CAR/LTV OCCUPANT FATALITIES

("A" = actual fatalities; "P" = potential fatalities without the vehicle safety technologies)

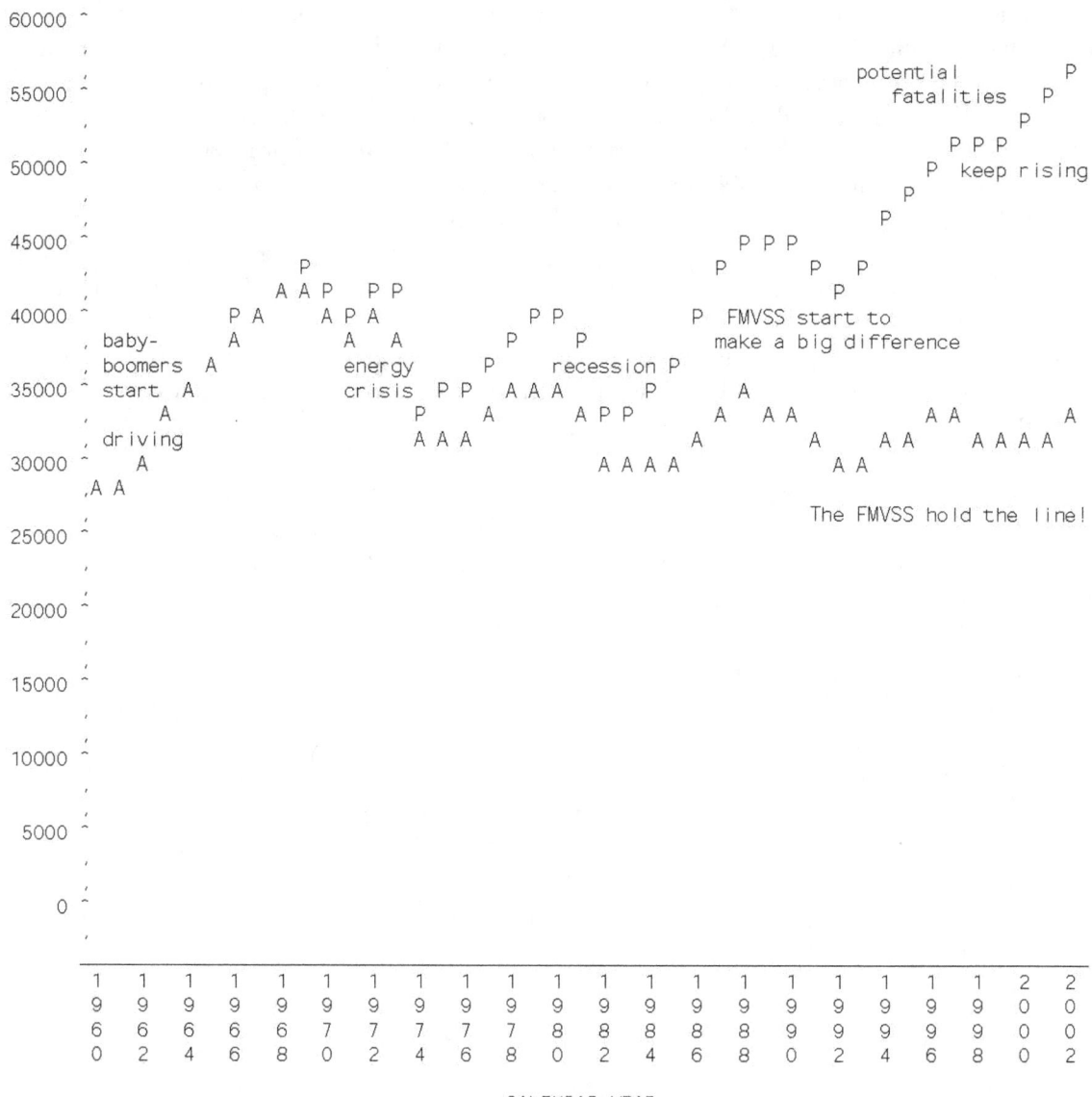

CALENDAR YEAR

FIGURE 2-3: PERCENT OF POTENTIAL FATALITIES SAVED BY VEHICLE SAFETY TECHNOLOGIES, 1960-2002

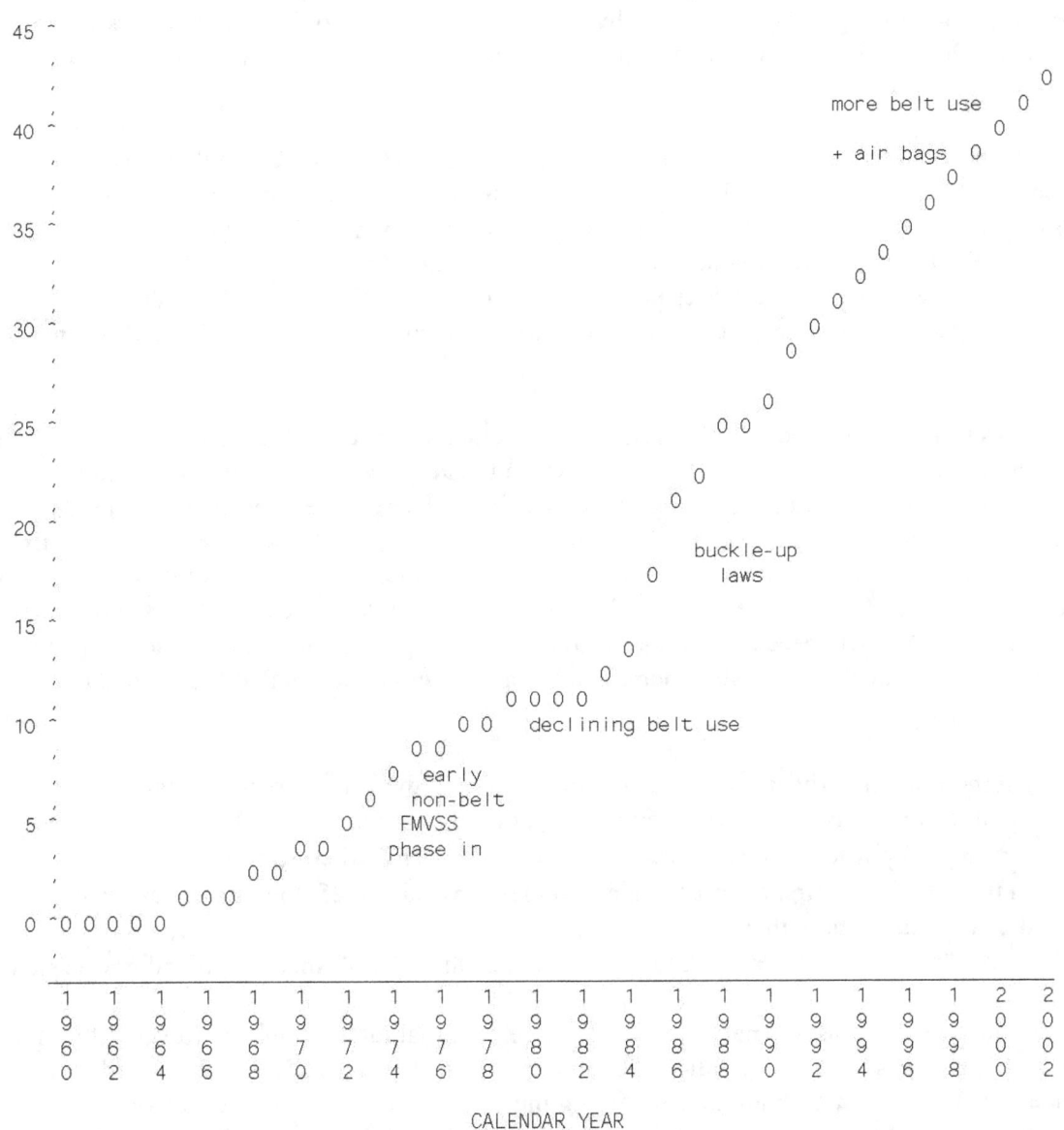

Car/LTV occupant fatalities per 100,000,000 VMT

A detailed analysis of fatality rates per 100,000,000 VMT comparing the strength of vehicle, behavioral, environmental and demographic factors is beyond the scope of this report, which concentrates on estimating the lives saved by vehicle safety technologies. Nevertheless, it is interesting to look at the trends in occupant fatality rates per 100,000,000 VMT – with and without the safety technologies.

Table 2-6 shows the VMT of passenger cars + LTVs in 1960-2002 and computes the rates of actual and potential fatalities per 100,000,000 miles in each calendar year.[3] America has been an increasingly mobile society, and VMT has grown every year except the energy-crisis years of 1974 and 1979-80, when it barely decreased. The actual fatality rate of car/LTV occupants has dramatically fallen from 4.26 fatalities per 100,000,000 miles in 1960 to 1.25 in 2002. Without the vehicle safety technologies, the potential fatality rates would have been 4.28 in 1960 and 2.18 in 2002.

Figure 2-4 shows the difference in the trends of the actual and potential fatality rates. From 1960 to the mid-1970's there was little difference between the actual and potential rates, because the vehicle technologies were not yet saving that many lives. There were large changes in the fatality rate for reasons unrelated to the FMVSS. From 1960 to 1964, the fatality rate actually increased as the first great cohort of baby-boomers, born immediately after World War 2, began to drive. From 1966 to 1977, the actual and potential rates fell dramatically (the potential rate from 4.61 to 2.70), partly because of major infrastructure improvements such as the Interstate Highway System, but also because of demographic and external factors that began to act at various times, such as:

- An increasingly urbanized and suburban population, with an increasing share of the VMT involved in relatively low-risk commuting and urban trips.
- A rising proportion of female drivers, who have lower fatal crash rates.
- In the 1970's, the first cohort of baby-boomers passed age 25, their most dangerous driving years behind them.
- Toward the end of that period, the energy crisis, the 55 mph speed limit and a recession.

From about 1983 onwards, the paradigm shifts. The potential fatality rate has stayed rather flat at about 2.1, whereas the actual fatality rate improved from 1.92 to 1.25. In other words, in the absence of vehicle safety technologies including increased belt use, if all other factors influencing the fatality rate had stayed the same, the fatality rate would also have stayed about the same.

---

[3] *Traffic Safety Facts 2002*, NHTSA Report No. DOT HS 809 620, Washington, 2004, p. 22 specifies VMT for cars in 1975-2002 and p. 24 for LTVs; add them to obtain the left column of Table 2-6. VMT estimates were obtained from the Federal Highway Administration (FHWA). P. 15 shows VMT for all vehicles on the road in 1966-2002, also obtained from FHWA. In 1975-80, cars + LTVs accounted for 92.4 percent of total VMT, and this percentage is assumed also for years before 1975. *Accident Facts 1974 Edition*, National Safety Council, Chicago, 1974, p. 59 tabulates VMT back to 1960 and earlier, also obtained from FHWA but with a slightly different definition. In 1966-73 the *Traffic Safety Facts* VMT is 99.6 percent of the *Accident Facts* VMT, and this percentage is also assumed for years before 1966.

TABLE 2-6: ACTUAL AND POTENTIAL OCCUPANT FATALITIES
IN CARS/LTVs PER 100,000,000 MILES VMT

CY	VMT (10^8 MILES)	ACTUAL FATALITIES		POTENTIAL FATALITIES W/O SAFETY TECHNOLOGIES	
		N	RATE	N	RATE
1960	6,617	28,183	4.26	28,298	4.28
1961	6,791	28,087	4.14	28,203	4.15
1962	7,058	30,544	4.33	30,679	4.35
1963	7,408	32,664	4.41	32,823	4.43
1964	7,794	35,603	4.57	35,805	4.59
1965	8,172	36,518	4.47	36,767	4.50
1966	8,560	39,131	4.57	39,465	4.61
1967	8,911	39,327	4.41	39,825	4.47
1968	9,392	41,019	4.37	41,818	4.45
1969	9,817	42,117	4.29	43,273	4.41
1970	10,260	39,556	3.86	40,972	3.99
1971	10,898	38,916	3.57	40,650	3.73
1972	11,647	40,103	3.44	42,280	3.63
1973	12,137	38,739	3.19	41,258	3.40
1974	11,841	31,145	2.63	33,608	2.84
1975	12,347	31,360	2.54	34,356	2.78
1976	13,040	32,222	2.47	35,397	2.71
1977	13,598	33,173	2.44	36,772	2.70
1978	14,259	34,988	2.45	38,952	2.73
1979	14,055	35,108	2.50	39,325	2.80
1980	14,025	35,098	2.50	39,554	2.82
1981	14,297	33,911	2.37	38,284	2.68
1982	14,679	29,855	2.03	33,834	2.30
1983	15,227	29,209	1.92	33,384	2.19
1984	15,850	30,177	1.90	34,934	2.20
1985	16,378	30,043	1.83	36,357	2.22
1986	16,941	32,380	1.91	40,828	2.41
1987	17,729	33,306	1.88	43,204	2.44
1988	18,725	34,217	1.83	45,407	2.42
1989	19,377	33,709	1.74	45,127	2.33
1990	19,828	32,830	1.66	44,470	2.24
1991	20,076	30,929	1.54	43,060	2.14
1992	20,784	29,542	1.42	41,965	2.02
1993	21,205	30,182	1.42	43,917	2.07
1994	21,707	30,979	1.43	46,075	2.12
1995	22,283	32,057	1.44	48,114	2.16
1996	22,864	32,534	1.42	50,289	2.20
1997	23,533	32,501	1.38	51,003	2.17
1998	24,179	31,940	1.32	51,263	2.12
1999	24,701	32,151	1.30	52,038	2.11
2000	25,233	32,233	1.28	53,968	2.14
2001	25,715	32,009	1.24	54,557	2.12
2002	26,248	32,737	1.25	57,243	2.18

FIGURE 2-4: ACTUAL AND POTENTIAL OCCUPANT FATALITIES
IN CARS/LTVs PER 100,000,000 MILES VMT

("A" = actual fatalities; "P" = potential fatalities without the vehicle safety technologies)

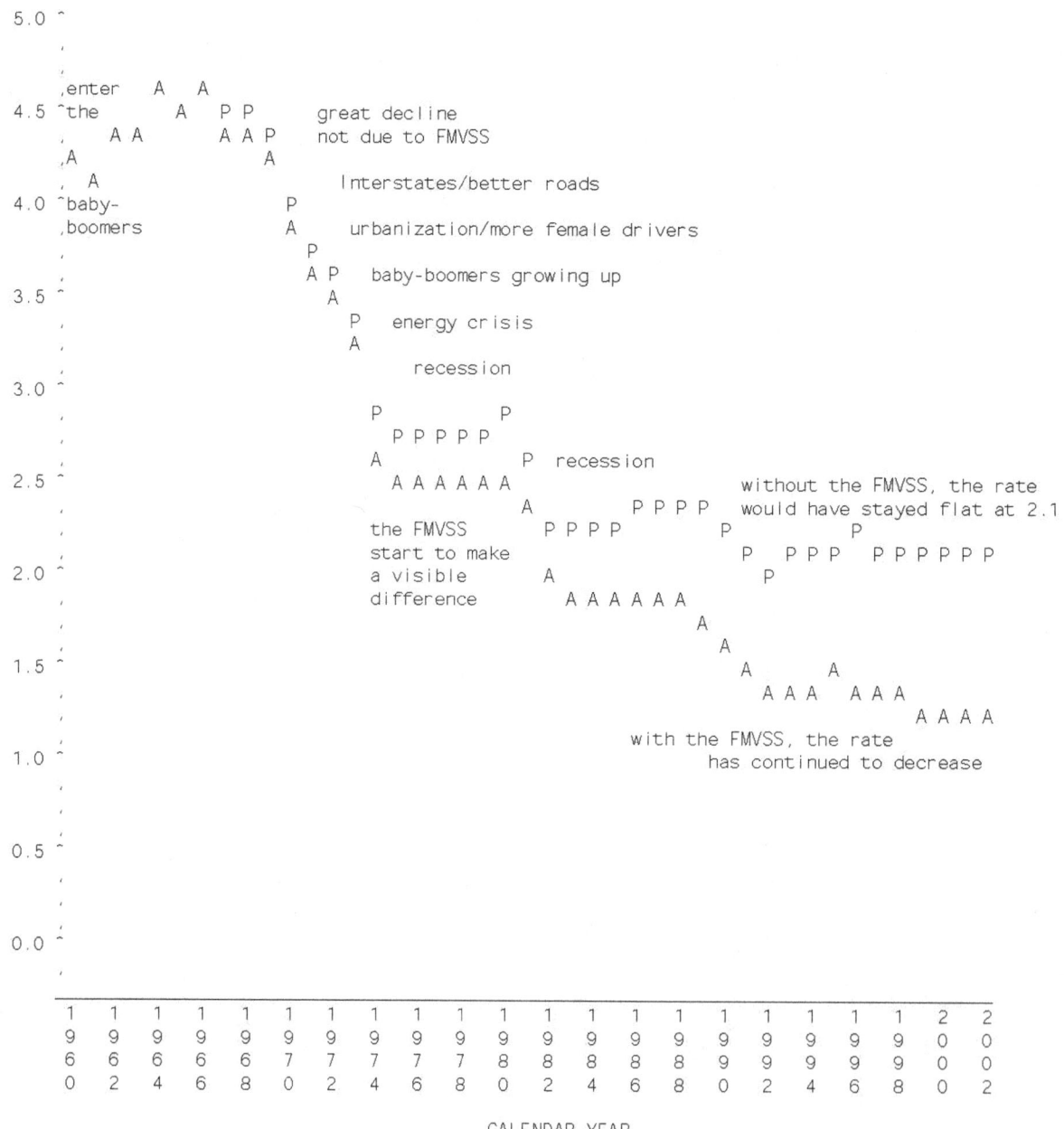

CALENDAR YEAR

For example, during 1983-2002, there were some factors that tended to reduce the fatality rate while other offsetting factors increased it:

- Effective programs to reduce drunk driving helped to reduce the fatality rate.
- A growing and increasingly mobile population of crash-prone, physically vulnerable older drivers contributed to increasing the rate.
- A reduction in the average mass of passenger cars on the road, as pre-downsized vehicles were retired, followed by an influx of rollover-prone and aggressive LTVs, may have contributed to increasing the rate.
- An increasingly prosperous society, with more recreational driving destinations and a greater hurry to get there, may have increased the rate.

But the steady increase in belt use, air bags, and other vehicle safety technology helped preserve the long-term downward trend in the fatality rate.

Estimates of lives saved by each FMVSS

Tables 2-7, 2-8 and 2-9 apportion the 328,551 lives saved in 1960-2002 among the various safety standards. Table 2-7 indicates that safety belts saved 168,524 occupants of cars and LTVs in 1960-2002, including 14,570 in 2002. The estimate comprises safety belts of all types – lap/shoulder, lap only and automatic 2-point – at all seat positions, front and rear, outboard and center. Lap belts were saving about 500 lives a year in the early 1970's. The transition to 3-point belts in the mid-1970's increased that to about 800. In the mid-1980's, buckle-up laws had a huge impact, raising lives saved to 6,000 per year by 1988-90 and, for the first time since 1970, putting belts far ahead of any other FMVSS in number of lives saved. Since 1990, continuing, successful programs to increase belt use, and the expanding "base" of VMT, vehicles and potentially fatal crashes has kept the "saved by the belt" number growing every year. Furthermore, belts are especially effective in LTVs (because many potentially fatal crashes are rollovers and/or ejections), and the growing SUV population means belts are saving more people. The 168,524 lives saved by belts are just over half of the total of 328,551; the 14,570 saved in 2002 are well over half of the total of 24,561 saved in that year.

Frontal air bags saved 12,074 lives in 1986-2002, including 2,473 in 2002. They are one of the most recent safety technologies to appear. In CY 2002, 133 million cars and LTVs had driver air bags, comprising 63 percent of all cars and LTVs on the road.[4] Vehicles with air bags are gradually replacing the older fleet without air bags, and the number of lives saved by air bags has been growing, year by year, as the fleet with air bags grows. When the proportion of vehicles with air bags gets closer to 100 percent, air bags will climb to clear 2nd place, behind safety belts but ahead of all the other FMVSS, in lives saved per year.

---

[4] *Traffic Safety Facts 2002 – Occupant Protection*, NHTSA Publication No. DOT HS 809 610, Washington, 2004 states there were 133 million vehicles with driver air bags; *Traffic Safety Facts 2002, op. cit.*, pp. 22 and 24 indicate there were 212 million cars and LTVs on the road in 2002.

TABLE 2-7: CAR + LTV OCCUPANTS SAVED BY SAFETY BELTS, FRONTAL AIR BAGS, AND CHILD SAFETY SEATS, 1960-2002

CY	SAFETY BELTS[5]	AIR BAGS[6]	CHILD SAFETY SEATS
1960	115	0	0
1961	116	0	0
1962	129	0	0
1963	140	0	0
1964	160	0	0
1965	174	0	0
1966	195	0	0
1967	205	0	0
1968	240	0	1
1969	298	0	1
1970	332	0	4
1971	379	0	9
1972	470	0	14
1973	546	0	19
1974	591	0	20
1975	843	0	36
1976	755	0	17
1977	791	0	34
1978	797	0	24
1979	707	0	49
1980	706	0	43
1981	666	0	70
1982	657	0	73
1983	823	0	112
1984	1,151	0	131
1985	2,370	0	166
1986	3,976	1	157
1987	5,066	2	209
1988	5,967	3	226
1989	6,187	8	239
1990	6,367	43	233
1991	6,912	75	252
1992	7,251	103	283
1993	8,222	207	278
1994	9,118	308	326
1995	9,443	538	331
1996	10,563	764	393
1997	11,017	984	372
1998	11,535	1,233	346
1999	11,687	1,485	396
2000	12,915	1,781	427
2001	13,373	2,065	328
2002	14,570	2,473	335
	===========	===========	===========
	168,524	12,074	5,954

---

[5] Includes lives saved by NCAP-related voluntary improvements for belted car drivers (1,024 total in 1979-2002).
[6] Net lives saved by air bags: benefits for adults and adolescents (age 13+) are to a small extent offset by increased risk to certain groups of child passengers (age 0-12); see Tables 2-38 – 2-40.

Child safety seats have been saving 300-400 lives per year since the early 1990's. They have been available and saving lives since the late 1960's. Benefits really began to increase after the States enacted child passenger protection laws in 1978-85, and followed up with campaigns to increase use. Benefits were lower in 2001-2002 than in 1996-2000, not because use declined, but because more people are putting their children in the back seat, a safer place, and are also driving more safely.

NHTSA's National Center for Statistics and Analysis (NCSA) generates the agency's official estimates of lives saved by safety belts, air bags and safety seats by the same general computational method.[7] NCSA's and this report's estimates are not identical because this report introduces slightly different computational procedures as it estimates the lives saved by all vehicle safety technologies, not just belts, air bags and safety seats. But the estimates are, on the average, very close – e.g., this report estimates 164,435 lives saved by belts in 1975-2002 while NCSA estimates 164,753.

Tables 2-8 and 2-9 estimate the lives saved by all the vehicle technologies other than safety belts, air bags and safety seats, grouping them under the eight FMVSS associated with them.

Importantly, the right column of Table 2-9 shows that, together, these "built-in," non-belt technologies saved an estimated 142,000 lives in 1960-2002, a substantial portion of the 328,551 lives saved by all the technologies. That column shows they were saving 3,000 lives per year by the late 1970's, by which time most cars of the pre-FMVSS era had been retired. The savings continued to grow, reaching 7,183 in 2002, as some new technologies were introduced, most pre-FMVSS LTVs were retired, and above all as the "base" of vehicles, VMT and potentially fatal crashes continued to grow. The eight individual FMVSS generally exhibit similar growth in lives saved during 1980-2002.

Braking improvements directly or indirectly associated with FMVSS 105 include dual master cylinders, implemented in the early-to-mid 1960's, and front disc brakes, introduced during 1965-1977. The first column of Table 2-8 shows these technologies began saving lives in 1962. Benefits grew rapidly in 1967-80 as vehicles with dual master cylinders replaced older vehicles without them in the on-road fleet. Benefits continued to grow after 1980 as the "base" of potentially fatal crashes kept expanding. They saved an estimated 13,053 lives in 1962-2002, including 538 in 2002. The people who owe their lives to these safety improvements are not limited to occupants of the cars and LTVs equipped with them. When a car or LTV, as a consequence of these improvements, avoids hitting another car/LTV, a pedestrian, bicyclist or motorcyclist, it may prevent a fatality among any of those groups.

The life-saving technology associated with FMVSS 108 is the red-and-white conspicuity tape on heavy truck trailers. Although this device is furnished on heavy trailers, not cars or LTVs, it is the occupants of cars and LTVs who primarily benefit, because it helps them avoid hitting the trailers in the dark. The tape began to appear on substantial numbers of trailers by 1991, and it has saved an estimated 1,105 car/LTV occupants, including 159 in 2002.

---

[7] *Traffic Safety Facts 2002 – Occupant Protection, op. cit.*

TABLE 2-8: LIVES SAVED BY TECHNOLOGIES ASSOCIATED WITH FMVSS 105, 108, 201, 203 AND 206, 1960-2002

CY	FMVSS 105 BRAKES (INCLUDES PEDS, MCs)	FMVSS 108 HEAVY TRAILER CONSPICUITY	FMVSS 201 INSTRUMENT PANEL	FMVSS 203 STEERING ASSEMBLY	FMVSS 206 DOOR LOCKS
1960	0	0	0	0	0
1961	0	0	0	0	0
1962	2	0	0	0	4
1963	4	0	0	0	15
1964	7	0	0	0	34
1965	9	0	0	0	60
1966	24	0	1	3	98
1967	54	0	14	68	141
1968	92	0	45	201	198
1969	129	0	88	346	258
1970	157	0	128	449	293
1971	189	0	179	563	337
1972	232	0	247	706	396
1973	262	0	305	808	427
1974	242	0	307	760	378
1975	278	0	366	894	385
1976	300	0	408	989	435
1977	330	0	458	1,113	566
1978	360	0	528	1,262	620
1979	377	0	548	1,334	753
1980	387	0	600	1,381	838
1981	380	0	602	1,351	795
1982	344	0	535	1,231	692
1983	338	0	522	1,210	702
1984	357	0	553	1,294	730
1985	369	0	605	1,469	763
1986	407	0	700	1,614	908
1987	428	0	733	1,718	1,012
1988	448	0	782	1,834	1,098
1989	442	0	792	1,877	1,036
1990	438	0	793	1,887	1,045
1991	420	9	767	1,821	1,026
1992	408	20	766	1,805	980
1993	427	32	792	1,924	992
1994	443	50	808	2,032	1,059
1995	462	69	851	2,162	1,110
1996	479	81	885	2,243	1,156
1997	485	113	916	2,306	1,153
1998	487	129	876	2,345	1,179
1999	492	154	832	2,330	1,238
2000	508	158	900	2,440	1,298
2001	516	132	882	2,592	1,292
2002	538	159	930	2,657	1,398
	===========	===========	===========	===========	===========
	13,053	1,105	21,043	53,017	28,902

TABLE 2-9: LIVES SAVED BY TECHNOLOGIES ASSOCIATED WITH
FMVSS 212, 214 AND 216, 1960-2002

CY	FMVSS212 WINDSHIELD BONDING	FMVSS214 SIDE IMPACT PROTECTION	FMVSS216 ROOF CRUSH STRENGTH	TOTAL FOR 8 NON-BELT FMVSS
1960	0	0	0	0
1961	0	0	0	0
1962	0	0	0	6
1963	0	0	0	20
1964	2	0	0	43
1965	8	0	0	78
1966	18	0	0	144
1967	28	0	0	304
1968	39	0	0	576
1969	53	7	0	880
1970	62	21	2	1,112
1971	72	40	5	1,387
1972	86	63	11	1,742
1973	94	96	18	2,010
1974	83	115	22	1,907
1975	85	142	29	2,179
1976	94	209	34	2,468
1977	104	218	59	2,847
1978	119	259	70	3,219
1979	121	308	103	3,543
1980	133	335	116	3,790
1981	133	343	115	3,719
1982	126	297	102	3,327
1983	131	311	98	3,312
1984	149	370	99	3,553
1985	169	370	108	3,853
1986	208	419	131	4,389
1987	238	429	137	4,695
1988	268	490	151	5,070
1989	259	504	144	5,053
1990	265	504	136	5,067
1991	262	512	138	4,956
1992	247	489	133	4,847
1993	263	525	134	5,088
1994	278	585	148	5,402
1995	294	708	147	5,804
1996	300	789	160	6,093
1997	298	767	150	6,187
1998	306	788	156	6,266
1999	315	860	154	6,374
2000	322	892	148	6,666
2001	333	944	148	6,839
2002	347	994	161	7,183
	===========	===========	===========	===========
	6,710	14,703	3,466	142,000

Improvements to mid- and lower instrument panels, not actually required by FMVSS 201 but historically and functionally associated with that standard to some extent, began to appear in cars in the later 1960's and in LTVs a few years after that. They have saved an estimated 21,043 right-front passengers in frontal crashes, including 930 in 2002.

Energy-absorbing steering assemblies meeting FMVSS 203 and 204 were one of the most important "built-in" early safety technologies. They significantly reduced the fatality risk of drivers in frontal crashes, and have saved an estimated 53,017 lives in cars and LTVs, including 2,657 in 2002. Energy-absorbing column and well-designed instrument panels continue to be important components of the overall occupant protection system in today's vehicles with air bags, and they continue to save lives. Specifically, as explained in the "Summary of the Estimation Method," our model first computes by how much fatalities would increase if the most recent technologies such as air bags were "removed" from today's vehicles, and then by how much more they would increase if the earlier technologies such as energy-absorbing columns were also "removed."

Improvements to door locks, latches, hinges and other components regulated by FMVSS 206 have significantly reduced door ejections in rollover crashes, saving an estimated 28,902 lives in 1962-2002, including 1,398 in 2002. Recent increases in the population of relatively rollover-prone LTVs have expanded the "base" of potentially fatal crashes addressed by this technology. FMVSS 212, windshield bonding, is the first standard addressed in Table 2-9. The introduction of adhesive windshield bonding extended from 1963 in some cars until approximately 1985 in some LTVs. As a result, the benefits for FMVSS 212 built up more gradually than some other standards. Adhesive bonding has significantly reduced occupant ejections via the windshield portal, saving an estimated 6,710 lives in 1964-2002, including 347 in 2002.

The estimates of benefits associated with FMVSS 214, side impact protection, are based on three technologies that entered at different times: side door beams for passenger cars (1969-73), voluntary TTI(d) reductions in 2-door cars before the dynamic test requirement went into effect (1986-93) and side door beams for LTVs (mostly 1994). Each new technology increased benefits, reaching 994 lives saved in 2002, and a total of 14,703 in 1969-2002. The most recent technologies – the 1994-97 upgrade in response to the new dynamic test, and side air bags – have not yet been evaluated and are not included in these estimates.

FMVSS 216, roof crush strength, is associated with the redesign of true hardtops as pillared hardtops or sedans during the 1970's. The elimination of these hardtops with low, flat, vulnerable roofs saved lives; if cars were still built that way there would have been 3,466 additional fatalities, including an estimated 161 in 2002.

Ranked by their estimates of lives saved in 1960-2002, the 11 groups of safety technologies line up as follows:

202

Lives Saved in 1960-2002

208/209/210	Safety belts	168,524
203/204	Energy-absorbing steering assemblies	53,017
206	Door locks, latches and hinges	28,902
201	Instrument panels	21,043
214	Side impact protection	14,703
105	Dual master cylinders/front disc brakes	13,053
208	Air bags (frontal)	12,074
212	Adhesive windshield bonding	6,710
213	Child safety seats	5,954
216	Roof crush resistance	3,466
108	Trailer conspicuity tape	1,105
		328,551

Safety belts are first by far, followed by three of the early occupant protection standards dating back to January 1, 1968 for passenger cars. However, some of the technologies were later arrivals and have not had as many years of accumulated benefits. Ranked by their estimates of lives saved in 2002 alone, the 11 FMVSS line up as follows:

Lives Saved in 2002

208/209/210	Safety belts	14,570
203/204	Energy-absorbing steering assemblies	2,657
208	Air bags (frontal)	2,473
206	Door locks, latches and hinges	1,398
214	Side impact protection	994
201	Instrument panels	930
105	Dual master cylinders/front disc brakes	538
212	Adhesive windshield bonding	347
213	Child safety seats	335
216	Roof crush resistance	161
108	Trailer conspicuity tape	159
		24,561

Safety belts and air bags account for a much large proportion of the total in recent years. Air bags moved up to 3rd place on the list and are likely to have moved to 2nd in 2003, as older vehicles without air bags continued to be retired.

As stated in the "Summary of the Estimation Method," our model comprises any technology where cars/LTVs produce and/or receive benefits: not only (1) the lives saved in cars and LTVs by car/LTV technologies (plus child safety seats), but also (2) the non-occupants and motorcyclists saved by car/LTV technologies and (3) the car/LTV occupants saved by technologies in other vehicles (heavy trailers). We can now estimate the sizes of these three groups: 99 percent of the benefits estimated by the model are car/LTV occupant lives saved by car/LTV technologies (plus child safety seats)[8]:

Car/LTV occupant lives saved by car/LTV technologies	325,266
Non-occupant/motorcyclist lives saved by car/LTV brake improvements	2,180
Car/LTV occupant lives saved by heavy-trailer conspicuity tape	1,105
	328,551

Benefits for occupants of passenger cars

Table 2-10 is a summary of the benefits of safety technologies for passenger car occupants. Actual fatalities in passenger cars increased from 24,689 in 1960 to an all-time high of 36,406 in 1969. They have rather steadily declined since then to 20,489 in 2002, reflecting cars' gradual loss of market share to LTVs. But safety improvements greatly contributed to the decline; without them, fatalities would only have fallen from 37,449 in 1969 to 34,664 in 2002. Safety technologies saved 232,255 car occupants in 1960-2002, including 14,175 in 2002. Voluntary improvements such as better instrument panels, TTI(d) reductions in 2-door cars and designs to new FMVSS before their effective date have been saving between 600 and 1,100 lives a year since 1968; however, most of the lives saved (200,156) are by safety technologies directly associated with FMVSS in effect at that time.

The proportion of potential fatalities saved by safety technologies has grown from 0.37 percent in 1960, when 92 car occupants were saved by lap belts, to 40.89 percent in 2002. The largest boost (from 14.32 to 25.35 percent) came in 1984-88, when buckle-up laws took effect in most of the States, but there have been continued, steady gains since with belt use increasing and air bags phasing in.

---

[8] The non-occupant/motorcyclist lives saved are the sum of the last two columns (1,782 + 398) on Table 2-4; the car/LTV occupant lives saved by trailer conspicuity tape is the total for FMVSS 108 in Table 2-8. (Trailer conspicuity tape also has the potential to save lives of motorcyclists and heavy-truck occupants, but these are not tallied by our model, as explained in the "Summary of the Estimation Method.")

TABLE 2-10: PASSENGER CAR SUMMARY – ACTUAL OCCUPANT FATALITIES, POTENTIAL FATALITIES WITHOUT THE SAFETY TECHNOLOGIES, AND LIVES SAVED

	CAR OCCUPANT FATALITIES			LIVES SAVED		
CY	ACTUAL	W/O SAFETY TECHS.	BY VOLUNTARY IMPROVEMENTS	BY FMVSS IN EFFECT	TOTAL	PERCENT SAVED
1960	24,689	24,782	92	0	92	0.37
1961	24,605	24,699	95	0	95	0.38
1962	26,757	26,867	110	0	110	0.41
1963	28,614	28,745	131	0	131	0.46
1964	31,189	31,357	168	0	168	0.54
1965	31,991	32,202	211	0	211	0.66
1966	34,346	34,634	288	0	288	0.83
1967	34,542	34,986	444	0	444	1.27
1968	35,720	36,439	614	105	719	1.97
1969	36,406	37,449	700	342	1,042	2.78
1970	34,051	35,331	711	568	1,279	3.62
1971	33,463	35,036	778	796	1,574	4.49
1972	34,444	36,427	899	1,084	1,984	5.45
1973	32,874	35,155	957	1,325	2,282	6.49
1974	26,103	28,322	809	1,411	2,220	7.84
1975	26,601	29,325	873	1,851	2,723	9.29
1976	26,803	29,655	864	1,988	2,852	9.62
1977	27,337	30,527	881	2,310	3,190	10.45
1978	28,438	31,940	878	2,624	3,501	10.96
1979	28,069	31,725	857	2,800	3,657	11.53
1980	27,709	31,557	839	3,009	3,848	12.20
1981	26,945	30,703	789	2,970	3,758	12.24
1982	23,390	26,783	686	2,708	3,394	12.67
1983	22,932	26,466	646	2,889	3,534	13.35
1984	23,596	27,538	626	3,317	3,943	14.32
1985	23,217	28,413	684	4,512	5,196	18.29
1986	24,966	31,794	826	6,001	6,827	21.47
1987	25,098	32,881	835	6,948	7,783	23.67
1988	25,720	34,453	879	7,854	8,733	25.35
1989	25,020	33,697	874	7,803	8,677	25.75
1990	24,089	32,773	859	7,825	8,684	26.50
1991	22,338	31,281	855	8,088	8,943	28.59
1992	21,303	30,310	866	8,141	9,007	29.72
1993	21,543	31,459	913	9,004	9,916	31.52
1994	21,949	32,575	962	9,664	10,626	32.62
1995	22,342	33,457	1,026	10,088	11,115	33.22
1996	22,495	34,571	1,082	10,994	12,076	34.93
1997	22,142	34,288	1,121	11,025	12,146	35.42
1998	21,210	33,535	1,072	11,253	12,325	36.75
1999	20,902	33,303	1,034	11,367	12,401	37.24
2000	20,712	33,765	1,094	11,958	13,052	38.66
2001	20,324	33,856	1,087	12,445	13,532	39.97
2002	20,489	34,664	1,083	13,092	14,175	40.89
	===========	===========	===========	===========	===========	
	1,137,471	1,369,726	32,099	200,156	232,255	

VMT data are available for passenger cars in 1975-2002.[9] They make it possible to compute occupant fatality rates per 100,000,000 VMT, as shown in Table 2-11 and Figure 2-5. Without the safety technologies, the fatality rate would potentially have remained nearly flat, close to 2.2, from about 1983 onward (with possibly a rising trend in the mid-1980's as new, downsized cars continued to replace the pre-downsized fleet). But thanks to the safety improvements, the actual fatality rate trended steadily downward from 1.93 in 1983 to 1.27 in 2002. Specifically, the buckle-up laws enacted in the mid-1980's more than cancelled any upward trend in the potential fatality rate.

At this point, we can review the lives saved by 22 individual safety technologies for passenger cars. Table 2-1 lists them in a chronological order based on the "median installation year": the first year that at least 50 percent of new cars had the technology. Our presentation will follow the same order, from oldest to newest (even though the computer models generating the estimates actually analyze the technologies in reverse chronological order). Table 2-12 covers the two earliest technologies, 208A – lap belts for front-outboard occupants (age 5 and older), and 208B – lap belt use (at any seat position) by children age 1-4.

Lap belts at the driver and right-front seats were offered in some new cars as early as 1956 and were retrofitted into some cars even older than that. (Part 1 of this report discusses the availability and effectiveness of each safety technology, in this case the section on "lap belts for front-seat occupants" in the chapter on FMVSS 208.) By model year 1962, over half of new cars were equipped with lap belts; States were passing laws requiring lap belts on new cars sold there. Lap belts reduce fatality risk by an average of 27 percent. FMVSS 208, like other initial safety standards, went into effect on January 1, 1968. Any lap-belt installation before January 1, 1968 – i.e., before mid-model year 1968 – is considered a "voluntary" installation, at least from a Federal point of view (although State laws might have required it), and any life saved by belts in a pre-MY 1968 or early-MY 1968 car is counted as a voluntary save in Table 2-12. Voluntary saves grow from 92 in 1960 to 176 in 1968 as more cars became equipped with belts. These cars, however, remained on the road after 1968 and their lap belts continued to accrue voluntary saves every year up to and including 2002, but in ever dwindling numbers as more and more were retired.

Most cars built from January 1, 1968 until August 31, 1973 had separate lap and shoulder belts. Here, our model counts fatality reductions among people who wore only the lap belt as "saved by the lap belt" and tallies them in the "Post-FMVSS" column of Table 2-12. But fatality reductions among people who wore both belts are counted with "saved by lap/shoulder belts" in Table 2-18, technology 208F. The number of post-FMVSS saves grew from 12 in 1968 to 233 in 1973 and generally declined after that as the MY 1968-73 cars were retired. The sum of voluntary and post-FMVSS saves, in the third column of Table 2-12, peaked at 337 in 1973 and totaled 4,312 in 1960-2002. Relatively low use and low effectiveness limited the number of lives saved by lap belts.

---

[9] *Traffic Safety Facts 2002, op. cit.*, pp. 22.

TABLE 2-11

CARS: ACTUAL AND POTENTIAL OCCUPANT FATALITIES
PER 100,000,000 MILES VMT

CY	VMT (10^8 MILES)	ACTUAL FATALITIES		POTENTIAL FATALITIES W/O SAFETY TECHNOLOGIES	
		N	RATE	N	RATE
1975	10,304	26,601	2.58	29,325	2.85
1976	10,707	26,803	2.50	29,655	2.77
1977	11,027	27,337	2.48	30,527	2.77
1978	11,365	28,438	2.50	31,940	2.81
1979	11,117	28,069	2.52	31,725	2.85
1980	11,071	27,709	2.50	31,557	2.85
1981	11,221	26,945	2.40	30,703	2.74
1982	11,458	23,390	2.04	26,783	2.34
1983	11,878	22,932	1.93	26,466	2.23
1984	12,265	23,596	1.92	27,538	2.25
1985	12,490	23,217	1.86	28,413	2.27
1986	12,776	24,966	1.95	31,794	2.49
1987	13,285	25,098	1.89	32,881	2.48
1988	13,840	25,720	1.86	34,453	2.49
1989	14,152	25,020	1.77	33,697	2.38
1990	14,272	24,089	1.69	32,773	2.30
1991	14,117	22,338	1.58	31,281	2.22
1992	14,360	21,303	1.48	30,310	2.11
1993	14,451	21,543	1.49	31,459	2.18
1994	14,592	21,949	1.50	32,575	2.23
1995	14,784	22,342	1.51	33,457	2.26
1996	14,991	22,495	1.50	34,571	2.31
1997	15,284	22,142	1.45	34,288	2.24
1998	15,559	21,210	1.36	33,535	2.16
1999	15,668	20,902	1.33	33,303	2.13
2000	15,805	20,712	1.31	33,765	2.14
2001	15,935	20,324	1.28	33,856	2.12
2002	16,085	20,489	1.27	34,664	2.16

FIGURE 2-5: ACTUAL AND POTENTIAL OCCUPANT FATALITIES IN CARS PER 100,000,000 MILES VMT

("A" = actual fatalities; "P" = potential fatalities without the vehicle safety technologies)

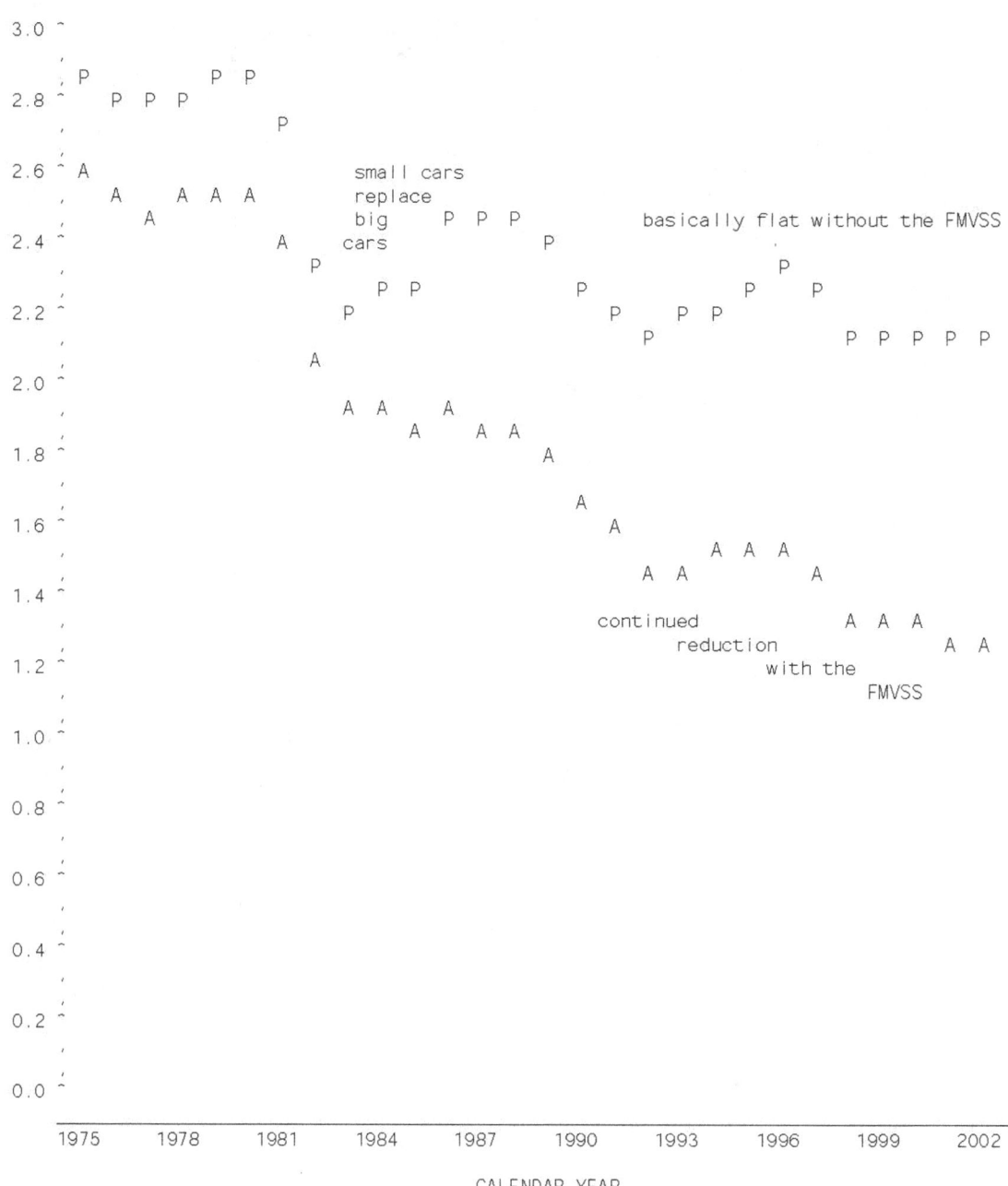

CALENDAR YEAR

TABLE 2-12: CAR OCCUPANT LIVES SAVED BY FRONT-OUTBOARD LAP BELTS AND LAP BELTS FOR CHILDREN AGE 1-4, 1960-2002

CY	208A: LAP BELTS FOR FRONT-OUTBOARD OCCS (MEDIAN INSTALL YR 1962, FMVSS 1/1/68)			208B: LAP BELT USE BY CHILDREN AGE 1-4 (MEDIAN INSTALL YR 1964, FMVSS 1/1/68)		
	VOLUNTARY	POST-FMVSS	TOTAL	VOLUNTARY	POST-FMVSS	TOTAL
1960	92	0	92	1	0	1
1961	94	0	94	1	0	1
1962	104	0	104	1	0	1
1963	114	0	114	1	0	1
1964	131	0	131	1	0	1
1965	142	0	142	1	0	1
1966	160	0	160	1	0	1
1967	169	0	169	1	0	1
1968	176	12	188	2	0	2
1969	168	48	216	1	0	2
1970	145	83	228	1	1	2
1971	130	118	249	1	1	2
1972	122	180	303	1	1	2
1973	104	233	337	1	1	2
1974	72	194	266	1	1	2
1975	64	188	253	0	1	1
1976	44	140	184	0	2	3
1977	33	109	142	0	3	3
1978	30	95	125	1	2	3
1979	21	62	83	1	3	4
1980	15	45	59	1	5	5
1981	7	39	46	0	4	4
1982	10	25	34	0	7	7
1983	7	22	28	0	7	8
1984	8	23	30	0	11	11
1985	11	35	46	0	14	14
1986	20	51	71	0	23	23
1987	17	41	58	0	27	27
1988	17	36	53	0	28	28
1989	16	23	40	0	30	30
1990	12	26	38	0	20	20
1991	8	19	27	0	19	19
1992	8	16	24	0	25	25
1993	9	14	23	0	26	26
1994	11	15	26	0	42	42
1995	4	12	16	0	30	30
1996	7	13	20	0	33	33
1997	5	8	13	0	29	29
1998	5	10	15	0	34	34
1999	6	9	15	0	18	18
2000	6	6	12	0	15	15
2001	5	6	11	0	22	22
2002	7	23	30	0	17	17
	===========	===========	===========	===========	===========	===========
	2,335	1,978	4,312	21	504	525

Technology 208B counts any child age 1-4 saved by belts (but not in a child safety seat) at any seat position. It counts them as "saved by lap belts" regardless of whether that seat was equipped with a lap belt only, a 3-point belt, or separate lap and shoulder belts, because the shoulder portion of the belt is of little value to a small child who is not in a safety seat. Lap belts reduce the fatality risk of 1-4 year old children by an average of 33 percent. They are not nearly as effective as a correctly used safety seat. Belts, without a safety seat, are assumed to have little or no value for infants (age 0); the model calculates no savings for infants. Lives saved in cars built before January 1, 1968 are "voluntary" saves, and they can accrue as long as such cars remain in service; savings in cars built on or after January 1, 1968 are counted as "post-FMVSS." This technology had quite limited benefits before States enacted child passenger protection laws in the early 1980's. Since then, belts have been saving 15-42 lives a year. Some State laws specifically permit a belt in lieu of a safety seat if a child has reached a certain age; in other cases, the adults transporting the child may simply decide on their own to use belts rather than a safety seat. Belts have saved a total of 525 children age 1-4 in cars.

Table 2-13 tallies savings by FMVSS 206 – improved door locks and by technology 208B – lap belts for rear-outboard occupants (age 5 and older). A series of improvements to door locks in approximately 1962-68 makes that the earliest non-belt technology evaluated by NHTSA. The evaluation estimated a 15 percent reduction of fatalities in rollovers. Benefits grew rapidly to about 500 lives saved per year as cars with the new technology replaced older cars; after that they gradually escalated to 704 in 2002 as the "base" of VMT and potentially fatal crashes grew larger and as smaller, more rollover-prone cars replaced full-size cars. As explained in the "Summary of the Estimation Method," our model first "removes" all safety technologies that came after FMVSS 206, most importantly 3-point belts. The 704 lives saved in 2002 are an estimate of the difference between a hypothetical fleet of cars equipped only with technologies lap belts for front-outboard occupants, lap belts for children and improved door locks (208A, 208B and 206) and another fleet that does not even have the improved door locks – specifically it is the benefit of FMVSS 206 in a fleet where nobody has 3-point belts. As with the other technologies, all lives saved in cars built before the January 1, 1968 effective date are "voluntary" saves. Improved door locks saved an estimated 19,504 car occupants in 1960-2002.

Lap belts were available at the rear-outboard seats of most new cars by 1965-66 and were required effective January 1, 1968. In the 1980's, 3-point belts increasingly superseded lap belts, and FMVSS 208 required lap/shoulder belts effective December 11, 1989. Lap belts are estimated to reduce fatality risk by an average of 32 percent for back-seat occupants age 5 and older. They achieved their largest benefit, 116 lives saved, in 1988, just before the shift to 3-point belts but after buckle-up laws, although primarily applicable to front-seat occupants, began to have a favorable influence on the back-seat occupants as well. These lap belts saved an estimated 1,360 lives in 1960-2002.

TABLE 2-13: CAR OCCUPANT LIVES SAVED BY IMPROVED DOOR LOCKS AND LAP BELTS FOR REAR-OUTBOARD OCCUPANTS, 1960-2002

CY	206: IMPROVED DOOR LOCKS (MEDIAN INSTALL YR 1965, FMVSS 1/1/68)			208C: LAP BELTS FOR REAR-OUTBOARD OCCS (MEDIAN INSTALL YR LATE 65, FMVSS 1/1/68)		
	VOLUNTARY	POST-FMVSS	TOTAL	VOLUNTARY	POST-FMVSS	TOTAL
1960	0	0	0	0	0	0
1961	0	0	0	0	0	0
1962	4	0	4	0	0	0
1963	13	0	13	0	0	0
1964	30	0	30	1	0	1
1965	52	0	52	2	0	2
1966	86	0	86	3	0	3
1967	123	0	123	4	0	4
1968	152	21	173	4	0	5
1969	157	67	224	4	1	6
1970	143	110	253	4	2	6
1971	139	151	291	4	3	7
1972	141	199	340	4	6	10
1973	129	232	361	4	9	13
1974	95	221	316	3	10	13
1975	74	248	322	5	12	17
1976	68	290	357	2	13	15
1977	69	392	461	4	21	25
1978	60	445	505	3	20	22
1979	53	536	589	0	12	12
1980	41	609	650	0	10	10
1981	29	590	619	1	13	13
1982	24	506	530	0	13	13
1983	20	511	532	0	29	30
1984	15	533	549	0	30	30
1985	15	553	568	0	65	65
1986	12	653	664	0	87	88
1987	10	711	721	1	88	89
1988	9	769	778	1	115	116
1989	5	703	708	1	108	109
1990	5	704	709	0	83	83
1991	5	676	681	0	77	77
1992	3	626	629	1	64	65
1993	2	630	632	0	65	65
1994	3	651	655	2	64	66
1995	2	679	681	1	67	68
1996	2	688	689	0	47	47
1997	2	644	646	0	42	42
1998	1	639	640	0	39	39
1999	1	667	668	0	18	18
2000	1	677	679	0	39	39
2001	1	669	670	0	15	15
2002	0	703	704	0	11	11
	===========	===========	===========	===========	===========	===========
	1,797	17,707	19,504	59	1,301	1,360

211

Table 2-14 considers adhesive windshield bonding and lap belts for front-center occupants. Adhesive windshield bonding is a technology associated with FMVSS 212 but not explicitly required by that standard. Its introduction extended from MY 1963 through the effective date of January 1, 1970 until approximately 1980, although the majority of new cars had it by MY 1966. Adhesive bonding significantly reduces the risk of ejection through the windshield portal. Lives saved in cars built before January 1, 1970 are counted as voluntary saves, all others as post-FMVSS saves. The number of lives saved has gradually risen over 200 per year as the "base" of VMT and potentially fatal crashes has increased. Adhesive bonding saved an estimated 5,248 lives in 1960-2002.

The front-center seat, like all designated seat positions, had to be equipped with lap belts by January 1, 1968 and often was equipped before that. Many cars do not have a front-center seat; when they do, it is usually unoccupied; and belt use is low. Nevertheless, over four decades, an estimated 81 car occupants can thank this technology for saving their lives.

Table 2-15 addresses dual master cylinders (105A) and lap belts for rear-center occupants (age 5+). Both of these technologies were required effective January 1, 1968 and often introduced a few years before that. Dual master cylinders reduce the risk of a fatal crash involvement by an estimated 0.7 percent. As the "base" of potentially fatal crashes grew, benefits also grew to 235 car occupants saved in 2002, and a total of 6,737 in 1960-2002.

Lap belts reduce the fatality risk of rear-center occupants by about 32 percent. Benefits have increased in recent years as belt use increased and as more children age 5-12 moved to the back seat. However, especially since 1999, manufacturers have been phasing in 3-point belts in anticipation of Anton's law, and the numbers for lap belts will dwindle. Lap belts saved an estimated 191 rear-center occupants in 1960-2002.

Table 2-16 analyzes two basic "built-in" protections for front-seat occupants in frontal crashes: energy-absorbing/telescoping steering assemblies to protect drivers (FMVSS 203/204), and voluntary modifications to mid and lower instrument panels, protecting passengers (associated with FMVSS 201). Energy-absorbing columns were introduced in the majority of cars in MY 1967 – i.e., late CY 1966 – and FMVSS 203 and 204 took effect on January 1, 1968. Thus, lives saved in the MY 1967 and early MY 1968 cars are counted as voluntary saves. The technology reduced drivers' fatality risk in frontal crashes by an estimated 12.1 percent. NHTSA's evaluation report, using 1978 as the baseline year, said FMVSS 203/204 would be saving 1,300 lives per year when all cars on the road met the standards. In fact, estimated benefits reached 1,300 in 1986 and since then have grown to 1,660 in 2002 as the base of VMT and potentially fatal crashes has expanded since 1978. Energy-absorbing and telescoping columns continue to be a critical component of the overall frontal energy-management and crash-protection system, even in today's cars with air bags and high belt use. With 41,545 lives saved in 1966-2002, energy-absorbing steering assemblies clearly rank first in cumulative benefits among the non-belt technologies for passenger cars.

TABLE 2-14: CAR OCCUPANT LIVES SAVED BY ADHESIVE WINDSHIELD BONDING AND LAP BELTS FOR FRONT-CENTER OCCUPANTS, 1960-2002

	212: ADHESIVE WINDSHIELD BONDING (MEDIAN INSTALL EARLY 66, FMVSS 1/1/70)			208D: LAP BELTS FOR FRONT-CENTER OCCS (MEDIAN INSTALL YR 1966, FMVSS 1/1/68)		
CY	VOLUNTARY	POST-FMVSS	TOTAL	VOLUNTARY	POST-FMVSS	TOTAL
1960	0	0	0	0	0	0
1961	0	0	0	0	0	0
1962	0	0	0	0	0	0
1963	0	0	0	0	0	0
1964	2	0	2	0	0	0
1965	8	0	8	0	0	0
1966	18	0	18	0	0	0
1967	28	0	28	0	0	0
1968	39	0	39	0	0	1
1969	53	0	53	1	0	1
1970	56	6	62	0	1	1
1971	55	17	72	0	1	2
1972	56	31	86	0	1	2
1973	52	41	94	0	2	2
1974	40	42	83	0	2	2
1975	37	48	85	1	3	4
1976	35	58	94	0	3	3
1977	33	71	104	0	2	2
1978	31	87	118	0	3	3
1979	23	92	115	0	1	1
1980	20	104	124	0	2	2
1981	12	110	122	0	1	1
1982	10	103	113	0	2	2
1983	8	107	115	0	1	1
1984	7	120	127	0	1	1
1985	5	133	138	0	3	3
1986	5	164	170	0	2	2
1987	4	183	187	0	2	2
1988	3	206	209	0	3	3
1989	2	196	197	0	4	5
1990	1	196	198	0	4	4
1991	1	194	195	0	3	3
1992	1	184	185	0	2	2
1993	1	193	194	0	3	3
1994	1	201	202	0	3	3
1995	1	210	211	0	3	3
1996	0	214	214	0	4	4
1997	0	208	208	0	4	4
1998	1	207	208	0	3	3
1999	0	209	210	0	2	2
2000	1	214	214	0	2	2
2001	0	221	221	0	2	2
2002	0	229	229	0	1	1
	===========	===========	===========	===========	===========	===========
	652	4,597	5,248	6	75	81

TABLE 2-15: CAR OCCUPANT LIVES SAVED BY DUAL MASTER CYLINDERS AND LAP BELTS FOR REAR-CENTER OCCUPANTS, 1960-2002

	105A: DUAL MASTER CYLINDERS (MEDIAN INSTALL YR LATE 66, FMVSS 1/1/68)			208E: LAP BELTS FOR REAR-CENTER OCCS (MEDIAN INSTALL EARLY 67, FMVSS 1/1/68)		
CY	VOLUNTARY	POST-FMVSS	TOTAL	VOLUNTARY	POST-FMVSS	TOTAL
1960	0	0	0	0	0	0
1961	0	0	0	0	0	0
1962	1	0	1	0	0	0
1963	3	0	3	0	0	0
1964	5	0	5	0	0	0
1965	6	0	6	0	0	0
1966	16	0	16	0	0	0
1967	37	0	37	0	0	0
1968	51	10	61	0	0	0
1969	53	32	85	0	0	0
1970	47	53	100	0	0	0
1971	46	73	119	0	0	0
1972	47	96	143	0	0	1
1973	45	112	157	0	0	1
1974	34	107	141	0	0	1
1975	32	132	164	0	0	0
1976	29	146	175	0	0	0
1977	23	163	186	0	1	1
1978	19	182	201	3	1	4
1979	15	188	203	0	0	1
1980	12	194	205	0	0	0
1981	9	192	201	0	0	0
1982	7	170	176	0	4	4
1983	5	170	176	0	1	1
1984	4	180	184	0	5	5
1985	4	186	190	0	6	6
1986	3	209	212	0	4	4
1987	2	218	220	0	10	10
1988	2	229	231	0	4	4
1989	2	225	227	0	10	10
1990	1	219	221	0	10	10
1991	1	210	211	0	7	7
1992	1	204	205	0	8	8
1993	1	212	213	0	5	5
1994	1	220	220	0	8	8
1995	0	226	226	0	10	10
1996	0	234	234	0	17	17
1997	0	232	233	0	14	14
1998	0	227	228	0	13	13
1999	0	226	226	0	10	10
2000	0	229	229	0	14	14
2001	0	230	230	0	9	9
2002	0	235	235	0	14	14
	===========	===========	===========	===========	===========	===========
	565	6,172	6,737	5	186	191

TABLE 2-16: CAR OCCUPANT LIVES SAVED BY ENERGY-ABSORBING STEERING ASSEMBLIES AND VOLUNTARY INSTRUMENT PANEL IMPROVEMENTS, 1960-2002

	203: ENERGY-ABSORBING STEERING ASSEMBLIES (MEDIAN INSTALL YEAR 1967, FMVSS 1/1/68)			201: VOLUNTARY INSTRUMENT PANEL IMPROVEMENTS (MEDIAN INSTALL YR 1968, FMVSS 1/1/68)
CY	VOLUNTARY	POST-FMVSS	TOTAL	TOTAL (VOLUNTARY)
1960	0	0	0	0
1961	0	0	0	0
1962	0	0	0	0
1963	0	0	0	0
1964	0	0	0	0
1965	0	0	0	0
1966	3	0	3	1
1967	68	0	68	14
1968	140	60	201	45
1969	154	192	346	87
1970	136	313	448	125
1971	131	431	562	172
1972	135	569	704	234
1973	133	663	795	286
1974	106	630	735	283
1975	100	764	864	339
1976	98	846	944	371
1977	78	964	1,043	410
1978	65	1,106	1,171	462
1979	54	1,168	1,222	475
1980	43	1,204	1,247	510
1981	31	1,178	1,209	510
1982	25	1,049	1,073	451
1983	19	1,029	1,047	439
1984	15	1,088	1,103	457
1985	13	1,211	1,224	502
1986	10	1,328	1,339	578
1987	8	1,388	1,396	593
1988	6	1,462	1,468	636
1989	6	1,460	1,466	626
1990	4	1,447	1,451	614
1991	5	1,373	1,377	591
1992	3	1,358	1,360	595
1993	2	1,428	1,430	612
1994	2	1,485	1,487	620
1995	2	1,533	1,535	645
1996	1	1,592	1,592	646
1997	2	1,601	1,603	677
1998	1	1,589	1,590	621
1999	1	1,544	1,545	587
2000	1	1,575	1,576	617
2001	0	1,657	1,658	608
2002	1	1,660	1,660	631
	===========	===========	===========	===========
	1,601	39,944	41,545	16,670

FMVSS 201 at one time proposed to regulate the force-deflection characteristics of mid and lower instrument panels, but those proposals were not included in the final rule. Nevertheless, manufacturers voluntarily remodeled the panels, using newly available materials, to absorb passenger impacts at a safer and more controlled force level. That has resulted in a 15.9 percent fatality reduction for unrestrained right-front passengers in frontal impacts. It saved an estimated 16,670 lives from late CY 1966 through 2002. The benefits for this technology are about one-third of the benefits for energy-absorbing steering assemblies principally because one-third of crash-involved cars have a right-front passenger, while all have a driver.

Table 2-17 addresses front disc brakes (105B) and side door beams (214A). Front disc brakes are well suited for meeting tests added to FMVSS 105 effective January 1, 1976, but were never explicitly required by that standard. Consumers preferred them to drum brakes. Their introduction extended from MY 1965 until 1977, but the majority of new cars had them by 1971. Front disc brakes reduce the risk of a fatal crash involvement by an estimated 0.17 percent. They have been saving approximately 50 car occupants per year, and a total of 1,256 in 1960-2002.

Side door beams were introduced in some cars as early as MY 1969, four years before the January 1, 1973 effective date of FMVSS 214. This technology aims to enhance crashworthiness and structural integrity in side impacts. While there are significant injury reductions in all types of side impacts, fatality reduction has been limited to single-vehicle crashes, such as side impacts with poles or trees after a car has run off the road and gone out of control. Side door beams reduce these fatalities by 14 percent, saving 450-500 lives per year, and an estimated total of 11,207 in 1960-2002.

Table 2-18 covers FMVSS 216, roof crush resistance and the big one: 3-point belts for front-outboard occupants, age 5+ (208F). The visible change associated with FMVSS 216 was the redesign of true hardtops as pillared hardtops or sedans. Nevertheless, many hardtops could and did meet FMVSS 216. The transition to pillared hardtops and sedans extended from approximately MY 1970 through the FMVSS 216 effective date of September 1, 1973 until about 1977. Pillared cars do a significantly better job of protecting non-ejected occupants in rollover crashes, saving an estimated 161 lives in 2002, and 3,466 in 1960-2002.

FMVSS 208 has required manual, integral 3-point belts at the driver's and right-front seats in all cars without automatic protection since September 1, 1973 and in all cars since September 1, 1997. From January 1, 1972 until August 31, 1973, FMVSS 208 required lap and shoulder belts at these positions in all cars, but manufacturers could choose between integral 3-point belts or separate lap and shoulder belts. From January 1, 1968 until December 31, 1971, this requirement applied only to cars where a lap belt was insufficient to prevent a dummy from contacting the windshield header (but, in reality, manufacturers equipped all their cars with separate or integral lap/shoulder belts).

TABLE 2-17: CAR OCCUPANT LIVES SAVED BY FRONT DISC BRAKES
AND SIDE DOOR BEAMS, 1960-2002

	105B: FRONT DISC BRAKES (MEDIAN INSTALL YEAR 1971, FMVSS 1/1/76)			214A: SIDE DOOR BEAMS (MEDIAN INSTALL EARLY 73, FMVSS 1/1/73)		
CY	VOLUNTARY	POST-FMVSS	TOTAL	VOLUNTARY	POST-FMVSS	TOTAL
1960	0	0	0	0	0	0
1961	0	0	0	0	0	0
1962	0	0	0	0	0	0
1963	0	0	0	0	0	0
1964	0	0	0	0	0	0
1965	0	0	0	0	0	0
1966	0	0	0	0	0	0
1967	1	0	1	0	0	0
1968	1	0	1	0	0	0
1969	3	0	3	7	0	7
1970	4	0	4	21	0	21
1971	7	0	7	40	0	40
1972	11	0	11	63	0	63
1973	15	0	15	75	21	96
1974	15	0	15	61	54	115
1975	19	0	19	66	76	142
1976	21	2	23	84	125	209
1977	21	6	27	65	153	218
1978	21	11	32	69	190	259
1979	19	16	35	68	240	308
1980	18	19	37	69	266	335
1981	15	22	38	53	290	343
1982	12	22	34	45	251	297
1983	11	24	35	41	270	311
1984	10	28	37	39	330	370
1985	8	32	40	23	347	370
1986	8	37	45	25	390	415
1987	6	42	47	20	398	418
1988	5	46	50	13	452	465
1989	3	46	50	9	456	465
1990	3	46	49	9	445	454
1991	2	45	47	6	448	454
1992	2	44	45	6	414	420
1993	1	46	47	4	422	426
1994	1	48	49	3	426	429
1995	1	50	50	3	481	484
1996	1	52	52	1	493	494
1997	1	51	52	1	454	454
1998	0	50	51	2	450	452
1999	0	50	51	1	456	457
2000	0	51	51	1	459	460
2001	0	51	51	0	465	466
2002	0	52	53	1	488	489
	===========	===========	===========	===========	===========	===========
	267	989	1,256	998	10,209	11,207

TABLE 2-18: CAR OCCUPANT LIVES SAVED BY ROOF CRUSH STRENGTH AND 3-POINT BELTS FOR FRONT-OUTBOARD OCCUPANTS, 1960-2002

	216: ROOF CRUSH STRENGTH - ELIMINATE HARDTOPS (MEDIAN INSTALL YR 1973, FMVSS 9/1/73)			208F: 3-POINT BELTS FOR FRONT-OUTBOARD OCCS (MEDIAN INSTALL YR 1974, FMVSS 9/1/73)		
CY	VOLUNTARY	POST-FMVSS	TOTAL	VOLUNTARY	POST-FMVSS	TOTAL
1960	0	0	0	0	0	0
1961	0	0	0	0	0	0
1962	0	0	0	0	0	0
1963	0	0	0	0	0	0
1964	0	0	0	0	0	0
1965	0	0	0	0	0	0
1966	0	0	0	0	0	0
1967	0	0	0	0	0	0
1968	0	0	0	2	0	2
1969	0	0	0	12	0	12
1970	2	0	2	23	0	23
1971	5	0	5	38	0	38
1972	11	0	11	59	0	59
1973	18	0	18	77	9	86
1974	15	7	22	64	144	208
1975	14	15	29	85	363	448
1976	14	20	34	83	344	426
1977	19	41	59	88	384	472
1978	16	54	70	71	427	499
1979	19	84	103	53	398	451
1980	19	97	116	53	449	502
1981	16	99	115	31	429	460
1982	12	90	102	33	452	486
1983	9	89	98	33	563	596
1984	8	92	99	28	776	804
1985	5	103	108	57	1,685	1,742
1986	7	125	131	77	2,801	2,878
1987	4	133	137	70	3,515	3,585
1988	3	148	151	54	4,072	4,127
1989	2	141	144	44	4,005	4,049
1990	2	134	136	45	3,824	3,869
1991	1	137	138	45	4,009	4,054
1992	1	132	133	34	4,066	4,100
1993	1	133	134	34	4,561	4,595
1994	1	148	148	26	4,791	4,817
1995	0	147	147	32	4,882	4,914
1996	0	159	160	22	5,401	5,424
1997	0	149	150	22	5,471	5,493
1998	0	155	156	28	5,562	5,591
1999	0	153	154	7	5,602	5,608
2000	0	148	148	16	5,883	5,899
2001	0	148	148	12	6,288	6,299
2002	0	160	161	7	6,554	6,561
	===========	===========	===========	===========	===========	===========
	225	3,241	3,466	1,466	87,710	89,176

The model developed for this study counts as "voluntary" saves by technology 208F:

- All people (age 5+) saved by integral 3-point belts installed before September 1, 1973 – primarily in European cars.
- Lives saved among occupants who used both the lap and the shoulder belts, in cars with separate lap and shoulder belts built before September 1, 1973.

However, as stated earlier, people who wore only the lap belt in cars with separate belts are counted as "saved by lap belts" (208A).

The model counts as "post-FMVSS" saves by technology 208F:

- All people (age 5+) saved by manual, integral 3-point belts installed on or after September 1, 1973, regardless of whether these integral belts were used correctly or incorrectly (e.g., wearing the shoulder portion of the belt behind the back, essentially obtaining protection from the lap portion only).
- Lives saved by automatic 3-point belts. Although they are automatic belts, they have had use patterns and effectiveness nearly identical to manual 3-point belts.

As stated above, any child age 1-4 saved by a belt, without a safety seat, is counted as "saved by lap belts" (208B), even if it was a 3-point belt, because the shoulder portion is intrinsically of little value to a child that small.

Three-point belts are the most important available occupant protection in most crash modes, reducing fatality risk by an average of 45 percent in passenger cars, when they are used. They were saving about 500 lives per year in 1977-83 as declining belt use more or less offset the growing proportion of the fleet equipped with 3-point belts. Buckle-up laws in most States phenomenally increased the benefits of belts to 4,127 lives saved by 1988. Continued increases in belt use and moderate growth in the "base" of VMT and potentially fatal crashes have raised that to 6,561 lives saved in 2002, and a cumulative 89,176 saved in 1960-2002. Both numbers are somewhat under half the overall benefits of the FMVSS in passenger cars.

Table 2-19 estimates lives saved by voluntary NCAP-related improvements and child safety seats (FMVSS 213). Since 1979, the New Car Assessment Program (NCAP) has advised the public about the injury performance of belted dummies in 35-mph frontal impact tests. During the early-to-mid 1980's, manufacturers substantially improved NCAP performance, modifying belt systems, steering assemblies, instrument panels and/or seat structures. For cars without air bags, NHTSA's evaluation estimated a 20 percent reduction of fatality risk for belted drivers in frontal impacts with other cars. NHTSA has not evaluated the relationship of NCAP scores with fatality risk in cars with air bags; the model limits its estimate of a benefit to cars without air bags. Lives saved by vehicle modifications associated with NCAP performance have saved an estimated 1,024 lives in 1983-2002; the estimates peaked at 79 in 1993 and have declined since then because almost all new cars have had driver air bags. Because NCAP is not a FMVSS, and any modifications to improve performance were voluntary on the part of the manufacturers, all benefits are counted as voluntary saves.

TABLE 2-19: CAR OCCUPANT LIVES SAVED BY VOLUNTARY NCAP-RELATED IMPROVEMENTS AND CHILD SAFETY SEATS, 1960-2002

	NCAP-RELATED VOLUNTARY IMPROVEMENTS (MEDIAN IMPLEMENTATION EARLY 1984)	213: CHILD SAFETY SEATS (USE EXCEEDED 50% IN 1985, FMVSS 4/1/71, STATE LAWS 1978-85)		
CY	TOTAL (VOLUNTARY)	VOLUNTARY	OBLIGATORY	TOTAL
1960	0	0	0	0
1961	0	0	0	0
1962	0	0	0	0
1963	0	0	0	0
1964	0	0	0	0
1965	0	0	0	0
1966	0	0	0	0
1967	0	0	0	0
1968	0	1	0	1
1969	0	1	0	1
1970	0	4	0	4
1971	0	8	0	8
1972	0	14	0	14
1973	0	18	0	18
1974	0	19	0	19
1975	0	34	0	34
1976	0	15	0	15
1977	0	32	0	32
1978	0	22	0	22
1979	0	48	0	48
1980	0	33	6	40
1981	0	66	2	69
1982	0	51	16	67
1983	2	42	65	107
1984	4	16	101	117
1985	16	7	138	145
1986	36	15	125	140
1987	52	12	164	175
1988	66	8	181	189
1989	65	15	194	209
1990	61	14	188	201
1991	71	10	198	208
1992	70	18	226	245
1993	79	4	238	242
1994	72	15	250	265
1995	68	17	243	260
1996	66	26	289	314
1997	67	20	268	288
1998	64	12	256	268
1999	50	22	273	295
2000	43	28	306	334
2001	41	30	205	236
2002	27	25	198	223
	===========	===========	===========	===========
	1,024	723	4,131	4,854

FMVSS 213 regulates the performance of child safety seats. However, the seats themselves are not part of a new vehicle. They must be purchased or acquired separately. At first, that was a voluntary decision for parents. During 1978-85, every State enacted laws requiring a safety seat for child passengers up to a certain age. The model counts as a "voluntary" save by child safety seats:

- Any child saved before a child passenger protection law was in effect in that State.
- Any child saved who was older than the range of ages included in that State's law.

The model counts as an "obligatory" save:

- Any child saved by a child safety seat in a State where a law was in effect, and that law required a child of that age to be in a safety seat.

However, any child saved by a belt, without a safety seat, is counted in one of the "saved by belts" categories, depending on the child's age and seat position, and is not counted as saved by a safety seat.

NHTSA's evaluation estimates that safety seats reduce fatality risk in passenger cars by 71 percent for infants and by 54 percent for toddlers. Safety seats saved an estimated 4,854 lives in 1960-2002, of which 4,131 were children obligated by a State law to be in a safety seat, while 723 were voluntary saves before State laws or involving older children. Benefits peaked at 334 in 2000, but dropped in 2001-2002 because the exposure "base" of children in potentially fatal crashes shrank – in part because more and more children are placed in the safer back seat to avoid exposure to air bags, and because many children are now transported in minivans or SUVs rather than cars.

Table 2-20 considers two of the more recent belt technologies: 3-point belts for rear-outboard occupants (208G) and automatic 2-point belts for front-outboard occupants (208H). FMVSS 208 has required lap/shoulder belts at rear-outboard seats since December 11, 1989[10], and quite a few make-models were voluntarily equipped with 3-point belts a year or two or even many years before that date. Three-point belts are quite effective for back-seat occupants, reducing fatality risk by an average of 44 percent (as compared to 32 percent for the lap belt only). Belt use, occupancy of the back seat and the proportion of cars on the road equipped with 3-point belts have steadily increased since 1990, pushing benefits to 297 lives saved in 2002, and a total of 2,224 in 1960-2002.

---

[10] A choice of 3-point belts or separate lap and shoulder belts was allowed from December 11, 1989 to August 31, 1990, whereas 3-point belts were required starting September 1, 1990; in reality, all cars have been equipped with 3-point belts since before December 11, 1989.

TABLE 2-20: CAR OCCUPANT LIVES SAVED BY 3-POINT BELTS FOR REAR-OUTBOARD OCCUPANTS AND AUTOMATIC 2-POINT BELTS, 1960-2002

	208G: 3-POINT BELTS FOR REAR-OUTBOARD OCCS (MEDIAN INSTALL YR 1989, FMVSS 12/11/89)			208H: AUTOMATIC 2-POINT BELTS (PEAK YEAR 1991, PHASE-IN BEGAN 9/1/86)		
CY	VOLUNTARY	POST-FMVSS	TOTAL	VOLUNTARY	POST-FMVSS	TOTAL
1960	0	0	0	0	0	0
1961	0	0	0	0	0	0
1962	0	0	0	0	0	0
1963	0	0	0	0	0	0
1964	0	0	0	0	0	0
1965	0	0	0	0	0	0
1966	0	0	0	0	0	0
1967	0	0	0	0	0	0
1968	0	0	0	0	0	0
1969	0	0	0	0	0	0
1970	0	0	0	0	0	0
1971	0	0	0	0	0	0
1972	0	0	0	0	0	0
1973	0	0	0	0	0	0
1974	0	0	0	0	0	0
1975	0	0	0	0	0	0
1976	0	0	0	1	0	1
1977	0	0	0	5	0	5
1978	0	0	0	6	0	6
1979	1	0	1	5	0	5
1980	0	0	0	5	0	5
1981	0	0	0	7	0	7
1982	0	0	0	5	0	5
1983	1	0	1	8	0	8
1984	3	0	3	11	0	11
1985	2	0	2	16	0	16
1986	8	0	8	16	0	16
1987	15	0	15	9	28	37
1988	13	0	13	16	102	117
1989	24	4	28	14	191	204
1990	23	18	41	16	417	433
1991	37	46	83	7	554	561
1992	27	58	85	15	616	631
1993	29	69	98	15	764	779
1994	23	115	139	15	906	921
1995	23	125	148	7	914	922
1996	29	149	178	5	970	975
1997	23	157	180	6	905	910
1998	32	212	244	10	854	864
1999	19	152	171	5	865	870
2000	22	216	238	5	812	817
2001	17	234	252	3	753	756
2002	14	282	297	9	730	739
	386	1,838	2,224	241	10,380	10,621

Automatic 2-point belts were introduced on a voluntary basis in the Volkswagen Rabbit (non-motorized) starting in 1975 and the Toyota Cressida (motorized) starting in 1981. Quite a few make-models were equipped with them during the MY 1987-89 phase-in period for automatic protection under FMVSS 208 and in the early 1990's. Consumers indicated a clear preference for the more effective combination of air bags and manual 3-point belts; automatic 2-point belts were gone from new cars by MY 1996, two years before FMVSS 208 required air bags and manual 3-point belts in all cars.

Automatic 2-point belts are less effective than 3-point belts, in part because the manual lap belt that accompanies them is often not used. Fatality reduction is an estimated 32 percent with automatic 2-point belts (vs. 45 percent with 3-point belts). During the 1980's the lower effectiveness was more than offset by a much higher use rate for the 2-point belts; by 2002, use of 3-point belts had largely caught up with the automatic use rate. Lives saved by automatic 2-point belts peaked at 975 in CY 1996 and subsequently declined as no new cars were equipped with them. Through 2002, automatic 2-point belts had saved an estimated 10,621 lives.

Table 2-21 assesses two of the most important recent safety technologies: voluntary TTI(d) reductions to improve side impact protection in 2-door cars (214B) and frontal air bags (208I). A dynamic test was added to FMVSS 214, with phase-in scheduled from September 1, 1993 to September 1, 1996, to improve occupant protection when a car is struck in the side by another vehicle. During the decade before the phase-in manufacturers tried out the test procedure and began to improve performance on the test criterion TTI(d) on a voluntary basis, especially in 2-door cars (that are the most vulnerable in side impacts). Side structure improvements in 1986-93 were similar to the additional improvements that many cars would receive during the phase-in period. If NHTSA had already evaluated the dynamic test standard itself, the model used in this report would have combined these voluntary improvements with the post-FMVSS improvements as a single technology; however, as of October 2004, NHTSA had only evaluated the fatality reduction associated with the voluntary improvements. They are estimated to reduce fatality risk in side impacts of 2-door cars by 23 percent over the time period they were introduced (1986-93). That amounts to 2,795 lives saved in 1960-2002, including 359 in 2002.

Frontal air bags are effective in reducing the fatality risk of adults and adolescents (age 13+) in frontal and partially frontal crashes. Benefits are to a small extent offset by increased risk to certain groups of child passengers (age 0-12). Table 2-21 presents *net* lives saved by air bags. At the end of this chapter, Tables 2-38 – 2-39 analyzes the effects of air bags separately by occupant age group and seat position. Mercedes offered air bags on a voluntary basis in MY 1985-86. After the FMVSS 208 phase-in of automatic protection began on September 1, 1986, other manufacturers began to offer driver or dual air bags, building up to 100 percent of new cars by MY 1997. Only the 1985-86 Mercedes air bags are included in the voluntary saves on Table 2-21. In all crashes, driver air bags reduce the fatality risk of belted occupants by an average of 11 percent, and unbelted occupants by 14 percent. The number of lives saved by air bags has grown rapidly, commensurate with the proportion of registered cars equipped with air bags, reaching 1,642 in 2002. It will continue to grow as older cars are retired, because even in 2002, 38 percent of the cars on the road still did not have any air bags, and 11 percent had only the driver air bag. Air bags saved an estimated 8,770 lives in passenger cars through 2002.

TABLE 2-21: CAR OCCUPANT LIVES SAVED BY VOLUNTARY TTI(d) REDUCTIONS IN 2-DOOR CARS AND BY FRONTAL AIR BAGS, 1960-2002

	214B: VOLUNTARY (PRE-1994) TTI(d) REDUCTIONS IN 2-DOOR CARS (MEDIAN IMPLEMENTATION YEAR 1993)	208I: FRONTAL AIR BAGS (MEDIAN INSTALL YR 1994, PHASE-IN BEGAN 9/1/86)		
CY	TOTAL (VOLUNTARY)	VOLUNTARY	POST-FMVSS	TOTAL
1960	0	0	0	0
1961	0	0	0	0
1962	0	0	0	0
1963	0	0	0	0
1964	0	0	0	0
1965	0	0	0	0
1966	0	0	0	0
1967	0	0	0	0
1968	0	0	0	0
1969	0	0	0	0
1970	0	0	0	0
1971	0	0	0	0
1972	0	0	0	0
1973	0	0	0	0
1974	0	0	0	0
1975	0	0	0	0
1976	0	0	0	0
1977	0	0	0	0
1978	0	0	0	0
1979	0	0	0	0
1980	0	0	0	0
1981	0	0	0	0
1982	0	0	0	0
1983	0	0	0	0
1984	0	0	0	0
1985	0	0	0	0
1986	4	1	0	1
1987	11	1	0	2
1988	25	0	2	3
1989	39	1	7	8
1990	49	0	43	43
1991	58	0	74	74
1992	69	0	99	99
1993	96	1	194	195
1994	141	0	283	283
1995	191	1	456	457
1996	247	0	613	613
1997	266	0	748	748
1998	263	1	904	905
1999	303	0	1,053	1,053
2000	321	1	1,246	1,247
2001	356	1	1,395	1,396
2002	359	0	1,642	1,642
	===========	===========	===========	===========
	2,795	11	8,759	8,770

Table 2-22 analyzes the two latest technologies that save lives of car occupants for which NHTSA has evaluated effectiveness: conspicuity tape for heavy trailers and 3-point belts for rear-center occupants. FMVSS 108 has required conspicuity tape on new trailers since December 1, 1993. The Federal Motor Carrier Safety Administration required all trailers on the road, even those built before December 1, 1993 to be equipped or retrofitted with tape by June 1, 2001. When trailers are equipped with tape, they are easier to see in the dark, and cars are 29 percent less likely to hit them.

The model[11] counts as a "voluntary" save by conspicuity tape:

- Any car occupant saved when his or her car, before June 1, 2001, avoided hitting a trailer built before December 1, 1993 yet equipped or retrofitted with tape.

The model counts as a "post-standard" save:

- Any car occupant saved when his or her car, at any time, avoided hitting a trailer built after December 1, 1993.
- Any car occupant saved when his or her car, after June 1, 2001, avoided hitting *any* trailer.

In 1991-93 conspicuity tape applied to trailers in advance of any Federal requirement began saving car occupants. From December 1, 1993 through May 31, 2001, lives were saved by a mix of trailers built after FMVSS 108 required the tape and older trailers voluntarily equipped or retrofitted with tape. After June 1, 2001, every save is a post-standard save, because FMCSA requires every trailer on the road to have the tape. This technology has saved a total of 683 car occupants through 2002, including 91 in 2002.

Anton's Law directs NHTSA to issue a regulation requiring the phase-in of 3-point belts at rear-center seats between September 1, 2005 and August 31, 2007. However, manufacturers have already furnished this technology voluntarily on numerous make-models, especially from MY 1999 onwards. Three-point belts reduce fatality risk, on the average, by 44 percent. It is estimated that they had saved 5 lives by the end of 2002. The safest place for a passenger to ride in a car is in the rear-center seat, protected by a 3-point belt, or by an age-appropriate safety seat if the passenger is a child too small for correct use of 3-point belts.

---

[11] Because FARS data do not report the MY of trailers, the model can only assess the probability of the following three scenarios based on the CY and month of the crash.

TABLE 2-22: CAR OCCUPANT LIVES SAVED BY CONSPICUITY TAPE ON HEAVY TRAILERS AND 3-POINT BELTS FOR REAR-CENTER OCCUPANTS, 1960-2002

CY	108: CONSPICUITY TAPE ON HEAVY TRAILERS (ON-ROAD FLEET EXCEEDED 50% IN 1996, FMVSS 12/1/93, FMCSA RETROFIT 6/1/2001)			208J: 3-POINT BELTS FOR REAR-CENTER OCCS (MEDIAN INSTALL YR 2001, FMVSS PHASE-IN TO BEGIN 9/2005)		
	VOLUNTARY	POST-STANDARD	TOTAL	VOLUNTARY	POST-FMVSS	TOTAL
1960	0	0	0	0	0	0
1961	0	0	0	0	0	0
1962	0	0	0	0	0	0
1963	0	0	0	0	0	0
1964	0	0	0	0	0	0
1965	0	0	0	0	0	0
1966	0	0	0	0	0	0
1967	0	0	0	0	0	0
1968	0	0	0	0	0	0
1969	0	0	0	0	0	0
1970	0	0	0	0	0	0
1971	0	0	0	0	0	0
1972	0	0	0	0	0	0
1973	0	0	0	0	0	0
1974	0	0	0	0	0	0
1975	0	0	0	0	0	0
1976	0	0	0	0	0	0
1977	0	0	0	0	0	0
1978	0	0	0	0	0	0
1979	0	0	0	0	0	0
1980	0	0	0	0	0	0
1981	0	0	0	0	0	0
1982	0	0	0	0	0	0
1983	0	0	0	0	0	0
1984	0	0	0	0	0	0
1985	0	0	0	0	0	0
1986	0	0	0	0	0	0
1987	0	0	0	0	0	0
1988	0	0	0	0	0	0
1989	0	0	0	0	0	0
1990	0	0	0	0	0	0
1991	6	0	6	0	0	0
1992	14	0	14	0	0	0
1993	22	0	22	0	0	0
1994	25	8	33	0	0	0
1995	27	18	45	0	0	0
1996	28	28	55	0	0	0
1997	30	40	70	0	0	0
1998	29	48	77	0	0	0
1999	30	60	91	0	0	0
2000	29	67	96	1	0	1
2001	10	74	84	2	0	2
2002	0	91	91	2	0	2
	===========	===========	===========	===========	===========	===========
	248	435	683	5	0	5

Benefits for occupants of LTVs

Table 2-23 is a summary of the benefits of safety technologies for LTV occupants. Actual fatalities in LTVs have rather steadily increased from 3,494 in 1960 to 12,248 in 2002, reflecting the growing market share of LTVs. But without the safety improvements they would have increased much more, from 3,516 to 22,579. Safety technologies saved 94,117 LTV occupants in 1960-2002, including 10,331 in 2002. Most of the lives saved (82,302) are by safety technologies directly associated with FMVSS in effect at that time. Voluntary improvements such as better instrument panels and compliance with FMVSS before their effective date have been saving about 400 lives a year since the mid-1990's.

The proportion of potential fatalities saved has grown from 0.63 percent in 1960, when 22 LTV occupants were saved by lap belts, to 45.75 percent in 2002. The latter is even higher than the corresponding proportion saved in cars (40.89 according to Table 2-10). That is primarily because safety belts are even more effective in LTVs than in cars – because fatal crashes of LTVs are more likely to involve rollover and/or ejection, and belts are especially effective in those situations.

Indeed, the largest boosts in the percent saved came in 1984-88 (from 11.02 to 22.43 percent), when buckle-up laws took effect in most of the States, and in 1995-2002 (from 33.72 to 45.75 percent) when belt use, historically lower in LTVs, especially in pickup trucks, rapidly caught up to the use rate in cars.

TABLE 2-23: LTV SUMMARY – ACTUAL OCCUPANT FATALITIES, POTENTIAL FATALITIES WITHOUT THE SAFETY TECHNOLOGIES, AND LIVES SAVED

| | LTV OCCUPANT FATALITIES | | | LIVES SAVED | | |
CY	ACTUAL	W/O SAFETY TECHS.	BY VOLUNTARY IMPROVEMENTS	BY FMVSS IN EFFECT	TOTAL	PERCENT SAVED
1960	3,494	3,516	22	0	22	0.63
1961	3,482	3,504	22	0	22	0.63
1962	3,787	3,812	25	0	25	0.65
1963	4,050	4,078	28	0	28	0.70
1964	4,414	4,448	34	0	34	0.76
1965	4,527	4,565	38	0	38	0.83
1966	4,785	4,831	46	0	46	0.95
1967	4,785	4,839	55	0	55	1.13
1968	5,299	5,379	80	0	80	1.49
1969	5,711	5,824	114	0	114	1.95
1970	5,505	5,641	136	0	136	2.41
1971	5,453	5,614	159	2	161	2.87
1972	5,659	5,853	173	21	194	3.31
1973	5,865	6,103	184	54	238	3.90
1974	5,042	5,286	172	72	244	4.61
1975	4,759	5,031	173	98	272	5.40
1976	5,419	5,742	199	125	324	5.64
1977	5,836	6,245	248	161	409	6.55
1978	6,550	7,012	279	183	462	6.59
1979	7,039	7,600	311	249	561	7.38
1980	7,389	7,997	350	258	608	7.60
1981	6,966	7,581	351	263	614	8.10
1982	6,465	7,051	332	253	585	8.30
1983	6,277	6,918	317	324	641	9.27
1984	6,581	7,396	308	507	815	11.02
1985	6,826	7,944	368	750	1,118	14.07
1986	7,414	9,034	364	1,256	1,620	17.93
1987	8,208	10,323	374	1,740	2,115	20.49
1988	8,497	10,954	399	2,058	2,457	22.43
1989	8,689	11,430	406	2,335	2,741	23.98
1990	8,741	11,697	402	2,555	2,957	25.28
1991	8,591	11,779	388	2,800	3,188	27.07
1992	8,239	11,655	393	3,024	3,417	29.31
1993	8,639	12,458	400	3,418	3,818	30.65
1994	9,030	13,500	443	4,026	4,469	33.11
1995	9,715	14,657	438	4,504	4,942	33.72
1996	10,039	15,718	468	5,212	5,679	36.13
1997	10,359	16,715	467	5,890	6,356	38.03
1998	10,730	17,728	475	6,523	6,998	39.48
1999	11,249	18,735	485	7,001	7,486	39.96
2000	11,521	20,203	473	8,209	8,682	42.97
2001	11,685	20,701	439	8,577	9,016	43.55
2002	12,248	22,579	478	9,853	10,331	45.75
	===========	===========	===========	===========	===========	
	305,559	399,675	11,814	82,302	94,117	

VMT data are available for LTVs in 1975-2002.[12] Occupant fatality rates per 100,000,000 VMT are shown in Table 2-24 and Figure 2-6. The trends are somewhat different from cars, because LTVs are a less homogeneous group of vehicles. Originally, the majority of LTVs were pickup trucks. The introduction of minivans in the mid-1980's, a new type of vehicle with exceptionally low fatality rates, helped to lower the average potential fatality rate for all LTVs from 2.32 in 1987 to 1.81 in 1992; the actual fatality rate fell even more sharply, from 1.85 to 1.28. After 1992, the market share for SUVs increased steadily relative to minivans and pickup trucks. Without the safety technologies, the potential fatality rate would have increased from 1.81 to 2.22 in 2002, back to mid-1980's levels. But thanks to safety improvements, above all increased belt use, the actual fatality rate dropped from 1.28 to 1.21. Essentially, increased belt use, air bags and other vehicle safety technologies made it possible to accommodate the shift to SUVs without an increase of actual fatality rates per 100,000,000 VMT.

Lives saved by 18 individual safety technologies for LTVs will now be discussed in chronological order as listed in Table 2-2, based on the "median installation year." Table 2-25 covers the two earliest technologies, 208A – lap belts for front-outboard occupants (age 5 and older), and 208B – lap belt use (at any seat position) by children age 1-4. By model year 1964, over half of new LTVs were equipped with lap belts; FMVSS 208 required them in LTVs as of July 1, 1971. Lap belts are more effective in LTVs than in cars, reducing fatality risk by 48 percent (vs. 27 percent in cars). Although some LTV make-models received 3-point belts in the early 1970's, most kept lap belts through MY 1975 and some even until 1980. Some of those LTVs were still on the road in 2002. Thus, the number of lives saved by lap belts reached a peak of 110 in 1986, after buckle-up laws increased use, and it did not decline that much in the 1990's. In all, lap belts have saved an estimated 2,612 front-outboard occupants age 5 or older.

Safety belts, without a safety seat, reduce the fatality risk of 1-4 year old children by an average of 48 percent in LTVs, and are assumed to have little or no value for infants (age 0). Belts have saved an estimated 295 children age 1-4, at all seat positions, in LTVs.

Table 2-26 tallies the benefits of dual master cylinders (105A) and improved door locks (206). Dual master cylinders were not required until FMVSS 105 was extended to LTVs, effective September 1, 1983. But they were introduced much earlier, presumably at about the same time as in cars, often a few years before 1968. They reduce the risk of a fatal crash involvement by an estimated 0.7 percent, saving a total of 2,379 LTV occupants in 1960-2002, including 157 in 2002. The number of lives saved has grown year by year as the "base" of registered LTVs, their VMT and potentially fatal crashes continues to grow.

---

[12] *Traffic Safety Facts 2002, op. cit.*, pp. 24.

TABLE 2-24

LTVs: ACTUAL AND POTENTIAL OCCUPANT FATALITIES
PER 100,000,000 MILES VMT

CY	VMT (10^8 MILES)	ACTUAL FATALITIES		POTENTIAL FATALITIES W/O SAFETY TECHNOLOGIES	
		N	RATE	N	RATE
1975	2,043	4,759	2.33	5,031	2.46
1976	2,334	5,419	2.32	5,742	2.46
1977	2,571	5,836	2.27	6,245	2.43
1978	2,895	6,550	2.26	7,012	2.42
1979	2,938	7,039	2.40	7,600	2.59
1980	2,955	7,389	2.50	7,997	2.71
1981	3,076	6,966	2.26	7,581	2.46
1982	3,220	6,465	2.01	7,051	2.19
1983	3,349	6,277	1.87	6,918	2.07
1984	3,586	6,581	1.84	7,396	2.06
1985	3,888	6,826	1.76	7,944	2.04
1986	4,165	7,414	1.78	9,034	2.17
1987	4,444	8,208	1.85	10,323	2.32
1988	4,884	8,497	1.74	10,954	2.24
1989	5,225	8,689	1.66	11,430	2.19
1990	5,557	8,741	1.57	11,697	2.11
1991	5,959	8,591	1.44	11,779	1.98
1992	6,424	8,239	1.28	11,655	1.81
1993	6,754	8,639	1.28	12,458	1.84
1994	7,115	9,030	1.27	13,500	1.90
1995	7,500	9,715	1.30	14,657	1.95
1996	7,873	10,039	1.28	15,718	2.00
1997	8,249	10,359	1.26	16,715	2.03
1998	8,620	10,730	1.24	17,728	2.06
1999	9,033	11,249	1.25	18,735	2.07
2000	9,429	11,521	1.22	20,203	2.14
2001	9,781	11,685	1.19	20,701	2.12
2002	10,164	12,248	1.21	22,579	2.22

FIGURE 2-6: ACTUAL AND POTENTIAL OCCUPANT FATALITIES
IN LTVs PER 100,000,000 MILES VMT

("A" = actual fatalities; "P" = potential fatalities without the vehicle safety technologies)

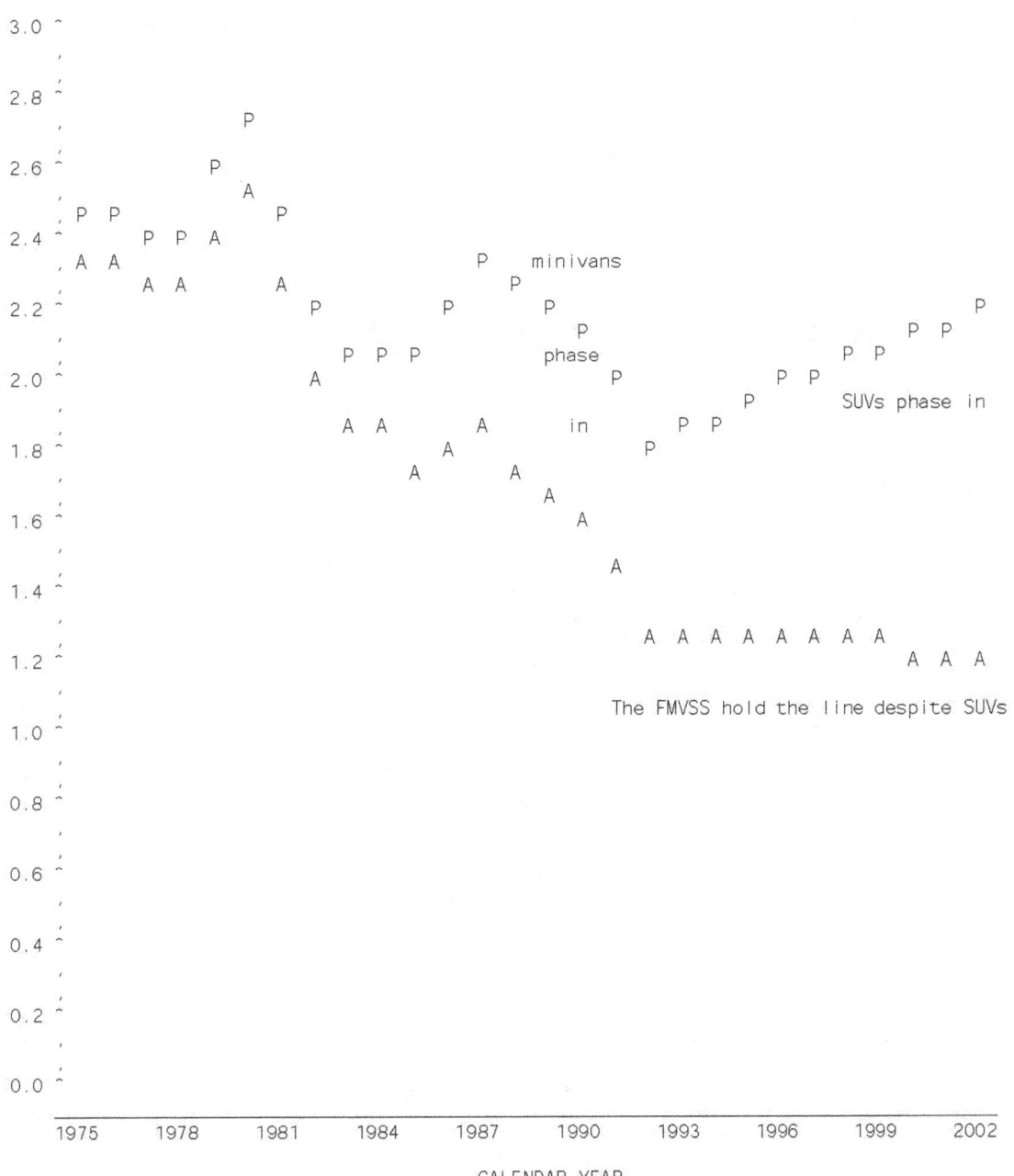

TABLE 2-25: LTV OCCUPANT LIVES SAVED BY FRONT-OUTBOARD LAP BELTS AND LAP BELTS FOR CHILDREN AGE 1-4, 1960-2002

	208A: LAP BELTS FOR FRONT-OUTBOARD OCCS (MEDIAN INSTALL YR 1964, FMVSS 7/1/71)			208B: LAP BELT USE BY CHILDREN AGE 1-4 (MEDIAN INSTALL YR 1966, FMVSS 7/1/71)		
CY	VOLUNTARY	POST-FMVSS	TOTAL	VOLUNTARY	POST-FMVSS	TOTAL
1960	22	0	22	0	0	0
1961	22	0	22	0	0	0
1962	24	0	24	0	0	0
1963	26	0	26	0	0	0
1964	28	0	28	0	0	0
1965	29	0	29	0	0	0
1966	30	0	30	0	0	0
1967	31	0	31	0	0	0
1968	42	0	42	0	0	0
1969	59	0	59	0	0	0
1970	68	0	68	0	0	0
1971	73	2	76	0	0	0
1972	71	15	86	0	0	0
1973	62	35	97	0	0	0
1974	46	43	89	0	0	0
1975	35	55	89	0	0	0
1976	26	58	84	0	0	1
1977	23	58	80	0	0	0
1978	20	53	73	0	0	0
1979	12	67	79	0	0	0
1980	8	42	50	0	1	1
1981	8	40	47	0	1	1
1982	12	38	50	0	0	0
1983	6	25	32	0	0	0
1984	14	40	54	0	4	4
1985	21	60	81	0	3	3
1986	22	88	110	0	9	9
1987	18	74	92	0	10	10
1988	26	84	109	1	10	11
1989	14	56	70	0	12	12
1990	22	50	72	0	9	9
1991	25	64	89	0	12	12
1992	17	56	73	1	11	12
1993	22	61	83	1	13	14
1994	19	47	67	0	21	21
1995	22	61	83	1	23	23
1996	20	45	65	0	25	25
1997	13	32	46	0	23	23
1998	14	31	45	0	27	27
1999	15	36	51	0	16	16
2000	13	25	38	0	26	26
2001	12	18	30	0	18	18
2002	7	34	41	0	13	13
	===========	===========	===========	===========	===========	===========
	1,118	1,494	2,612	7	288	295

TABLE 2-26: LTV OCCUPANT LIVES SAVED BY DUAL MASTER CYLINDERS AND IMPROVED DOOR LOCKS, 1960-2002

	105A: DUAL MASTER CYLINDERS (MEDIAN INSTALL YR LATE 66, FMVSS 9/1/83)			206: IMPROVED DOOR LOCKS (MEDIAN INSTALL YR 1967, FMVSS 1/1/72)		
CY	VOLUNTARY	POST-FMVSS	TOTAL	VOLUNTARY	POST-FMVSS	TOTAL
1960	0	0	0	0	0	0
1961	0	0	0	0	0	0
1962	0	0	0	1	0	1
1963	1	0	1	2	0	2
1964	1	0	1	5	0	5
1965	1	0	1	8	0	8
1966	3	0	3	13	0	13
1967	7	0	7	17	0	17
1968	12	0	12	25	0	25
1969	17	0	17	34	0	34
1970	20	0	20	39	0	39
1971	23	0	23	46	0	46
1972	26	0	26	51	5	56
1973	30	0	30	48	18	66
1974	28	0	28	35	27	63
1975	28	0	28	22	41	63
1976	33	0	33	23	55	78
1977	38	0	38	23	82	105
1978	43	0	43	25	90	115
1979	48	0	48	27	137	164
1980	52	0	52	25	163	188
1981	49	0	49	24	152	176
1982	46	0	46	20	142	162
1983	46	0	46	19	152	171
1984	43	6	49	18	163	181
1985	40	13	53	14	181	195
1986	39	22	61	13	231	244
1987	38	33	70	15	276	291
1988	35	40	75	12	307	319
1989	31	47	78	11	317	328
1990	28	52	80	8	328	336
1991	26	55	81	10	335	345
1992	23	58	80	7	344	351
1993	22	64	86	6	354	360
1994	20	73	93	7	397	404
1995	18	83	101	6	423	429
1996	16	93	109	5	462	467
1997	14	102	116	5	502	507
1998	12	111	123	4	535	539
1999	11	119	130	3	567	570
2000	9	131	141	3	617	620
2001	7	137	144	2	620	622
2002	7	151	157	2	692	694
	===========	===========	===========	===========	===========	===========
	989	1,390	2,379	681	8,717	9,398

FMVSS 206 took effect on January 1, 1972 in LTVs, four years later than in cars. The model assumes that door lock improvements in LTVs started at about the same time as in cars, MY 1962, but extended until MY 1972, the year the standard took effect. In cars, NHTSA's evaluation estimated a 15 percent reduction of fatalities in rollovers. However, NCSS data show that the proportion of ejections that are through the door portal is smaller in LTVs than in cars. The effectiveness in LTVs is estimated to be 10 percent. Despite that lower effectiveness, the benefits of improved door locks in LTVs are estimated to be substantial, because so many potentially fatal LTV crashes involve rollover and ejection: 9,398 lives saved in 1960-2002, including 694 in 2002.

Table 2-27 looks at the benefits of lap belts for occupants age 5+ at the rear-outboard (208C) and front-center (208D) seat positions. Lap belts were required at all designated seat positions in LTVs on July 1, 1971. FMVSS 208 required 3-point belts at the rear-outboard seats in LTVs starting September 1, 1991. Even lap belts alone are highly effective in the back seats of LTVs, reducing fatality risk by an average of 63 percent. They achieved their largest benefit, 80 lives saved, in 1993, just after the shift to 3-point belts, and have saved an estimated 978 lives in 1960-2002.

The front-center seat is more frequently occupied in pickup trucks, especially those that have a wide bench seat in the front and no back seats at all, than in cars. As a consequence, here is one technology that has saved more lives in LTVs (177) than in cars (81) during 1960-2002.

Table 2-28 considers lap belts for rear-center occupants (208E) and front disc brakes (105B). Rear-center seats are occupied fairly often in vans and large SUVs, and lap belts are quite effective, reducing fatality risk by an estimated 63 percent. The belts saved 403 lives in 1960-2002.

The model assumes front disc brakes were introduced in LTVs at about the same time as in cars – i.e., during MY 1965-77 – even though FMVSS 105 did not apply to LTVs until September 1, 1983.[13] Front disc brakes reduce the risk of a fatal crash involvement by an estimated 0.17 percent. They saved an estimated 501 LTV occupants in 1960-2002.

Table 2-29 analyzes voluntary modifications to mid and lower instrument panels and energy-absorbing steering assemblies, two basic "built-in" protections in frontal crashes. Instrument panels of LTVs were redesigned over several years, from about MY 1969 to 1976. That resulted in a 15.9 percent fatality reduction for unrestrained right-front passengers in frontal impacts. It saved an estimated 4,373 lives through CY 2002.

FMVSS 203 and 204 were not extended to LTVs until September 1, 1981. However, energy-absorbing columns were introduced in some LTVs as early as 1970, and in the majority by MY 1975-76, but in some not until MY 1982. Energy-absorbing steering assemblies have saved 11,472 lives through CY 2002, including 997 in 2002. They rank first in cumulative benefits among the non-belt technologies for LTVs, as well as for passenger cars.

---

[13] *Ward's Automotive Yearbooks* (Ward's Communications, Detroit, annual publication) show nearly identical percentages of U.S. cars, U.S. trucks and imported cars with front disc brakes in MY 1975-77 and have no information for U.S. trucks and imported cars before MY 1975.

TABLE 2-27: LTV OCCUPANT LIVES SAVED BY LAP BELTS FOR REAR-OUTBOARD OCCUPANTS AND LAP BELTS FOR FRONT-CENTER OCCUPANTS, 1960-2002

	208C: LAP BELTS FOR REAR-OUTBOARD OCCS (MEDIAN INSTALL YR 1968, FMVSS 7/1/71)			208D: LAP BELTS FOR FRONT-CENTER OCCS (MEDIAN INSTALL YR 1968, FMVSS 7/1/71)		
CY	VOLUNTARY	POST-FMVSS	TOTAL	VOLUNTARY	POST-FMVSS	TOTAL
1960	0	0	0	0	0	0
1961	0	0	0	0	0	0
1962	0	0	0	0	0	0
1963	0	0	0	0	0	0
1964	0	0	0	0	0	0
1965	0	0	0	0	0	0
1966	0	0	0	0	0	0
1967	0	0	0	0	0	0
1968	0	0	0	0	0	0
1969	0	0	0	0	0	0
1970	1	0	1	0	0	0
1971	1	0	1	0	0	0
1972	1	0	1	0	0	0
1973	1	1	1	0	0	1
1974	1	1	1	0	0	1
1975	2	1	3	1	1	2
1976	0	10	10	0	1	1
1977	0	3	3	1	0	1
1978	0	4	4	0	1	1
1979	0	6	6	0	1	1
1980	0	1	1	0	0	0
1981	0	2	2	0	1	1
1982	0	2	2	0	0	0
1983	0	1	1	0	0	0
1984	0	9	10	0	0	0
1985	0	13	13	0	5	5
1986	0	25	25	0	1	1
1987	0	46	46	0	10	10
1988	0	52	52	0	8	8
1989	0	46	46	0	11	11
1990	1	58	59	0	5	5
1991	0	67	67	0	5	5
1992	0	63	63	0	10	10
1993	0	80	80	0	8	8
1994	0	63	63	1	12	12
1995	0	48	48	1	12	13
1996	0	43	43	0	11	11
1997	0	76	76	0	14	14
1998	0	59	59	0	10	10
1999	0	50	50	0	7	7
2000	0	64	64	0	9	9
2001	0	45	45	0	13	13
2002	0	32	32	0	14	14
	===========	===========	===========	===========	===========	===========
	9	969	978	6	171	177

TABLE 2-28: LTV OCCUPANT LIVES SAVED BY LAP BELTS FOR REAR-CENTER OCCUPANTS AND FRONT DISC BRAKES, 1960-2002

	208E: LAP BELTS FOR REAR-CENTER OCCS (MEDIAN INSTALL YR 1968, FMVSS 7/1/71)			105B: FRONT DISC BRAKES (MEDIAN INSTALL YEAR 1971, FMVSS 9/1/83)		
CY	VOLUNTARY	POST-FMVSS	TOTAL	VOLUNTARY	POST-FMVSS	TOTAL
1960	0	0	0	0	0	0
1961	0	0	0	0	0	0
1962	0	0	0	0	0	0
1963	0	0	0	0	0	0
1964	0	0	0	0	0	0
1965	0	0	0	0	0	0
1966	0	0	0	0	0	0
1967	0	0	0	0	0	0
1968	0	0	0	0	0	0
1969	0	0	0	1	0	1
1970	0	0	0	1	0	1
1971	0	0	0	2	0	2
1972	0	0	0	2	0	2
1973	0	0	1	3	0	3
1974	0	0	1	4	0	4
1975	0	0	0	4	0	4
1976	1	0	1	5	0	5
1977	0	4	4	7	0	7
1978	0	7	7	8	0	8
1979	0	2	2	9	0	9
1980	0	0	0	10	0	10
1981	0	0	0	10	0	10
1982	0	0	0	9	0	9
1983	0	0	0	9	0	9
1984	0	0	0	9	1	10
1985	0	2	2	8	3	11
1986	0	3	3	8	5	13
1987	0	5	5	8	8	15
1988	4	10	14	7	9	17
1989	0	11	11	7	11	18
1990	0	10	10	6	12	18
1991	0	18	18	5	13	18
1992	0	14	14	5	13	18
1993	0	13	13	5	15	20
1994	0	16	16	4	17	21
1995	0	18	18	4	19	23
1996	0	19	19	4	22	25
1997	0	39	39	3	24	27
1998	0	36	36	3	26	28
1999	0	31	31	2	28	30
2000	0	65	65	2	30	32
2001	0	30	30	2	32	33
2002	0	44	44	1	35	37
	7	396	403	178	323	501

TABLE 2-29: LTV OCCUPANT LIVES SAVED BY VOLUNTARY INSTRUMENT PANEL IMPROVEMENTS AND ENERGY-ABSORBING STEERING ASSEMBLIES, 1960-2002

CY	201: VOLUNTARY INSTRUMENT PANEL IMPROVEMENTS (MEDIAN INSTALL YR 1972, FMVSS 9/1/81) TOTAL (VOLUNTARY)	203: ENERGY-ABSORBING STEERING ASSEMBLIES (MEDIAN INSTALL EARLY 76, FMVSS 9/1/81) VOLUNTARY	POST-FMVSS	TOTAL
1960	0	0	0	0
1961	0	0	0	0
1962	0	0	0	0
1963	0	0	0	0
1964	0	0	0	0
1965	0	0	0	0
1966	0	0	0	0
1967	0	0	0	0
1968	0	0	0	0
1969	1	0	0	0
1970	4	1	0	1
1971	7	2	0	2
1972	13	2	0	2
1973	20	13	0	13
1974	23	25	0	25
1975	27	30	0	30
1976	37	45	0	45
1977	47	71	0	71
1978	66	91	0	91
1979	73	112	0	112
1980	90	133	0	133
1981	92	141	1	142
1982	84	136	22	158
1983	83	114	48	162
1984	96	97	93	191
1985	103	108	137	245
1986	122	96	180	275
1987	140	89	232	322
1988	146	86	280	366
1989	166	84	327	411
1990	179	72	364	436
1991	176	65	379	444
1992	171	61	383	444
1993	180	63	430	493
1994	187	57	488	545
1995	206	51	576	627
1996	239	47	603	651
1997	239	43	660	703
1998	255	33	722	755
1999	245	32	754	785
2000	283	25	838	863
2001	274	21	912	934
2002	299	22	975	997
	===========	===========	===========	===========
	4,373	2,066	9,407	11,472

Table 2-30 estimates the benefits of 3-point belts for front-outboard occupants (208F) and adhesive windshield bonding (212). FMVSS 208 required 3-point belts at the front-outboard seats of most LTVs on January 1, 1976 and on some not until September 1, 1981. However, some make-models received 3-point belts well before January 1, 1976. Safety belt use is critically important in LTVs, because a high proportion of potentially fatal crashes involve rollover and/or ejection, and because belts are also highly effective in the other crash modes. Three-point belts reduce fatality risk by an average of 60 percent in LTVs, when they are used. Fortunately, belt use is as high in SUVs and vans as in cars.[14] Three-point belts saved an estimated 52,037 drivers and right-front passengers in 1960-2002, including 6,199 in 2002. Both of these numbers are more than half the overall benefits of all the LTV safety technologies in the same time period, and so are the estimates for each individual year from 1988 to 2002.

Adhesive windshield bonding gradually superseded the older installation method using rubber gaskets during approximately 1978-85 in LTVs. The model assumes that adhesive bonding has the same effect on ejection fatalities in LTVs as in cars: NCSS data show that the proportion of ejections that are via the windshield portal was about the same in LTVs with rubber gaskets as in cars with rubber gaskets. Adhesive bonding has saved an estimated 1,462 lives in 1960-2002, including 118 in 2002.

Table 2-31 considers child safety seats (213) and 3-point belts for rear-outboard occupants (208G). NHTSA's evaluation estimates that safety seats reduce fatality risk in LTVs by 58 percent for infants and by 59 percent for toddlers. Safety seats saved an estimated 1,100 lives in 1960-2002, of which 934 were children obligated by a State law to be in a safety seat, while 165 were voluntary saves before State laws or involving older children.

FMVSS 208 has required lap/shoulder belts at rear-outboard seats of LTVs since September 1, 1991, but most LTVs were voluntarily equipped with 3-point belts a few years before that date. Three-point belts are highly effective for back-seat occupants, reducing fatality risk by an average of 73 percent. Belt use, occupancy of the back seat and the proportion of LTVs on the road equipped with 3-point belts have increased rapidly, pushing benefits to 526 lives saved in 2002, and a total of 2,499 in 1960-2002.

Table 2-32 addresses side door beams (214A) and frontal air bags (208I). The side door strength test of FMVSS 214 was extended to LTVs effective September 1, 1993. Most LTVs were first equipped with side door beams in MY 1994, but on a few models they were voluntarily introduced in 1991-93. Side door beams reduce fatality risk in single-vehicle side impacts by 19 percent. They saved an estimated 701 lives in 1960-2002; annual benefits have been rising as new LTVs with side door beams replace the pre-standard fleet.

---

[14] Glassbrenner, D., *Safety Belt Use in 2003*, NHTSA Technical Report No. DOT HS 809 646, Washington, 2003.

TABLE 2-30: LTV OCCUPANT LIVES SAVED BY 3-POINT BELTS FOR FRONT-OUTBOARD OCCUPANTS AND ADHESIVE WINDSHIELD BONDING, 1960-2002

	208F: 3-POINT BELTS FOR FRONT-OUTBOARD OCCS (MEDIAN INSTALL YR 1976, FMVSS 1/1/76)			212: ADHESIVE WINDSHIELD BONDING (MEDIAN INSTALL YR 1980, FMVSS 9/1/78)		
CY	VOLUNTARY	POST-FMVSS	TOTAL	VOLUNTARY	POST-FMVSS	TOTAL
1960	0	0	0	0	0	0
1961	0	0	0	0	0	0
1962	0	0	0	0	0	0
1963	0	0	0	0	0	0
1964	0	0	0	0	0	0
1965	0	0	0	0	0	0
1966	0	0	0	0	0	0
1967	0	0	0	0	0	0
1968	0	0	0	0	0	0
1969	1	0	1	0	0	0
1970	2	0	2	0	0	0
1971	3	0	3	0	0	0
1972	4	0	4	0	0	0
1973	5	0	5	0	0	0
1974	9	0	9	0	0	0
1975	23	0	23	0	0	0
1976	26	0	26	0	0	0
1977	37	13	50	0	0	0
1978	23	28	51	1	0	1
1979	28	34	61	2	3	5
1980	27	43	70	2	7	9
1981	25	56	82	1	9	11
1982	20	33	53	1	12	13
1983	35	81	116	1	15	16
1984	27	157	184	1	21	22
1985	67	289	356	1	29	30
1986	63	637	701	1	38	39
1987	64	965	1,029	1	50	51
1988	72	1,166	1,237	1	58	59
1989	83	1,410	1,493	1	60	61
1990	76	1,571	1,648	1	66	67
1991	45	1,746	1,791	1	66	67
1992	62	1,939	2,002	1	62	62
1993	50	2,220	2,270	1	68	69
1994	80	2,640	2,720	0	76	76
1995	49	2,924	2,973	0	83	84
1996	63	3,467	3,530	0	85	86
1997	55	3,789	3,845	0	90	90
1998	41	4,217	4,258	0	99	99
1999	55	4,391	4,445	0	105	105
2000	48	5,271	5,319	0	108	108
2001	35	5,447	5,482	0	112	112
2002	39	6,160	6,199	0	118	118
	===========	===========	===========	===========	===========	===========
	1,342	50,695	52,037	20	1,441	1,462

239

TABLE 2-31: LTV OCCUPANT LIVES SAVED BY CHILD SAFETY SEATS
AND 3-POINT BELTS FOR REAR-OUTBOARD OCCUPANTS, 1960-2002

	213: CHILD SAFETY SEATS (USE EXCEEDED 50% IN 1985, FMVSS 4/1/71, STATE LAWS 1978-85)			208G: 3-POINT BELTS FOR REAR-OUTBOARD OCCS (MEDIAN INSTALL YR 1990, FMVSS 9/1/91)		
CY	VOLUNTARY	OBLIGATORY	TOTAL	VOLUNTARY	POST-FMVSS	TOTAL
1960	0	0	0	0	0	0
1961	0	0	0	0	0	0
1962	0	0	0	0	0	0
1963	0	0	0	0	0	0
1964	0	0	0	0	0	0
1965	0	0	0	0	0	0
1966	0	0	0	0	0	0
1967	0	0	0	0	0	0
1968	0	0	0	0	0	0
1969	0	0	0	0	0	0
1970	0	0	0	0	0	0
1971	0	0	0	0	0	0
1972	1	0	1	0	0	0
1973	1	0	1	0	0	0
1974	1	0	1	0	0	0
1975	2	0	2	0	0	0
1976	3	0	3	0	0	0
1977	2	0	2	0	0	0
1978	2	0	2	0	0	0
1979	1	0	1	0	0	0
1980	3	0	3	0	0	0
1981	1	0	1	0	0	0
1982	3	3	6	0	0	0
1983	3	2	5	0	0	0
1984	3	10	13	0	0	0
1985	5	16	20	0	0	0
1986	0	17	17	0	0	0
1987	2	32	34	0	0	0
1988	2	35	37	8	0	8
1989	3	27	30	6	0	6
1990	3	29	32	6	0	6
1991	4	39	44	28	0	28
1992	9	29	38	26	43	69
1993	2	34	36	24	57	80
1994	5	56	61	22	105	127
1995	8	63	71	16	88	105
1996	9	70	79	12	96	108
1997	15	68	84	16	208	224
1998	7	70	77	31	202	233
1999	17	85	101	35	287	322
2000	12	82	93	9	304	313
2001	13	79	92	14	332	346
2002	24	88	112	10	515	526
	===========	===========	===========	===========	===========	===========
	165	934	1,100	263	2,236	2,499

240

TABLE 2-32: LTV OCCUPANT LIVES SAVED BY SIDE DOOR BEAMS AND FRONTAL AIR BAGS, 1960-2002

| CY | 214A: SIDE DOOR BEAMS (MEDIAN INSTALL YEAR 1994, FMVSS 9/1/93) | | | 208I: FRONTAL AIR BAGS (MEDIAN INSTALL YR 1995, PHASE-IN BEGAN 9/1/94) | | |
	VOLUNTARY	POST-FMVSS	TOTAL	VOLUNTARY	POST-FMVSS	TOTAL
1960	0	0	0	0	0	0
1961	0	0	0	0	0	0
1962	0	0	0	0	0	0
1963	0	0	0	0	0	0
1964	0	0	0	0	0	0
1965	0	0	0	0	0	0
1966	0	0	0	0	0	0
1967	0	0	0	0	0	0
1968	0	0	0	0	0	0
1969	0	0	0	0	0	0
1970	0	0	0	0	0	0
1971	0	0	0	0	0	0
1972	0	0	0	0	0	0
1973	0	0	0	0	0	0
1974	0	0	0	0	0	0
1975	0	0	0	0	0	0
1976	0	0	0	0	0	0
1977	0	0	0	0	0	0
1978	0	0	0	0	0	0
1979	0	0	0	0	0	0
1980	0	0	0	0	0	0
1981	0	0	0	0	0	0
1982	0	0	0	0	0	0
1983	0	0	0	0	0	0
1984	0	0	0	0	0	0
1985	0	0	0	0	0	0
1986	0	0	0	0	0	0
1987	0	0	0	0	0	0
1988	0	0	0	0	0	0
1989	0	0	0	0	0	0
1990	0	0	0	0	0	0
1991	0	0	0	0	0	0
1992	1	0	1	4	0	4
1993	3	1	3	12	0	12
1994	4	12	16	24	1	25
1995	4	29	33	37	44	81
1996	2	46	47	39	112	151
1997	1	46	47	43	193	236
1998	3	70	73	52	277	329
1999	4	96	100	45	387	432
2000	3	108	111	48	486	534
2001	3	120	123	50	619	669
2002	4	143	146	59	772	831
	===========	===========	===========	===========	===========	===========
	31	670	701	414	2,890	3,304

Frontal air bags are effective in reducing the fatality risk of adults and adolescents (age 13+) in frontal and partially frontal crashes. Benefits are to a small extent offset by increased risk to certain groups of child passengers (age 0-12). Furthermore, in pickup trucks with dual air bags and on-off switches, turning the switch "off" can prevent harm to a child passenger, but takes away the benefit of air bags from an adult passenger. Table 2-32 presents *net* lives saved by air bags in LTVs. At the end of this chapter, Tables 2-38 – 2-40 analyze the effects of air bags separately by occupant age group and seat position and investigates the effects of on-off switches.

The FMVSS 208 phase-in period for automatic protection in LTVs extended from September 1, 1994 to August 30, 1997. LTVs equipped with driver or dual air bags before September 1, 1994 account for the voluntary saves on Table 2-32. NHTSA's evaluation showed air bags are about equally effective in cars and LTVs: driver air bags reduce the fatality risk of belted occupants by an average of 11 percent, and unbelted occupants by 14 percent. Air bags saved 831 lives in 2002; in that year, 41 percent of the LTVs on the road still did not have any air bags, and 11 percent had only the driver air bag. The number of lives saved by air bags will continue to grow as older LTVs are retired. Air bags saved an estimated 3,304 lives in LTVs through 2002.

Table 2-33 analyzes conspicuity tape for heavy trailers (108) and 3-point belts for rear-center occupants (208J). When heavy trailers are equipped with the red-and-white tape, they are easier to see in the dark, and LTVs are 29 percent less likely to hit them. The tape has saved a total of 422 LTV occupants through 2002, including 68 in 2002.

Three-point belts began to appear on the rear-center seats of LTVs before 2002.[15] Anton's Law will require them on all rear-center seats by MY 2008. The model estimates 3 lives saved in LTVs by the end of 2002.

---

[15] In a typical minivan, the two seats in the second row are considered "outboard" seats and have been equipped with 3-point belts since MY 1992 or earlier. Only the middle seat in the third row is considered a "center" seat.

TABLE 2-33: LTV OCCUPANT LIVES SAVED BY CONSPICUITY TAPE ON HEAVY TRAILERS AND 3-POINT BELTS FOR REAR-CENTER OCCUPANTS, 1960-2002

	108: CONSPICUITY TAPE ON HEAVY TRAILERS (ON-ROAD FLEET EXCEEDED 50% IN 1996, FMVSS 12/1/93, FMCSA RETROFIT 6/1/2001)			208J: 3-POINT BELTS FOR REAR-CENTER OCCS (MEDIAN INSTALL YR AFTER 2001, FMVSS PHASE-IN TO BEGIN 9/2005)		
CY	VOLUNTARY	POST-STANDARD	TOTAL	VOLUNTARY	POST-FMVSS	TOTAL
1960	0	0	0	0	0	0
1961	0	0	0	0	0	0
1962	0	0	0	0	0	0
1963	0	0	0	0	0	0
1964	0	0	0	0	0	0
1965	0	0	0	0	0	0
1966	0	0	0	0	0	0
1967	0	0	0	0	0	0
1968	0	0	0	0	0	0
1969	0	0	0	0	0	0
1970	0	0	0	0	0	0
1971	0	0	0	0	0	0
1972	0	0	0	0	0	0
1973	0	0	0	0	0	0
1974	0	0	0	0	0	0
1975	0	0	0	0	0	0
1976	0	0	0	0	0	0
1977	0	0	0	0	0	0
1978	0	0	0	0	0	0
1979	0	0	0	0	0	0
1980	0	0	0	0	0	0
1981	0	0	0	0	0	0
1982	0	0	0	0	0	0
1983	0	0	0	0	0	0
1984	0	0	0	0	0	0
1985	0	0	0	0	0	0
1986	0	0	0	0	0	0
1987	0	0	0	0	0	0
1988	0	0	0	0	0	0
1989	0	0	0	0	0	0
1990	0	0	0	0	0	0
1991	3	0	3	0	0	0
1992	6	0	6	0	0	0
1993	11	0	11	0	0	0
1994	12	4	17	0	0	0
1995	14	9	23	0	0	0
1996	13	13	26	0	0	0
1997	19	25	43	0	0	0
1998	19	32	52	0	0	0
1999	21	42	63	0	0	0
2000	19	43	62	0	0	0
2001	5	43	48	0	0	0
2002	0	68	68	3	0	3
	===========	===========	===========	===========	===========	===========
	142	280	422	3	0	3

Benefits for pedestrians, bicyclists and other non-occupants

Table 2-34 is a summary of the non-occupant lives saved by car/LTV safety technologies, specifically braking improvements. The computation is straightforward: dual master cylinders reduce fatal crash involvements, including collisions with non-occupants by 0.70 percent, front disc brakes by 0.17 percent. Thus, when all cars and LTVs on the road are equipped with dual master cylinders and front disc brakes, 0.87 percent of potential non-occupant fatalities will be avoided. Indeed, the right column of Table 2-34 shows that the proportion of potential fatalities saved had reached 0.85 percent by 1987 and was converging on 0.87 percent by 2000-2002. For example, in 2002, there were 5,011 actual pedestrian, bicyclist and other non-occupant fatalities in collisions with cars or LTVs. Without the braking improvements there would have been 5,055; the brakes prevented 44 fatalities, 0.87 percent of 5,055.

Non-occupant fatalities in collisions with cars/LTVs peaked at 9,277 in 1972 and have greatly declined since then. That is primarily due to demographic factors: an increasingly suburban, aging and affluent population is less likely to walk, bike or ride a horse to work or to other destinations on public roads. Braking improvements in cars and LTVs have played a relatively small role in the decline. They have prevented an estimated 1,782 of 304,025 potential fatalities. Furthermore, the absolute number of lives saved has decreased from a peak of 65 in 1980 to 44 in 2002 because the "base" of non-occupant exposure to potentially fatal crash situations had shrunk. Table 2-35 apportions the non-occupant lives saved between the two brake technologies: dual master cylinders saved an estimated 1,496 lives in 1960-2002; front disc brakes, 286.

Benefits for motorcyclists

Table 2-36 summarizes motorcyclist lives saved by car/LTV braking improvements. The computation is similar to Table 2-34: when all cars and LTVs on the road are equipped with dual master cylinders and front disc brakes, 0.87 percent of potential motorcyclist fatalities in collisions between cars/LTVs and motorcycles will be avoided. In 2002, for example, 1,366 motorcyclists were actually killed in collisions with cars or LTVs. Without the braking improvements in the cars/LTVs there would have been 1,378; the brakes saved 12 lives, 0.87 percent of 1,378.

As the popularity of motorcycles has risen and fallen over the years, so have motorcyclist fatalities in collisions with cars/LTVs: rising from 388 in 1960 to 2,285 in 1980, falling to 1,000 in 1996-97, and returning to 1,366 in 2002 as many baby-boomers have taken up cycling again. Motorcycle helmets (a safety technology outside the scope of this report) also have a major impact on fatalities. Braking improvements in cars and LTVs have played a relatively small role. They have prevented an estimated 398 of 58,349 potential fatalities. Table 2-37 shows that dual master cylinders in cars/LTVs saved an estimated 332 motorcyclists in 1960-2002; front disc brakes, 66.

TABLE 2-34: PEDESTRIAN/BICYLIST SUMMARY
ACTUAL NON-OCCUPANT FATALITIES IN COLLISIONS WITH CARS/LTVs, POTENTIAL FATALITIES WITHOUT CAR/LTV BRAKING IMPROVEMENTS, AND LIVES SAVED BY CAR/LTV BRAKING IMPROVEMENTS

| | NON-OCCUPANT FATALITIES | | LIVES SAVED | | | |
CY	ACTUAL	W/O CAR/LTV BRAKE TECHS.	BY VOLUNTARY IMPROVEMENTS	BY FMVSS IN EFFECT	TOTAL	PERCENT SAVED
1960	6,447	6,447	0	0	0	0.00
1961	6,407	6,407	0	0	0	0.00
1962	6,604	6,604	0	0	0	0.00
1963	6,918	6,919	1	0	1	0.01
1964	7,587	7,588	1	0	1	0.02
1965	7,469	7,470	2	0	2	0.02
1966	7,901	7,905	4	0	4	0.05
1967	7,941	7,949	9	0	9	0.11
1968	8,334	8,349	13	2	15	0.18
1969	8,334	8,355	15	6	21	0.25
1970	8,805	8,834	17	11	28	0.32
1971	9,002	9,037	18	17	35	0.39
1972	9,277	9,319	21	22	42	0.46
1973	9,159	9,207	23	26	48	0.52
1974	7,626	7,671	20	25	45	0.59
1975	7,882	7,935	22	31	53	0.67
1976	7,599	7,653	21	33	54	0.70
1977	7,828	7,886	22	36	58	0.73
1978	7,776	7,836	22	38	60	0.77
1979	8,088	8,152	22	42	64	0.78
1980	8,077	8,142	22	43	65	0.80
1981	7,879	7,943	20	45	64	0.81
1982	7,472	7,533	20	42	62	0.82
1983	6,835	6,891	18	39	57	0.82
1984	7,116	7,176	17	43	60	0.83
1985	6,911	6,969	16	43	59	0.84
1986	6,975	7,035	13	47	59	0.84
1987	6,982	7,042	12	48	60	0.85
1988	7,117	7,179	10	51	61	0.85
1989	6,706	6,763	9	49	58	0.85
1990	6,692	6,750	8	49	58	0.86
1991	6,084	6,137	7	46	53	0.86
1992	5,746	5,796	5	45	50	0.86
1993	5,974	6,026	5	47	52	0.86
1994	5,729	5,778	5	45	50	0.86
1995	5,908	5,960	4	47	51	0.86
1996	5,681	5,730	3	46	50	0.86
1997	5,621	5,670	3	46	49	0.86
1998	5,477	5,524	2	45	48	0.87
1999	5,185	5,230	2	43	45	0.86
2000	4,963	5,006	2	42	43	0.87
2001	5,119	5,164	1	44	45	0.87
2002	5,011	5,055	1	43	44	0.87
	===========	===========	===========	===========	===========	
	302,243	304,025	456	1,326	1,782	

TABLE 2-35: NON-OCCUPANT LIVES SAVED BY
CAR/LTV DUAL MASTER CYLINDERS AND FRONT DISC BRAKES, 1960-2002

	105A: DUAL MASTER CYLINDERS (MEDIAN INSTALLATION YEAR LATE 1966, FMVSS 1/1/68 [CARS], 9/1/83 [LTVs])			105B: FRONT DISC BRAKES (MEDIAN INSTALLATION YEAR 1971, FMVSS 1/1/76 [CARS], 9/1/83 [LTVs])		
CY	VOLUNTARY	POST-FMVSS	TOTAL	VOLUNTARY	POST-FMVSS	TOTAL
1960	0	0	0	0	0	0
1961	0	0	0	0	0	0
1962	0	0	0	0	0	0
1963	1	0	1	0	0	0
1964	1	0	1	0	0	0
1965	2	0	2	0	0	0
1966	4	0	4	0	0	0
1967	9	0	9	0	0	0
1968	13	2	15	0	0	0
1969	15	6	21	1	0	1
1970	16	11	27	1	0	1
1971	16	17	33	2	0	2
1972	17	22	39	3	0	3
1973	18	26	44	4	0	4
1974	16	25	40	5	0	5
1975	16	31	47	6	0	6
1976	14	33	47	6	0	7
1977	15	35	50	7	1	8
1978	15	36	51	7	2	9
1979	16	38	54	7	3	10
1980	15	39	54	7	4	11
1981	14	40	54	6	5	11
1982	14	37	51	6	5	11
1983	13	34	47	5	5	10
1984	12	37	49	5	6	11
1985	12	36	48	4	7	11
1986	9	39	48	3	8	11
1987	9	40	49	3	8	11
1988	7	42	50	3	9	11
1989	7	40	47	2	9	11
1990	6	40	47	2	9	11
1991	5	37	43	1	9	10
1992	4	36	40	1	8	10
1993	4	38	42	1	9	10
1994	4	37	40	1	9	10
1995	3	38	42	1	9	10
1996	3	37	40	1	9	10
1997	2	37	40	1	9	9
1998	2	37	39	0	9	9
1999	2	35	36	0	8	9
2000	1	34	35	0	8	8
2001	1	35	36	0	8	9
2002	1	34	35	0	8	8
	===========	===========	===========	===========	===========	===========
	353	1,143	1,496	103	183	286

TABLE 2-36: MOTORCYLIST SUMMARY
ACTUAL MOTORCYCLIST FATALITIES IN COLLISIONS WITH CARS/LTVs, POTENTIAL FATALITIES WITHOUT CAR/LTV BRAKING IMPROVEMENTS, AND LIVES SAVED BY CAR/LTV BRAKING IMPROVEMENTS

	MOTORCYCLIST FATALITIES		LIVES SAVED			
CY	ACTUAL	W/O CAR/LTV BRAKE TECHS.	BY VOLUNTARY IMPROVEMENTS	BY FMVSS IN EFFECT	TOTAL	PERCENT SAVED
1960	388	388	0	0	0	0.00
1961	362	362	0	0	0	0.00
1962	414	414	0	0	0	0.00
1963	465	465	0	0	0	0.01
1964	569	569	0	0	0	0.02
1965	827	827	0	0	0	0.02
1966	1,111	1,112	1	0	1	0.05
1967	1,034	1,035	1	0	1	0.11
1968	982	984	2	0	2	0.18
1969	1,008	1,011	2	1	2	0.24
1970	1,189	1,193	2	1	4	0.31
1971	1,137	1,142	2	2	4	0.37
1972	1,396	1,402	3	3	6	0.44
1973	1,603	1,611	4	4	8	0.51
1974	1,654	1,664	5	5	9	0.57
1975	1,569	1,579	5	6	10	0.65
1976	1,602	1,613	5	6	11	0.69
1977	1,957	1,971	6	8	14	0.72
1978	2,136	2,152	7	9	16	0.75
1979	2,238	2,256	7	10	18	0.78
1980	2,285	2,303	7	11	18	0.79
1981	2,188	2,205	7	11	18	0.80
1982	2,019	2,036	6	10	16	0.81
1983	1,864	1,879	6	10	15	0.81
1984	2,040	2,057	6	11	17	0.83
1985	2,003	2,020	5	12	17	0.84
1986	2,017	2,034	4	13	17	0.83
1987	1,803	1,818	3	12	15	0.84
1988	1,682	1,696	3	11	14	0.85
1989	1,385	1,396	2	9	12	0.85
1990	1,448	1,460	2	10	12	0.86
1991	1,219	1,229	2	9	11	0.85
1992	1,043	1,052	1	8	9	0.86
1993	1,125	1,135	1	8	10	0.86
1994	1,034	1,043	1	8	9	0.86
1995	1,041	1,050	1	8	9	0.87
1996	1,000	1,009	1	8	9	0.86
1997	1,000	1,009	1	8	9	0.87
1998	1,004	1,013	1	8	9	0.87
1999	1,068	1,078	0	9	9	0.87
2000	1,271	1,282	0	11	11	0.87
2001	1,405	1,417	0	12	12	0.87
2002	1,366	1,378	0	12	12	0.87
	===========	===========	===========	===========	===========	
	57,951	58,349	114	284	398	

TABLE 2-37: MOTORCYCLIST LIVES SAVED BY
CAR/LTV DUAL MASTER CYLINDERS AND FRONT DISC BRAKES, 1960-2002

CY	105A: DUAL MASTER CYLINDERS (MEDIAN INSTALLATION YEAR LATE 1966, FMVSS 1/1/68 [CARS], 9/1/83 [LTVs])			105B: FRONT DISC BRAKES (MEDIAN INSTALLATION YEAR 1971, FMVSS 1/1/76 [CARS], 9/1/83 [LTVs])		
	VOLUNTARY	POST-FMVSS	TOTAL	VOLUNTARY	POST-FMVSS	TOTAL
1960	0	0	0	0	0	0
1961	0	0	0	0	0	0
1962	0	0	0	0	0	0
1963	0	0	0	0	0	0
1964	0	0	0	0	0	0
1965	0	0	0	0	0	0
1966	1	0	1	0	0	0
1967	1	0	1	0	0	0
1968	2	0	2	0	0	0
1969	2	1	2	0	0	0
1970	2	1	4	0	0	0
1971	2	2	4	0	0	0
1972	3	3	6	0	0	0
1973	3	4	7	1	0	1
1974	4	5	9	1	0	1
1975	4	6	9	1	0	1
1976	4	6	10	1	0	1
1977	4	8	12	2	0	2
1978	5	9	14	2	1	2
1979	5	10	15	2	1	3
1980	5	10	15	2	1	3
1981	5	10	15	2	1	3
1982	5	9	14	2	1	3
1983	4	8	13	2	1	3
1984	4	10	14	2	1	3
1985	4	10	14	1	2	3
1986	3	11	14	1	2	3
1987	2	10	13	1	2	3
1988	2	9	12	1	2	3
1989	2	8	10	1	2	2
1990	2	8	10	0	2	2
1991	1	7	9	0	2	2
1992	1	6	7	0	1	2
1993	1	7	8	0	2	2
1994	1	7	7	0	2	2
1995	1	7	7	0	2	2
1996	1	6	7	0	2	2
1997	1	6	7	0	2	2
1998	0	7	7	0	2	2
1999	0	7	8	0	2	2
2000	0	9	9	0	2	2
2001	0	10	10	0	2	2
2002	0	9	10	0	2	2
	===========	===========	===========	===========	===========	===========
	88	244	332	26	40	66

Effect of air bags, by seat position, occupant age and on-off switch availability

The analysis model developed for this report permits more detailed estimate of the *net* effect of air bags on various population subgroups. Tables 2-7, 2-21 and 2-32 show that air bags had saved a net total of 12,074 lives by the end of CY 2002, including 8,770 in passenger cars and 3,304 in LTVs. Table 2-38 shows that 10,349 of the lives saved were drivers. The computation of benefits is simpler for drivers than for right-front passengers. Even though NHTSA's Special Crash Investigations (SCI) show individual cases where air bags have harmed drivers, the statistical analysis of FARS data estimates a net fatality reduction for every age/gender group and vehicle type. The model applies these effectiveness estimates – ranging from 33 percent for unbelted drivers in directly frontal impacts (12:00) down to 5 percent for belted drivers in side impacts partially frontal force (10:00 or 2:00) to zero in entirely non-frontal impacts – to the actual FARS cases of driver fatalities in vehicles equipped with air bags, inflating these cases to determine how many additional fatalities there would have been without air bags. Table 2-38 estimates that air bags saved 7,394 car drivers and 2,955 LTV drivers. The number of lives saved has grown year by year, as vehicles with driver or dual air bags replace older vehicles without air bags.

Drivers account for a very large proportion (10,349 of 12,074) of the lives saved by air bags because:

- The right-front seat is occupied in only 1 of every 3 vehicles, but the driver's seat is always occupied.
- There are many vehicles with driver air bags only, and they have been in service longer than the more recent vehicles with dual air bags.

Table 2-39 estimates the effects of passenger air bags in vehicles without on-off switches for those air bags, including cars, SUVs, minivans, most large vans, and some pickup trucks with full crew cabs. Table 2-39 has two sections. The upper half considers adults and adolescents age 13 or older. Here, air bags are beneficial; in fact, they are slightly more effective than for drivers. The computation is similar to the analysis of drivers in Table 2-38. Passenger air bags saved an estimated 1,743 adults and adolescents, in vehicles without on-off switches, in 1987-2002.

The lower half of Table 2-39 estimates the *net* effect of air bags on child passengers age 0-12. In this group, SCI has identified 154 individual cases (CY 1993-2002 crashes reported to SCI as of October 1, 2004) where air bags resulted in fatal injuries to children in vehicles without on-off switches, but there conceivably have been other crashes where the air bags prevented a fatality. A negative number in Table 2-39 indicates a fatality increase. The computations in Table 2-39 are:

249

TABLE 2-38

LIVES SAVED BY DRIVER AIR BAGS IN CARS AND LTVs, 1984-2002

LIVES SAVED BY DRIVER AIR BAGS

CY	TOTAL	IN CARS	IN LTVs
1984	0	0	0
1985	0	0	0
1986	1	1	0
1987	2	2	0
1988	3	3	0
1989	8	8	0
1990	43	43	0
1991	72	71	0
1992	101	97	4
1993	199	187	12
1994	291	266	25
1995	478	399	79
1996	692	543	149
1997	851	630	221
1998	1,052	752	300
1999	1,261	877	384
2000	1,494	1,026	468
2001	1,749	1,153	595
2002	2,051	1,334	717
	===========	===========	===========
	10,349	7,394	2,955

TABLE 2-39

ESTIMATED EFFECT OF PASSENGER AIR BAGS ON ADULTS (AGE 13+) AND CHILDREN (AGE 0-12) IN CARS AND LTVs *WITHOUT* ON-OFF SWITCHES, 1987-2002

ADULTS (AGE 13+) SAVED BY PASSENGER AIR BAGS

CY	TOTAL	IN CARS	IN LTVs W/O SWITCHES
1987	0	0	0
1988	0	0	0
1989	0	0	0
1990	0	0	0
1991	3	3	0
1992	2	2	0
1993	8	8	0
1994	21	20	1
1995	67	64	2
1996	92	84	8
1997	144	131	13
1998	192	167	25
1999	222	189	33
2000	277	234	43
2001	305	258	46
2002	411	322	89
	===========	===========	===========
	1,743	1,482	261

NET EFFECT ON CHILD PASSENGERS (AGE 0-12)
(Negative numbers indicate a net increase in fatalities)

CY	TOTAL	IN CARS	IN LTVs W/O SWITCHES
1987	0	0	0
1988	0	0	0
1989	0	0	0
1990	0	0	0
1991	0	0	0
1992	0	0	0
1993	-0	-0	0
1994	-4	-3	-1
1995	-6	-6	0
1996	-19	-14	-5
1997	-13	-13	-1
1998	-18	-14	-4
1999	-15	-13	-1
2000	-17	-13	-4
2001	-19	-16	-3
2002	-18	-14	-4
	===========	===========	===========
	-130	-106	-23

251

- Strictly based on the same type of statistical model used to compute the benefits of air bags for people age 13 or older. In other words,
  - The [often negative] effect of air bags was estimated by double-pair comparison analysis of FARS data, and it depends on the age of the child, restraint use, and vehicle impact location.
  - The negative effectiveness estimates are applied to the actual FARS cases of child passenger fatalities in vehicles equipped with dual air bags, deflating these cases to determine how many fewer fatalities there would have been without air bags.
- Information from SCI cases was not used at all in generating these estimates (except, of course, many of the SCI cases are also FARS cases).
- Because NHTSA has not yet statistically evaluated the redesign of air bags circa 1998, this analysis assumes the same effectiveness for frontal air bags before and after the redesign. If the later air bags are less harmful for children, the numbers for 1998-2002 may be overstated, and the numbers for 1994-97 understated.

Based on this statistical approach, the model estimates that passenger air bags are associated with a net increase of 130 FARS-reportable child passenger fatalities through 2002, in vehicles without on-off switches. "FARS-reportable" means, among other things, that the crash happened on a public road. This statistical estimate is quite compatible with the count of 154 SCI cases, considering that the count includes some crashes on private property such as parking lots, and makes no allowance for lives that air bags may have saved in other crashes (it is not a net effect).

Table 2-39 allows a year-by-year comparison of the estimated fatality reduction for adults and adolescents (age 13+) and the fatality increase for children (age 0-12). In 1996, air bags saved 92 adults, but child passenger fatalities increased by 19. That is a ratio of 4.8 adults saved per child passenger fatality. In 2002, a much larger number of vehicles with dual air bags on the road saved 411 adult lives, but child passenger fatalities had decreased to 18. That is a ratio of 22.8 adults saved per child passenger fatality. Because our statistical model assumes the same effects for earlier and later air bags – it does not adjust the effectiveness for the air bag redesign circa 1998 – the enormous improvement in the ratio can be attributed to one factor alone: the public, to a large extent, has moved child passengers to the back seat, and the exposure of small children to air bags has been greatly reduced.

Table 2-40 estimates the effects of passenger air bags in pickup trucks with on-off switches, and also the effect of the switches. Like Table 2-39, it has two sections. Ideally, switches would always be "on" for passengers age 13 and older, securing for them the benefits of air bags, and they would always be "off" for child passengers up to age 12, assuring that children would never be exposed to air bags. If that were the case, the first column of the top half of Table 2-40 would be identical to the second column, showing air bags achieving their full potential to protect adults (147 lives saved); the first column of the lower section of Table 2-40 would show air bags having no effect at all on child passengers (because they are always turned off) and it would be entirely filled with zeros.

TABLE 2-40

ESTIMATED EFFECT OF PASSENGER AIR BAGS ON ADULTS (AGE 13+) AND CHILDREN (AGE 0-12) IN PICKUP TRUCKS *WITH* ON-OFF SWITCHES, 1995-2002

(Positive numbers indicate lives saved; negative numbers indicate net increases in fatalities.
Block print indicates actual effects of air bags with switches 'on';
italics indicate hypothetical effects;
bold italics indicate children saved by turning switches 'off')

EFFECT ON ADULT PASSENGERS (AGE 13+)

CY	ACTUAL LIVES SAVED BY AIR BAGS WITH SWITCHES 'ON'	*LIVES SAVED IF ALL SWITCHES HAD BEEN 'ON'*	*POTENTIAL SAVINGS LOST BECAUSE SWITCHES WERE TURNED OFF*
1995	0	*0*	*0*
1996	0	*0*	*0*
1997	2	*3*	*-0*
1998	9	*11*	*-2*
1999	17	*21*	*-4*
2000	27	*35*	*-7*
2001	32	*40*	*-8*
2002	31	*38*	*-7*
	===========	===========	===========
	117	*147*	*-30*

EFFECT ON CHILD PASSENGERS (AGE 0-12)

CY	NET EFFECT ON CHILDREN BY AIR BAGS BECAUSE SWITCHES WERE 'ON'	*NET EFFECT IF ALL SWITCHES HAD BEEN 'ON'*	***CHILDREN SAVED BECAUSE SWITCHES WERE TURNED OFF***
1995	0	*0*	***0***
1996	0	*0*	***0***
1997	-0	*-1*	***0***
1998	-1	*-6*	***4***
1999	-1	*-2*	***2***
2000	-1	*-2*	***1***
2001	-1	*-6*	***4***
2002	-2	*-8*	***6***
	===========	===========	===========
	-6	*-24*	***18***

In reality, though, NHTSA's survey in 2000 indicated that switches were turned off, for example, for 22 percent of 13-15 year olds, 15 percent of adults age 20-59, and 56 percent of people age 70 or older. Switches were turned on for 14 percent of infants, 26 percent of 1-6 year olds, and 70 percent of 11-12 year olds. The statistical model for this report factors in the probability that the switch was on and the effectiveness of air bags, given the occupant's age and the vehicle's impact location, to estimate how the actual number of fatalities in FARS would have changed:

- If the trucks had passenger air bags and no switches – or if they did have switches, but they were left on all the time (these are the same).
- If the trucks did not have passenger air bags – or if they did have air bags and the switches were left off all the time (these are the same).

The upper half of Table 2-40 estimates that passenger air bags in pickup trucks with switches actually saved 117 occupants age 13 or older. In other words, the switches were left on for these occupants; the air bags deployed and saved their lives. However, if in these same trucks the switch had been correctly left on for every occupant age 13 or older (or if pickup trucks had passenger air bags and no switches), the number of lives (age 13+) saved by air bags would have been 147 rather than 117. In other words, 30 of the 147 potential saves were lost because an air bag was turned off for a person age 13 or older. There are two ways of looking at these statistics. On the one hand, it could be said that inappropriate use of the switches cost 30 lives. On the other, it could be argued that, without some device such as the switches to protect child passengers, dual air bags simply could not have been mandated in pickup trucks without a back seat; in that sense, the switches saved 117 adults who, otherwise, would not have had the protection of air bags.

In the lower half of Table 2-40, the statistical method used by our model estimates air bags net-increased child passenger fatalities by 6 in trucks with the switches left on. In fact, on October 1, 2004, there were 5 SCI cases of children with fatal injuries from air bags in pickup trucks where the switches were left on.

The statistical model also considers the hypothetical situation that all switches had been left on (or that all the trucks had air bags but no switches). The model estimates that air bags would have increased child passenger fatalities by 24 rather than 6. In other words, 18 potential fatalities to children were avoided because air bags were turned off. Essentially, the switch was turned off in three-quarters of the situations (18 of 24) where it needed to be off – i.e., especially for infants and toddlers.

The statistics for adults and children – the two halves of Table 2-40 – can be compared:

- As actually used by the public in 1995-2002, the combination of passenger air bags with on-off switches saved an estimated 117 adults at a cost of 6 child passenger fatalities. That is a ratio of 19.5 adult lives saved per child passenger fatality: not too different from the 22.8 ratio, in 2002, in cars, vans and SUVs (where children can escape harm from air bags simply by riding in the back seat, an option not available in pickup trucks without back seats).

- Air bags without switches, hypothetically, would have saved an estimated 147 adults at a cost of 24 child passenger fatalities. That would have been a ratio of 6.1 adult lives saved per child passenger fatality: not too different from the 4.8 ratio, in 1996, in cars, vans and SUVs, before the public became well aware of the risks of air bags to small children.
- If pickup trucks had not been equipped with passenger air bags, the statistics would have been, of course, zero adults saved and zero children harmed.
- If the public had been fully aware of the correct way to use the switches, or if the manual switches were superseded by advanced systems that automatically trigger or suppress deployment depending on the size and location of the occupant, the statistics could have been 147 adults saved and zero children harmed.

Finally, the statistics in Table 2-39 for vehicles other than pickup trucks can be combined with the estimates for pickup trucks in Table 2-40.

- In all vehicles, passenger air bags saved 1,743 + 117 = 1,860 people age 13 or older in 1987-2002.
- Child passenger fatalities increased by 130 + 6 = 136.
- Net lives saved by passenger air bags were 1,860 − 136 = 1,724.

REFERENCES

Accident Analysis and Prevention, Vol. 32, No. 2, March 2000 (entire issue on head restraints and occupant injury in rear impacts).

Accident Facts. National Safety Council, Chicago, annual publication.

ACTS, Automotive Coalition for Traffic Safety, Inc., State Laws Database. www.actsinc.org.

"Air Bag Chronology," *USA Today*, July 13, 1999.

Air Bags & On-Off Switches. NHTSA Publication No. DOT HS 808 629, Washington, 1997.

Air Bag & Seat Belt Safety Campaign, Air Bag & Seat Belt Safety Tips, National Safety Council, Chicago, November 17-30, 2003 (http://www.nsc.org/partners/safetips.htm).

Analysis of Effects of Proposed Changes to Passenger Car Requirements of FMVSS 208. NHTSA Docket No. 74-14-N01-104, Washington, 1974.

Anton's Law. Public Law 107-318, December 4, 2002.

Arehart, C., Radlinski, R. and Hiltner, E. *Light Vehicle ABS Performance Evaluation – Phase II.* NHTSA Technical Report No. DOT HS 807 924, Washington, 1992.

B.01.04 Door Latch Integrity. NHTSA Research & Development, www-nrd.nhtsa.dot.gov/departments/nrd-01/summaries/b0104.html, June 2000.

Ballard, C. and Andrade, D. *Systems and Hardware Effects of FMVSS 105-75.* Paper No. 760216, Society of Automotive Engineers, Warrendale, PA, 1976.

Bayer, A.R., Jr. and Peterson, B.S. *Child Restraint Systems Testing.* NHTSA Report No. DOT HS 803 408, Washington, 1978.

Bloch, B. "Advanced Designs for Side Impact and Rollover Protection," *Sixteenth International Technical Conference on the Enhanced Safety of Vehicles.* NHTSA Report No. DOT HS 808 759, Washington, 1998, p. 1780.

Bowman, B.L. and Rounds, D.A. *Restraint System Usage in the Traffic Population 1987 Annual Report.* NHTSA Report No. DOT HS 807 342, Washington, 1988.

_____. *Restraint System Usage in the Traffic Population 1988 Annual Report.* NHTSA Report No. DOT HS 807 447, Washington, 1989.

Buckle Up America. NHTSA Publication No. DOT HS 808 628, Washington, 1997.

Buckle-Up America Child Passenger Safety Week, February 10-16, 2002, Talking Points, NHTSA, (http://www.nhtsa.dot.gov/people/injury/airbags/buckleplan/ CPS%20Week%20Planner_files/talking1.html).

Buckle Up: Avoid the Summertime Blues. HSL Publication No. DOT HS 041 653, NHTSA, Washington, 1991.

Burger, W.J., et al. *Improved Commercial Vehicle Conspicuity and Signalling Systems, Task I – Accident Analysis and Functional Requirements*. NHTSA Technical Report No. DOT HS 806 100, Washington, 1981.

_____. *Improved Commercial Vehicle Conspicuity and Signalling Systems, Task III – Field Test Evaluation of Vehicle Reflectorization Effectiveness*. NHTSA Technical Report No. DOT HS 806 923, Washington, 1985.

Buying a Safer Car for Child Passengers 2003. NHTSA Publication No. DOT HS 809 545, Washington, 2003.

Buying a Safer Car 2000. NHTSA Publication No. DOT HS 809 046, Washington, 2000.

Buying a Safer Car 2001. NHTSA Publication No. DOT HS 809 152, Washington, 2000.

Buying a Safer Car 2002. NHTSA Publication No. DOT HS 809 409, Washington, 2002.

Buying a Safer Car 2003. NHTSA Publication No. DOT HS 809 546, Washington, 2003.

Campbell, B.J. *A Study of Injuries Related to Padding on Instrument Panels*. HSL Publication No. 00427812, Report No. VJ-1823-R2, Cornell Aeronautical Laboratory, Buffalo, 1963.

Child Restraint Laws. Insurance Institute for Highway Safety, Arlington, VA, 2004 (http://www.hwysafety.org/safety%5Ffacts/state%5Flaws/restrain2.htm).

Clark, C., Blechschmidt, C. and Gordon, F. "Impact Protection with the 'Airstop' Restraint System," *Eighth Stapp Car Crash and Field Demonstration Conference*. Wayne State University Press, Detroit, 1966.

"'Click It Or Ticket' Expands beyond North Carolina," *Status Report*, Vol. 36, November 15, 2001, p. 4, Insurance Institute for Highway Safety, Arlington, VA.

Code of Federal Regulations, Title 49. Government Printing Office, Washington, 2002.

Cole, B.L., Dain, S.J. and Fisher, A.J. *Study of Motor Vehicle Signal Systems*. HSL Publication No. DOT HS 022 690, Road Safety Information Service, Melbourne, 1977.

Consumer Reports, Vol. 37, August 1972, p. 484 (30 mph sled test of child safety seats).

Dang, J.N. *Preliminary Results Analyzing the Effectiveness of Electronic Stability Control (ESC) Systems.* NHTSA Evaluation Note No. DOT HS 809 790, Washington, 2004.

Daniel, R.P. "Vehicle Interior Safety Constraint Systems," *1970 International Automobile Safety Conference Compendium.* Paper No. 700423, Society of Automotive Engineers, New York, 1970.

Datta, T.K. and Guzek, P. *Restraint System Use in 19 U.S. Cities 1989 Annual Report.* NHTSA Report No. DOT HS 807 595, Washington, 1990.

_____. *Restraint System Use in 19 U.S. Cities 1990 Annual Report.* NHTSA Report No. DOT HS 808 147, Washington, 1991.

_____. *Restraint System Use in 19 U.S. Cities 1991 Annual Report.* NHTSA Report No. DOT HS 808 148, Washington, 1992.

Decina, L.E. and Knoebel, K.Y. "Child Safety Seat Misuse Patterns in Four States," *Accident Analysis and Prevention,* Vol. 29, January 1997, pp. 125-132.

Digges, K.H., Nicholson, R.M. and Rouse, E.J. *The Technical Basis for the Center High Mounted Stoplamp.* Paper No. 851240, Society of Automotive Engineers, Warrendale, PA, 1985.

Eby, D.W., Vivoda, J.M. and Fordyce, T.A. *Direct Observation of Safety Belt Use in Michigan: Fall 1999.* Report No. UMTRI-99-33, University of Michigan Transportation Research Institute, Ann Arbor, 1999.

Evaluation of the Effectiveness of Occupant Protection, Interim Report. NHTSA Report No. DOT HS 807 843, 1992.

Evaluation Plan for Center High-Mounted Stop Lamps. NHTSA Docket No. 81-02-N02-002, 1983.

Evaluation Program Plan, 1998-2002. NHTSA Report No. DOT HS 808 709, Washington, 1998.

Evans, L. "Double Pair Comparison - A New Method to Determine How Occupant Characteristics Affect Fatality Risk in Traffic Crashes," *Accident Analysis and Prevention,* Vol. 18, June 1986, pp. 217-227.

_____. "The Effectiveness of Safety Belts in Preventing Fatalities," *Accident Analysis and Prevention,* Vol. 18, June 1986, pp. 229-241.

_____. *Traffic Safety and the Driver.* Van Nostrand Reinhold, New York, 1991.

Fargo, R.B. *Windshield Glazing as an Injury Factor in Automobile Accidents.* Cornell Aeronautical Laboratory, Buffalo, 1968.

Farmer, C.M. "New Evidence Concerning Fatal Crashes of Passenger Vehicles Before and After Adding Antilock Braking Systems," *Accident Analysis and Prevention*, Vol. 33, 2001, pp. 361-369.

Farmer, C.M., Lund, A.K., Trempel, R.E. and Braver, E.R. *Fatal Crashes of Passenger Vehicles Before and After Adding Antilock Brake Systems.* Insurance Institute for Highway Safety, Arlington, VA, 1996.

Federal Register Notices:

31 (March 8, 1966): 4091, GSA proposes safety standards for Federally purchased vehicles.

31 (July 15, 1966): 9631, GSA Final Rule, safety standards for Federally purchased vehicles.

31 (December 3, 1966): 15212, NPRM for the initial FMVSS.

32 (February 3, 1967): 2414, Final Rule for the initial FMVSS, including among others FMVSS Nos. 103, 105, 108, 201, 203, 204, 205, 206, 207, 208, 209, 210 and 301.

32 (October 13, 1967): 14278, ANPRM to consider regulations to limit roof crush and intrusion.

32 (December 16, 1967): 18033, Final Rule extending FMVSS 108 to cars and LTVs.

32 (December 28, 1967): 20865, NPRM to establish FMVSS 202 for passenger cars.

33 (February 14, 1968): 2945, Final Rule establishing FMVSS 202 for passenger cars.

33 (August 16, 1968): 11652, Final Rule establishing FMVSS 212 for passenger cars.

33 (October 5, 1968): 14971, ANPRM announcing the intention to regulate side door strength.

33 (December 11, 1968): 18386, first NPRM proposing FMVSS 214.

34 (January 24, 1969): 1150, Final Rule extending FMVSS 206 to trucks, buses and multipurpose passenger vehicles.

34 (July 2, 1969): 11148, initial ANPRM to consider air bags or other automatic protection.

35 (January 21, 1970): 813, second NPRM proposing FMVSS 214.

35 (April 23, 1970): 6513, third NPRM proposing FMVSS 214.

35 (May 7, 1970): 7187, initial NPRM for automatic occupant protection, did not become a Final Rule.

35 (September 23, 1970): 14778, Final Rule establishing FMVSS 213.

35 (September 30, 1970): 15222, Final Rule extending the original FMVSS 208 to LTVs.

35 (September 25, 1970): 14936, NPRM proposing to add a knee impact test to FMVSS 201, never became a Final Rule.

35 (October 30, 1970): 16801, Final Rule establishing FMVSS 214 (side door strength, passenger cars).

35 (November 11, 1970): 17345, original NPRM to establish FMVSS 105a for passenger cars, effective October 1, 1972.

36 (January 6, 1971): 166, NPRM to establish FMVSS 216 for passenger cars.

36 (March 10, 1971): 4600, Final Rule amending FMVSS 208 to require shoulder belts and warning buzzers in all cars, and to permit automatic occupant protection.

36 (December 2, 1971): 22902, Final Rule extending FMVSS 207 to LTVs.

36 (December 8, 1971): 23299, Final Rule establishing FMVSS 216 for passenger cars.

37 (September 2, 1972): 17970, original Final Rule establishing FMVSS 105a in passenger cars and LTVs, effective September 1, 1974 (subsequently revised).

38 (May 18, 1973): 13017, delays FMVSS 105a to September 1, 1975 (subsequently revised).

38 (June 20, 1973): 16072, Final Rule amending FMVSS 208 to require integral 3-point belts and ignition interlocks in passenger cars.

39 (February 22, 1974): 6708, renames FMVSS 105a to FMVSS 105-75 and limits it to passenger cars (subsequently revised).

39 (March 19, 1974): 10268, NPRM proposing to upgrade FMVSS 202, including fixed head restraints for drivers, never became a Final Rule.

39 (March 21, 1974): 10588, Final Rule adding lateral and rear impact tests to FMVSS 301 and extending the standard to LTVs.

39 (October 31, 1974): 38380, Final Rule amending FMVSS 208 to delete the interlock requirement.

39 (November 21, 1974): 40857, minor revisions to FMVSS 301 Final Rule of March 21, 1974.

39 (December 6, 1974): 42692, Final Rule amending FMVSS 208 to require a 4-8 second audible warning on belt use.

40 (June 9, 1975): 24525, Final Rule for FMVSS 105-75 in passenger cars, effective January 1, 1976.

40 (July 9, 1975): 28805, Final Rule amending FMVSS 208, extending 3-point belts to LTVs.

40 (August 6, 1975): 33036, revises fuel spillage requirements in FMVSS 301 Final Rule of March 21, 1974.

40 (October 10, 1975): 47790, revises effective dates in FMVSS 301 Final Rule of March 21, 1974.

41 (August 30, 1976): 36493, Final Rule extending FMVSS 212 to LTVs and also amending the standard, allowing NHTSA to test at a wider range of temperatures.

42 (July 5, 1977): 34288, postpones until September 1, 1978 the effective date of the extension of FMVSS 212 to LTVs.

42 (July 5, 1977): 34289, Final Rule requiring automatic occupant protection, rescinded in 1981 before its effective date.

44 (November 29, 1979): 68470, Final Rule extending FMVSS 201, 203 and 204 to LTVs.

44 (December 13, 1979): 72131, Final Rule adding a 30 mph test to FMVSS 213.

46 (January 2, 1981): 55, Final Rule extending FMVSS 105 to LTVs

46 (January 8, 1981): 2064, Final Rule amending FMVSS 208 to add comfort and convenience standards for safety belts.

46 (February 19, 1981): 13193, Executive Order 12291 – Federal Regulation.

46 (October 29, 1981): 53419, rescission of 1977 Final Rule requiring automatic occupant protection.

48 (October 18, 1983): 48235, Final Rule amending FMVSS 108 to require Center High Mounted Stop Lamps on passenger cars.

48 (November 16, 1983): 52065, Final Rule amending FMVSS 205 to permit the optional use of glass-plastic glazing.

49 (July 17, 1984): 28962, Final Rule amending FMVSS 208 to require automatic occupant protection in passenger cars.

50 (August 23, 1985): 34152, Final Rule amending FMVSS 208 to postpone belt comfort and convenience standards to September 1, 1986.

54 (June 14, 1989): 25275, Final Rule amending FMVSS 208 to require 3-point belts at the rear-outboard seats of cars.

54 (September 25, 1989): 39183, Final Rule extending FMVSS 202 to LTVs.

54 (November 2, 1989): 46257, Final Rule amending FMVSS 208 to require 3-point belts at the rear-outboard seats of LTVs.

55 (January 17, 1990): 1586, NHTSA evaluation plan for automatic occupant protection.

55 (October 30, 1990): 45752, Final Rule upgrading FMVSS 214, adding a dynamic side impact test for passenger cars.

56 (March 26, 1991): 12472, Final Rule amending FMVSS 208 to extend automatic protection to LTVs.

56 (April 17, 1991): 15510, Final Rule extending FMVSS 216 to LTVs.

56 (April 19, 1991): 16015, Final Rule amending FMVSS 108 to require Center High Mounted Stop Lamps on LTVs.

56 (June 14, 1991): 27427, Final Rule extending FMVSS 214 side door strength requirement to LTVs.

57 (December 10, 1992): 58406, Final Rule amending FMVSS 108 to require conspicuity tape on heavy trailers.

58 (September 2, 1993): 46551, Final Rule amending FMVSS 208 to require manual 3-point belts and dual air bags in cars and LTVs.

58 (October 4, 1993): 51735, Executive Order 12866 – Regulatory Planning and Review.

59 (January 4, 1994): 281, ANPRM asking for information about the effectiveness and potential benefits of ABS technologies.

60 (April 12, 1995): 18566, ANPRM to consider upgrading FMVSS 301.

60 (May 23, 1995): 27233, Final Rule amending FMVSS 208 to allow on-off switches for passenger air bags in pickup trucks and some other vehicles.

60 (July 12, 1995): 35889, Notice of public meeting to discuss potential upgrades to FMVSS 206.

60 (July 28, 1995): 38749, Final Rule extending the FMVSS 214 dynamic side impact test to LTVs.

60 (August 18, 1995): 43031, Final Rule upgrading FMVSS 201.

60 (September 28, 1995): 50124, Final Rule extending FMVSS 206 to the back doors of vehicles.

61 (July 12, 1996): 36698, ANPRM deferring indefinitely the ABS requirement.

62 (March 19, 1997): 12960, Final Rule modifying the FMVSS 208 test to permit air bags that deploy less forcefully.

62 (November 21, 1997): 62406, sets up procedures enabling the public to obtain aftermarket on-off switches for air bags.

63 (August 4, 1998): 41451, Final Rule reducing the FMVSS 201 test speed from 15 mph to 12 mph on target areas where a head air bag is stored.

64 (March 5, 1999): 10786, Final Rule establishing FMVSS 225 and amending FMVSS 213 to require upper and lower tethers on child safety seats.

64 (March 31, 1999): 15587, Federal Motor Carrier Safety Regulation (Final Rule) requiring all heavy trailers on the road on or after June 1, 2001 to be equipped or retrofitted with conspicuity tape.

65 (May 12, 2000): 30679, Final Rule amending FMVSS 208 to require advanced air bags.

65 (November 13, 2000): 67693, NPRM to upgrade the rear- and lateral-impact test procedures of FMVSS 301.

66 (January 4, 2001): 967, NPRM to upgrade FMVSS 202.

66 (January 12, 2001): 3388, establishes NHTSA's rollover resistance ratings based on the Static Stability Factor.

68 (December 1, 2003): 67068, Final Rule upgrading the rear- and lateral-impact test procedures of FMVSS 301.

Fifth/Sixth Report to Congress – Effectiveness of Occupant Protection Systems and Their Use. NHTSA Report No. DOT HS 809 442, Washington, 2001.

51 Young Lives. HSL Report No. DOT HS 034 896, Tennessee Highway Patrol, Nashville, 1983.

Final Regulatory Evaluation, Anti-Lacerative Glazing, FMVSS 205. NHTSA Plans and Policy, Washington, 1983.

Final Regulatory Evaluation, Rear Seat Lap Shoulder Belts in Passenger Cars. NHTSA Docket No. 87-08-N03-001, 1989.

Final Regulatory Impact Analysis, Amendment to Federal Motor Vehicle Safety Standard 208, Passenger Car Front Seat Occupant Protection. NHTSA Report No. DOT HS 806 572, Washington, 1984.

Final Regulatory Impact Analysis, Amendment to FMVSS 108 to Require Center High-Mounted Stop Lamps on Light Trucks and Buses. NHTSA Docket No. 81-02-N10-001, 1991.

Final Regulatory Impact Analysis, Extension of the Automatic Restraint Requirements of FMVSS 208 to Trucks, Buses and Multi-Purpose Passenger Vehicles. NHTSA Docket No. 74-14-N70-001, 1990.

Final Regulatory Impact Analysis, Federal Motor Vehicle Safety Standard 108, Center High-Mounted Stop Lamps. NHTSA Docket No. 81-02-N02-001, 1983.

Flora, J.D., Beitler, P., Bromberg, J., Goldstein, N., and O'Day, J. *An Evaluation of FMVSS 301 – Fuel System Integrity.* Report No. UM-HSRI-79-42, UMTRI, University of Michigan, Ann Arbor, 1979.

Fourth Report to Congress – Effectiveness of Occupant Protection Systems and Their Use. NHTSA Report No. DOT HS 808 919, Washington, 1999.

Garrett, J.W. *Comparison of Door Opening Frequency in 1967-68 Cars with Earlier Model U.S. Cars.* NHTSA Technical Report No. DOT HS 800 231, Washington, 1969.

_____. *The Safety Performance of 1962-63 Automobile Door Latches and Comparison with Earlier Latch Designs.* Report No. VJ-1823-R7, Cornell Aeronautical Laboratory, Buffalo, 1964.

Garrett, J.W. and Hendricks, D.L. "Factors Influencing the Performance of the Energy Absorbing Steering Column in Accidents," *Report on the Fifth International Technical Conference on Experimental Safety Vehicles.* NHTSA, Washington, 1975.

Get It Together. NHTSA Publication No. DOT HS 806 254, Washington, 1982.

Gloyns, P.F. and Mackay, G.M. "Impact Performance of Some Designs of Steering Assembly in Real Accidents and under Test Conditions," *Proceedings of Eighteenth Stapp Car Crash Conference.* Society of Automotive Engineers, Warrendale, PA, 1974.

Glassbrenner, D. *Safety Belt and Helmet Use in 2002 – Overall Results.* NHTSA Technical Report No. DOT HS 809 500, Washington, 2002

_____. *Safety Belt Use in 2004 – Overall Results.* NHTSA Research Note No. DOT HS 809 646, Washington, 2004.

_____. *Safety Belt Use in 2003.* NHTSA Technical Report No. DOT HS 809 646, Washington, 2003.

_____. *The Use of Child Restraints in 2002.* NHTSA Research Note No. DOT HS 809 555, Washington, 2003.

Goryl, M.E. *Restraint System Usage in the Traffic Population.* NHTSA Report No. DOT HS 806 987, Washington, 1986.

Goryl, M.E. and Bowman, B.L. *Restraint System Usage in the Traffic Population.* NHTSA Report No. DOT HS 807 080, Washington, 1987.

Goryl, M.E. and Cynecki, M.J. *Restraint System Usage in the Traffic Population.* NHTSA Report No. DOT HS 806 714, Washington, 1985.

Government Performance and Results Act of 1993. Public Law 103-62, August 3, 1993.

Hackney, J. and Ellison, C. "A Review of the Effects of Belts Systems, Steering Assemblies, and Structural Design on the Safety Performance of Vehicles in the New Car Assessment Program," *Tenth International Technical Conference on Experimental Safety Vehicles.* NHTSA Report No. DOT HS 806 916, Washington, 1983, pp. 380-413.

Hackney, J.R. and Kahane, C.J. *The New Car Assessment Program: Five Star Rating System and Vehicle Safety Performance Characteristics.* Technical Paper No. 950888, Society of Automotive Engineers, Warrendale, PA, 1995.

Hackney, J. and Quarles, V. "The New Car Assessment Program – Status and Effect," *Ninth International Technical Conference on Experimental Safety Vehicles.* NHTSA Report No. DOT HS 806 341, Washington, 1983, pp. 809-824.

Harless, D.W. and Hoffer, G.E. "The Antilock Braking System Anomaly: A Drinking Driver Problem?," *Accident Analysis and Prevention,* Vol. 34, 2002, pp. 333-341.

Head Restraints – Identification of Issues Relevant to Regulation, Design, and Effectiveness. NHTSA Performance Standards, Washington, 1996 (www.nhtsa.dot.gov/cars/rules/crashworthy/ headrest/status9/status9.html#28).

Hedeen, C.E. and Campbell, D.D. *Side Impact Structures.* Paper No. 690003, Society of Automotive Engineers, New York, 1969.

Hertz, E. *Analysis of the Crash Experience of Vehicles Equipped with All Wheel Antilock Braking Systems (ABS) – A Second Update Including Vehicles with Optional ABS.* NHTSA Technical Report No. DOT HS 809 144, Washington, 2000.

_____. *Revised Estimates of Child Restraint Effectiveness.* NHTSA NCSA Research Note, Washington, 1996.

Hertz, E., Hilton, J. and Johnson, D.M. *An Analysis of the Crash Experience of Light Trucks Equipped with Antilock Braking Systems.* NHTSA Technical Report No. DOT HS 808 278, Washington, 1995.

_____. *An Analysis of the Crash Experience of Passenger Cars Equipped with Antilock Braking Systems.* NHTSA Technical Report No. DOT HS 808 279, Washington, 1995.

Hight, P.V., Siegel, A.W. and Nahum, A.M. "Injury Mechanisms in Rollover Collisions," *Proceedings of Sixteenth Stapp Car Crash Conference.* Society of Automotive Engineers, New York, 1972.

Hill, A. *Steering Wheel Oscillations and Vertical Movements in 30 mph Barrier Impacts.* NHTSA Technical Report No. DOT HS 803 606, Washington, 1978.

Hiltner, E., Arehart, C. and Radlinski, R. *Light Vehicle ABS Performance Evaluation.* NHTSA Technical Report No. DOT HS 807 813, Washington, 1991.

Horsch, J.D., Petersen, K.R. and Viano, D.C. *Laboratory Study of Factors Influencing the Performance of Energy Absorbing Steering Systems.* Paper No. 820475, Society of Automotive Engineers, Warrendale, PA, 1982.

Huelke, D.F. *Accident Investigations of Performance Characteristics of Energy Absorbing Steering Columns.* Paper No. 690184, Society of Automotive Engineers, New York, 1969.

Huelke, D.F. and Compton, C. "Injury Frequency and Severity in Rollover Car Crashes as Related to Occupant Ejection, Contacts and Roof Damage," *Accident Analysis and Prevention,* Vol. 15, October 1983.

Huelke, D.F. and Gikas, P.W. "Causes of Death in Automobile Accidents," *Police,* Vol. 13, November/December 1968, pp. 81-89.

_____. "How Do They Die? Medical Engineering Data from On-Scene Investigation of Fatal Automobile Accidents", *Highway Vehicle Safety*. Society of Automotive Engineers, New York, 1968.

Huelke, D.F., Grabb, W.C., Dingman, R.O. and Oneal, R.M. "The New Automotive Windshield and its Effectiveness in Reducing Facial Lacerations," *Plastic and Reconstructive Surgery*, Vol. 41, No. 6, June 1968.

Huelke, D.F., Marsh, J.C. IV and Sherman, H.W. "Analysis of Rollover Accident Factors and Injury Causation," *Proceedings of the Sixteenth Conference of the American Association for Automotive Medicine*. American Association for Automotive Medicine, Morton Grove, IL, 1972.

Hunter, W.W. et al. *Analysis of Occupant Restraint Issues from State Accident Data*. HSL Publication No. 00578847, Report No. HSRC-TR75, Highway Safety Research Center, University of North Carolina, Chapel Hill, 1990.

Kahane, C.J. *Addendum to "Seat Belt Effectiveness Estimates Using Data Adjusted for Damage Type."* NHTSA Docket No. 74-14-N35-229-05, 1984.

_____. *Correlation of NCAP Performance with Fatality Risk in Actual Head-On Collisions*. NHTSA Technical Report No. DOT HS 808 061, Washington, 1994.

_____. *The Effectiveness of Center High Mounted Stop Lamps: A Preliminary Evaluation*. NHTSA Technical Report No. DOT HS 807 076, Washington, 1987.

_____. *Estimates of Fatality Reduction for Air Bags and Lap/Shoulder Belts Based on Case by Case Analysis of Unrestrained Fatalities*. NHTSA Docket No. 74-14-N35-012, 1984.

_____. *An Evaluation of Center High Mounted Stop Lamps Based on 1987 Data*. NHTSA Technical Report No. DOT HS 807 442, Washington, 1989.

_____. *An Evaluation of Child Passenger Safety: The Effectiveness and Benefits of Safety Seats*. NHTSA Technical Report No. DOT HS 806 890, Washington, 1986.

_____. *An Evaluation of Door Locks and Roof Crush Resistance of Passenger Cars – Federal Motor Vehicle Safety Standards 206 and 216*. NHTSA Technical Report No. DOT HS 807 489, Washington, 1989.

_____. *An Evaluation of Federal Motor Vehicle Safety Standards for Passenger Car Steering Assemblies*. NHTSA Technical Report No. DOT HS 805 705, Washington, 1981.

_____. *Evaluation of FMVSS 214 - Side Impact Protection: Dynamic Performance Requirement; Phase 1: Correlation of TTI(d) with Fatality Risk in Actual Side Impact Collisions of Model Year 1981-1993 Passenger Cars*. NHTSA Technical Report No. DOT HS 809 004, Washington, 1999.

_____. *An Evaluation of Head Restraints - Federal Motor Vehicle Safety Standard 202.* NHTSA Technical Report No. DOT HS 806 108, Washington, 1982.

_____. *An Evaluation of Occupant Protection in Frontal Interior Impact for Unrestrained Front Seat Occupants of Cars and Light Trucks.* NHTSA Technical Report No. DOT HS 807 203, Washington, 1988.

_____. *An Evaluation of Side Marker Lamps for Cars, Trucks and Buses.* NHTSA Technical Report No. DOT HS 806 430, Washington, 1983.

_____. *An Evaluation of Side Structure Improvements in Response to Federal Motor Vehicle Safety Standard 214.* NHTSA Technical Report No. DOT HS 806 314, Washington, 1982.

_____. *An Evaluation of Windshield Glazing and Installation Methods for Passenger Cars.* NHTSA Technical Report No. DOT HS 806 693, Washington, 1985.

_____. "Fatality and Injury Reducing Effectiveness of Lap Belts for Back Seat Occupants," *Restraint Technologies: Rear Seat Occupant Protection.* Paper No. 870486, Society of Automotive Engineers Publication No. SP-691, Warrendale, PA, 1987.

_____. *Fatality Reduction by Air Bags: Analyses of Accident Data through Early 1996.* NHTSA Technical Report No. DOT HS 808 470, Washington, 1996.

_____. *Fatality Reduction by Safety Belts for Front-Seat Occupants of Cars and Light Trucks.* NHTSA Technical Report No. DOT HS 809 199, Washington, 2000.

_____. *A Preliminary Evaluation of Seat Back Locks.* NHTSA Technical Report No. DOT HS 807 067, Washington, 1987.

_____. *Preliminary Evaluation of the Effectiveness of Antilock Brake Systems for Passenger Cars.* NHTSA Technical Report No. DOT HS 808 206, Washington, 1994.

_____. *Preliminary Evaluation of the Effectiveness of Rear-Wheel Antilock Brake Systems for Light Trucks.* NHTSA Docket No. 70-27-GR-026, Washington, 1993.

_____. *A Preliminary Evaluation of Two Braking Improvements for Passenger Cars.* NHTSA Technical Report No. DOT HS 806 359, Washington, 1983.

_____. *Usage and Effectiveness of Seat and Shoulder Belts in Rural Pennsylvania Accidents.* NHTSA Technical Note No. DOT HS 801 398, Washington, 1974.

Kahane, C.J. and Hertz, E. *The Long-Term Effectiveness of Center High Mounted Stop Lamps in Passenger Cars and Light Trucks.* NHTSA Technical Report No. DOT HS 808 696, Washington, 1998.

Kahane, C.J. and Ichter, K.D. *Statistical Evaluation of Brake Safety Improvements for Passenger Cars.* Paper No. 841236, Society of Automotive Engineers, Warrendale, PA, 1984.

Kahane, C.J., Smith, R.A. and Tharp, K.J. "The National Crash Severity Study," *Report on the Sixth International Technical Conference on Experimental Safety Vehicles.* NHTSA Report No. DOT HS 802 501, Washington, 1977, pp. 493-515.

Kelleher, B.J. and Walsh, M.J. "Sled Test Comparisons of Child Restraint Performance," *Twenty-Second Stapp Car Crash Conference.* Paper No. 780903, Society of Automotive Engineers, Warrendale, PA, 1978.

Kelleher, B.J., Walsh, M.J , Dance, D.M. and Gardner, W.T. "An Experimental Study of the Effects of Child Restraint Improper Installation and Crash Protection for Larger Size Children," *SAE Child Injury and Restraint Conference Proceedings.* Paper No. 831602, Society of Automotive Engineers Publication No. P-135, Warrendale, PA, 1983, pp. 31-51

Kindelberger, J. and Starnes, M. *Moving Children from the Front Seat to the Back Seat: The Influence of Child Safety Campaigns.* NHTSA Research Note No. DOT HS 809 698, Washington, 2003.

Kohl, J.S. and Baker, C. *Field Test Evaluation of Rear Lighting Systems.* NHTSA Technical Report No. DOT HS 803 467, Washington, 1978.

Kullgren, A., Lie A. and Tingvall, C. "The Effectiveness of ABS in Real Life Accidents," *Proceedings of the Fourteenth International Technical Conference on the Enhanced Safety of Vehicles.* NHTSA, Washington, 1994, Paper No. 94-S4-O-07.

Lap/Shoulder Belt Kits for Rear Seats. NHTSA Publication No. DOT HS 807 881, Washington, 1992.

Lawless, E.W. and Siani, T.A. "The State of the Art of Child Passenger Safety Legislation in North America," *SAE Child Injury and Restraint Conference Proceedings.* Paper No. 831650, Society of Automotive Engineers Publication No. P-135, Warrendale, PA, 1983, pp. 229-242..

Levine, D.N. and Campbell, B.J. *Effectiveness of Lap Seat Belts and the Energy Absorbing Steering Column in the Reduction of Injuries.* HSL Publication No. 00221617, Highway Safety Research Center, University of North Carolina, Chapel Hill, 1971.

Ludtke, N.F. "1980 and 1979 Ford F-150 Weight and Cost Analysis," *Third Automotive Fuel Economy Research Contractors' Coordination Meeting – Summary Report.* NHTSA Technical Report No. DOT HS 805 875, Washington, 1980.

Marquis, D.P. *The General Motors Energy Absorbing Column.* Paper No. 670039, Society of Automotive Engineers, New York, 1967.

McLean, R.F., Eckel, C. and Cowan, D. *Cost Evaluation for Four Federal Motor Vehicle Standards, Volume I*. NHTSA Report No. DOT HS 803 871, Washington, 1978.

Mertz, H.J., Jr., and Patrick, L.M. "Investigation of the Kinematics and Kinetics of Whiplash," *Proceedings of Eleventh Stapp Car Crash Conference*. Society of Automotive Engineers, New York, 1967.

Morgan, C. *Effectiveness of Lap/Shoulder Belts in the Back Outboard Seating Positions*. NHTSA Technical Report No. DOT HS 808 945, Washington, 1999.

_____. *The Effectiveness of Retroreflective Tape on Heavy Trailers*. NHTSA Technical Report No. DOT HS 809 222, Washington, 2001.

_____. *Evaluation of Rear Window Defrosting and Defogging Systems*. NHTSA Technical Report No. DOT HS 809 724, Washington, 2004.

_____. *Results of the Survey on the Use of Passenger Air Bag On-Off Switches*. NHTSA Technical Report No. DOT HS 809 689, Washington, 2003.

The Motor Vehicle Information and Cost Savings Act of 1973. Public Law 92-513, as amended, 15 *United States Code* 1912-2012.

Multidisciplinary Accident Investigation Data File, Editing Manual and Reference Information, Vol. 2. NHTSA Technical Report No. DOT HS 802 412, Washington, 1977.

National Child Passenger Safety Week: One-Minute Child Safety Seat Facts. (http://www.nhtsa.dot.gov/people/outreach/safesobr/OPlanner/ncpsw/kidseat3.html).

National Highway Traffic Safety Administration Evaluation Program Plan, Calendar Years 2004-2007. NHTSA Report No. DOT HS 809 699, Washington, 2004.

National Occupant Protection Use Survey: Controlled Intersection Study. NHTSA NCSA Research Note, Washington, 1995.

National Occupant Protection Use Survey – 1996: Controlled Intersection Study. NHTSA NCSA Research Note, Washington, 1997.

NCSA Special Crash Investigations, Downloadable Files. NHTSA, Washington (www-nrd.nhtsa.dot.gov/departments/nrd-30/SCI.html).

New Car Assessment Program: Response to the NCAP FY 1992 Congressional Requirements. Report to the Congress, NHTSA, Washington, 1993

NHTSA Child Seat Safety Information. NHTSA, Washington, 2004 (http://www.nhtsa.dot.gov/people/injury/childps/csr2001/csrhtml/safetyFeatures.html#forward).

1978 Model Year Passenger Car and Truck Accident Investigator's Manual. Motor Vehicle Manufacturers Association, Detroit, 1978.

Observed Safety Belt Use, Fall 2000 National Occupant Protection Use Survey. NHTSA (NCSA) Research Note, Washington, 2001.

Observed Safety Belt Use in 1996. NHTSA (NCSA) Research Note, Washington, 1997.
A Parent's Guide to Booster Seats. NHTSA Publication No. DOT HS 808 671, Washington, 1998.

Parsons, G.G. *An Evaluation of the Effects of Glass-Plastic Windshield Glazing in Passenger Cars.* NHTSA Technical Report No. DOT HS 808 062, Washington, 1993.

_____. *Evaluation of Federal Motor Vehicle Standard 301-75, Fuel System Integrity: Passenger Cars.* NHTSA Technical Report No. DOT HS 806 335, Washington, 1983.

_____. *Motor Vehicle Fires in Traffic Crashes and the Effects of the Fuel System Integrity Standard.* NHTSA Technical Report No. DOT HS 807 675, Washington, 1990.

Partyka, S.C. "Belt Effectiveness in Pickup Trucks and Passenger Cars by Crash Direction and Accident Year (May 1988)," *Papers on Adult Seat Belts - Effectiveness and Use.* NHTSA Technical Report No. DOT HS 807 285, Washington, 1988.

_____. *Belt Use in Serious Impacts Estimated from Fatality Data.* NHTSA Technical Report No. 807 519, 1989.

_____. *Lives Saved by Child Restraints from 1982 through 1987.* NHTSA Technical Report No. DOT HS 807 371, Washington, 1988.

_____. *Restraint Use and Fatality Risk for Infants and Toddlers.* NHTSA National Center for Statistics and Analysis, Washington, 1984.

Patrick, L.M. and Daniel, B.P. "Comparison of Standard and Experimental Windshields," *Eighth Stapp Car Crash and Field Demonstration Conference.* Wayne State University Press, Detroit, 1966.

Performance of Lap Belts in 26 Frontal Crashes. National Transportation Safety Board Report No. NTSB/SS-86/03, Publication No. PB86-917006, Washington, 1986.

Perkins, D.D., Cynecki, M.J. and Goryl, M.E. *Restraint System Usage in the Traffic Population.* NHTSA Report No. DOT HS 806 582, Washington, 1984.

Phillips, B.M. *Restraint System Usage in the Traffic Population.* NHTSA Report No. DOT HS 806 424, Washington, 1983.

_____. *Safety Belt Usage Among Drivers*. NHTSA Report No. DOT HS 805 398, Washington, 1980.

Preliminary Evaluation of the Proposed Extension of Standards No. 201, 203 and 204 to Light Trucks, Buses, and Multipurpose Vehicles. NHTSA Plans and Policy, Washington, 1978.

Preliminary Regulatory Evaluation – Proposed Amendment to FMVSS No. 108 to Require Retroreflective Material on the Side and Rears of Heavy Trailers. Submission to NHTSA Docket No. 80-02-N09, NHTSA, 1991.

Radovich, V.G. "Development of Infant and Child Restraint Regulations and Their Application," *SAE Child Injury and Restraint Conference Proceedings*. Paper No. 831655, Society of Automotive Engineers Publication No. P-135, Warrendale, PA, 1983, pp. 101-111.

Rausch, A., Wong, Jackson I. and Kirkpatrick, M. *A Field Test of Two Single Center High-Mounted Brake Light Systems*. Submission to NHTSA Docket No. 81-02-N01-031, Insurance Institute for Highway Safety, Arlington, VA, 1981.

Recent Air Bag Field Performance. Research Note, NHTSA National Center for Statistics and Analysis, 1986.

Reilly, R.E., Kurke, D.S. and Buckenmaier, C.C., Jr. *Validation of the Reduction of Rear End Collisions by a High Mounted Auxiliary Stoplamp*. NHTSA Technical Report No. DOT HS 805 360, Washington, 1980.

Reinfurt, D.W. *A Statistical Evaluation of the Effectiveness of FMVSS 301: Fuel System Integrity*. NHTSA Report No. DOT HS 805 969, Washington, 1981.

Reinfurt, D.W., Silva, C.Z. and Seila, A.F. *A Statistical Analysis of Seat Belt Effectiveness in 1973-75 Model Cars Involved in Towaway Crashes, Volume 1*. NHTSA Report No. DOT HS 802 035, Washington, 1976.

Rieser, R.G. and Michaels, G.E. "Factors in the Development and Evaluation of Safer Glazing," *Ninth Stapp Car Crash Conference*. University of Minnesota, Minneapolis, 1966.

SAE Handbook. Society of Automotive Engineers, Warrendale, PA, annual publication.

Safety Belt Use Laws. Insurance Institute for Highway Safety, Arlington, VA, 2004 (http://www.hwysafety.org/safety%5Ffacts/state%5Flaws/restrain3.htm).

Safety Belt Usage – A Review of Effectiveness Studies – Suggestions for State Programs. NHTSA Report No. DOT HS 801 988, Washington, 1976.

Safety Code for Safety Glazing Materials for Glazing Motor Vehicles Operating on Land Highways, Standard Z26.1. American National Standards Institute, New York, 1966

SAS/STAT® User's Guide, Vol. 1, Version 6, 4ᵗʰ Ed. SAS Institute, Cary, NC, 1990.

Schmidt, R.A., Young, D.E. and Ayres, T.J. "Automobile Seat Belts: Usage Patterns in Automatic Belt Systems," *Human Factors*, Vol. 40, March 1998, pp. 126-135.

"70 by 92: Increased Safety-Belt Use," *Journal of the American Medical Association*, Vol. 268, July 15, 1992, p. 318.

Severy, D.M., Brink, H.M. and Baird, J.D. *Backrest and Head Restraint Design for Rear-End Collision Protection*. Paper No. 680079, Society of Automotive Engineers, New York, 1968.

Severy, D.M., Brink, H.M., Baird, J.D. and Blaisdell, D.M. "Safer Seat Designs," *Thirteenth Stapp Car Crash Conference*. Society of Automotive Engineers, New York, 1969.

Side Impact Conference. NHTSA Report No. DOT HS 805 614, Washington, 1980.

Sikora, J.J. *Relative Risk of Death for Ejected Occupants in Fatal Traffic Accidents*. NHTSA Report No. DOT HS 807 096, Washington, 1986.

Smith, R.A. and Kahane, C.J. *1974 Accident Experience with Air Cushion Restraint Systems*. NHTSA Technical Note No. DOT HS 801 565, Paper No. 750190, Society of Automotive Engineers, Warrendale, PA, 1975.

Standard No. 208 – Passive Restraint Amendment, Explanation of Rulemaking Action. NHTSA Docket No. 74-14-N10-011, NHTSA Report No. DOT HS 802 523, Washington, 1977.

States, J.D., Korn, M.W. and Massengill, J.B. "The Enigma of Whiplash Injuries," *Proceedings of the Thirteenth Annual Conference of the American Association of Automotive Medicine*. American Association of Automotive Medicine, Morton Grove, IL, 1969.

Stowell, C. and Bryant, J. *Safety Belt Usage: Survey of the Traffic Population*. NHTSA Report No. DOT HS 803 354, Washington, 1978.

Streff, F.M., Molnar, L.J. and Christoff, C. "Automatic Safety Belt Use in Michigan: A Two-Year Follow-Up," *Journal of Safety Research*, Vol. 25, Winter 1994, pp. 215-219.

Third Report to Congress – Effectiveness of Occupant Protection Systems and Their Use. NHTSA Report No. DOT HS 808 537, Washington, 1996.

Tourin, B. and Garrett, J.W. *Safety Belt Effectiveness in Rural California Automobile Accidents*. Cornell Aeronautical Laboratory, Buffalo, 1960.

Traffic Safety Facts 1999 – Children. NHTSA Publication No. DOT HS 809 087, Washington, 2000.

Traffic Safety Facts 2002. NHTSA Report No. DOT HS 809 620, Washington, 2004.

Traffic Safety Facts 2002 – Occupant Protection. NHTSA Publication No. DOT HS 809 610, Washington, 2003.

Treat, J.R. et al. *A Tri-Level Study of the Causes of Traffic Accidents*, Vol. 1. NHTSA Technical Report No. DOT HS 805 085, Washington, 1977.

Voight, G. and Wilfert, K. *Mechanisms of Injury to Unrestrained Drivers in Head-On Collisions.* Paper No. 690811, Society of Automotive Engineers, New York, 1969.

Walz, M.C. *The Effectiveness of Head Restraints in Light Trucks.* Technical Report No. DOT HS 809 247, Washington, 2001.

_____. *Evaluation of FMVSS 214 Side Impact Protection for Light Trucks: Crush Resistance Requirements for Side Doors.* NHTSA Technical Report No. DOT HS 809 719, Washington, 2004.

_____. *NCAP Test Improvements with Pretensioners and Load Limiters.* NHTSA Technical Report No. DOT HS 809 562, Washington, 2003.

Ward's Automotive Yearbook. Ward's Communications, Detroit, annual publication.

Widman, J.C. *Recent Developments in Penetration Resistance of Windshield Glass.* Paper No. 650474, Society of Automotive Engineers, New York, 1965.

Williams, A.F., Wells, J.K., Lund, A.K. and Teed, N. "Observed Use of Automatic Seat Belts in 1987 Cars," *Accident Analysis and Prevention*, Vol. 21, October 1989, pp. 427-433.

Williams, A.F., Wong, J. and O'Neill, B. "Occupant Protection in Interior Impacts – An Analysis of FMVSS No. 201," *Proceedings of the Twenty-Third Conference of the American Association for Automotive Medicine.* American Association for Automotive Medicine, Morton Grove, IL, 1979.

Willke, D., Summers, S., Wang, J., Lee, J., Partyka, S. and Duffy, S. *Ejection Mitigation Using Advanced Glazing: Status Report II.* NHTSA, Washington, 1999 (www-nrd.nhtsa.dot.gov/pdf/nrd-11/glazing0999.pdf).

Wilson, R.A. "Evaluating Knee-to-Instrument Panel Impacts," *Thirteenth Stapp Car Crash Conference.* Paper No. 690801, Society of Automotive Engineers, New York, 1969.

Ziedman, K., et al. *Improved Commercial Vehicle Conspicuity and Signalling Systems, Task II – Analyses, Experiments and Design Recommendations.* NHTSA Technical Report No. DOT HS 806 098, Washington, 1981.

APPENDIX A

SAS PROGRAMS USED TO ESTIMATE LIVES SAVED BY THE FMVSS AND OTHER VEHICLE SAFETY TECHNOLOGIES, 1960-2002

OVERVIEW

LS2004CR is the program that creates a census file called *LSM2004* of the 1,241,796 fatality cases on the Fatality Analysis Reporting System (FARS) for calendar years 1975 through 2002, on file as of February 27, 2004.[1] It includes all fatalities, non-occupants as well as vehicle occupants, occupants of heavy trucks, unknown vehicles, and other types as well as occupants of cars and LTVs. It includes necessary data elements about the person who was the fatality and the crash. It contains data elements about a vehicle: the vehicle occupied by the fatality, if the fatality was an occupant, or the vehicle that hit the non-occupant. Vehicle data include an analysis of the VIN, based on a series of programs developed and maintained by NHTSA staff since 1991 for use in FMVSS evaluations and other vehicle safety analyses. If the crash is a 2-vehicle crash, data elements are obtained on the "other" vehicle.

LS2004 is the main program that:

- Selects the fatality cases from *LSM2004* that were occupants of cars, occupants of LTVs, non-occupants struck by cars /LTVs, or motorcyclists struck by cars/LTVs.
- Adjusts for missing data on vehicle type or vehicle model year.
- Replaces each case with unknown impact type, seat position or belt/safety seat use by a set of cases, one for each possible value of those variables, weighted by the probability of occurrence of that value – based on analyses of the distribution of these variables, when they are known, as a function of other variables on the file.
- "Removes" the safety technologies from each case vehicle, one-by-one, in the reverse chronological order that they were installed, calculating the increase in the fatality risk, if any, as a consequence.
- Tallies, for the entire file, the number of additional fatalities there would have been if none of the vehicles had been equipped with any safety technologies, and apportions the additional fatalities by safety standard/technology, by vehicle type, by calendar year (1975-2002), and by whether the technology was mandated by a standard already in effect or was voluntarily furnished by the manufacturer at the time the vehicle was built.
- Generates information about CY 1975-80 crashes needed to extend the model to 1960-74, and stores it in three files named *FARS7580, PED7580* and *MC7580*.
- Prints out, for each model year, the proportion of cars and LTVs equipped with various safety technologies.

LS2004 includes both the "preprocessor" and the "main model" described in the Summary chapter of Part 2 of this report. Before LS2004 can be successfully operated, it is necessary to

---

[1] On that day FARS had been complete for the full year 2002 for some time, but selected cases may still be revised in the future.

run nine auxiliary programs that provide the information the preprocessor needs to adjust or distribute the unknowns.

LS_UNK4 finds the proportion of cases in each calendar year where the case vehicle type and/or its model year are unknown, and it develops the factors to adjust the weights of the remaining cases after these cases are deleted from the analysis.

L_CRSH4 creates four files, *CRSH1*, *CRSH2*, *CRSH3* and *CRSH4* that state the probability distribution of fatalities by crash type/impact location as a function of other variables. LS2004 will use these files to impute crash type/impact locations on *LSM2004* cases where they are unknown. For example in single-vehicle crashes, a multinomial logistic regression (CATMOD) calibrates the probability that the crash type/impact location was frontal, side, rollover or rear/other as a function of the specific object struck, the vehicle type, the occupant's age, and the occupant's ejection status, based on cases where everything is known, and it stores these probabilities in *CRSH1*.

L_SEAT4 creates a file, *SEAT1* that states the probability distribution, calibrated by CATMOD, of passenger fatalities by seat position as a function of the passenger's age, the vehicle type and the crash type/impact location. LS2004 will use this file to impute seat positions when they are completely unknown, or when they are just partially known (e.g., front seat, unknown location).

L3PT4 is a binomial logistic regression (LOGIST) of belt use by front-outboard occupants age 6 years or older in vehicles with integral 3-point belts (including automatic 3-point belts) at the front-outboard positions, for fatality cases where the belt use is not reported as "unknown." It calibrates belt use (yes or no) as a function of the calendar year, the occupant's age and gender, the vehicle type, vehicle age, crash type/impact location and seat position (driver or front-right). The regression equation is copied into LS2004, where it will impute the belt use of occupants for whom it is unknown on FARS.

LBK3PT4 similarly calibrates belt use for rear-outboard and rear-center occupants age 6+, in vehicles with integral 3-point belts at the applicable seat position, as a function of the calendar year, the occupant's age and gender, the vehicle type, vehicle age, crash type/impact location and seat position (outboard or center).

L2PT4 calibrates use of automatic 2-point shoulder belts at the front-outboard seats for occupants age 6+, as a function of the type of automatic 2-point belt (motorized or non-motorized), the occupant's age and gender, the vehicle age, crash type/impact location and seat position (driver or front-right). The analysis does not investigate whether the manual lap belt, if one exists, was also used.

L_LAP4 calibrates belt use of occupants age 6+ at seat positions equipped with lap belts only or with separate lap and shoulder belts (front-center in all model years; front-outboard, rear-outboard and rear-center before the shift to integral 3-point belts), as a function of the calendar year, the occupant's age and gender, the vehicle type, vehicle age, crash type/impact location and seat position (driver, front-center, front-right, rear-outboard or rear-center). In the cars with separate lap and shoulder belts, it merely calibrates if any belt was used, and it does not distinguish between the lap belt, the shoulder belt, or both.

LKIDRES4 comprises two logistic regressions and performs two calibrations of restraint systems use by child passengers age 0-5. The first regression, excluding only the passengers whose restraint use was completely unknown (codes 9 and 99), calibrates the proportion that used any type of restraint (belts, safety seats, or "used, type not specified") versus no restraint at all. The second regression, limited to restrained children where the type of restraint was also known, calibrates the proportion in safety seats versus belts. Both regression equations, as well as those from the three preceding programs, are copied into LS2004.

L_UNR_EJ tabulates the percentages of unrestrained fatalities that were ejected, by vehicle type, age group, seat position and impact location/crash type. These percentages are copied into the belt and safety-seat effectiveness analyses of LS2004.

A preliminary run of LS2004, and another program, NCNA2084, supplies information on the median installation years for the various safety technologies. (NCNA2084 shows the proportion of vehicles on NCSS and NASS files equipped with lap belts or lap/shoulder belts, by seat position, vehicle type and model year.) If any of the safety technologies on the current version of LS2004 are not in the right reverse-chronological order, rearrange the program and run it again. LS2004 generates a complete listing of lives saved by the safety technologies in 1975-2002 and also generates the files named *FARS7580*, *PED7580* and *MC7580* used to describe the distribution of fatalities in 1975-80 and make inferences about the distribution before 1975.

OLDFARS1 and OLDFA14 analyze files named *FARS7580*, *PED7580* and *MC7580* to find the CY 1975-80 distributions of:

- Car occupant fatalities by age of the case vehicle; LTV occupant fatalities by age of the case vehicle; non-occupant fatalities by age of the striking vehicle, when it was a car or LTV; motorcyclist fatalities in collisions with cars or LTVs, by age of the car/LTV.
- Distribution of car and LTV occupant fatalities by impact location/crash type.
- Distribution of car and LTV occupant fatalities by seat position and occupant age group.
- Proportion of fatalities that were ejected, by vehicle type, impact location/crash type, and belt use.
- Belt use by vehicle type, seat position and model year.
- Safety seat use and lap belt use by infants and toddlers.

OLDFA24 is the postprocessor that estimates lives saved in CY 1960-74. It uses information from *Accident Facts* and OLDFA14 to estimate:

- How many fatalities there would have been on FARS in each calendar year, from 1960 to 1974, if FARS had existed.
- How many of them would have been car occupants, LTV occupants, or non-occupants and motorcyclists struck by cars or LTVs.

- The model-year distribution of the cars and LTVs in these crashes, and for the occupant fatalities, their distribution by the vehicle's impact location/crash type, and whether or not the occupant was ejected.
- How many of the occupants were belted, and how many of the vehicles were equipped with the various safety technologies that existed before 1975.

Having created these hypothetical populations of fatal crash cases, OLDFA24 runs through the same model as LS2004, removing the safety technologies from the case vehicles in the reverse chronological order that they were installed, calculating the increase in the fatality risk, if any, as a consequence, and tallying the number of additional fatalities there would have been if none of the vehicles had been equipped with any safety technologies in 1960-74.

Finally, LIFE6002 combines the estimates from LS2004 and OLDFA24 to provide a single printout of the lives saved, year-by-year, in 1960-2002, apportioned by safety standard/ technology, by vehicle type, and by whether the technology was mandated by a standard already in effect or was voluntarily furnished by the manufacturer at the time the vehicle was built. TRND6002 computes occupant fatality rates per 100,000,000 VMT, based on VMT totals derived from *Traffic Safety Facts 2002* and *Accident Facts*.[2]

This overview is followed by complete listings, with commentary, of the long programs LS2004 and OLDFA24. All the other programs are much shorter, generally self-explanatory, or are discussed within the commentary on LS2004 and OLDFA24.

---

[2] *Traffic Safety Facts 2002*, NHTSA Report No. DOT HS 809 620, Washington, 2004; *Accident Facts*, National Safety Council, Chicago, annual publication.

DESCRIPTION OF THE MAIN ANALYSIS PROGRAM LS2004

Files used to run this program:

```
LIBNAME LIBRARY 'O:\FARSSAS\FORMATS\WINFMT91.610';
LIBNAME LSM2004 'C:\...';
LIBNAME CRSH1 'C:\...';
LIBNAME CRSH2 'C:\...';
LIBNAME CRSH3 'C:\...';
LIBNAME CRSH4 'C:\...';
LIBNAME SEAT1 'C:\...';
```

Files created by this program

```
LIBNAME FARS7580 'C:\...';
LIBNAME PED7580 'C:\...';
LIBNAME MC7580 'C:\...';

OPTION NOCENTER NOFMTERR OBS=5000000 LS=153 PAGESIZE=47;
/*  PROGRAM NAME: LS2004.SAS            UPDATED: 04/14/2004    */
/*  LIVES SAVED BY FMVSS IN 1975-2002, CARS, LIGHT TRUCKS, PEDS, MCs  */
```

Codes for impact location/crash type (CRSH):

```
PROC FORMAT; VALUE CRSH 1='FRONTAL' 2='SIDE IMPACT'
  3='ROLLOVER (PRIMARY)' 4='REAR & OTHER' 9='UNKNOWN';
```

Codes for vehicle type:

```
PROC FORMAT; VALUE VTYP 1='PASSENGER CAR' 2='LIGHT TRUCK' 3='BIG TRUCK, BUS'
  4='MOTORCYCLE' 5='OTHER' 9='UNKNOWN';

PROC FORMAT; VALUE CYGP 1='1975-81' 2='1982-90' 3='1991-2002';
PROC FORMAT; VALUE MY 1900-1954='PRE-1955' 9999='UNKNOWN';
```

Codes for streamlined seat position variable (SEAT2):

```
PROC FORMAT; VALUE NEWSEAT 11='DRIVER' 12='CENTER FRONT' 13='RIGHT FRONT' 18='OTHER FRONT'
  19='UNK FRONT' 21='OUTBOARD REAR' 22='CENTER REAR' 28='OTHER REAR'
  29='UNK REAR' 51='OTHER ENCLOSED' 52='UNENCLOSED' 99='UNK PASSENGER';

PROC FORMAT; VALUE NEWVTYP 1='CONVERTIBLE' 2='2-DOOR CAR' 4='4-DOOR CAR'
  6='STATION WAGON' 9='CAR, UNK STYLE'
  11='PICKUP TRUCK' 12='S U V' 13='VAN' 19='TRUCK, UNK STYLE';
```

Codes for streamlined restraint use variables (REST2 and REST3):

```
PROC FORMAT; VALUE NEWREST 0='UNRESTRAINED' 1='SHOULDER BELT ONLY' 2='LAP BELT ONLY'
  3='LAP+SHOULDER BELT' 4='CHILD SAFETY SEAT' 7='2-PT AUTOMATIC BELT'
  8='USED, TYPE NOT SPECIFIED' 99='UNKNOWN IF USED';
```

Special codes for analysis of air bags and child passengers:

```
PROC FORMAT; VALUE KCRSH 1='12:00 IMPACT1 OR 2' 2='11,1:00' 3='10,2:00' 4='NONFRONTAL';
PROC FORMAT; VALUE RESTGP 1='INFANT IN CRD' 2='INFANT NOT IN CRD'
  3='AGE 1-5 UNRESTRAINED' 4='AGE 1-5 IN CRD' 5='AGE 1-5 BELTED'
  6='AGE 6-10 UNRESTRAINED' 7='AGE 6-10 BELTED'
  8='AGE 11-12 UNRESTRAINED' 9='AGE 11-12 BELTED';
RUN;
```

The preprocessor section of LS2004 starts here. Selects from *LSM2004*, the census of fatality cases, those fatalities that were occupants of cars or LTVs (OCC1), non-occupants struck by cars or LTVs (PED1) or motorcyclists struck by cars or LTVs (MC1).

```
/* ----------------------------------------------------------------------- */
/* ----------------------------------------------------------------------- */
/*     SUBDIVIDES THE FATALITIES INTO (1) OCCUPANTS OF CARS/LTVs            */
/* (2) NONOCCUPANTS KILLED BY CARS/LTVS (3) MOTORCYCLISTS KILLED BY CAR/LTVs */
/* ADJUSTS ALL NUMBERS UPWARD FOR MISSING DATA ON VEHICLE TYPE OR MODEL YEAR */
/* MY = 0-9 IN CY < 1982 ARE HIGHLY SUSPECT AND WILL BE CONSIDERED MISSING  */
/* ----------------------------------------------------------------------- */
/* ----------------------------------------------------------------------- */

DATA OCC1(KEEP=ORIGWT WEIGHTFA CY MY VTYP IMPACT1 IMPACT2 HARM_EV MAN_COLL
         BOD2 TRKTYP BODY_TYP V1-V8 AGE SEX VE_FORMS CYGP NEWVTYP EJECTION
         SEAT_POS REST_USE AUT_REST PER_TYP LGT_COND CRSH PASSIVE MAK2 MM2 CG
         MAKE MAK_MOD PER_NO OVTYP OIMPACT2 OVCONFIG STATE M_HARM TAPECASE)
     PED1(KEEP=ORIGWT WEIGHTFA CY MY VTYP)
     MC1(KEEP=ORIGWT WEIGHTFA CY OMY OVTYP);
SET LSM2004.LSM2004;
```

Deletes cases with unknown vehicle type, unknown model year, or that did not involve a car/LTV.

```
IF VTYP=. THEN VTYP=9;
IF VTYP IN (1,2) OR (VTYP=4 AND OVTYP IN (1,2));
IF VTYP IN (1,2) AND (MY GT CY+1 OR MY=9999 OR 0 LE MY LE 1909) THEN DELETE;
IF VTYP=4 AND (OMY GT CY+1 OR OMY=9999 OR OMY=. OR 0 LE MY LE 1909) THEN DELETE;

IF VTYP=4 THEN GOTO MCWATE;
IF PER_TYP IN (1,2,9) THEN GOTO OCCWATE;

/* DELETE (AND ADJUST FOR) PARKED-VEHICLE OCC SINCE VTYP/MY ALWAYS UNKNOWN   */
IF (CY LE 1981 AND PER_TYP=5) OR (CY GE 1982 AND PER_TYP=3) THEN DELETE;

/* ----------------------------------------------------------------------- */
/* ADJUSTMENT FACTORS FOR NONOCCUPANTS - OBTAINED BY RUNNING LS_UNK4         */
/* ----------------------------------------------------------------------- */
```

Creates the two weight-factor variables, ORIGWT and WEIGHTFA that will stay with each fatality cases as it proceeds through the model. ORIGWT is the original weight assigned to each fatality case. WEIGHTFA is the inflated weight denoting how many fatalities there would have been if safety technologies were removed from the vehicle. Throughout the preprocessor, WEIGHTFA = ORIGWT.

Having deleted all the cases with unknown vehicle type or model year, must give all the remaining cases an ORIGWT higher than 1, so that the ORIGWTs will add up to the original fatality count, including an allowance for all cases deleted due to missing MY, and an allowance for a portion of the cases with unknown vehicle type – as computed by the program LS_UNK4.

For example, suppose there are 10,000 fatality cases in cars/LTVs with known MY, 100 cases in cars/LTVs with unknown MY, and 200 fatality cases in vehicles of unknown type. Assume also that, when the vehicle type is known, 80 percent are cars/LTVs and 20 percent are heavy trucks, motorcycles or other known types. Then the inflation factor computed by LS_UNK4 would be:

$$[10,000 + 100 + (.80 \times 200)] / 10,000 = 1.026$$

The vehicle's model year or type is unknown in 9-19 percent of pedestrian crashes, primarily because many of them are hit-and-run.

```
IF CY=1975 THEN WEIGHTFA=1.15080; ELSE IF CY=1976 THEN WEIGHTFA=1.16331;
  ELSE IF CY=1977 THEN WEIGHTFA=1.16595; ELSE IF CY=1978 THEN WEIGHTFA=1.16998;
  ELSE IF CY=1979 THEN WEIGHTFA=1.17921; ELSE IF CY=1980 THEN WEIGHTFA=1.18698;
  ELSE IF CY=1981 THEN WEIGHTFA=1.20951; ELSE IF CY=1982 THEN WEIGHTFA=1.13142;
  ELSE IF CY=1983 THEN WEIGHTFA=1.11566; ELSE IF CY=1984 THEN WEIGHTFA=1.11156;
  ELSE IF CY=1985 THEN WEIGHTFA=1.12026; ELSE IF CY=1986 THEN WEIGHTFA=1.11781;
  ELSE IF CY=1987 THEN WEIGHTFA=1.11672; ELSE IF CY=1988 THEN WEIGHTFA=1.11157;
  ELSE IF CY=1989 THEN WEIGHTFA=1.12342; ELSE IF CY=1990 THEN WEIGHTFA=1.11187;
  ELSE IF CY=1991 THEN WEIGHTFA=1.11680; ELSE IF CY=1992 THEN WEIGHTFA=1.10650;
  ELSE IF CY=1993 THEN WEIGHTFA=1.10809; ELSE IF CY=1994 THEN WEIGHTFA=1.10696;
  ELSE IF CY=1995 THEN WEIGHTFA=1.10145; ELSE IF CY=1996 THEN WEIGHTFA=1.10478;
  ELSE IF CY=1997 THEN WEIGHTFA=1.10029; ELSE IF CY=1998 THEN WEIGHTFA=1.09377;
  ELSE IF CY=1999 THEN WEIGHTFA=1.09198; ELSE IF CY=2000 THEN WEIGHTFA=1.09535;
  ELSE IF CY=2001 THEN WEIGHTFA=1.10633; ELSE IF CY=2002 THEN WEIGHTFA=1.11733;
  ORIGWT=WEIGHTFA;
  OUTPUT PED1; RETURN;

/* ------------------------------------------------------------------------ */
/* ADJUSTMENT FACTORS FOR MOTORCYCLISTS - OBTAINED BY RUNNING LS_UNK4        */
/* ------------------------------------------------------------------------ */

MCWATE: IF CY=1975 THEN WEIGHTFA=1.03029; ELSE IF CY=1976 THEN WEIGHTFA=1.03235;
  ELSE IF CY=1977 THEN WEIGHTFA=1.02838; ELSE IF CY=1978 THEN WEIGHTFA=1.01218;
  ELSE IF CY=1979 THEN WEIGHTFA=1.00780; ELSE IF CY=1980 THEN WEIGHTFA=1.00795;
  ELSE IF CY=1981 THEN WEIGHTFA=1.01466; ELSE IF CY=1982 THEN WEIGHTFA=1.01772;
  ELSE IF CY=1983 THEN WEIGHTFA=1.01923; ELSE IF CY=1984 THEN WEIGHTFA=1.02307;
  ELSE IF CY=1985 THEN WEIGHTFA=1.02295; ELSE IF CY=1986 THEN WEIGHTFA=1.01927;
  ELSE IF CY=1987 THEN WEIGHTFA=1.02442; ELSE IF CY=1988 THEN WEIGHTFA=1.01736;
  ELSE IF CY=1989 THEN WEIGHTFA=1.01430; ELSE IF CY=1990 THEN WEIGHTFA=1.02179;
  ELSE IF CY=1991 THEN WEIGHTFA=1.02575; ELSE IF CY=1992 THEN WEIGHTFA=1.00806;
  ELSE IF CY=1993 THEN WEIGHTFA=1.01340; ELSE IF CY=1994 THEN WEIGHTFA=1.01456;
  ELSE IF CY=1995 THEN WEIGHTFA=1.01144; ELSE IF CY=1996 THEN WEIGHTFA=1.01047;
  ELSE IF CY=1997 THEN WEIGHTFA=1.01546; ELSE IF CY=1998 THEN WEIGHTFA=1.01122;
  ELSE IF CY=1999 THEN WEIGHTFA=1.01471; ELSE IF CY=2000 THEN WEIGHTFA=1.01027;
  ELSE IF CY=2001 THEN WEIGHTFA=1.01655; ELSE IF CY=2002 THEN WEIGHTFA=1.01231;
  ORIGWT=WEIGHTFA;
  OUTPUT MC1; RETURN;

/* ------------------------------------------------------------------------ */
/* ADJUSTMENT FACTORS FOR CAR/LTV OCCUPANTS - OBTAINED BY RUNNING LS_UNK4    */
/* INCLUDES OCCUPANTS OF PARKED VEHICLES                                     */
/* ------------------------------------------------------------------------ */
```

In recent years, the vehicle type and model year have been known in well over 99 percent of the occupant fatality cases on FARS.

```
OCCWATE: IF CY=1975 THEN WEIGHTFA=1.02596; ELSE IF CY=1976 THEN WEIGHTFA=1.02452;
  ELSE IF CY=1977 THEN WEIGHTFA=1.02496; ELSE IF CY=1978 THEN WEIGHTFA=1.01685;
  ELSE IF CY=1979 THEN WEIGHTFA=1.01236; ELSE IF CY=1980 THEN WEIGHTFA=1.01104;
  ELSE IF CY=1981 THEN WEIGHTFA=1.01537; ELSE IF CY=1982 THEN WEIGHTFA=1.01228;
  ELSE IF CY=1983 THEN WEIGHTFA=1.00685; ELSE IF CY=1984 THEN WEIGHTFA=1.00814;
  ELSE IF CY=1985 THEN WEIGHTFA=1.01133; ELSE IF CY=1986 THEN WEIGHTFA=1.00817;
  ELSE IF CY=1987 THEN WEIGHTFA=1.00883; ELSE IF CY=1988 THEN WEIGHTFA=1.00627;
  ELSE IF CY=1989 THEN WEIGHTFA=1.00802; ELSE IF CY=1990 THEN WEIGHTFA=1.00643;
  ELSE IF CY=1991 THEN WEIGHTFA=1.00580; ELSE IF CY=1992 THEN WEIGHTFA=1.00448;
  ELSE IF CY=1993 THEN WEIGHTFA=1.00658; ELSE IF CY=1994 THEN WEIGHTFA=1.00539;
  ELSE IF CY=1995 THEN WEIGHTFA=1.00559; ELSE IF CY=1996 THEN WEIGHTFA=1.00640;
  ELSE IF CY=1997 THEN WEIGHTFA=1.00659; ELSE IF CY=1998 THEN WEIGHTFA=1.00474;
  ELSE IF CY=1999 THEN WEIGHTFA=1.00456; ELSE IF CY=2000 THEN WEIGHTFA=1.00404;
  ELSE IF CY=2001 THEN WEIGHTFA=1.00437; ELSE IF CY=2002 THEN WEIGHTFA=1.00905;
  ORIGWT=WEIGHTFA;
```

```
/* DEFINES THE CRASH CONFIGURATION                                    */
```

The CRSH variable (1 = frontal, 2 = side, 3 = rollover, 4 = rear/other) is critically important because the effectiveness of many of the safety technologies varies by crash type. First-harmful-event fires, immersions and other non-collision types are included in the rear/other category. If the principal impact (IMPACT2) is unknown, rely on the initial impact (IMPACT1).

```
IF IMPACT2=13 OR HARM_EV=1      THEN CRSH=3;
  ELSE IF 2 LE HARM_EV LE 6     THEN CRSH=4;
  ELSE IF 5 LE IMPACT2 LE 7     THEN CRSH=4;
  ELSE IF 8 LE IMPACT2 LE 10    THEN CRSH=2;
  ELSE IF 2 LE IMPACT2 LE 4     THEN CRSH=2;
  ELSE IF IMPACT2 IN (1,11,12)  THEN CRSH=1;
  ELSE IF HARM_EV=7             THEN CRSH=4;
  ELSE IF IMPACT1=13            THEN CRSH=3;
  ELSE IF 5 LE IMPACT1 LE 7     THEN CRSH=4;
  ELSE IF 8 LE IMPACT1 LE 10    THEN CRSH=2;
  ELSE IF 2 LE IMPACT1 LE 4     THEN CRSH=2;
  ELSE IF IMPACT1 IN (1,11,12)  THEN CRSH=1;

/* (HEAD-ON COLLISION INVOLVMENTS ARE ASSUMED TO BE FRONTAL IMPACTS) */
  ELSE IF MAN_COLL=2            THEN CRSH=1;

  ELSE                              CRSH=9;

/* DEFINES THE CALENDAR YEAR GROUPS (MAJOR CHANGES IN FARS DATA DEFINITIONS) */
IF 1975 LE CY LE 1981 THEN CYGP=1;
  ELSE IF 1982 LE CY LE 1990 THEN CYGP=2;
  ELSE IF CY GE 1991 THEN CYGP=3;

/* DEFINES THE VEHICLE BODY TYPE (N OF CAR DOORS, PICKUP, SUV, VAN)    */
IF VTYP=1 THEN DO;
IF BOD2 IN (1,2,4,6,9) THEN NEWVTYP=BOD2;
  ELSE IF BOD2 IN (3,7) THEN NEWVTYP=2;
  ELSE IF BOD2 IN (5,8) THEN NEWVTYP=4;
  ELSE IF CYGP=1 THEN DO;
    IF BODY_TYP IN (1,2,6,9) THEN NEWVTYP=BODY_TYP;
    ELSE IF BODY_TYP=3 THEN NEWVTYP=4; ELSE NEWVTYP=9; END;
  ELSE IF CYGP IN (2,3) THEN DO;
    IF BODY_TYP IN (1,2,4,6,9) THEN NEWVTYP=BODY_TYP;
    ELSE IF BODY_TYP=3 THEN NEWVTYP=2;
    ELSE IF BODY_TYP=5 THEN NEWVTYP=4; ELSE NEWVTYP=9; END; END;

ELSE IF VTYP=2 THEN DO;
IF TRKTYP IN (1,2) THEN NEWVTYP=11;
  ELSE IF TRKTYP IN (3,4) THEN NEWVTYP=12;
  ELSE IF TRKTYP IN (5,6) THEN NEWVTYP=13;
  ELSE IF TRKTYP IN (7,8,9) THEN NEWVTYP=19;
  ELSE IF CYGP=1 THEN DO;
    IF BODY_TYP=50 THEN NEWVTYP=11;
    ELSE IF BODY_TYP IN (43,52) THEN NEWVTYP=12;
    ELSE IF BODY_TYP=51 THEN NEWVTYP=13; ELSE NEWVTYP=19; END;
  ELSE IF CYGP=2 THEN DO;
    IF BODY_TYP IN (50,51) THEN NEWVTYP=11;
    ELSE IF BODY_TYP IN (12,55,56) THEN NEWVTYP=12;
    ELSE IF BODY_TYP=40 THEN NEWVTYP=13; ELSE NEWVTYP=19; END;
  ELSE IF CYGP=3 THEN DO;
    IF BODY_TYP IN (30,31,32,39) THEN NEWVTYP=11;
    ELSE IF BODY_TYP IN (14,15,16,19) THEN NEWVTYP=12;
    ELSE IF BODY_TYP IN (20,21) THEN NEWVTYP=13; ELSE NEWVTYP=19; END; END;

OUTPUT OCC1;
RUN;
```

Next is the main analysis section for non-occupant fatalities: pedestrians, bicyclists and other non-motorists that were struck by cars or LTVs. Two safety technologies, dual master cylinders (105A) and front disc brakes (105B) saved non-occupant lives because they enabled car/LTV drivers to avoid hitting the non-occupants. The model is simpler here because it involves just those two technologies, but the setup is basically the same as for car and LTV occupant fatalities.

```
/* ------------------------------------------------------------------- */
/* ------------------------------------------------------------------- */
/*  ANALYZES EFFECT OF BRAKE IMPROVEMENTS ON NONOCCUPANT FATALITIES    */
/* ------------------------------------------------------------------- */
/* ------------------------------------------------------------------- */

DATA PED2; SET PED1;
/* IMPLEMENTATION OF FRONT DISC BRAKES IN THE VEHICLES THAT HIT THESE PEDESTRIANS */
```

Proceed in reverse chronological order: disc brakes first because they were installed later. Ideally, we should determine if the vehicle (that hit the pedestrian) was equipped with front disc brakes. However, we don't know whether specific vehicles were equipped or not equipped. We only know, based on the vehicle's model year, what percent of vehicles were equipped in that model year. PS105B + PV105B is the percentage of vehicles, in any given model year, that had front disc brakes; PS105B are the installations after FMVSS 105 was amended, effective 1/1/1976 in cars and 9/1/1983 in LTVs, to include performance tests most easily met with disc brakes; PV105B are any disc-brake installations before the standard was amended. PS = "proportion, standard" PV = "proportion voluntary"

```
IF MY GE 1984 THEN DO; PS105B=1; PV105B=0; END;
  ELSE IF MY LE 1964 THEN DO; PS105B=0; PV105B=0; END;
  ELSE IF MY=1965 THEN DO; PS105B=0; PV105B=.02; END;
  ELSE IF MY=1966 THEN DO; PS105B=0; PV105B=.03; END;
  ELSE IF MY=1967 THEN DO; PS105B=0; PV105B=.06; END;
  ELSE IF MY=1968 THEN DO; PS105B=0; PV105B=.13; END;
  ELSE IF MY=1969 THEN DO; PS105B=0; PV105B=.28; END;
  ELSE IF MY=1970 THEN DO; PS105B=0; PV105B=.41; END;
  ELSE IF MY=1971 THEN DO; PS105B=0; PV105B=.63; END;
  ELSE IF MY=1972 THEN DO; PS105B=0; PV105B=.74; END;
  ELSE IF MY=1973 THEN DO; PS105B=0; PV105B=.86; END;
  ELSE IF MY=1974 THEN DO; PS105B=0; PV105B=.84; END;
  ELSE IF MY=1975 THEN DO; PS105B=0; PV105B=.93; END;
  ELSE IF MY=1976 AND CY=1975 THEN DO; PS105B=0; PV105B=.99; END;
  ELSE IF MY=1976 AND VTYP=1 THEN DO; PS105B=.50; PV105B=.49; END;
  ELSE IF MY=1976 AND VTYP=2 THEN DO; PS105B=0; PV105B=.99; END;
  ELSE IF 1977 LE MY LE 1983 AND VTYP=1 THEN DO; PS105B=1; PV105B=0; END;
  ELSE IF 1977 LE MY LE 1983 AND VTYP=2 THEN DO; PS105B=0; PV105B=1; END;
```

Thus, for example, in MY 1971, PV105B = .63 because 63 percent of cars (and we assume also LTVs) were equipped with front disc brakes, voluntarily, before the amendment to FMVSS 105. Note that PV105B > 0 even in MY 1976, because nearly half of MY 1976 cars were produced before 1/1/1976.

```
/* IMPLEMENTATION OF DUAL MASTER CYLINDERS IN THE VEHICLES THAT HIT THESE PEDESTRIANS */
  IF MY GE 1984 THEN DO; PS105A=1; PV105A=0; END;
    ELSE IF MY LE 1961 THEN DO; PS105A=0; PV105A=0; END;
    ELSE IF MY IN (1962,1963) THEN DO; PS105A=0; PV105A=.09; END;
    ELSE IF MY IN (1964,1965) THEN DO; PS105A=0; PV105A=.07; END;
    ELSE IF MY=1966 THEN DO; PS105A=0; PV105A=.54; END;
    ELSE IF MY=1967 THEN DO; PS105A=0; PV105A=1; END;
    ELSE IF MY=1968 AND VTYP=1 THEN DO; PS105A=.5; PV105A=.5; END;
    ELSE IF MY=1968 AND VTYP=2 THEN DO; PS105A=0; PV105A=1; END;
```

```
ELSE IF 1969 LE MY LE 1983 AND VTYP=1 THEN DO; PS105A=1; PV105A=0; END;
ELSE IF 1969 LE MY LE 1983 AND VTYP=2 THEN DO; PS105A=0; PV105A=1; END;
```

Here is the basic routine for estimating lives saved by a safety technology, specifically non-occupant lives saved by front disc brakes. All effectiveness estimates are derived from NHTSA evaluation reports and discussed in Part 1 of this report. NHTSA's evaluation estimated that front disc brakes reduced fatal crash involvements by 0.17 percent. Effectiveness, E = .0017. P = PS105B + PV105B is the probability that the case vehicle was equipped with front disc brakes (based on its model year). Up to this point in the model, this case has a weight of WEIGHTFA fatalities, renamed OLDWTFA. There would have been OLDWTFA / (1 – E x P) fatalities if the vehicle had no front disc brakes at all rather than a P probability of disc brakes. That becomes the new value of WEIGHTFA. S = OLDWTFA x E x P / (1 – E x P) is the difference between the new WEIGHTFA and the OLDWTFA, and it is the increase in fatalities that would have occurred in the complete absence of front disc brakes. S is apportioned between lives saved by post-standard installations of disc brakes (PLS105B) and voluntary installations of disc brakes (PLV105B). The new value of WEIGHTFA here will become the old value (OLDWTFA) in the next step (dual master cylinders).

PLV105B: "P" = pedestrian, "L" = lives saved, "V" = voluntary, "105B" = front disc brakes

```
/* PEDESTRIAN LIVES SAVED BY FRONT DISC BRAKES */
  OLDWTFA=WEIGHTFA;
  IF PS105B GT 0 OR PV105B GT 0 THEN DO;
    E=.0017;
    P=PS105B+PV105B;
    S=OLDWTFA*E*P / (1 - E*P);
    PLS105B=S*PS105B/P;
    PLV105B=S*PV105B/P;
    WEIGHTFA=OLDWTFA+PLS105B+PLV105B; END;
  ELSE DO; PLS105B=0; PLV105B=0; END;
```

NHTSA's evaluation estimated that dual master cylinders reduced fatal crash involvements by 0.7 percent.

```
/* PEDESTRIAN LIVES SAVED BY DUAL MASTER CYLINDERS */
  OLDWTFA=WEIGHTFA;
  IF PS105A GT 0 OR PV105A GT 0 THEN DO;
    E=.007;
    P=PS105A+PV105A;
    S=OLDWTFA*E*P / (1 - E*P);
    PLS105A=S*PS105A/P;
    PLV105A=S*PV105A/P;
    WEIGHTFA=OLDWTFA+PLS105A+PLV105A; END;
  ELSE DO; PLS105A=0; PLV105A=0; END;
  PLS=PLS105A+PLS105B; PLV=PLV105A+PLV105B;
  PL105A=PLS105A+PLV105A; PL105B=PLS105B+PLV105B; PL=PLS+PLV;
```

At this point, the model has "removed" all applicable safety technologies from the vehicles, and WEIGHTFA indicates the implicit number of non-occupant fatalities in the absence of front disc brakes and dual master cylinders on the case car/LTV.

```
RUN;
```

Adds up the actual fatality cases (ORIGWT), the estimate of how many fatalities there would have been if all safety technologies had been removed from the vehicles (WEIGHTFA), and the lives saved by each technology, by calendar year. PLV = total non-occupant lives saved (voluntary installations). PLS = total non-occupant lives saved (post-standard installations). PL = PLV + PLS = total non-occupant lives saved by all technologies.

```
PROC MEANS SUM NOPRINT DATA=PED2; BY CY;
  VAR ORIGWT WEIGHTFA PLV105B PLS105B PL105B PLV105A PLS105A PL105A PLV PLS PL;
  OUTPUT OUT=PED3
    SUM=P_ORIGWT P_WTFA PLV105B PLS105B PL105B PLV105A PLS105A PL105A PLV PLS PL;
```

Prints out the totals for each calendar year, and the sum for all years, 1975-2002.

```
PROC PRINT DATA=PED3;
  FORMAT P_ORIGWT P_WTFA PLV105B PLS105B PL105B PLV105A PLS105A PL105A PLV PLS PL 9.0;
  ID CY; VAR P_ORIGWT P_WTFA PLV105B PLS105B PL105B PLV105A PLS105A PL105A PLV PLS PL;
  SUM P_ORIGWT P_WTFA PLV105B PLS105B PL105B PLV105A PLS105A PL105A PLV PLS PL;
TITLE1 'PEDESTRIANS SAVED BY CAR/LIGHT TRUCK DISC BRAKES AND DUAL MASTER CYLINDERS, 1975-2002';
TITLE2 ' ';
TITLE3 '...105B = FRONT DISC BRAKES (MEDIAN INSTALLATION YEAR 1971, FMVSS EFFECTIVE 1/1/76 OR
9/1/83)';
TITLE4 '...105A = DUAL MASTER CYLINDERS (MEDIAN INSTALLATION YEAR LATE 1966, FMVSS EFFECTIVE 1/1/68
OR 9/1/83)';
TITLE5 '...V = VOLUNTARY INSTALLATIONS, BEFORE EFFECTIVE DATE';
TITLE6 '...S = STANDARD INSTALLATIONS, AFTER EFFECTIVE DATE';
TITLE7 'P_ORIGWT = ACTUAL PED/BIKE/NONMOTORIST FATALITIES';
TITLE8 'P_WTFA = FATALITIES THAT WOULD HAVE OCCURRED WITHOUT THESE SAFETY IMPROVEMENTS';
TITLE9 'PL = OVERALL PED/BIKE/NONMOTORIST LIVES SAVED BY BRAKE IMPROVEMENTS IN CARS/LIGHT TRUCKS';
RUN;


/* ------------------------------------------------------------------ */
/* ------------------------------------------------------------------ */
/*  ANALYZES EFFECT OF BRAKE IMPROVEMENTS ON MOTORCYCLIST FATALITIES  */
/* ------------------------------------------------------------------ */
/* ------------------------------------------------------------------ */
```

Next is the main analysis for motorcyclist fatalities in collisions with cars/LTVs. The model is exactly the same as for non-occupants, except lives-saved estimates are named "MLV105B" instead of "PLV105B," etc. ("M" = motorcyclist) OMY = the model year of the car/LTV that hit the motorcycle.

```
DATA MC2; SET MC1;
/* IMPLEMENTATION OF FRONT DISC BRAKES IN THE VEHICLES THAT HIT THESE MOTORCYCLES */
  IF OMY GE 1984 THEN DO; MS105B=1; MV105B=0; END;
    ELSE IF OMY LE 1964 THEN DO; MS105B=0; MV105B=0; END;
    ELSE IF OMY=1965 THEN DO; MS105B=0; MV105B=.02; END;
    ELSE IF OMY=1966 THEN DO; MS105B=0; MV105B=.03; END;
    ELSE IF OMY=1967 THEN DO; MS105B=0; MV105B=.06; END;
    ELSE IF OMY=1968 THEN DO; MS105B=0; MV105B=.13; END;
    ELSE IF OMY=1969 THEN DO; MS105B=0; MV105B=.28; END;
    ELSE IF OMY=1970 THEN DO; MS105B=0; MV105B=.41; END;
    ELSE IF OMY=1971 THEN DO; MS105B=0; MV105B=.63; END;
    ELSE IF OMY=1972 THEN DO; MS105B=0; MV105B=.74; END;
    ELSE IF OMY=1973 THEN DO; MS105B=0; MV105B=.86; END;
    ELSE IF OMY=1974 THEN DO; MS105B=0; MV105B=.84; END;
    ELSE IF OMY=1975 THEN DO; MS105B=0; MV105B=.93; END;
    ELSE IF OMY=1976 AND CY=1975 THEN DO; MS105B=0; MV105B=.99; END;
    ELSE IF OMY=1976 AND OVTYP=1 THEN DO; MS105B=.50; MV105B=.49; END;
    ELSE IF OMY=1976 AND OVTYP=2 THEN DO; MS105B=0; MV105B=.99; END;
    ELSE IF 1977 LE OMY LE 1983 AND OVTYP=1 THEN DO; MS105B=1; MV105B=0; END;
    ELSE IF 1977 LE OMY LE 1983 AND OVTYP=2 THEN DO; MS105B=0; MV105B=1; END;
/* IMMLEMENTATION OF DUAL MASTER CYLINDERS IN THE VEHICLES THAT HIT THESE MOTORCYCLES */
  IF OMY GE 1984 THEN DO; MS105A=1; MV105A=0; END;
    ELSE IF OMY LE 1961 THEN DO; MS105A=0; MV105A=0; END;
    ELSE IF OMY IN (1962,1963) THEN DO; MS105A=0; MV105A=.09; END;
```

```
        ELSE IF OMY IN (1964,1965) THEN DO; MS105A=0; MV105A=.07; END;
        ELSE IF OMY=1966 THEN DO; MS105A=0; MV105A=.54; END;
        ELSE IF OMY=1967 THEN DO; MS105A=0; MV105A=1; END;
        ELSE IF OMY=1968 AND OVTYP=1 THEN DO; MS105A=.5; MV105A=.5; END;
        ELSE IF OMY=1968 AND OVTYP=2 THEN DO; MS105A=0; MV105A=1; END;
        ELSE IF 1969 LE OMY LE 1983 AND OVTYP=1 THEN DO; MS105A=1; MV105A=0; END;
        ELSE IF 1969 LE OMY LE 1983 AND OVTYP=2 THEN DO; MS105A=0; MV105A=1; END;
/* MOTORCYLIST LIVES SAVED BY FRONT DISC BRAKES */
  OLDWTFA=WEIGHTFA;
  IF MS105B GT 0 OR MV105B GT 0 THEN DO;
    E=.0017;
    P=MS105B+MV105B;
    S=OLDWTFA*E*P / (1 - E*P);
    MLS105B=S*MS105B/P;
    MLV105B=S*MV105B/P;
    WEIGHTFA=OLDWTFA+MLS105B+MLV105B; END;
  ELSE DO; MLS105B=0; MLV105B=0; END;
/* MOTORCYCLIST LIVES SAVED BY DUAL MASTER CYLINDERS */
  OLDWTFA=WEIGHTFA;
  IF MS105A GT 0 OR MV105A GT 0 THEN DO;
    E=.007;
    P=MS105A+MV105A;
    S=OLDWTFA*E*P / (1 - E*P);
    MLS105A=S*MS105A/P;
    MLV105A=S*MV105A/P;
    WEIGHTFA=OLDWTFA+MLS105A+MLV105A; END;
  ELSE DO; MLS105A=0; MLV105A=0; END;
  MLS=MLS105A+MLS105B; MLV=MLV105A+MLV105B;
  ML105A=MLS105A+MLV105A; ML105B=MLS105B+MLV105B; ML=MLS+MLV;
RUN;

PROC MEANS SUM NOPRINT DATA=MC2; BY CY;
  VAR ORIGWT WEIGHTFA MLV105B MLS105B ML105B MLV105A MLS105A ML105A MLV MLS ML;
  OUTPUT OUT=MC3
    SUM=M_ORIGWT M_WTFA MLV105B MLS105B ML105B MLV105A MLS105A ML105A MLV MLS ML;
PROC PRINT DATA=MC3;
  FORMAT M_ORIGWT M_WTFA MLV105B MLS105B ML105B MLV105A MLS105A ML105A MLV MLS ML 9.0;
  ID CY; VAR M_ORIGWT M_WTFA MLV105B MLS105B ML105B MLV105A MLS105A ML105A MLV MLS ML;
  SUM M_ORIGWT M_WTFA MLV105B MLS105B ML105B MLV105A MLS105A ML105A MLV MLS ML;
TITLE1 'MOTORCYCLISTS SAVED BY CAR/LIGHT TRUCK DISC BRAKES AND DUAL MASTER CYLINDERS, 1975-2002';
TITLE2 ' ';
TITLE3 '...105B = FRONT DISC BRAKES (MEDIAN INSTALLATION YEAR 1971, FMVSS EFFECTIVE 1/1/76 OR
9/1/83)';
TITLE4 '...105A = DUAL MASTER CYLINDERS (MEDIAN INSTALLATION YEAR LATE 1966, FMVSS EFFECTIVE 1/1/68
OR 9/1/83)';
TITLE5 '...V = VOLUNTARY INSTALLATIONS, BEFORE EFFECTIVE DATE';
TITLE6 '...S = STANDARD INSTALLATIONS, AFTER EFFECTIVE DATE';
TITLE7 'M_ORIGWT = ACTUAL MOTORCYLIST FATALITIES IN COLLISIONS WITH A CAR OR LTV';
TITLE8 'M_WTFA = FATALITIES THAT WOULD HAVE OCCURRED WITHOUT THESE SAFETY IMPROVEMENTS';
TITLE9 'ML = OVERALL MOTORCYLIST LIVES SAVED BY BRAKE IMPROVEMENTS IN CARS/LIGHT TRUCKS';
RUN;
```

The program now turns to car/LTV occupant fatalities, and it returns to the pre-processor section, splitting cases with unknown crash mode, seat position or restraint use into two or more cases with known, imputed values for these variables.

```
/* ---------------------------------------------------------------- */
/* ---------------------------------------------------------------- */
/*   THE ANALYSIS OF OCCUPANT FATALITIES BEGINS HERE                */
/*   FIRST TASK IS TO IMPUTE ALL UNKOWN CRASHMODES, SEAT POS, BELT USE */
/* ---------------------------------------------------------------- */
/* ---------------------------------------------------------------- */

proc freq data=occ1; weight origwt; format crsh crsh.; tables crsh;
```

288

```
TITLE 'UNKNOWN CRSH IN OCC1';
run;
```

Splits the file OCC1 into OCC2 and UNKCRSH1. OCC2 are the cases with known impact
location/crash type (CRSH) and are left alone. UNKCRSH1 contains the cases with unknown
crash mode (CRSH = 9).

```
DATA OCC2 UNKCRSH1; SET OCC1;
IF CRSH=9 THEN OUTPUT UNKCRSH1; ELSE OUTPUT OCC2;
RUN;
/* ----------------------------------------------------------------------- */
/* ----------------------------------------------------------------------- */
/*   DISTRIBUTES AND IMPUTES CRASH MODES WHEN THEY ARE UNKNOWN           */
/*   SEPARATE IMPUTATIONS FOR FIXED OBJECT, ANGLE, REAR-END AND          */
/*   OTHER CRASHES BASED ON CATMOD MODELS DEVELOPED IN L_CRSH4.SAS       */
/*   (HEAD-ON COLLISION INVOLVMENTS ARE ASSUMED TO BE FRONTAL IMPACTS)   */
/* ----------------------------------------------------------------------- */
/* ----------------------------------------------------------------------- */
```

UNKCRSH1, in turn, is split into FIXOBJ, ANGLE, REAR, OTH; the first consists of single-
vehicle crashes with unknown crash mode; the second, multivehicle crashes described as "angle
collisions" (MAN_COLL = 1,3); the third, rear-end collisions; the fourth, all other crash
involvements. Defines various subgroups of first harmful event and seat position/age/gender.
These groupings, together with vehicle type and occupant ejection, are good predictors of whether
the impact was frontal, side, rollover, or rear/other.

```
DATA FIXOBJ ANGLE REAR OTH;
SET UNKCRSH1;
IF HARM_EV IN (10,11,12,14,18,24,30,34,35,42,43,99) THEN EVENT1=HARM_EV;
  ELSE IF HARM_EV IN (8,9,15,44,47) THEN EVENT1=99;
  ELSE IF HARM_EV=13 THEN EVENT1=12;
  ELSE IF HARM_EV IN (16,19,22,48) THEN EVENT1=34;
  ELSE IF HARM_EV=17 THEN EVENT1=18;
  ELSE IF HARM_EV IN (20,23,26) THEN EVENT1=24;
  ELSE IF HARM_EV IN (21,25,32,36,37,38,39,46) THEN EVENT1=35;
  ELSE IF HARM_EV IN (27,29,40) THEN EVENT1=42;
  ELSE IF HARM_EV=28 THEN EVENT1=30;
  ELSE IF HARM_EV IN (31,33,41) THEN EVENT1=43;
  ELSE IF HARM_EV=45 THEN EVENT1=14;
  ELSE IF HARM_EV=49 THEN EVENT1=11;
  ELSE EVENT1=99;
IF EVENT1=12 THEN DO;
    IF MAN_COLL IN (1,3) THEN EVENT2=12.1;
    ELSE IF MAN_COLL=4 THEN EVENT2=12.2;
    ELSE IF MAN_COLL IN (5,6,7) THEN EVENT2=12.3;
    ELSE EVENT2=99; END;
  ELSE EVENT2=EVENT1;
IF EJECTION IN (1,2) THEN EJECT=1; ELSE EJECT=0;
IF PER_TYP=2 THEN DO; AGEGP3=1;
    IF 0 LE AGE LE 15 THEN AGEGP4=1; ELSE AGEGP4=2; END;
  ELSE DO; IF 14 LE AGE LE 29 THEN AGEGP3=2;
      ELSE IF 30 LE AGE LE 49 THEN AGEGP3=3;
      ELSE IF 50 LE AGE LE 97 THEN AGEGP3=4;
      ELSE AGEGP3=2;
    IF SEX=1 THEN AGEGP4=3; ELSE IF SEX=2 THEN AGEGP4=4; END;
IF 14 LE EVENT2 LE 43 THEN OUTPUT FIXOBJ;
  ELSE IF EVENT2 IN (10,12.2) THEN OUTPUT ANGLE;
  ELSE IF EVENT2=12.1 THEN OUTPUT REAR;
  ELSE OUTPUT OTH;
RUN;

/* ----------------------------------------------------------------- */
```

```
/*   CRSH1.CRSH1 ASSIGNS FIXED-OBJECT CRASHES TO FRONTAL, SIDE, ROLLOVER   */
/*   OR REAR BASED ON FIRST HARMFUL EVENT, EJECTION, DRV/PAS AGE, CAR/LTRK */
/* ------------------------------------------------------------------- */
```

CRSH1 was generated by the program L_CRSH4 by a CATMOD analysis of the cases with known crash modes. For each combination of EVENT2, EJECT, AGEGP3 and VTYP, CATMOD calibrates the proportions of fatalities that are frontal, side, rollover and rear/other, and it stores these proportions in *CRSH1*. For example, a typical record on *CRSH1* could look like:

Event2	Eject	Agegp3	Vtyp	Frontal	Side	Roll	Rearothr
42 (tree)	0	2 (young drv)	1 (car)	.58	.35	.05	.02

In other words, when a young car driver initially hits a tree, and is fatally injured but not ejected, 58 percent of the crashes with known crash type are frontals, 35 percent are side impacts, 5 percent are classified as rollovers for the purpose of this model because the principal impact location is the top of the car (due to a subsequent rollover), and 2 percent are rear/other.

```
PROC SORT DATA=FIXOBJ; BY EVENT2 EJECT AGEGP3 VTYP;
```

The information from *CRSH1* is added to each case.

```
DATA FIXOBJ2; MERGE FIXOBJ(IN=OCC) CRSH1.CRSH1;
  BY EVENT2 EJECT AGEGP3 VTYP; IF OCC;
```

Now each original case with unknown crash mode is split into four separate cases with known crash modes. For example, if the original case had a weight factor (ORIGWT = WEIGHTFA) of 1, and it was a young non-ejected car driver who hit a tree, create four cases, with the same values as the original case on all variables except CRSH, ORIGWT and WEIGHTFA. The first case will have CRSH = 1 (frontal) and ORIGWT = WEIGHTFA = .58. The second case will have CRSH = 2 (side) and ORIGWT = WEIGHTFA = .35. The third case will have CRSH = 3 (rollover) and ORIGWT = WEIGHTFA = .05. The fourth case will have CRSH = 4 (rear/other) and ORIGWT = WEIGHTFA = .02. The four cases, together will have ORIGWTs adding up to 1, like the initial case.

```
DATA FIXOBJ3; SET FIXOBJ2;
O1=ORIGWT; W1=WEIGHTFA;
CRSH=1; ORIGWT=FRONTAL * O1; WEIGHTFA=FRONTAL * W1; OUTPUT;
CRSH=2; ORIGWT=SIDE    * O1; WEIGHTFA=SIDE    * W1; OUTPUT;
CRSH=3; ORIGWT=ROLL    * O1; WEIGHTFA=ROLL    * W1; OUTPUT;
CRSH=4; ORIGWT=REAROTHR* O1; WEIGHTFA=REAROTHR* W1; OUTPUT;

/* ------------------------------------------------------------------- */
/*   CRSH2.CRSH2 ASSIGNS MULTIVEH ANGLE COLLISIONS TO FRONTAL, SIDE, ROLL */
/*   OR REAR BASED ON FIRST HARMFUL EVENT, EJECTION, DRV/PAS AGE, CAR/LTRK */
/* ------------------------------------------------------------------- */

PROC SORT DATA=ANGLE; BY EVENT2 EJECT AGEGP4 VTYP;
DATA ANGLE2; MERGE ANGLE(IN=OCC) CRSH2.CRSH2;
  BY EVENT2 EJECT AGEGP4 VTYP; IF OCC;
DATA ANGLE3; SET ANGLE2;
O1=ORIGWT; W1=WEIGHTFA;
CRSH=1; ORIGWT=FRONTAL * O1; WEIGHTFA=FRONTAL * W1; OUTPUT;
CRSH=2; ORIGWT=SIDE    * O1; WEIGHTFA=SIDE    * W1; OUTPUT;
CRSH=3; ORIGWT=ROLL    * O1; WEIGHTFA=ROLL    * W1; OUTPUT;
CRSH=4; ORIGWT=REAROTHR* O1; WEIGHTFA=REAROTHR* W1; OUTPUT;
```

```
/* ------------------------------------------------------------------- */
/*  CRSH3.CRSH3 ASSIGNS MULTIVEH REAR-END COLLSNS TO FRONTAL, SIDE, ROLL */
/*  OR REAR BASED ON EJECTION, DRV/PAS AGE, CAR/LTRK                     */
/* ------------------------------------------------------------------- */

PROC SORT DATA=REAR; BY EVENT2 EJECT AGEGP4 VTYP;
DATA REAR2; MERGE REAR(IN=OCC) CRSH3.CRSH3;
  BY EVENT2 EJECT AGEGP4 VTYP; IF OCC;
DATA REAR3; SET REAR2;
O1=ORIGWT; W1=WEIGHTFA;
CRSH=1; ORIGWT=FRONTAL * O1; WEIGHTFA=FRONTAL * W1; OUTPUT;
CRSH=2; ORIGWT=SIDE    * O1; WEIGHTFA=SIDE    * W1; OUTPUT;
CRSH=3; ORIGWT=ROLL    * O1; WEIGHTFA=ROLL    * W1; OUTPUT;
CRSH=4; ORIGWT=REAROTHR* O1; WEIGHTFA=REAROTHR* W1; OUTPUT;


/* ------------------------------------------------------------------- */
/*  CRSH4.CRSH4 ASSIGNS OTHER-EVENT CRASHES TO FRONTAL, SIDE, ROLLOVER  */
/*  OR REAR BASED ON FIRST HARMFUL EVENT, EJECTION                     */
/* ------------------------------------------------------------------- */

PROC SORT DATA=OTH; BY EVENT2 EJECT;
DATA OTH2; MERGE OTH(IN=OCC) CRSH4.CRSH4;
  BY EVENT2 EJECT; IF OCC;
DATA OTH3; SET OTH2;
O1=ORIGWT; W1=WEIGHTFA;
CRSH=1; ORIGWT=FRONTAL * O1; WEIGHTFA=FRONTAL * W1; OUTPUT;
CRSH=2; ORIGWT=SIDE    * O1; WEIGHTFA=SIDE    * W1; OUTPUT;
CRSH=3; ORIGWT=ROLL    * O1; WEIGHTFA=ROLL    * W1; OUTPUT;
CRSH=4; ORIGWT=REAROTHR* O1; WEIGHTFA=REAROTHR* W1; OUTPUT;
```

UNKCRSH2 contains all the newly created cases with known crash modes. It is appended to
OCC2, the original cases with known crash modes, creating OCC3, a file on which all cases have
known crash modes.

```
DATA UNKCRSH2; SET FIXOBJ3 ANGLE3 REAR3 OTH3;
DROP EVENT1 EVENT2 AGEGP3 AGEGP4 FRONTAL SIDE ROLL REAROTHR EJECT O1 W1 N;
RUN;

DATA OCC3; SET OCC2 UNKCRSH2;
```

The next step is to replace unknown seat positions with known seat positions. All drivers
(PER_TYP = 1) are assumed to be in the left-front seat. Passengers (PER_TYP = 2) might have an
unknown seat position. A special case is vehicles where multiple people were ejected and killed,
and it is unknown who was the driver (PER_TYP = 9).

```
/* WHEN THERE ARE MULTIPLE (EJECTED) OCCUPANTS, UNKNOWN WHO WAS THE DRIVER,   */
/* ARBITRARILY SAY PER_NO 1 IS THE DRIVER, SINCE IT HARDLY AFFECTS OUR MODEL  */
IF PER_TYP=9 THEN DO;
  IF 0 LE AGE LE 14 THEN PER_TYP=2;
    ELSE IF SEAT_POS NE 1 AND SEAT_POS NE 11 AND SEAT_POS NE 19 AND
      SEAT_POS NE 99 THEN PER_TYP=2;
    ELSE IF PER_NO=1 THEN PER_TYP=1;
    ELSE PER_TYP=2; END;
```

Before the imputation of seat positions, some simplifications of the SEAT_POS variable on FARS
are needed. (For example, before 1982, FARS coded seat positions 1, 2, 3, 4, 5, 6 rather than 11,
12, 13, 21, 22, 23.)

```
/* --------------------------------------------------------------------------- */
/* STREAMLINES SEAT_POS CATEGORIES AND MAKES UNIFORM ACROSS 3 GENERATIONS OF FARS */
/* IN VANS, ETC., 3rd AND 4th ROW OF SEATS ARE TREATED AS REAR SEATS            */
```

```
/* ---------------------------------------------------------------------------- */
IF PER_TYP=1 AND O LE AGE LE 9 THEN DO; PER_TYP=2; SEAT_POS=17; END;
  ELSE IF PER_TYP=1 THEN SEAT_POS=11;
  ELSE IF PER_TYP=2 AND
    ((CYGP IN (2,3) AND SEAT_POS=11) OR (CYGP=1 AND SEAT_POS=1)) THEN SEAT_POS=17;
  ELSE IF PER_TYP=2 AND CYGP=1 THEN DO;
    IF 2 LE SEAT_POS LE 3 THEN SEAT_POS=SEAT_POS+10;
    ELSE IF 4 LE SEAT_POS LE 6 THEN SEAT_POS=SEAT_POS+17;
    ELSE IF 7 LE SEAT_POS LE 9 THEN SEAT_POS=SEAT_POS+24;
    ELSE IF SEAT_POS=10 THEN SEAT_POS=18;
    ELSE IF SEAT_POS=11 THEN SEAT_POS=28;
    ELSE IF SEAT_POS=12 THEN SEAT_POS=38;
    ELSE IF SEAT_POS=13 THEN SEAT_POS=51;
    ELSE IF SEAT_POS=14 THEN SEAT_POS=50;
    ELSE IF SEAT_POS=15 THEN SEAT_POS=55; END;

IF VTYP=1 THEN DO; IF SEAT_POS IN (11,12,13,18,19,51,99) THEN SEAT2=SEAT_POS;
  ELSE IF SEAT_POS=17 THEN SEAT2=18;
  ELSE IF SEAT_POS IN (21,23) THEN SEAT2=21;
  ELSE IF SEAT_POS IN (22,31,32,33) THEN SEAT2=22;
  ELSE IF SEAT_POS IN (28,38,41,42,43,48,49) THEN SEAT2=28;
  ELSE IF SEAT_POS IN (29,39) THEN SEAT2=29;
  ELSE IF SEAT_POS IN (50,52,53,54,55) THEN SEAT2=52; END;
ELSE DO; IF SEAT_POS IN (11,12,13,18,19,51,99) THEN SEAT2=SEAT_POS;
  ELSE IF SEAT_POS=17 THEN SEAT2=18;
  ELSE IF SEAT_POS IN (21,23,31,33,41,43) THEN SEAT2=21;
  ELSE IF SEAT_POS IN (22,32,42) THEN SEAT2=22;
  ELSE IF SEAT_POS IN (28,38,48) THEN SEAT2=28;
  ELSE IF SEAT_POS IN (29,39,49) THEN SEAT2=29;
  ELSE IF SEAT_POS IN (50,52,53,54,55) THEN SEAT2=52; END;
RUN;
```

The SEAT2 codes are: 11=driver 12=center front 13=right front 18=other front 19=unknown front 21=outboard rear 22=center rear 28=other rear 29=unknown rear 51=other enclosed 52=unenclosed 99=unknown passenger. Note: a child sitting on the driver's lap is coded 18, "other front."

```
proc freq data=occ3; weight origwt; format seat2 newseat.; tables seat2;
TITLE 'UNKNOWN SEAT2 IN OCC3';
run;
```

The imputation of seat position is quite similar to the procedure for imputing crash modes. The file OCC3 is split into OCC4 and UNKSEAT1. OCC4 are the cases with known seat position (SEAT2) and are left alone. UNKSEAT1 contains the cases with unknown seat position (SEAT2 = 19, 29, 99).

```
DATA OCC4 UNKSEAT1; SET OCC3;
IF SEAT2 IN (19,29,99) THEN OUTPUT UNKSEAT1; ELSE OUTPUT OCC4;
RUN;
```

```
/* ------------------------------------------------------------------ */
/*  DISTRIBUTES AND IMPUTES PASSENGER SEAT POSITIONS WHEN UNKNOWN.     */
/*  SEPARATE IMPUTATIONS FOR TOTALLY UNKNOWN, FRONT-UNKNOWN AND        */
/*  REAR-UNKNOWN PASSENGERS, BASED ON CATMOD MODEL IN LSEAT4.SAS       */
/* ------------------------------------------------------------------ */
```

UNKSEAT1 is further split into the cases where the seat position is entirely unknown (SEAT2 = 99) and the cases where it is known that the occupant was in the front seat, or in a back seat, but the exact location is unknown.

```
DATA USEAT119 USEAT129 USEAT199; SET UNKSEAT1;
```

```
IF AGE=0 THEN AGEGP2=1;
  ELSE IF   1 LE AGE LE   4 THEN AGEGP2=2;
  ELSE IF   5 LE AGE LE  10 THEN AGEGP2=3;
  ELSE IF  11 LE AGE LE  15 THEN AGEGP2=4;
  ELSE IF  16 LE AGE LE  19 THEN AGEGP2=5;
  ELSE IF  20 LE AGE LE  29 THEN AGEGP2=6;
  ELSE IF  30 LE AGE LE  97 THEN AGEGP2=7; ELSE AGEGP2=6;
IF SEAT2=19 THEN OUTPUT USEAT119;
  ELSE IF SEAT2=29 THEN OUTPUT USEAT129;
  ELSE IF SEAT2=99 THEN OUTPUT USEAT199;
RUN;
```

The program L_SEAT4 uses CATMOD to calibrate the distribution of passengers' known seat positions by detailed vehicle type (NEWVTYP), passengers' age group (AGEGP2), and CRSH. On a file *SEAT1*, it stores the proportions of passengers who are in the center-front seat, the right-front seat, and at any of the other known seat positions included in SEAT2, and it stores them in the data elements SEAT12, SEAT13, etc. These are the proportions that will be imputed if the seat position is completely unknown.

However, if we know the passenger was in the front seat, but we don't know exactly where, we can impute them by the proportions SEAT12 / (SEAT12 + SEAT13 + SEAT18), etc.

```
DATA PSEAT19; SET SEAT1.SEAT1;
P1=SEAT12+SEAT13+SEAT18;
SEAT12=SEAT12/P1; SEAT13=SEAT13/P1; SEAT18=SEAT18/P1;
KEEP NEWVTYP AGEGP2 CRSH SEAT12 SEAT13 SEAT18;

DATA PSEAT29; SET SEAT1.SEAT1;
P2=SEAT21+SEAT22+SEAT28;
SEAT21=SEAT21/P2; SEAT22=SEAT22/P2; SEAT28=SEAT28/P2;
KEEP NEWVTYP AGEGP2 CRSH SEAT21 SEAT22 SEAT28;
RUN;
```

Just as in the imputation of crash modes, split each case with unknown seat position into multiple cases with known seat position, each weighted by the probability that the passenger was in that position.

```
PROC SORT DATA=USEAT119; BY NEWVTYP AGEGP2 CRSH;
DATA USEAT219; MERGE USEAT119(IN=OCC) PSEAT19;
  BY NEWVTYP AGEGP2 CRSH; IF OCC;
DATA USEAT319; SET USEAT219;
O1=ORIGWT; W1=WEIGHTFA;
SEAT2=12; ORIGWT=SEAT12 * O1; WEIGHTFA=SEAT12 * W1; OUTPUT;
SEAT2=13; ORIGWT=SEAT13 * O1; WEIGHTFA=SEAT13 * W1; OUTPUT;
SEAT2=18; ORIGWT=SEAT18 * O1; WEIGHTFA=SEAT18 * W1; OUTPUT;

PROC SORT DATA=USEAT129; BY NEWVTYP AGEGP2 CRSH;
DATA USEAT229; MERGE USEAT129(IN=OCC) PSEAT29;
  BY NEWVTYP AGEGP2 CRSH; IF OCC;
DATA USEAT329; SET USEAT229;
O1=ORIGWT; W1=WEIGHTFA;
SEAT2=21; ORIGWT=SEAT21 * O1; WEIGHTFA=SEAT21 * W1; OUTPUT;
SEAT2=22; ORIGWT=SEAT22 * O1; WEIGHTFA=SEAT22 * W1; OUTPUT;
SEAT2=28; ORIGWT=SEAT28 * O1; WEIGHTFA=SEAT28 * W1; OUTPUT;

PROC SORT DATA=USEAT199; BY NEWVTYP AGEGP2 CRSH;
DATA USEAT299; MERGE USEAT199(IN=OCC) SEAT1.SEAT1;
  BY NEWVTYP AGEGP2 CRSH; IF OCC;
DATA USEAT399; SET USEAT299;
O1=ORIGWT; W1=WEIGHTFA;
```

```
SEAT2=12; ORIGWT=SEAT12 * O1; WEIGHTFA=SEAT12 * W1; OUTPUT;
SEAT2=13; ORIGWT=SEAT13 * O1; WEIGHTFA=SEAT13 * W1; OUTPUT;
SEAT2=18; ORIGWT=SEAT18 * O1; WEIGHTFA=SEAT18 * W1; OUTPUT;
SEAT2=21; ORIGWT=SEAT21 * O1; WEIGHTFA=SEAT21 * W1; OUTPUT;
SEAT2=22; ORIGWT=SEAT22 * O1; WEIGHTFA=SEAT22 * W1; OUTPUT;
SEAT2=28; ORIGWT=SEAT28 * O1; WEIGHTFA=SEAT28 * W1; OUTPUT;
SEAT2=51; ORIGWT=SEAT51 * O1; WEIGHTFA=SEAT51 * W1; OUTPUT;
SEAT2=52; ORIGWT=SEAT52 * O1; WEIGHTFA=SEAT52 * W1; OUTPUT;
```

UNKSEAT2 contains all the newly created cases with known seat positions. It is appended to
OCC4, the original cases with known seat positions, creating OCC5, a file on which all cases have
known seat positions (and crash modes).

```
DATA UNKSEAT2; SET USEAT319 USEAT329 USEAT399;
DROP AGEGP2 SEAT12 SEAT13 SEAT18 SEAT21 SEAT22 SEAT28 SEAT51 SEAT52 O1 W1;
RUN;
```

The next step is to replace unknown belt (or safety seat) use with known restraint use. Before the
imputation of restraint use, it is necessary to determine exactly what sort(s) of belt systems were
installed at each seat. For front-outboard seats of cars starting in MY 1981 and LTVs in 1985, a
detailed VIN analysis has already been performed on the FARS cases, and the information is stored
in the variable PASSIVE that is already on the file *LSM2004* and carried over to OCC5. For the
other vehicles/seat positions, restraint availability is determined from the VIN-based make-model
codes MAK2, CG and MM2, or from the FARS codes MAKE and MAK_MOD. The information
is encoded in the variables FTLS (for the front-outboard seats), BKLS (for the rear-outboard seats),
and CBKLS (for the rear-center seats). In vans, the "rear" seats include the 3rd and 4th rows as well
as the 2nd. The "lap belts only" code really means "lap belts only, if anything." Front-center seats
and third seats in station wagons are assumed to have lap belts only.

```
/* ---------------------------------------------------------------- */
/* FTLS: 3=FRONT OUTBOARD SEATS HAVE 3-POINT BELTS                   */
/*       2=AUTOMATIC 2-POINT BELTS                                   */
/*       1=LAP BELTS ONLY                                            */
/*       4=3-POINT BELTS FOR DRIVER, AUTOMATIC 2-POINT BELTS FOR RF  */
/*       5=SEPARATE MANUAL LAP AND TORSO BELTS                       */
/*       6=MIX OF 3-POINT BELTS AND LAP-ONLY BELTS                   */
/*       7=MIX OF SEPARATE LAP AND TORSO BELTS AND LAP-ONLY BELTS    */
/*       8=MIX OF 3-POINT BELTS AND 2-POINT AUTOMATIC BELTS          */
/*                                                                   */
/* BKLS: 3=OUTBOARD BACK SEATS HAVE 3-POINT BELTS                    */
/*       1=LAP BELTS ONLY                                            */
/*                                                                   */
/* CBKLS: 3=CENTER BACK SEATS HAVE 3-POINT BELTS                     */
/*        1=LAP BELTS ONLY                                           */
/*                                                                   */
/* CENTER FRONT SEATS AND STATION-WAGON 3RD SEATS: LAP BELT ONLY     */
/* OTHER (NONSTANDARD) SEATS HAVE NO BELTS AT ALL                    */
/* KIDS UNDER 6 CAN ONLY USE SAFETY SEATS OR LAP BELTS               */
/* PEOPLE AGE 10 OR OLDER ARE NEVER IN CHILD SAFETY SEATS            */
/* ---------------------------------------------------------------- */

DATA OCC5; SET OCC4 UNKSEAT2;
```

The program now edits the original restraint use code on FARS, REST_USE in three steps,
generating REST1, then REST2, then REST3, comprising the following codes: 0=unrestrained
1=shoulder belt only 2=lap belt only 3=lap + shoulder belt 4=child safety seat 7=2-pt automatic
belt 8=used, type not specified 99=unknown if used.

294

REST1 is a slight streamlining of REST_USE, changing codes used in earlier versions of FARS to current codes. Children age 0-5 will be considered protected by "lap belts only" when FARS says they wore lap and shoulder belts.

```
IF REST_USE=13 AND PER_TYP=2 AND 0 LE AGE LE 5 THEN REST1=2;
  ELSE IF REST_USE=13 THEN REST1=8;
  ELSE IF REST_USE=14 THEN REST1=4;
  ELSE IF REST_USE=9 THEN REST1=99; ELSE REST1=REST_USE;
IF (1977 LE CY LE 1979 AND AUT_REST=5) OR
  (1980 LE CY LE 1989 AND AUT_REST=1) THEN REST1=7;
```

REST2 edits REST1, taking into account what types of belts were actually in the vehicle and modifying implausible codes. Anybody age 10 or older will be considered to have unknown restraint use if FARS says they were in a child safety seat. "Motorcycle helmet" will always be recoded "unknown" for car/LTV occupants. People in non-designated seat positions (e.g., unenclosed areas) are assumed to be unrestrained.

```
IF SEAT2 IN (17,18,28,51,52) THEN REST2=0;
  ELSE IF REST1=5 THEN REST2=99;
  ELSE IF REST1=4 AND AGE GE 10 THEN REST2=99;
  ELSE REST2=REST1;
```

Belt type installed in front-outboard seats, passenger cars. In 1969-73, VW, Audi, BMW, Mercedes and Volvo are assumed to have 3-point belts, while other makes had separate lap and shoulder belts.

```
IF VTYP=1 THEN DO;
  IF PASSIVE IN (303,404,606,707,1303,1404,1606,1313) THEN FTLS=2;
    ELSE IF PASSIVE IN (1003,1004,1006) THEN FTLS=4;
    ELSE IF MAK_MOD IN (3006,3036,30036) AND 1975 LE MY LE 1980 THEN FTLS=8;
    ELSE IF 1974 LE MY LE 2002 THEN FTLS=3;
    ELSE IF 1969 LE MY LE 1973 AND MAKE IN (30,32,34,42,51) THEN FTLS=3;
    ELSE IF 1965 LE MY LE 1968 AND MAKE=51 THEN FTLS=3;
    ELSE IF 1969 LE MY LE 1973 THEN FTLS=5;
    ELSE IF MY=1968 AND MAKE IN (30,32,34,42) THEN FTLS=6;
    ELSE IF MY=1968 THEN FTLS=7;
    ELSE FTLS=1; END;
```

Belt type installed in front-outboard seats, LTVs. All LTVs had 3-point belts from MY 1981 onwards. All pickup trucks from 1977 onwards. All LTVs had only lap belts (if anything) up to 1968; SUVs and vans up to 1973. In 1969-80, LTVs could have lap belts, or 3-point belts, or a mix of the two, depending on the make, MY and vehicle type. In the absence of specific evidence, a mix of the two is assumed.

```
ELSE IF VTYP=2 THEN DO;
  IF 1981 LE MY LE 2002 THEN FTLS=3;
    ELSE IF MY LE 1968 THEN FTLS=1;
    ELSE IF MAKE=30 THEN FTLS=3;
    ELSE IF NEWVTYP=11 THEN DO;
      IF 1977 LE MY LE 1980 THEN FTLS=3;
        ELSE IF 2 LE MAKE LE 29 AND 1974 LE MY LE 1976 THEN FTLS=6;
        ELSE IF 2 LE MAKE LE 29 THEN FTLS=1;
        ELSE IF 31 LE MAKE LE 84 AND 1975 LE MY LE 1976 THEN FTLS=3;
        ELSE IF 31 LE MAKE LE 84 AND 1972 LE MY LE 1974 THEN FTLS=6;
        ELSE IF 31 LE MAKE LE 84 THEN FTLS=1;
        ELSE IF 1972 LE MY LE 1976 THEN FTLS=6; ELSE FTLS=1; END;
    ELSE IF NEWVTYP=12 THEN DO;
      IF 1969 LE MY LE 1973 THEN FTLS=1;
        ELSE IF MAKE IN (1,2,3,29,84) THEN FTLS=6;
```

```
        ELSE IF MAKE IN (7,9,20,23) AND 1974 LE MY LE 1976 THEN FTLS=6;
        ELSE IF MAKE IN (7,9,20,23) THEN FTLS=3;
        ELSE IF MAKE=12 AND 1974 LE MY LE 1977 THEN FTLS=6;
        ELSE IF MAKE=12 THEN FTLS=3;
        ELSE IF MAKE=49 AND 1974 LE MY LE 1975 THEN FTLS=1;
        ELSE IF MAKE=49 THEN FTLS=3; ELSE FTLS=6; END;
    ELSE IF 1969 LE MY LE 1973 THEN FTLS=1;
    ELSE IF 1974 LE MY LE 1979 THEN FTLS=6;
    ELSE IF MY=1980 THEN FTLS=3; END;
```

TY = first model year with 3-point belts at the back outboard seats. These dates are documented in NHTSA's evaluation of back seat belts.

```
IF MAK2=51 OR (MAK2=. AND MAKE=51) THEN TY=1971;
    ELSE IF VTYP=1 AND (MAK2=42 OR (MAK2=. AND MAKE=42)) THEN TY=1974;
    ELSE IF MM2=39032 THEN TY=1977;
    ELSE IF VTYP=1 AND
     (MAK2 IN (34,44,45,47) OR MAK2=. AND MAKE IN (34,44,45,47)) THEN TY=1981;
    ELSE IF MM2 IN (37032,32035) THEN TY=1982;
    ELSE IF MM2=30041 OR MAK2=32 OR (MAK2=. AND MAKE=32) THEN TY=1983;
    ELSE IF MAK2=56 OR (MAK2=. AND MAKE=56) THEN TY=1986;
    ELSE IF MM2=30040 THEN TY=1985;
    ELSE IF MM2=54032 THEN TY=1986;
    ELSE IF MM2 IN (35039,54032,61031,39031) THEN TY=1987;
    ELSE IF MM2 IN (6018,7018,12017,14017,13005,18005,19005,19014,21005,20019,
        30042,30044,54031,37031,37033,41035,41037,41043,41044,46044,49040,49035,
        60031,22031) OR (MM2=19003 AND CG=18042) OR
        (CG=18052 AND MM2 IN (18002,21002,22002)) THEN TY=1988;
    ELSE IF MM2 IN (7017,9017,7019,9019,12004,14004,18016,20016,22016,20009,22009,
        18017,20017,21017,22017,18003,19003,21003,18018,21018,22018,18020,20020,21020,
        22020,35042,35032,41034,49032,49034,49033,49038,7034,9034,52035,6035,52031,
        7044,9044,52034,10034,20034,53034,55033) OR CG IN (18039,18040,18048)
        THEN TY=1989;
  ELSE IF VTYP=1 THEN TY=1990;
    ELSE IF CG=38301 AND V7='8' THEN TY=1987;
    ELSE IF CG IN (18402,18403,38301,38303,38304) THEN TY=1988;
    ELSE IF CG IN (18301,18303,18304,18401,41401,49301,49302,49401,52301,53302,53303)
        THEN TY=1989;
    ELSE IF CG IN (18302,18404,18405,35301) THEN TY=1990;
    ELSE IF CG IN (1303,1304,1305,6301,6402,6403,6405,6406,18305,49303,49402)
        THEN TY=1991;
  ELSE IF VTYP=2 THEN TY=1992;
IF MY GE TY THEN BKLS=3; ELSE BKLS=1;
```

CTY = first model year with 3-point belts at the center-rear seat. This is documented in NHTSA brochures from MY 1999 onwards. There is limited information on earlier vehicles in NHTSA crash files. All cars and LTVs will be required to have 3-point belts by MY 2008.

```
IF MAK2 IN (47,51) THEN CTY=1994;
  ELSE IF (VTYP=1 AND MAK2 IN (13,58)) OR
     MM2 IN (19014,19017,20032,30046,32042,34035,35039,37032,37402,39032,
            41035,41037,42048,49032,49040,49043,52040,54035,54036,55033,
            59031,59032,59034,62307) OR CG IN (12039,49303) THEN CTY=1999;
  ELSE IF MM2 IN (12037,18002,19003,20002,30040,30042,32044,35043,35047,
            39034,42303,42307,48034,49045,52034,64032) THEN CTY=2000;
  ELSE IF MAK2 IN (42,48) OR (VTYP=1 AND MAK2 IN (6,7,9)) OR
     MM2 IN (6302,21022,22002,22352,22353,24005,24006,32047,49046,49322,
            49323,49342,49343,49352,49353,54323,59035,59332,59333) THEN CTY=2001;
  ELSE IF MAK2=62 OR MM2 IN (7214,7215,20036,32043,32046,34037,34303,37031,52046,
            52336,52337,55035,55038,63034) OR

/* ----------------------------------------------------------------------- */
/* ADD SATURN VUE TO THE 2002 LIST AFTER YOU DEFINE ITS CG AND MM2          */
```

```
/* ADD MERCURY MARAUDER ETC. TO THE 2003 LIST AFTER YOU DEFINE ITS CG AND MM2   */
/* ---------------------------------------------------------------------------- */

      CG IN (6305,18314,18315,18316,37303) THEN CTY=2002;
  ELSE IF MAK2 IN (13,19,24,32) OR (VTYP=1 AND MAK2 IN (41,49)) OR
    MM2 IN (12016,12017,12300,12301,12302,12303,12306,12307,12308,
            12312,12313,13312,13313,14016,14017,14302,14303,14308,
            20016,22016,22032,37322,49302,49303,52333,53033,55036) OR
    CG IN (18210,18311,18312) THEN CTY=2003;
  ELSE CTY=2008;                               /* FINAL DATE IN ANTON'S LAW */
IF MY GE CTY THEN CBKLS=3; ELSE CBKLS=1;

DROP TY CTY;
RUN;


proc freq data=occ5; weight origwt; format rest2 newrest.; tables rest2;
TITLE 'UNKNOWN REST2 IN OCC5';
run;
```

OCC5 is split into OCC6 and UNKBELT1. OCC6 are the cases with known restraint use (REST2) and are left alone. UNKBELT1 contains the cases with unknown restraint use (REST2 = 99), plus child passengers known to be restrained, but unknown if they were in a safety seat or just belts.

```
DATA OCC6 UNKBELT1; SET OCC5;
IF REST2=99 THEN OUTPUT UNKBELT1;
  ELSE IF O LE AGE LE 5 AND PER_TYP=2 AND REST2=8 THEN OUTPUT UNKBELT1;
  ELSE OUTPUT OCC6;
RUN;
```

UNKBELT1 is further subdivided into: occupants of front-outboard seats equipped with 3-point belts (UNK3PT1), with 2-point automatic belts (UNK2PT1), back seats equipped with 3-point belts (UNKBK31), any seats equipped with lap belts or separate lap and shoulder belts (UNKLAP1), child passengers age 0-5 with completely unknown restraint use (UNKKID1), and child passengers age 0-5 known to be restrained, but unknown what type of restraint (UNKKID2). FTLS = 8 is primarily 3-point belts with some 2-point, and is sent to UNK3PT1; other mixes are sent to UNKLAP1.

```
/* ---------------------------------------------------------------------- */
/*  DISTRIBUTES AND IMPUTES BELT USE WHEN UNKNOWN; SEPARATELY FOR         */
/*  FRONT-SEAT 3-POINT, AUTOMATIC 2-POINT, BACK-SEAT 3-POINT, LAP,        */
/*  INFANT-TODDLER (IN 2 STEPS: USED/NOT-USED BELT/SAFETY-SEAT)           */
/*  BASED ON LOGISTIC REGRESSION MODELS FROM L3PT4, L2PT4, LBK3PT4,       */
/*  L_LAP4, & LKIDRES4                                                    */
/* ---------------------------------------------------------------------- */

DATA UNK3PT1 UNK2PT1 UNKBK31 UNKLAP1 UNKKID1 UNKKID2;
SET UNKBELT1;
GENDER=SEX; IF GENDER NE 2 THEN GENDER=1;
IF O LE AGE LE 97 THEN NEWAGE=AGE; ELSE NEWAGE=30;
  IF PER_TYP=1 AND AGE LT 14 THEN NEWAGE=30;
IF O LE AGE LE 5 AND PER_TYP=2 AND REST2=8 THEN OUTPUT UNKKID2;
ELSE IF SEAT2=11 THEN DO;
  IF FTLS=2 THEN OUTPUT UNK2PT1;
  ELSE IF FTLS IN (3,4,8) THEN OUTPUT UNK3PT1;
  ELSE OUTPUT UNKLAP1; END;
ELSE IF SEAT2=13 THEN DO;
  IF O LE AGE LE 5 THEN OUTPUT UNKKID1;
  ELSE IF FTLS IN (2,4) THEN OUTPUT UNK2PT1;
  ELSE IF FTLS IN (3,8) THEN OUTPUT UNK3PT1;
  ELSE OUTPUT UNKLAP1; END;
ELSE IF SEAT2=21 THEN DO;
```

```
     IF O LE AGE LE 5 THEN OUTPUT UNKKID1;
     ELSE IF BKLS=3 THEN OUTPUT UNKBK31;
     ELSE OUTPUT UNKLAP1; END;
   ELSE IF SEAT2=22 THEN DO;
     IF O LE AGE LE 5 THEN OUTPUT UNKKID1;
     ELSE IF CBKLS=3 THEN OUTPUT UNKBK31;
     ELSE OUTPUT UNKLAP1; END;
   ELSE IF SEAT2=12 THEN DO;
     IF O LE AGE LE 5 THEN OUTPUT UNKKID1;
     ELSE OUTPUT UNKLAP1; END;
   RUN;
```

The program L3PT4 is a logistic regression that calibrates belt use of front-outboard occupants in vehicles with 3-point belts, based on cases where belt use is known, as a function of the calendar year, the occupant's age and gender, the vehicle type, vehicle age, crash mode and seat position. The code here defines the independent variables used in L3PT4 and copies the regression equation obtained there, in order to impute the belt use of occupants for whom it is unknown on FARS.

```
/* IMPUTES FRONT-SEAT 3-POINT BELT USE BASED ON LOGISTIC REGRESSION EQUATION */

DATA UNK3PT2; SET UNK3PT1;
IF NEWAGE GE 17 THEN FEMALE=GENDER-1;
   ELSE IF 6 LE NEWAGE LE 13 THEN FEMALE=.5;
   ELSE IF 14 LE NEWAGE LE 16 THEN FEMALE=.5*GENDER-.25;
PRETEEN=0; TEEN=0; ADULT=0; OLD=0;
   IF 6 LE NEWAGE LE 12 THEN PRETEEN=1;
   ELSE IF 13 LE NEWAGE LE 20 THEN TEEN=20-NEWAGE;
   ELSE IF 30 LE NEWAGE LE 79 THEN ADULT=NEWAGE-30;
   ELSE IF 80 LE NEWAGE LE 97 THEN OLD=1;
CONVRTBL=0; TWODOOR=0; STAWAGON=0; UNK_CAR=0; PICKUP=0; SUV=0; VAN=0; UNK_LTRK=0;
   IF NEWVTYP=1 THEN CONVRTBL=1;
   ELSE IF NEWVTYP=2 THEN TWODOOR=1;
   ELSE IF NEWVTYP=6 THEN STAWAGON=1;
   ELSE IF NEWVTYP=9 THEN UNK_CAR=1;
   ELSE IF NEWVTYP=11 THEN PICKUP=1;
   ELSE IF NEWVTYP=12 THEN SUV=1;
   ELSE IF NEWVTYP=13 THEN VAN=1;
   ELSE IF NEWVTYP=19 THEN UNK_LTRK=1;
CY7576=0; CY7778=0; CY7980=0; CY81=0; CY82=0; CY83=0; CY84=0;
   CY85=0; CY86=0; CY87=0; CY88=0; CY89=0; CY90=0; CY91=0;
   CY92=0; CY93=0; CY94=0; CY95=0; CY96=0; CY97=0; CY98=0;
   CY2000=0; CY2001=0; CY2002=0;
IF CY IN (1975,1976) THEN CY7576=1;
   ELSE IF CY IN (1977,1978) THEN CY7778=1;
   ELSE IF CY IN (1979,1980) THEN CY7980=1;
   ELSE IF CY=1981 THEN CY81=1; ELSE IF CY=1982 THEN CY82=1;
   ELSE IF CY=1983 THEN CY83=1; ELSE IF CY=1984 THEN CY84=1;
   ELSE IF CY=1985 THEN CY85=1; ELSE IF CY=1986 THEN CY86=1;
   ELSE IF CY=1987 THEN CY87=1; ELSE IF CY=1988 THEN CY88=1;
   ELSE IF CY=1989 THEN CY89=1; ELSE IF CY=1990 THEN CY90=1;
   ELSE IF CY=1991 THEN CY91=1; ELSE IF CY=1992 THEN CY92=1;
   ELSE IF CY=1993 THEN CY93=1; ELSE IF CY=1994 THEN CY94=1;
   ELSE IF CY=1995 THEN CY95=1; ELSE IF CY=1996 THEN CY96=1;
   ELSE IF CY=1997 THEN CY97=1; ELSE IF CY=1998 THEN CY98=1;
   ELSE IF CY=2000 THEN CY2000=1; ELSE IF CY=2001 THEN CY2001=1;
   ELSE IF CY=2002 THEN CY2002=1;
RTFRONT=0; IF SEAT2=13 THEN RTFRONT=1;
SIDE=0; ROLL=0; REAROTHR=0;
   IF CRSH=2 THEN SIDE=1; ELSE IF CRSH=3 THEN ROLL=1; ELSE IF CRSH=4 THEN REAROTHR=1;
VEH_AGE=CY-MY; IF VEH_AGE LT O THEN VEH_AGE=0;

Z=-.2954+.7226*PRETEEN+.0292*TEEN+.0202*ADULT+.9100*OLD+.4728*FEMALE;
Z=Z-.1273*CONVRTBL-.2593*TWODOOR+.1047*STAWAGON-.0414*UNK_CAR;
```

```
Z=Z-.7998*PICKUP-.3697*SUV-.1141*VAN-.4421*UNK_LTRK-.0647*VEH_AGE;
Z=Z-2.4205*CY7576-2.9491*CY7778-3.2598*CY7980-3.3202*CY81-3.2030*CY82;
Z=Z-2.9786*CY83-2.6736*CY84-1.8587*CY85-1.3004*CY86-1.0471*CY87;
Z=Z-.9215*CY88-.8555*CY89-.7852*CY90-.6234*CY91-.4971*CY92-.3323*CY93;
Z=Z-.2301*CY94-.2192*CY95-.1221*CY96-.0760*CY97-.0047*CY98;
Z=Z+.1102*CY2000+.1544*CY2001+.2124*CY2002;
Z=Z+.4169*SIDE-.4605*ROLL+.1482*REAROTHR-.0332*RTFRONT;
```

Based on the logistic regression equation, BELTED is the probability that the occupant was belted; UNBELTED is the probability they were unrestrained. Each case with unknown belt use is split into two cases with known belt use: one yes, weighted by the probability that the occupant was belted; the other no, weighted by UNBELTED.

```
BELTED=EXP(Z)/(1+EXP(Z)); UNBELTED=1-BELTED;
O1=ORIGWT; W1=WEIGHTFA;
REST2=0; ORIGWT=UNBELTED* O1; WEIGHTFA=UNBELTED* W1; OUTPUT;
REST2=8; ORIGWT= BELTED * O1; WEIGHTFA= BELTED * W1; OUTPUT;
RUN;

/* IMPUTES AUTOMATIC 2-POINT BELT USE BASED ON LOGISTIC REGRESSION EQUATION */

DATA UNK2PT2; SET UNK2PT1;
IF PASSIVE IN (606,707,1606) THEN MOTOR=0; ELSE MOTOR=1;
IF NEWAGE GE 17 THEN FEMALE=GENDER-1;
  ELSE IF 6 LE NEWAGE LE 13 THEN FEMALE=.5;
  ELSE IF 14 LE NEWAGE LE 16 THEN FEMALE=.5*GENDER-.25;
PRETEEN=0; TEEN=0; ADULT=0; OLD=0;
  IF 6 LE NEWAGE LE 12 THEN PRETEEN=1;
  ELSE IF 13 LE NEWAGE LE 20 THEN TEEN=20-NEWAGE;
  ELSE IF 30 LE NEWAGE LE 79 THEN ADULT=NEWAGE-30;
  ELSE IF 80 LE NEWAGE LE 97 THEN OLD=1;
RTFRONT=0; IF SEAT2=13 THEN RTFRONT=1;
SIDE=0; ROLL=0; REAROTHR=0;
  IF CRSH=2 THEN SIDE=1; ELSE IF CRSH=3 THEN ROLL=1; ELSE IF CRSH=4 THEN REAROTHR=1;
VEH_AGE=CY-MY; IF VEH_AGE LT 0 THEN VEH_AGE=0;

Z=-.2911+.6194*MOTOR+.5819*PRETEEN+.0323*TEEN+.0289*ADULT+1.4860*OLD+.4928*FEMALE;
Z=Z-.0199*VEH_AGE+.1496*SIDE-.7183*ROLL-.0317*REAROTHR-.0732*RTFRONT;
BELTED=EXP(Z)/(1+EXP(Z)); UNBELTED=1-BELTED;
O1=ORIGWT; W1=WEIGHTFA;
REST2=0; ORIGWT=UNBELTED* O1; WEIGHTFA=UNBELTED* W1; OUTPUT;
REST2=8; ORIGWT= BELTED * O1; WEIGHTFA= BELTED * W1; OUTPUT;
RUN;

/* IMPUTES BACK-SEAT 3-POINT BELT USE BASED ON LOGISTIC REGRESSION EQUATION */

DATA UNKBK32; SET UNKBK31;
IF NEWAGE GE 17 THEN FEMALE=GENDER-1;
  ELSE IF 6 LE NEWAGE LE 13 THEN FEMALE=.5;
  ELSE IF 14 LE NEWAGE LE 16 THEN FEMALE=.5*GENDER-.25;
PRETEEN=0; TEEN=0; ADULT=0; OLD=0;
  IF 6 LE NEWAGE LE 12 THEN PRETEEN=1;
  ELSE IF 13 LE NEWAGE LE 20 THEN TEEN=20-NEWAGE;
  ELSE IF 30 LE NEWAGE LE 79 THEN ADULT=NEWAGE-30;
  ELSE IF 80 LE NEWAGE LE 97 THEN OLD=1;
TWODOOR=0; VAN=0; OTHR_TRK=0;
  IF NEWVTYP IN (1,2) THEN TWODOOR=1;
  ELSE IF NEWVTYP=9 THEN TWODOOR=.5;
  ELSE IF NEWVTYP IN (11,12,19) THEN OTHR_TRK=1;
  ELSE IF NEWVTYP=13 THEN VAN=1;
CY7589=0; CY9094=0; CY9599=0;
IF 1975 LE CY LE 1989 THEN CY7589=1;
  ELSE IF 1990 LE CY LE 1994 THEN CY9094=1;
  ELSE IF 1995 LE CY LE 1999 THEN CY9599=1;
```

```
SIDE=0; ROLL=0; REAROTHR=0;
  IF CRSH=2 THEN SIDE=1; ELSE IF CRSH=3 THEN ROLL=1; ELSE IF CRSH=4 THEN REAROTHR=1;
VEH_AGE=CY-MY; IF VEH_AGE LT 0 THEN VEH_AGE=0;
CENREAR=0; IF SEAT2=22 THEN CENREAR=1;

Z=-1.4715+1.5495*PRETEEN+.1042*TEEN+.0294*ADULT+1.3595*OLD+.3314*FEMALE;
Z=Z-.1506*TWODOOR+.0012*VAN-.3215*OTHR_TRK-.0303*VEH_AGE;
Z=Z-.7777*CY7589-.4001*CY9094-.2469*CY9599;
Z=Z+.2107*SIDE-.6045*ROLL+.2894*REAROTHR-.6754*CENREAR;
BELTED=EXP(Z)/(1+EXP(Z)); UNBELTED=1-BELTED;
O1=ORIGWT; W1=WEIGHTFA;
REST2=0; ORIGWT=UNBELTED* O1; WEIGHTFA=UNBELTED* W1; OUTPUT;
REST2=8; ORIGWT= BELTED * O1; WEIGHTFA= BELTED * W1; OUTPUT;
RUN;

/* IMPUTES LAP BELT USE BASED ON LOGISTIC REGRESSION EQUATION */

DATA UNKLAP2; SET UNKLAP1;
PRE_STD=0;
  IF SEAT2 IN (11,13) AND MY LT 1964 THEN PRE_STD=1;
    ELSE IF SEAT2=21 AND MY LT 1966 THEN PRE_STD=1;
    ELSE IF SEAT2 IN (12,22) AND MY LT 1968 THEN PRE_STD=1;
IF NEWAGE GE 17 THEN FEMALE=GENDER-1;
  ELSE IF 6 LE NEWAGE LE 13 THEN FEMALE=.5;
  ELSE IF 14 LE NEWAGE LE 16 THEN FEMALE=.5*GENDER-.25;
PRETEEN=0; TEEN=0; ADULT=0; OLD=0;
  IF 6 LE NEWAGE LE 12 THEN PRETEEN=1;
  ELSE IF 13 LE NEWAGE LE 20 THEN TEEN=20-NEWAGE;
  ELSE IF 30 LE NEWAGE LE 79 THEN ADULT=NEWAGE-30;
  ELSE IF 80 LE NEWAGE LE 97 THEN OLD=1;
TWODOOR=0; PICKUP=0; SUV_VAN=0;
  IF NEWVTYP IN (1,2) THEN TWODOOR=1;
  ELSE IF NEWVTYP=9 THEN TWODOOR=.63;
  ELSE IF NEWVTYP=11 THEN PICKUP=1;
  ELSE IF NEWVTYP IN (12,13) THEN SUV_VAN=1;
  ELSE IF NEWVTYP=19 THEN DO; PICKUP=.72; SUV_VAN=.28; END;
CY7579=0; CY8084=0; CY8589=0; CY9094=0; CY9599=0;
IF 1975 LE CY LE 1979 THEN CY7579=1;
  ELSE IF 1980 LE CY LE 1984 THEN CY8084=1;
  ELSE IF 1985 LE CY LE 1989 THEN CY8589=1;
  ELSE IF 1990 LE CY LE 1994 THEN CY9094=1;
  ELSE IF 1995 LE CY LE 1999 THEN CY9599=1;
CENFRONT=0; RTFRONT=0; OUTBRD_R=0; CENREAR=0;
  IF SEAT2=12 THEN CENFRONT=1;
  ELSE IF SEAT2=13 THEN RTFRONT=1;
  ELSE IF SEAT2=21 THEN OUTBRD_R=1;
  ELSE IF SEAT2=22 THEN CENREAR=1;
SIDE=0; ROLL=0; REAROTHR=0;
  IF CRSH=2 THEN SIDE=1; ELSE IF CRSH=3 THEN ROLL=1; ELSE IF CRSH=4 THEN REAROTHR=1;
VEH_AGE=CY-MY; IF VEH_AGE LT 0 THEN VEH_AGE=0;

Z=-.3189+1.1862*PRETEEN+.0323*TEEN+.0204*ADULT+1.1997*OLD;
Z=Z-.2945*PRE_STD-1.2292*CENFRONT-.1795*RTFRONT-.7338*OUTBRD_R-1.6522*CENREAR;
Z=Z+.4191*FEMALE-.0717*TWODOOR-.8322*PICKUP-.0642*SUV_VAN-.0535*VEH_AGE;
Z=Z-3.0443*CY7579-3.2209*CY8084-1.4020*CY8589-.7785*CY9094-.3090*CY9599;
Z=Z+.0432*SIDE-.1522*ROLL+.0675*REAROTHR;
BELTED=EXP(Z)/(1+EXP(Z)); UNBELTED=1-BELTED;
O1=ORIGWT; W1=WEIGHTFA;
REST2=0; ORIGWT=UNBELTED* O1; WEIGHTFA=UNBELTED* W1; OUTPUT;
REST2=8; ORIGWT= BELTED * O1; WEIGHTFA= BELTED * W1; OUTPUT;
RUN;
```

The imputation procedure for child passengers age 0-5 is performed in two steps. First, each case with completely unknown restraint use is split into two cases based on the first logistic regression equation in LKIDRES4: one unrestrained, one restrained, type unknown. Next, the restrained cases generated from this procedure, plus the already existing cases of children that were restrained, type unknown are split into two cases based on the second equation in LKIDRES4: one in a safety seat, one in a lap belt only.

```
/* IMPUTES RESTRAINED AND UNRESTRAINED INFANTS/TODDLERS */

DATA UNKKID3 UNKKID4; SET UNKKID1 UNKKID2(IN=KID2);
INFANT=0; AGE1=0; AGE2=0; AGE3=0; AGE4=0;
  IF NEWAGE=0 THEN INFANT=1; ELSE IF NEWAGE=1 THEN AGE1=1;
  ELSE IF NEWAGE=2 THEN AGE2=1; ELSE IF NEWAGE=3 THEN AGE3=1;
  ELSE IF NEWAGE=4 THEN AGE4=1;
IF VTYP=2 THEN LTRK=1; ELSE LTRK=0;
CY7576=0; CY7778=0; CY7980=0; CY81=0; CY82=0; CY83=0; CY84=0;
  CY85=0; CY86=0; CY87=0; CY88=0; CY89=0; CY90=0; CY91=0;
  CY92=0; CY93=0; CY94=0; CY95=0; CY96=0; CY97=0; CY98=0;
  CY2000=0; CY2001=0; CY2002=0;
IF CY IN (1975,1976) THEN CY7576=1;
  ELSE IF CY IN (1977,1978) THEN CY7778=1;
  ELSE IF CY IN (1979,1980) THEN CY7980=1;
  ELSE IF CY=1981 THEN CY81=1; ELSE IF CY=1982 THEN CY82=1;
  ELSE IF CY=1983 THEN CY83=1; ELSE IF CY=1984 THEN CY84=1;
  ELSE IF CY=1985 THEN CY85=1; ELSE IF CY=1986 THEN CY86=1;
  ELSE IF CY=1987 THEN CY87=1; ELSE IF CY=1988 THEN CY88=1;
  ELSE IF CY=1989 THEN CY89=1; ELSE IF CY=1990 THEN CY90=1;
  ELSE IF CY=1991 THEN CY91=1; ELSE IF CY=1992 THEN CY92=1;
  ELSE IF CY=1993 THEN CY93=1; ELSE IF CY=1994 THEN CY94=1;
  ELSE IF CY=1995 THEN CY95=1; ELSE IF CY=1996 THEN CY96=1;
  ELSE IF CY=1997 THEN CY97=1; ELSE IF CY=1998 THEN CY98=1;
  ELSE IF CY=2000 THEN CY2000=1; ELSE IF CY=2001 THEN CY2001=1;
  ELSE IF CY=2002 THEN CY2002=1;
IF SEAT2 IN (12,13,21,22);
  CENFRONT=0; RTFRONT=0; CENREAR=0;
  IF SEAT2=12 THEN CENFRONT=1;
  ELSE IF SEAT2=13 THEN RTFRONT=1;
  ELSE IF SEAT2=22 THEN CENREAR=1;
SIDE=0; ROLL=0; REAROTHR=0;
  IF CRSH=2 THEN SIDE=1; ELSE IF CRSH=3 THEN ROLL=1; ELSE IF CRSH=4 THEN REAROTHR=1;

IF KID2 THEN DO; OUTPUT UNKKID4; RETURN; END;
ELSE DO;
Z=.3711+.7837*INFANT+.8726*AGE1+.5193*AGE2+.1485*AGE3+.1024*AGE4;
Z=Z-.0472*LTRK-3.4784*CY7576-3.5735*CY7778-3.0285*CY7980;
Z=Z-2.7536*CY81-2.4664*CY82-2.0078*CY83-1.5532*CY84-1.1660*CY85;
Z=Z-1.1936*CY86-1.0132*CY87-1.0726*CY88-1.0026*CY89-.9834*CY90;
Z=Z-.7092*CY91-.5285*CY92-.6801*CY93-.3309*CY94-.2090*CY95;
Z=Z-.0455*CY96-.0201*CY97+.0808*CY98;
Z=Z+.2306*CY2000+.1098*CY2001+.1525*CY2002;
Z=Z+.2874*SIDE-.5541*ROLL-.2060*REAROTHR;
Z=Z-1.4496*CENFRONT-.7022*RTFRONT-.7946*CENREAR;
BELTED=EXP(Z)/(1+EXP(Z)); UNBELTED=1-BELTED;
O1=ORIGWT; W1=WEIGHTFA;
REST2=0; ORIGWT=UNBELTED* O1; WEIGHTFA=UNBELTED* W1; OUTPUT UNKKID3;
REST2=8; ORIGWT= BELTED * O1; WEIGHTFA= BELTED * W1; OUTPUT UNKKID4; END;
RUN;

/* IMPUTES IF THE RESTRAINED INFANTS/TODDLERS WERE IN CHILD SEATS OR LAP BELTS */

DATA UNKKID5; SET UNKKID4;
Z=-1.6869+5.6002*INFANT+4.5930*AGE1+3.4650*AGE2+2.7264*AGE3+1.2914*AGE4;
Z=Z-.0439*LTRK-1.2118*CY7576-1.5049*CY7778-1.0482*CY7980;
Z=Z-.9440*CY81-.5881*CY82-.2602*CY83-.3824*CY84-.3179*CY85;
Z=Z-.9094*CY86-.8220*CY87-1.1240*CY88-.7704*CY89-.4048*CY90;
Z=Z-.2422*CY91-.1682*CY92-.6909*CY93-.7306*CY94-.4925*CY95;
```

```
Z=Z-.4550*CY96-.3514*CY97-.6729*CY98;
Z=Z+.1257*CY2000-.1710*CY2001+.0451*CY2002;
Z=Z+.0723*SIDE+.0479*ROLL+.2250*REAROTHR;
Z=Z-.7766*CENFRONT-1.0595*RTFRONT+.0752*CENREAR;
KIDSEAT=EXP(Z)/(1+EXP(Z)); LAPBELT=1-KIDSEAT;
O1=ORIGWT; W1=WEIGHTFA;
REST2=2; ORIGWT=LAPBELT * O1; WEIGHTFA=LAPBELT * W1; OUTPUT;
REST2=4; ORIGWT=KIDSEAT * O1; WEIGHTFA=KIDSEAT * W1; OUTPUT;
RUN;
```

UNKBELT2 contains all the newly created cases with known restraint use. It is appended to OCC6, the original cases with known restraint use, creating OCC7, a file on which all cases have known restraint use (and seat positions and crash modes).

```
DATA UNKBELT2;
SET UNK3PT2 UNK2PT2 UNKBK32 UNKLAP2 UNKKID3 UNKKID5;
DROP GENDER NEWAGE Z BELTED UNBELTED O1 W1 PRETEEN TEEN ADULT OLD FEMALE
    CONVRTBL TWODOOR STAWAGON UNK_CAR PICKUP SUV VAN UNK_LTRK
    VEH_AGE CY7576 CY7778 CY7980 CY81-CY98 CY2000 CY2001 SIDE ROLL REAROTHR RTFRONT
    MOTOR OTHR_TRK CY7589 CY9094 PRE_STD CENFRONT OUTBRD_R CENREAR SUV_VAN
    CY7579 CY8084 CY8589 KIDSEAT LAPBELT;
RUN;
```

REST3 edits REST2, assigning a specific restraint type to every restrained occupant on the file (and REST3 = 0 for unrestrained occupants).

- Every restrained child age 0-5 is either in a child safety seat (REST3 = 4) or in a lap belt (REST3 = 2), without distinguishing if the seat was correctly used; even if the belt is a 3-point belt, it will be considered "lap belt only" for children that age.
- Children age 6-9 that FARS explicitly states were in child safety seats (REST_USE = 4) will be considered in booster seats (REST3 = 4). All other children age 6-9 will be included with "adults" in the remaining categories.
- Every restrained adult in a seat equipped with integral (manual or automatic) 3-point belts will be considered as restrained with a 3-point belt (REST3 = 3), without distinguishing that some of these may have been incorrectly worn and, in effect, acted like just a lap belt.
- Every restrained adult in a seat equipped with automatic 2-point belts will be considered as restrained with a 2-point belt (REST3 = 7), without distinguishing between people who also wore the manual lap belt, if one was available, and those who didn't.
- In seats equipped with separate lap and shoulder belts, restrained adults will only be considered as having worn both belts (REST3 = 3) if FARS explicitly stated that the lap and shoulder belts were used (REST_USE = 3). All other restrained adults, including the imputed cases will be assumed to have worn only the lap belt (REST3 = 2). These assumptions are most consistent with observational surveys that showed separate shoulder belts were only used by about 15-25 percent of the people who buckled the lap belt.
- Restrained adults in 1975-80 VW Rabbits – vehicles where it is impossible to tell if they had 3-point or 2-point belts, but the 3-point belts were more common – will only be considered as 2-point belted if REST_USE = 1,8 or AUT_REST = 1,5.
- And, of course, in seats equipped only with lap belts, restrained adults will be considered lap-belted.

```
/* ------------------------------------------------------------ */
/*  EVERY OCCUPANT CASE IS NOW ASSIGNED A SPECIFIC BELT TYPE     */
/*  (OR UNRESTRAINED) ACCORDING TO REST2, SEAT2, FTLS, BKLS, & CBKLS  */
```

```
/* ------------------------------------------------------------------ */

DATA OCC7; SET OCC6 UNKBELT2;
IF REST2 IN (0,4) THEN REST3=REST2;
  ELSE IF 0 LE AGE LE 5 THEN REST3=2;
  ELSE IF SEAT2=12 THEN REST3=2;
  ELSE IF SEAT2=22 AND CBKLS=3 THEN REST3=3;
  ELSE IF SEAT2=22 THEN REST3=2;
  ELSE IF SEAT2=21 AND BKLS=3 THEN REST3=3;
  ELSE IF SEAT2=21 THEN REST3=2;
  ELSE IF FTLS=3 THEN REST3=3;
  ELSE IF FTLS=2 THEN REST3=7;
  ELSE IF FTLS=1 THEN REST3=2;
  ELSE IF FTLS=4 AND SEAT2=11 THEN REST3=3;
  ELSE IF FTLS=4 AND SEAT2=13 THEN REST3=7;
  ELSE IF FTLS IN (5,6,7) AND REST2=3 THEN REST3=3;
  ELSE IF FTLS IN (5,6,7) THEN REST3=2;
  ELSE IF FTLS=8 AND REST2 IN (1,7,8) THEN REST3=7;
  ELSE IF FTLS=8 THEN REST3=3;
```

The two variables RESTGP and KCRSH are used for estimating the effect of air bags on the risk of a child passenger.

```
/* ------------------------------------------------------------------ */
/*  CLASSIFICATION OF RF CHILD PASSENGERS EXPOSED TO AIR BAGS         */
/* ------------------------------------------------------------------ */

IF 0 LE AGE LE 12 AND SEAT2=13 AND PASSIVE IN (2,3,1313) THEN DO;
  IF AGE=0 AND REST3=4 THEN RESTGP=1;
    ELSE IF AGE=0 AND REST3 IN (0,2,3,7) THEN RESTGP=2;
    ELSE IF 1 LE AGE LE 5 AND REST3=0 THEN RESTGP=3;
    ELSE IF 1 LE AGE LE 5 AND REST3=4 THEN RESTGP=4;
    ELSE IF 1 LE AGE LE 5 AND REST3 IN (2,3,7) THEN RESTGP=5;
    ELSE IF 6 LE AGE LE 10 AND REST3=0 THEN RESTGP=6;
    ELSE IF 6 LE AGE LE 10 AND REST3 IN (2,3,4,7) THEN RESTGP=7;
    ELSE IF 11 LE AGE LE 12 AND REST3=0 THEN RESTGP=8;
    ELSE IF 11 LE AGE LE 12 AND REST3 IN (2,3,7) THEN RESTGP=9;

  IF IMPACT2=12 OR IMPACT1=12 THEN KCRSH=1;
      ELSE IF IMPACT2 IN (1,11) OR IMPACT1 IN (1,11) THEN KCRSH=2;
      ELSE IF IMPACT2 IN (2,10) OR IMPACT1 IN (2,10) THEN KCRSH=3;
      ELSE KCRSH=4; END;
RUN;
```

This completes the preprocessor. OCC7 is a file in which every record has non-missing values for all the variables needed to estimate lives saved by safety technologies, and whose ORIGWTs (and WEIGHTFAs) add up to the total number of car and LTV occupant fatalities in 1975-2002.

```
/* ------------------------------------------------------------------ */
/*  SPLIT THE CAR AND LIGHT-TRUCK OCCUPANT CASES INTO SEPARATE FILES  */
/* ------------------------------------------------------------------ */
```

OCC7 is split into CAR1 and TRK1 because the main model is run separately for cars and for LTVs (because the safety technologies were introduced in a different chronological order, and the effectiveness is sometimes different in cars and LTVs). Also, the preprocessor changed the order of the cases because the imputed cases were, at each step, appended at the end of the file. In order to tally up lives saved by calendar year, it is necessary to sort CAR1 and TRK1 by CY.

```
DATA CAR1 TRK1; SET OCC7;
IF VTYP=1 THEN OUTPUT CAR1; ELSE OUTPUT TRK1;
RUN;
```

```
PROC SORT DATA=CAR1; BY CY;
PROC SORT DATA=TRK1; BY CY;
RUN;

/* ---------------------------------------------------------------- */
/* ---------------------------------------------------------------- */
/*   THE OCCUPANT CASES ARE NOW FREE OF UNKNOWNS ON ANY VARIABLES    */
/*   AND WEIGHTED TO TOTAL UP TO THE ORIGINAL SAMPLE SIZE            */
/*                                                                  */
/*   WE MAY NOW CALCULATE THE LIVES SAVED BY THE VEHICLE SAFETY      */
/*   STANDARDS IN THE REVERSE ORDER OF THEIR IMPLEMENTATION          */
/*                                                                  */
/*   PASSENGER CARS FIRST                                            */
```

When the notes mention "NHTSA evaluations" and "effectiveness estimates," they are generally discussed in Part 1 of this report, in the chapter for that FMVSS and subsection for that technology. Footnotes/references are provided here only for analyses not discussed in Part 1.

```
/* ---------------------------------------------------------------- */
/* ---------------------------------------------------------------- */
```

Three "tags" of information accompany each case as it processes through the model:

- ORIGWT, the original case weight defined by the preprocessor, remains unchanged in the main model.
- WEIGHTFA, the inflated case weight, grows as safety technologies are "removed" from the vehicle. The existence of ORIGWT fatalities in a vehicle equipped with safety technologies implies that there would have been WEIGHTFA fatalities if the vehicle had no safety technologies at all, in the sense that the combined effectiveness of these technologies would reduce WEIGHTFA fatalities to ORIGWT fatalities.
- EJECT2, the probability that an occupant was ejected. Initially, EJECT2 is either 1 (if the occupant was ejected) or 0 (if not ejected or unknown[1]). However, some safety technologies, such as belts, can reduce the probability of ejection, whereas other technologies, such as improved door locks, are effective only if the occupant was ejected. Thus, it is necessary to track the occupant's changing probability of ejection as technologies are removed from the vehicle one-by-one.

```
DATA CAR2; SET CAR1;
IF EJECTION IN (1,2) THEN EJECT2=1; ELSE EJECT2=0;
```

Three-point belts for rear-center occupants, already widely available in cars in anticipation of Anton's Law, are the most recent safety technology for which an effectiveness estimate is available, so they will be the first technology that the model "removes" from the car. The basic procedure of the model was discussed in detail, above, in the analysis the effect of brake improvements on pedestrian crashes.

```
/* ---------------------------------------------------------------- */
/*   208J 3-POINT BELTS FOR BACK-SEAT CENTER OCCUPANTS (ANTON'S LAW) */
/*   MEDIAN INSTALLATION YEAR: 2001                                 */
/* ---------------------------------------------------------------- */
```

---

[1] Tabulation of EJECTION by calendar year suggests that, in the early years, especially 1975-76, "unknown" essentially means not ejected. From 1985 onward, well under 1 percent are unknown if ejected.

Unlike the situation with brakes, we know what specific occupants are protected by rear-center 3-point belts. In other words, PV208J or PS208J are not probabilities between 0 and 1, but are either exactly 0 or exactly 1. Moreover, because it is still before the effective date of the regulation specified by Anton's Law, PS208J = 0 in 1975-2002, and only PV208J can equal 1. PV208J = 1 if the occupant was age 6+, sat in a rear-center seat equipped with 3-point belts, and wore those belts, and PV208J = 0 for all other occupants.

```
/* IDENTIFIES BACK-SEAT OCCUPANTS WEARING 3-POINT BELTS  */
  PS208J=0; PV208J=0;
  IF SEAT2=22 AND REST3=3 THEN DO; IF MY LE 2008 THEN PV208J=1; ELSE PS208J=1; END;
/* LIVES SAVED AND EJECTIONS PREVENTED BY BACK-SEAT CENTER 3-POINT BELTS */
  OLDWTFA=WEIGHTFA;
  IF PS208J GT 0 OR PV208J GT 0 THEN DO;
```

The data in NHTSA's evaluation of 3-point belts in the rear-outboard seats indicate fatality reduction of 29 percent in frontals, 42 percent in side impacts, 77 percent in rollovers and 31 percent in rear/other impacts. We will assume the same effectiveness for 3-point belts in the rear-center seat. Moreover, when we "remove" the belts from the vehicle, the occupant becomes unrestrained and, in general, 26 percent of unrestrained rear-center occupant fatalities in frontal crashes were ejected, 33 percent in side impacts, 64 percent in rollovers and 31 percent in rear/other impacts. One of these becomes the new values of EJECT2 as this now-unrestrained occupant fatality case proceeds through the rest of the model.

CLV208J: "C" = car, "L" = lives saved, "V" = voluntary, "208J" = rear-center 3-point belt

```
    IF CRSH=1 THEN DO; E=.29; EJECT2=.26; END;
      ELSE IF CRSH=2 THEN DO; E=.42; EJECT2=.33; END;
      ELSE IF CRSH=3 THEN DO; E=.77; EJECT2=.64; END;
      ELSE IF CRSH=4 THEN DO; E=.31; EJECT2=.31; END;
    P=PS208J+PV208J;
    S=OLDWTFA*E*P / (1 - E*P);
    CLS208J=S*PS208J/P;
    CLV208J=S*PV208J/P;
    WEIGHTFA=OLDWTFA+CLS208J+CLV208J; END;
  ELSE DO; CLS208J=0; CLV208J=0; END;
  CL208J=CLS208J+CLV208J;
  CLS=CLS208J; CLV=CLV208J; CL=CL208J;
```

By the way, when this model "removes" a 3-point belt system from the vehicle, it removes the entire system, not just the shoulder harness. The occupant from this point onwards is unrestrained, not lap-belted. The lives saved by the 3-point system are the benefits of that system relative to being unrestrained, not the incremental benefit of a 3-point system relative to lap belts only. This model includes nine belt technologies (208A, 208B, 208C, 208D, 208E, 208F, 208G, 208H and 208J), but no more than one can apply to any specific occupant case.[2]

```
/* ------------------------------------------------------------------ */
/*  108 TRAILER CONSPICUITY TAPE (LIVES SAVED IN CARS NOT HITTING THEM) */
/*  50% OF TRAILERS ON THE ROAD HAD THEM IN: 1996                      */
```

---

[2] It would also have been theoretically possible to estimate the effectiveness of 3-point belts relative to lap belts only (rather than relative to being unrestrained), and then have the model subdivide the lives saved by 3-point belts into (1) lives that would have been saved even by just a lap belt (and count them under technology 208A) and (2) lives that were saved only because there was also a shoulder harness (and count only these in technology 208F). This approach would probably be confusing, since it would attribute a large number of saves to lap belts who were, in fact, people that used 3-point belts.

```
/*   FMCSA RETROFIT REQUIREMENT ISSUED 3/31/1999, TOOK EFFECT 6/1/2001   */
/* ------------------------------------------------------------------ */

/* IDENTIFIES SIDE/REAR IMPACTS INTO TRAILERS IN THE DARK AND % OF TRAILERS WITH TAPE */
   PS108=0; PV108=0;
```

Identifies cases where the front of a car hit the side or rear of a heavy trailer in the dark.

```
IF CY GE 1991 AND (TAPECASE=1 OR (OVTYP=3 AND
((1991 LE CY LE 1994 AND OVCONFIG=5) OR (CY GE 1995 AND OVCONFIG=6))
AND 2 LE LGT_COND LE 5 AND
  (OIMPACT2 IN (14,16) AND MAN_COLL IN (1,4,5) AND IMPACT2 IN (11,12,1))))
THEN DO;
```

Unlike all the other safety technologies, the tape is not on the case vehicle (the car) but on the other vehicle (the heavy trailer). The model year of the case vehicle is irrelevant. The model year of the trailer is not reported on FARS. It is unknown (before June 1, 2001, the effective date of the FMCSA retrofit requirement) if this specific trailer was equipped with tape, but we can infer, from the calendar year of the crash, the probability that it was equipped with tape. PS108 is the percentage of trailers on the road that FMVSS 108 requires to be equipped with tape when new (if built after 12/1/1993) plus the percent required by the FMCSA retrofit rule (after 6/1/2001). PV108 is the percentage of trailers on the road that were built before 12/1/1993 and voluntarily fitted or retrofitted with tape, before 6/1/2001.

```
IF CY=1991 THEN DO; PS108=0; PV108=.09; END;
 ELSE IF CY=1992 THEN DO; PS108=0; PV108=.18; END;
 ELSE IF CY=1993 THEN DO; PS108=0; PV108=.27; END;
 ELSE IF CY=1994 THEN DO; PS108=.09; PV108=.27; END;
 ELSE IF CY=1995 THEN DO; PS108=.18; PV108=.27; END;
 ELSE IF CY=1996 THEN DO; PS108=.27; PV108=.27; END;
 ELSE IF CY=1997 THEN DO; PS108=.36; PV108=.27; END;
 ELSE IF CY=1998 THEN DO; PS108=.45; PV108=.27; END;
 ELSE IF CY=1999 THEN DO; PS108=.54; PV108=.27; END;
 ELSE IF CY=2000 THEN DO; PS108=.63; PV108=.27; END;
 ELSE IF CY=2001 THEN DO; PS108=.8775; PV108=.1125; END;
 ELSE IF CY GE 2002 THEN DO; PS108=1; PV108=0; END; END;
/* CAR OCCUPANT LIVES SAVED BY CONSPICUITY TAPE ON TRUCK TRAILERS */
 OLDWTFA=WEIGHTFA;
 IF PS108 GT 0 OR PV108 GT 0 THEN DO;
```

The tape is estimated to reduce fatal impacts into the side/rear of trailers, in the dark, by 29 percent.

```
E=.29;
P=PS108+PV108;
S=OLDWTFA*E*P / (1 - E*P);
CLS108=S*PS108/P;
CLV108=S*PV108/P;
 WEIGHTFA=OLDWTFA+CLS108+CLV108; END;
ELSE DO; CLS108=0; CLV108=0; END;
CL108=CLS108+CLV108;
CLS=CLS+CLS108; CLV=CLV+CLV108; CL=CL+CL108;


/* ------------------------------------------------------------------ */
/*  208I FRONTAL AIR BAGS FOR DRIVERS AND RF PASSENGERS AGE 13+       */
/*  MEDIAN INSTALLATION YEAR: 1994                                    */
/* ------------------------------------------------------------------ */

/* IDENTIFIES CARS WITH AIR BAGS  */
 PS208I=0; PV208I=0; UDAB=0; UPAB=0;
```

"PASSIVE," the variable that indicates the types of occupant protection at the front-outboard seats, is only defined if FARS has at least a partial VIN to decode. However, even when there is no VIN at all, we can often infer from the model year and the make if the car had driver and/or passenger air bags. All cars from MY 1997 onwards have had dual air bags. Some makes had driver air bags on all of their cars (UDAB = 1) or passenger air bags (UPAB = 1) before 1997.

```
/* CARS WITH UNKNOWN "PASSIVE" (BAD VINs) CLASSIFIED BY MAKE AND MODEL YEAR */

  IF CY GE 1985 AND MY GE 1986 AND (PASSIVE=. OR PASSIVE=9999) AND SEAT2 IN (11,13) THEN DO;
    IF SEAT2=11 THEN DO;
      IF (MAKE=42 AND MY GE 1986)
      OR (MAKE IN (13,38,45,47,51,59) AND MY GE 1990)
      OR (MAKE IN (6,32,34) AND MY GE 1991)
      OR (MAKE=37 AND MY GE 1992)
      OR (MAKE IN (19,24,39,49) AND MY GE 1993)
      OR (MAKE IN (18,21,54,58) AND MY GE 1994)
      OR (MAKE IN (9,10,12,14,20,22,30,35,41,48,52,53) AND MY GE 1995)
      OR (MAKE IN (55,63) AND MY GE 1996)
      OR MY GE 1997 THEN UDAB=1; END;

    ELSE IF SEAT2=13 THEN DO;
      IF (MAKE=45 AND MY GE 1990)
      OR (MAKE=13 AND MY GE 1993)
      OR (MAKE IN (19,32,37,42,51,54,58,59) AND MY GE 1994)
      OR (MAKE IN (6,10,12,14,24,30,34,35,39,41,47,48,52,53) AND MY GE 1995)
      OR (MAKE IN (9,22,49,55,63) AND MY GE 1996)
      OR MY GE 1997 THEN UPAB=1; END; END;
```

These are the codes of PASSIVE, based on VIN analysis, that indicate a car is equipped with frontal air bags and the driver or right-front seats, respectively. PS208I includes any installation from MY 1987 onwards, because the phase-in of automatic occupant protection began on September 1, 1986.

```
  IF CY GE 1984 AND MY GE 1985 AND
    (SEAT2=11 AND
        (PASSIVE IN (1,1003,1004,1006,2,1090,1099,1303,1313,1404,1505,1606) OR UDAB=1))
    OR (SEAT2=13 AND (PASSIVE IN (2,1313) OR UPAB=1)) THEN DO;
      IF 1985 LE MY LE 1986 THEN PV208I=1;
        ELSE IF MY GE 1987 THEN PS208I=1; END;

/* ------------------------------------------------------------ */
/*   SPLITS CASES INTO SCENARIOS DEPENDING ON AIR BAG TYPE,       */
/*   SEAT POSITION, OCCUPANT AGE, RESTRAINT USE                   */
/*   COMPUTES LIVES SAVED BY AIR BAGS IN EACH SCENARIO            */
/* ------------------------------------------------------------ */
```

For car occupants, there are five different effectiveness scenarios – belted drivers, unbelted drivers, child passengers, belted adult (age 13+) passengers, and unbelted adult passengers – plus the default scenario, no air bag. Within each scenario, there are several effectiveness estimates.

```
  IF PS208I=0 AND PV208I=0 THEN GOTO CNOBAG;        /* NO AIR BAG */
    ELSE IF SEAT2=11 AND REST3 GT 0 THEN GOTO CDRVBELT;   /* BELTED DRIVER */
    ELSE IF SEAT2=11 THEN GOTO CDRVUNR;              /* UNRESTRAINED DRIVER */
    ELSE IF 0 LE AGE LE 12 THEN GOTO CKID;           /* CHILD RF NO SWITCH */
    ELSE IF REST3 GT 0 THEN GOTO CRFBELT;            /* BELTED ADULT RF NO SWITCH */
    ELSE GOTO CRFUNR;                                /* UNBELTED ADULT RF NO SWITCH */
```

In addition to the basic statistics, WEIGHTFA, CLS208I, CLV208I and CL208I, the model compiles more detailed information on air bags:

- CABDRV = car driver lives saved by air bags
- CABRF = adult (age 13+) passenger lives saved by air bags in cars
- CABKID = effect of air bags on child (age 0-12) passengers in cars; if this is a fatality increase, it will show up as a negative number.

```
CNOBAG:     CLS2081=0; CLV2081=0; CL2081=0;
            CABDRV=0; CABRF=0; CABKID=0; GOTO C214;

CDRVBELT:   OLDWTFA=WEIGHTFA;
```

Air bag effectiveness: the data from NHTSA's evaluation indicates that air bags reduce fatalities:

- In all crashes, for all drivers by 12.4 percent, for belted drivers by 10.8 percent, and for unbelted drivers by 14.0 percent.
- In frontal and partially frontal crashes, without a most-harmful-event rollover, fire or immersion, for all drivers, by 29.0 percent if the principal impact is 12:00, by 15.2 percent if it is 11:00 or 1:00, and by 5.8 percent if it is 10:00 or 2:00
- In 12:00 crashes, for all adult right-front passengers, by 31.9 percent.

There are not enough data to get precise individual estimates such as, for belted RF passengers in 11:00 or 1:00 crashes. Instead, ratios of the preceding estimates are used. For example, the effect of air bags for belted drivers in 12:00 crashes is estimated to be (.108/.124) x .290 = .25258. For belted RF passengers it would be (.319/.290) x (.108/.124) x .290 = .27784.

```
            IF 1 LE M_HARM LE 6 THEN E=0;
              ELSE IF IMPACT2=12 THEN E=.25258;
              ELSE IF IMPACT2 IN (1,11) THEN E=.13239;
              ELSE IF IMPACT2 IN (2,10) THEN E=.05052; ELSE E=0;
            P=PS2081+PV2081;
            S=OLDWTFA*E*P / (1 - E*P);
            CLS2081=S*PS2081/P;
            CLV2081=S*PV2081/P;
            WEIGHTFA=OLDWTFA+CLS2081+CLV2081;
            CL2081=CLS2081+CLV2081;
            CLS=CLS+CLS2081; CLV=CLV+CLV2081; CL=CL+CL2081;
            CABDRV=CL2081; CABRF=0; CABKID=0; GOTO C214;

CDRVUNR:    OLDWTFA=WEIGHTFA;
            IF 1 LE M_HARM LE 6 THEN E=0;
              ELSE IF IMPACT2=12 THEN E=.32742;
              ELSE IF IMPACT2 IN (1,11) THEN E=.17161;
              ELSE IF IMPACT2 IN (2,10) THEN E=.06548; ELSE E=0;
            P=PS2081+PV2081;
            S=OLDWTFA*E*P / (1 - E*P);
            CLS2081=S*PS2081/P;
            CLV2081=S*PV2081/P;
            WEIGHTFA=OLDWTFA+CLS2081+CLV2081;
            CL2081=CLS2081+CLV2081;
            CLS=CLS+CLS2081; CLV=CLV+CLV2081; CL=CL+CL2081;
            CABDRV=CL2081; CABRF=0; CABKID=0; GOTO C214;

CRFBELT:    OLDWTFA=WEIGHTFA;
            IF 1 LE M_HARM LE 6 THEN E=0;
              ELSE IF IMPACT2=12 THEN E=.27784;
              ELSE IF IMPACT2 IN (1,11) THEN E=.14563;
              ELSE IF IMPACT2 IN (2,10) THEN E=.05557; ELSE E=0;
```

```
        P=PS208I+PV208I;
        S=OLDWTFA*E*P / (1 - E*P);
        CLS208I=S*PS208I/P;
        CLV208I=S*PV208I/P;
        WEIGHTFA=OLDWTFA+CLS208I+CLV208I;
        CL208I=CLS208I+CLV208I;
        CLS=CLS+CLS208I; CLV=CLV+CLV208I; CL=CL+CL208I;
        CABDRV=0; CABRF=CL208I; CABKID=0; GOTO C214;
CRFUNR:     OLDWTFA=WEIGHTFA;
        IF 1 LE M_HARM LE 6 THEN E=0;
          ELSE IF IMPACT2=12 THEN E=.36016;
          ELSE IF IMPACT2 IN (1,11) THEN E=.18877;
          ELSE IF IMPACT2 IN (2,10) THEN E=.07203; ELSE E=0;
        P=PS208I+PV208I;
        S=OLDWTFA*E*P / (1 - E*P);
        CLS208I=S*PS208I/P;
        CLV208I=S*PV208I/P;
        WEIGHTFA=OLDWTFA+CLS208I+CLV208I;
        CL208I=CLS208I+CLV208I;
        CLS=CLS+CLS208I; CLV=CLV+CLV208I; CL=CL+CL208I;
        CABDRV=0; CABRF=CL208I; CABKID=0; GOTO C214;

CKID:       OLDWTFA=WEIGHTFA;
```

As stated in Part 1 of this report, in 12:00 impacts, air bags increased the fatality risk of infants in rear-facing seats by about 450 percent, for unrestrained children age 0-5 by about 100 percent, and for restrained children age 1-5 and unrestrained children age 6-10 by about 70 percent. In partially frontal crashes (10, 11, 1, or 2:00), the effect of air bags is about 38 percent as large as in 12:00 impacts. These estimates are based on MY 1985-2000 vehicles in 1986-2000 FARS data. Most, but not all of the vehicles pre-dated the 1998-99 redesign of air bags to reduce risk to out-of-position occupants.

```
        IF KCRSH=1 AND RESTGP=1 THEN E=-4.5;
          ELSE IF KCRSH=1 AND RESTGP IN (2,3) THEN E=-1;
          ELSE IF KCRSH=1 AND RESTGP IN (4,5,6) THEN E=-.70;
          ELSE IF KCRSH IN (2,3) AND RESTGP=1 THEN E=-1.71;
          ELSE IF KCRSH IN (2,3) AND RESTGP IN (2,3) THEN E=-.38;
          ELSE IF KCRSH IN (2,3) AND RESTGP IN (4,5,6) THEN E=-.27;
          ELSE E=0;
        P=PS208I+PV208I;
        S=OLDWTFA*E*P / (1 - E*P);
        CLS208I=S*PS208I/P;
        CLV208I=S*PV208I/P;
        WEIGHTFA=OLDWTFA+CLS208I+CLV208I;
        CL208I=CLS208I+CLV208I;
        CLS=CLS+CLS208I; CLV=CLV+CLV208I; CL=CL+CL208I;
        CABDRV=0; CABRF=0; CABKID=CL208I; GOTO C214;

/* -------------------------------------------------------------- */
/*  214B VOLUNTARY TTI(d) REDUCTIONS IN 2-DOOR CARS               */
/*  MEDIAN IMPLEMENTATION YEAR: 1993                              */
/* -------------------------------------------------------------- */

/* IDENTIFIES 2-DOOR CARS WITH VOLUNTARY TTI(d) REDUCTIONS   */
C214:  PS214B=0; PV214B=0;
   IF CY GE 1985 AND MY GE 1986 AND BOD2 IN (1,2,3) AND SEAT2 IN (11,13) THEN DO;
```

NHTSA's evaluation indicates that TTI(d) performance substantially improved in 2-door cars from MY 1985 to 1994, before the FMVSS 214 upgrade took effect, with most of the improvement coming in the last two years. (Substantial additional improvements occurred in both 2- and 4-door cars during and after the phase-in of FMVSS 214, but they have not yet been evaluated.)

```
    IF 1986 LE MY LE 1992 THEN PV214B=.06*(MY-1985);
      ELSE IF MY=1993 THEN PV214B=.68;
      ELSE IF MY GE 1994 THEN PV214B=1; END;
/* LIVES SAVED IN SIDE IMPACTS BY VOLUNTARY TTI(d) REDUCTIONS */
  OLDWTFA=WEIGHTFA;
  IF PS214B GT 0 OR PV214B GT 0 THEN DO;
```

The average TTI(d) improvement in 2-door cars from 1985 to 1994 was associated with a 23 percent reduction of fatality risk for front-seat occupants in side impacts.

```
    IF IMPACT2 IN (2,3,4,8,9,10) THEN E=.23; ELSE E=0;
    P=PS214B+PV214B;
    S=OLDWTFA*E*P / (1 - E*P);
    CLS214B=S*PS214B/P;
    CLV214B=S*PV214B/P;
    WEIGHTFA=OLDWTFA+CLS214B+CLV214B; END;
  ELSE DO; CLS214B=0; CLV214B=0; END;
  CL214B=CLS214B+CLV214B;
  CLS=CLS+CLS214B; CLV=CLV+CLV214B; CL=CL+CL214B;


/* ----------------------------------------------------------------- */
/*  208H AUTOMATIC 2-POINT BELTS                                      */
/*  PEAK INSTALLATION YEAR: 1991                                     */
/* ----------------------------------------------------------------- */

/* IDENTIFIES OCCUPANTS WEARING AUTOMATIC 2-POINT BELTS  */
  PS208H=0; PV208H=0;
```

"REST3 = 7" indicates use of 2-point automatic belts. PS208H includes any installation from MY 1987 onwards, because the phase-in of automatic occupant protection began on September 1, 1986.

```
  IF REST3=7 THEN DO; IF MY LE 1986 THEN PV208H=1; ELSE PS208H=1; END;
/* LIVES SAVED AND EJECTIONS PREVENTED BY AUTOMATIC 2-POINT BELTS */
  OLDWTFA=WEIGHTFA;
  IF PS208H GT 0 OR PV208H GT 0 THEN DO;
```

Two-point automatic belts reduce fatalities by an estimated 30 percent in frontals, 18 percent in side impacts, 62 percent in rollovers and 68 percent in rear/other impacts. In general, 21 percent of unrestrained front-outboard occupant fatalities in frontal crashes were ejected, 23 percent in side impacts, 66 percent in rollovers and 35 percent in rear/other impacts.

```
   IF CRSH=1 THEN DO; E=.30; EJECT2=.21; END;
     ELSE IF CRSH=2 THEN DO; E=.18; EJECT2=.23; END;
     ELSE IF CRSH=3 THEN DO; E=.62; EJECT2=.66; END;
     ELSE IF CRSH=4 THEN DO; E=.68; EJECT2=.35; END;
    P=PS208H+PV208H;
    S=OLDWTFA*E*P / (1 - E*P);
    CLS208H=S*PS208H/P;
    CLV208H=S*PV208H/P;
    WEIGHTFA=OLDWTFA+CLS208H+CLV208H; END;
  ELSE DO; CLS208H=0; CLV208H=0; END;
  CL208H=CLS208H+CLV208H;
  CLS=CLS+CLS208H; CLV=CLV+CLV208H; CL=CL+CL208H;
```

```
/* ----------------------------------------------------------------- */
/*   208G 3-POINT BELTS FOR BACK-SEAT OUTBOARD OCCUPANTS              */
/*   MEDIAN INSTALLATION YEAR: 1989                                  */
/* ----------------------------------------------------------------- */

/* IDENTIFIES BACK-SEAT OCCUPANTS WEARING 3-POINT BELTS */
  PS208G=0; PV208G=0;
  IF SEAT2=21 AND REST3=3 THEN DO; IF MY LE 1989 THEN PV208G=1; ELSE PS208G=1; END;
/* LIVES SAVED AND EJECTIONS PREVENTED BY BACK-SEAT 3-POINT BELTS */
  OLDWTFA=WEIGHTFA;
  IF PS208G GT 0 OR PV208G GT 0 THEN DO;
```

Three-point belts in the rear-outboard seats reduce fatalities by 29 percent in frontals, 42 percent in side impacts, 77 percent in rollovers and 31 percent in rear/other impacts. In general, 23 percent of unrestrained rear-outboard occupant fatalities in frontal crashes were ejected, 28 percent in side impacts, 61 percent in rollovers and 27 percent in rear/other impacts.

```
   IF CRSH=1 THEN DO; E=.29; EJECT2=.23; END;
     ELSE IF CRSH=2 THEN DO; E=.42; EJECT2=.28; END;
     ELSE IF CRSH=3 THEN DO; E=.77; EJECT2=.61; END;
     ELSE IF CRSH=4 THEN DO; E=.31; EJECT2=.27; END;
   P=PS208G+PV208G;
   S=OLDWTFA*E*P / (1 - E*P);
   CLS208G=S*PS208G/P;
   CLV208G=S*PV208G/P;
    WEIGHTFA=OLDWTFA+CLS208G+CLV208G; END;
  ELSE DO; CLS208G=0; CLV208G=0; END;
  CL208G=CLS208G+CLV208G;
  CLS=CLS+CLS208G; CLV=CLV+CLV208G; CL=CL+CL208G;

/* ----------------------------------------------------------------- */
/*   213 CHILD SAFETY SEATS                                          */
/*   USE RATE WENT OVER 50% IN: 1985                                 */
/* ----------------------------------------------------------------- */

/* IDENTIFIES CHILD PASSENGERS IN SAFETY SEATS */
  PS213=0; PV213=0;
  IF REST3=4 THEN DO;
```

"REST3 = 4" indicates a child passenger in a safety seat (not necessarily correctly used). FMVSS 213 regulates the performance of safety seats but does not mandate their use. Therefore, the distinction between PV213 and PS213 will not be based on the effective date of FMVSS 213 (April 1, 1971) but on whether the child was covered by a State law for child passenger protection. That depends on the State and the age of the child. Here are when the State laws took effect, and for what age groups:

```
  IF STATE=47 AND CY GE 1978 AND AGE LE 3 THEN PS213=1;
    ELSE IF STATE=44 AND CY GE 1980 AND AGE LE 5 THEN PS213=1;
    ELSE IF STATE=54 AND CY GE 1981 AND AGE LE 2 THEN PS213=1;
    ELSE IF STATE=25 AND CY GE 1982 AND AGE LE 5 THEN PS213=1;
    ELSE IF STATE IN (9,10,20,21,26,36,37) AND CY GE 1982 AND AGE LE 3 THEN PS213=1;
    ELSE IF STATE IN (32,34,35) AND CY GE 1983 AND AGE LE 4 THEN PS213=1;
    ELSE IF STATE IN (1,6,11,12,17,28,33,39,45,51,55) AND CY GE 1983 AND AGE LE 3 THEN PS213=1;
    ELSE IF STATE=15 AND CY GE 1983 AND AGE LE 2 THEN PS213=1;
    ELSE IF STATE IN (4,5,46,50) AND CY GE 1984 AND AGE LE 4 THEN PS213=1;
    ELSE IF STATE IN (8,18,23,24,27,29,31,40,41,42) AND CY GE 1984 AND AGE LE 3 THEN PS213=1;
    ELSE IF STATE IN (13,38,53) AND CY GE 1984 AND AGE LE 2 THEN PS213=1;
    ELSE IF STATE IN (30,49) AND CY GE 1984 AND AGE LE 1 THEN PS213=1;
    ELSE IF STATE=56 AND CY GE 1985 AND AGE LE 4 THEN PS213=1;
    ELSE IF STATE IN (2,16) AND CY GE 1985 AND AGE LE 3 THEN PS213=1;
    ELSE IF STATE IN (19,22) AND CY GE 1985 AND AGE LE 2 THEN PS213=1;
```

311

```
        ELSE IF STATE=48 AND CY GE 1985 AND AGE LE 1 THEN PS213=1;
        ELSE PV213=1; END;
/* LIVES SAVED AND EJECTIONS PREVENTED BY CHILD SAFETY SEATS */
  OLDWTFA=WEIGHTFA;
  IF PS213 GT 0 OR PV213 GT 0 THEN DO;
```

NHTSA's evaluation estimates that safety seats reduce fatality risk by 71 percent for infants and by 54 percent for toddlers. Effectiveness is not estimated separately by crash mode. In general, 19 percent of unrestrained child passenger fatalities in frontal crashes were ejected, 26 percent in side impacts, 65 percent in rollovers and 34 percent in rear/other impacts.

```
    IF AGE=0 THEN E=.71; ELSE E=.54;
    IF CRSH=1 THEN EJECT2=.19;
      ELSE IF CRSH=2 THEN EJECT2=.26;
      ELSE IF CRSH=3 THEN EJECT2=.65;
      ELSE IF CRSH=4 THEN EJECT2=.34;
    P=PS213+PV213;
    S=OLDWTFA*E*P / (1 - E*P);
    CLS213=S*PS213/P;
    CLV213=S*PV213/P;
    WEIGHTFA=OLDWTFA+CLS213+CLV213; END;
  ELSE DO; CLS213=0; CLV213=0; END;
  CL213=CLS213+CLV213;
  CLS=CLS+CLS213; CLV=CLV+CLV213; CL=CL+CL213;
```

NCAP-related crashworthiness improvement is the only component of the model that is not associated with a specific FMVSS or even a single, specific technology. Nevertheless, cars became safer, saving lives. Starting in 1979, NHTSA's New Car Assessment Program (NCAP) tested the injury performance of belted dummies in 35-mph frontal impacts, and advised the public about the comparative performance of make-models. Starting about 1982, rapidly at first and more gradually in 1984-87, manufacturers modified or redesigned their cars, resulting in substantial improvements on the NCAP test. Modifications included the belt system, the steering assembly, the instrument panel and the seat structure, taking into account how dummies interacted with those systems in 35 mph tests. NHTSA's evaluation showed a 20 percent reduction of fatality risk, for MY 1983-86 vs. MY 1979-82, for belted drivers in head-on collisions with other cars. NHTSA's evaluation did not consider crashes with other types of vehicles or with fixed objects, and it did not study the fatality risk of passengers; the data were limited almost entirely to cars without air bags. We will assume a 20 percent fatality reduction, for belted drivers only (not passengers), when a car frontally impacts another car (but not limited to head-on collisions). Conservatively, we will not assume any benefit for other frontal impacts, such as with an LTV, a heavy truck, or a fixed object; we will limit the benefit to cars without air bags.

```
/* -------------------------------------------------------------------- */
/*   NCAP VOLUNTARY FRONTAL CRASHWORTHINESS IMPROVEMENTS                 */
/*   MEDIAN IMPLEMENTATION YEAR: EARLY 1984                              */
/* -------------------------------------------------------------------- */

/* IDENTIFIES NON-AIR-BAG, 3-POINT-BELT CARS WITH IMPROVED NCAP SCORES   */
  PSNCAP=0; PVNCAP=0;
  IF CY GE 1981 AND MY GE 1982 AND PASSIVE IN (0,505) AND SEAT2=11 THEN DO;
      IF MY=1982 THEN PVNCAP=.25;
        ELSE IF MY=1983 THEN PVNCAP=.50;
        ELSE IF MY=1984 THEN PVNCAP=.67;
        ELSE IF MY=1985 THEN PVNCAP=.83;
        ELSE IF MY GE 1986 THEN PVNCAP=1; END;
/* LIVES SAVED BY NCAP-RELATED IMPROVEMENTS: BELTED DRIVERS IN HEAD-ON CRASHES WITH CARS */
  OLDWTFA=WEIGHTFA;
  IF PSNCAP GT 0 OR PVNCAP GT 0 THEN DO;
    IF REST3=3 AND IMPACT2 IN (1,11,12) AND OVTYP=1 THEN E=.20; ELSE E=0;
    P=PSNCAP+PVNCAP;
    S=OLDWTFA*E*P / (1 - E*P);
    CLSNCAP=S*PSNCAP/P;
    CLVNCAP=S*PVNCAP/P;
    WEIGHTFA=OLDWTFA+CLSNCAP+CLVNCAP; END;
  ELSE DO; CLSNCAP=0; CLVNCAP=0; END;
  CLNCAP=CLSNCAP+CLVNCAP;
  CLS=CLS+CLSNCAP; CLV=CLV+CLVNCAP; CL=CL+CLNCAP;

/* -------------------------------------------------------------------- */
/*   208F 3-POINT BELTS FOR FRONT-SEAT OUTBOARD OCCUPANTS               */
/*   MEDIAN INSTALLATION YEAR: 1974                                     */
/* -------------------------------------------------------------------- */

/* IDENTIFIES FRONT-SEAT OCCUPANTS WEARING 3-POINT BELTS  */
  PS208F=0; PV208F=0;
```

"REST3 = 3" includes use of manual 3-point belts, use of automatic 3-point belts, and use of both belts at seats with separate lap and shoulder belts (i.e., for the separate belts only, FARS must say REST_USE = 3; for the 3-point systems, REST_USE can be 1, 2, 3, 8, 13 or imputed). The FMVSS 208 requirement for integral 3-point belts took effect on September 1, 1973 (model year 1974); all lap/shoulder belt use (integral or separate) before MY 1974 will be credited to PV208F.

```
  IF SEAT2 IN (11,13) AND REST3=3 THEN DO; IF MY LE 1973 THEN PV208F=1; ELSE PS208F=1; END;
/* LIVES SAVED AND EJECTIONS PREVENTED BY FRONT-OUTBOARD 3-POINT BELTS */
  OLDWTFA=WEIGHTFA;
  IF PS208F GT 0 OR PV208F GT 0 THEN DO;
```

With the abundant data for evaluating lap/shoulder belt effectiveness, it is possible to fine-tune the estimates more than for other belt systems: 60 percent fatality reduction in single-vehicle frontals; 42 percent in multivehicle frontals; 21 percent in single-vehicle nearside; 5 percent in multivehicle nearside; 46 percent in single-vehicle farside; 35 percent in multivehicle farside; 74 percent in rollovers; 56 percent in rear/other impacts. The proportion of unrestrained fatalities that were ejectees is also computed separately for each of those crash types.

```
    IF CRSH=1 AND VE_FORMS=1 THEN DO; E=.60; EJECT2=.31; END;
      ELSE IF CRSH=1 THEN DO; E=.42; EJECT2=.12; END;
      ELSE IF CRSH=2 THEN DO;
        IF VE_FORMS=1 AND
          ((SEAT2=11 AND IMPACT2 IN (8,9,10)) OR (SEAT2=13 AND IMPACT2 IN (2,3,4)))
          THEN DO; E=.21; EJECT2=.33; END;
        ELSE IF ((SEAT2=11 AND IMPACT2 IN (8,9,10)) OR (SEAT2=13 AND IMPACT2 IN (2,3,4)))
          THEN DO; E=.05; EJECT2=.15; END;
        ELSE IF VE_FORMS=1 THEN DO; E=.46; EJECT2=.36; END;
```

```
       ELSE DO; E=.35; EJECT2=.20; END; END;
      ELSE IF CRSH=3 THEN DO; E=.74; EJECT2=.66; END;
      ELSE IF CRSH=4 THEN DO; E=.56; EJECT2=.35; END;
    P=PS208F+PV208F;
    S=OLDWTFA*E*P / (1 - E*P);
    CLS208F=S*PS208F/P;
    CLV208F=S*PV208F/P;
     WEIGHTFA=OLDWTFA+CLS208F+CLV208F; END;
   ELSE DO; CLS208F=0; CLV208F=0; END;
   CL208F=CLS208F+CLV208F;
   CLS=CLS+CLS208F; CLV=CLV+CLV208F; CL=CL+CL208F;

/* ------------------------------------------------------------------ */
/*  216 B-PILLARS FOR HARDTOP CARS AND OTHER ROOF STRENGTHENING        */
/*  MEDIAN INSTALLATION YEAR: 1973                                     */
/* ------------------------------------------------------------------ */

/* ROOF STRENGTH IMPROVED GRADUALLY FROM 1970 TO 1977          */
  IF MY LE 1969 THEN DO; PS216=0; PV216=0; END;
    ELSE IF 1970 LE MY LE 1973 THEN DO; PS216=0; PV216=.125*(MY-1969); END;
    ELSE IF 1974 LE MY LE 1976 THEN DO; PS216=.125*(MY-1969); PV216=0; END;
    ELSE IF MY GE 1977 THEN DO; PS216=1; PV216=0; END;
/* REDUCTION IN NONEJECTION ROLLOVER FATALITY RISK WITH IMPROVED ROOF STRENGTH */
  OLDWTFA=WEIGHTFA;
```

Roof crush resistance can benefit any non-ejected occupant (EJECT2 NE 1) within the vehicle (SEAT2 NE 52) in a rollover crash. NHTSA's evaluation showed an overall 7.4 percent reduction in non-ejection rollover fatalities during the 1970-77 implementation period associated with FMVSS 216. This is one of the FMVSS where it was important to track the changes in EJECT2 in previous steps of the model. But EJECT2 also has to be recomputed at the end of this step EJECT2 = EJECT2/(1+REL_S), because "removing" roof crush strength increases non-ejection fatalities, and thus increases the proportion of fatalities that are not ejected

```
  IF (PS216 GT 0 OR PV216 GT 0) AND (CRSH=3 OR M_HARM=1)
     AND SEAT2 NE 52 AND EJECT2 NE 1 THEN DO;
   E=.074;
   P=PS216+PV216;
   REL_S=(1-EJECT2)*E*P / (1 - E*P);
   S=OLDWTFA*REL_S;
   CLS216=S*PS216/P;
   CLV216=S*PV216/P;
    WEIGHTFA=OLDWTFA+CLS216+CLV216;
    EJECT2=EJECT2/(1+ REL_S); END;
  ELSE DO; CLS216=0; CLV216=0; END;
  CL216=CLS216+CLV216;
  CLS=CLS+CLS216; CLV=CLV+CLV216; CL=CL+CL216;
```

```
/* ---------------------------------------------------------------- */
/*   214A SIDE DOOR BEAMS                                            */
/*   MEDIAN INSTALLATION YEAR: EARLY 1973                            */
/* ---------------------------------------------------------------- */

/* IDENTIFIES CARS WITH SIDE DOOR BEAMS                  */
```

We know exactly when each make-model was equipped with side door beams. PS214A and PV214A can be set to 0 or 1 (or possibly to .5 in MY 1973 because FMVSS 214 took effect in the middle of the model year). Identification of make-models is based on the VIN, if available, and on the FARS MAK_MOD code, otherwise.

```
    IF MY GE 1974 THEN DO; PS214A=1; PV214A=0; GOTO SAVE214; END;
    IF MY LE 1968 THEN DO; PS214A=0; PV214A=0; GOTO SAVE214; END;
    IF 1969 LE MY LE 1972 THEN PS214A=0;
      ELSE IF MY=1973 THEN PS214A=.5;
    IF MAKE IN (1,7,9,12,13,14,18,19,20,21,22) THEN GOTO SOMEVOL;
      ELSE GOTO NOVOL;
SOMEVOL: V13=V1||V2||V3;
    IF V13 IN ('999','000','   ') THEN GOTO NOVIN;
    IF MY=1969 THEN DO;
      IF MAKE IN (1,7,9,12,13,14) THEN PV214A=0;
        ELSE IF MAKE IN (18,20,21,22) AND V2 IN ('5','6','7','8') THEN PV214A=1;
        ELSE IF MAKE=19 AND V1 NE 'H' THEN PV214A=1;
        ELSE PV214A=0; GOTO SAVE214; END;
    IF MY=1970 THEN DO;
      IF MAKE IN (1,12,13,14) THEN PV214A=0;
        ELSE IF MAKE IN (18,20,21,22) AND V2 IN ('3','4','5','6','7','8') THEN PV214A=1;
        ELSE IF MAKE=19 AND V1 NE 'H' THEN PV214A=1;
        ELSE IF MAKE IN (20,22) AND V2='2' THEN PV214A=.5;
        ELSE IF MAKE=7 AND V1='J' THEN PV214A=1;
        ELSE IF MAKE=9 AND V1='B' THEN PV214A=1;
        ELSE PV214A=0; GOTO SAVE214; END;
    IF MY=1971 THEN DO;
      IF MAKE IN (18,19,21,22) THEN PV214A=1;
        ELSE IF MAKE=20 AND V2 IN ('2','3','4','5','6') THEN PV214A=1;
        ELSE IF MAKE=12 AND V3 IN ('0','5','6','7') THEN PV214A=1;
        ELSE IF MAKE=13 AND V4 IN ('1','2') THEN PV214A=1;
        ELSE IF MAKE=14 AND V3 IN ('4','5','6','7','9') THEN PV214A=1;
        ELSE IF MAKE=7 AND V1='J' THEN PV214A=1;
        ELSE IF MAKE=9 AND V1='B' THEN PV214A=1;
        ELSE IF MAKE=1 AND V4 IN ('3','7') THEN PV214A=1;
        ELSE PV214A=0; GOTO SAVE214; END;
    IF MY=1972 THEN DO;
      IF MAKE IN (13,18,19,21) THEN PV214A=1;
        ELSE IF MAKE=20 AND V2 NE 'X' AND V2 NE 'Z' THEN PV214A=1;
        ELSE IF MAKE=22 AND V2 NE 'Y' THEN PV214A=1;
        ELSE IF MAKE=12 AND V3 IN ('0','2','3','4','5','6','7') THEN PV214A=1;
        ELSE IF MAKE=14 AND V3 NE '3' THEN PV214A=1;
        ELSE IF MAKE=7 AND V1='J' THEN PV214A=1;
        ELSE IF MAKE=9 AND V1='B' THEN PV214A=1;
        ELSE IF MAKE=1 AND V4='7' THEN PV214A=1;
        ELSE PV214A=0; GOTO SAVE214; END;
    IF MY=1973 THEN DO;
      IF MAKE IN (1,13,14,18,19,20,21,22) THEN PV214A=.5;
        ELSE IF MAKE=12 AND V3 NE '8' THEN PV214A=.5;
        ELSE IF MAKE=7 AND V1='J' THEN PV214A=.5;
        ELSE IF MAKE=9 AND V1='B' THEN PV214A=.5;
        ELSE PV214A=0; GOTO SAVE214; END;
NOVIN: IF MY=1969 THEN DO;
      IF MAKE IN (1,7,9,12,13,14) THEN PV214A=0;
        ELSE IF MAK_MOD IN (1802,18002,1803,18003,1903,19003,2002,20002,
                            2102,21002,2103,21003,2202,22002,2210,22010) THEN PV214A=1;
        ELSE PV214A=0; GOTO SAVE214; END;
```

```
 IF MY=1970 THEN DO;
   IF MAKE IN (1,12,13,14) THEN PV214A=0;
     ELSE IF MAK_MOD IN (1802,18002,1803,18003,1903,19003,2002,20002,
                         2102,21002,2103,21003,2202,22002,2210,22010,
                         1801,18001,2001,20001,2101,21001,2201,22001,
                         2010,20010,705,7005,905,9005) THEN PV214A=1;
     ELSE IF MAK_MOD IN (2009,20009,2209,22009) THEN PV214A=.5;
     ELSE PV214A=0; GOTO SAVE214; END;
 IF MY=1971 THEN DO;
   IF MAKE IN (18,19,21,22) THEN PV214A=1;
     ELSE IF MAK_MOD IN (2002,20002,2001,20001,2010,20010,705,7005,905,9005,
                         2009,20009,2011,20011,
                         1203,12003,1206,12006,1301,13001,1404,14004,1406,14006,
                         105,1005,106,1006) THEN PV214A=1;
     ELSE PV214A=0; GOTO SAVE214; END;
 IF MY=1972 THEN DO;
   IF MAKE IN (13,18,19,21) THEN PV214A=1;
     ELSE IF MAK_MOD IN (2002,20002,2001,20001,2010,20010,705,7005,905,9005,
                         2009,20009,2011,20011,
                         1203,12003,1206,12006,1202,12002,1404,14004,1406,14006,
                         1402,14002,105,1005,106,1006) THEN PV214A=1;
     ELSE IF MAKE=22 AND MAK_MOD NE 2208 AND MAK_MOD NE 22008 THEN PV214A=1;
     ELSE PV214A=0; GOTO SAVE214; END;
 IF MY=1973 THEN DO;
   IF MAKE IN (1,13,14,18,19,20,21,22) THEN PV214A=.5;
     ELSE IF MAKE=12 AND MAK_MOD NE 1204 AND MAK_MOD NE 12004 THEN PV214A=.5;
     ELSE IF MAK_MOD IN (705,7005,905,9005) THEN PV214A=.5;
     ELSE PV214A=0; GOTO SAVE214; END;
NOVOL: IF 1969 LE MY LE 1972 THEN DO; PS214A=0; PV214A=0; GOTO SAVE214; END;
   ELSE IF MY=1973 THEN DO; PS214A=.5; PV214A=0; GOTO SAVE214; END;
/* REDUCTION IN SINGLE-VEHICLE SIDE-IMPACT FATALITIES WITH SIDE DOOR BEAMS    */
SAVE214:  OLDWTFA=WEIGHTFA;
```

Side door beams reduce fatality risk by 14 percent for front- and rear-outboard occupants in single-vehicle crashes. Whenever possible (CY GE 1979), also exclude collisions with parked vehicles from the "single" vehicle crashes.

```
 IF (PS214A GT 0 OR PV214A GT 0) AND IMPACT2 IN (2,3,4,8,9,10) AND
     SEAT2 IN (11,13,21) AND
     (VE_FORMS=1 OR (CY GE 1979 AND M_HARM NE 12 AND M_HARM NE 13 AND M_HARM NE 14))
     THEN DO;
   E=.14;
   P=PS214A+PV214A;
   S=OLDWTFA*E*P / (1 - E*P);
   CLS214A=S*PS214A/P;
   CLV214A=S*PV214A/P;
   WEIGHTFA=OLDWTFA+CLS214A+CLV214A; END;
 ELSE DO; CLS214A=0; CLV214A=0; END;
 CL214A=CLS214A+CLV214A;
 CLS=CLS+CLS214A; CLV=CLV+CLV214A; CL=CL+CL214A;


/* ----------------------------------------------------------------- */
/*  105B FRONT DISC BRAKES                                           */
/*  MEDIAN INSTALLATION YEAR: 1971 (INSTALLATION COMPLETED 1977)     */
/* ----------------------------------------------------------------- */
```

As discussed in the analysis of non-occupant fatalities, front disc brakes were gradually introduced into passenger cars during 1965-77 and they reduce fatal crash involvements of all types by an estimated 0.17 percent.

```
/* IMPLEMENTATION OF FRONT DISC BRAKES */
 IF MY GE 1977 THEN DO; PS105B=1; PV105B=0; END;
   ELSE IF MY LE 1964 THEN DO; PS105B=0; PV105B=0; END;
```

316

```
        ELSE IF MY=1965 THEN DO; PS105B=0; PV105B=.02; END;
        ELSE IF MY=1966 THEN DO; PS105B=0; PV105B=.03; END;
        ELSE IF MY=1967 THEN DO; PS105B=0; PV105B=.06; END;
        ELSE IF MY=1968 THEN DO; PS105B=0; PV105B=.13; END;
        ELSE IF MY=1969 THEN DO; PS105B=0; PV105B=.28; END;
        ELSE IF MY=1970 THEN DO; PS105B=0; PV105B=.41; END;
        ELSE IF MY=1971 THEN DO; PS105B=0; PV105B=.63; END;
        ELSE IF MY=1972 THEN DO; PS105B=0; PV105B=.74; END;
        ELSE IF MY=1973 THEN DO; PS105B=0; PV105B=.86; END;
        ELSE IF MY=1974 THEN DO; PS105B=0; PV105B=.84; END;
        ELSE IF MY=1975 THEN DO; PS105B=0; PV105B=.93; END;
        ELSE IF MY=1976 AND CY=1975 THEN DO; PS105B=0; PV105B=.99; END;
        ELSE IF MY=1976 THEN DO; PS105B=.50; PV105B=.49; END;
/* CAR OCCUPANT LIVES SAVED BY FRONT DISC BRAKES */
  OLDWTFA=WEIGHTFA;
  IF PS105B GT 0 OR PV105B GT 0 THEN DO;
    E=.0017;
    P=PS105B+PV105B;
    S=OLDWTFA*E*P / (1 - E*P);
    CLS105B=S*PS105B/P;
    CLV105B=S*PV105B/P;
    WEIGHTFA=OLDWTFA+CLS105B+CLV105B; END;
  ELSE DO; CLS105B=0; CLV105B=0; END;
  CL105B=CLS105B+CLV105B;
  CLS=CLS+CLS105B; CLV=CLV+CLV105B; CL=CL+CL105B;


/* ---------------------------------------------------------------- */
/* 201 VOLUNTARY INSTRUMENT PANEL IMPROVEMENTS (FMVSS-201 INSPIRED)  */
/* MEDIAN IMPLEMENTATION YEAR: 1968                                  */
/* ---------------------------------------------------------------- */

/* IDENTIFIES CARS WITH VOLUNTARY INSTRUMENT PANEL IMPROVEMENTS    */
  PS201=0; PV201=0;
```

Part 1 of this report describes how manufacturers significantly modified mid and lower instrument panels in the years before/after FMVSS 201 took effect, even though the standard as finally issued only regulated the upper part of the instrument panel. The technology benefits right-front passengers (SEAT2 = 13).

```
  IF MY GE 1967 AND SEAT2=13 THEN DO;
      IF MY GE 1973 THEN PV201=1;
        ELSE IF MY=1967 THEN PV201=.25; ELSE IF MY=1968 THEN PV201=.5;
        ELSE IF MY=1969 THEN PV201=.6;  ELSE IF MY=1970 THEN PV201=.7;
        ELSE IF MY=1971 THEN PV201=.8;  ELSE IF MY=1972 THEN PV201=.9; END;
/* LIVES SAVED IN FRONTAL IMPACTS BY INSTRUMENT PANEL IMPROVEMENTS */
  OLDWTFA=WEIGHTFA;
```

The evaluation indicated a 15.9 percent fatality reduction for unrestrained right-front passengers in frontal impacts. Since 3-point belts, automatic belts and child safety seats have already been "removed" from the vehicle, REST3 NE 2 (i.e., not lap-belted) identifies unrestrained people.

```
  IF PS201 GT 0 OR PV201 GT 0 THEN DO;
    IF CRSH=1 AND REST3 NE 2 THEN E=.159; ELSE E=0;
    P=PS201+PV201;
    S=OLDWTFA*E*P / (1 - E*P);
    CLS201=S*PS201/P;
    CLV201=S*PV201/P;
    WEIGHTFA=OLDWTFA+CLS201+CLV201; END;
  ELSE DO; CLS201=0; CLV201=0; END;
  CL201=CLS201+CLV201;
  CLS=CLS+CLS201; CLV=CLV+CLV201; CL=CL+CL201;
```

```
/* ------------------------------------------------------------------ */
/*  203 ENERGY-ABSORBING AND TELESCOPING STEERING ASSEMBLIES          */
/*  MEDIAN IMPLEMENTATION YEAR: 1967                                  */
/* ------------------------------------------------------------------ */

/* IDENTIFIES CARS WITH ENERGY-ABSORBING STEERING ASSEMBLIES     */
  PS203=0; PV203=0;
```

FMVSS 203 and 204 took effect on January 1, 1968, but energy-absorbing steering assemblies had already been installed on AMC, Chrysler and GM cars in MY 1967 and all other cars in MY 1968.

```
  IF MY GE 1967 AND SEAT2=11 THEN DO;
     IF MY GE 1969 THEN PS203=1;
        ELSE IF MY=1968 THEN DO; PS203=.5; PV203=.5; END;
        ELSE IF MY=1967 AND MAKE IN (1,6,7,8,9,18,19,20,21,22) THEN PV203=1;
        ELSE IF MY=1967 THEN PV203=0; END;
/* LIVES SAVED IN FRONTAL IMPACTS BY INSTRUMENT PANEL IMPROVEMENTS */
  OLDWTFA=WEIGHTFA;
  IF PS203 GT 0 OR PV203 GT 0 THEN DO;
```

Energy-absorbing steering assemblies reduce the fatality risk of drivers by 12.1 percent in frontal impacts.

```
    IF IMPACT2 IN (11,12,1) THEN E=.121; ELSE E=0;
    P=PS203+PV203;
    S=OLDWTFA*E*P / (1 - E*P);
    CLS203=S*PS203/P;
    CLV203=S*PV203/P;
    WEIGHTFA=OLDWTFA+CLS203+CLV203; END;
  ELSE DO; CLS203=0; CLV203=0; END;
  CL203=CLS203+CLV203;
  CLS=CLS+CLS203; CLV=CLV+CLV203; CL=CL+CL203;
```

```
/* ------------------------------------------------------------------ */
/*   208E LAP BELTS FOR BACK-SEAT CENTER OCCUPANTS                     */
/*   MEDIAN INSTALLATION YEAR: EARLY 1967                              */
/* ------------------------------------------------------------------ */

/* IDENTIFIES BACK-CENTER-SEAT OCCUPANTS WEARING LAP BELTS   */
  PS208E=0; PV208E=0;
  IF SEAT2=22 AND REST3=2 AND (5 LE AGE LE 99 OR AGE=.) THEN DO;
    IF MY GE 1969 THEN PS208E=1;
      ELSE IF MY LE 1967 THEN PV208E=1;
```

Lap belts (as a minimum) were required at all designated seating positions of cars and LTVs effective January 1, 1968, but were already installed in many vehicles well before that date. Since 1/1/1968 is in the middle of the 1968 model year, PV = .5 and PS = .5 for MY 1968 for all of the lap belt technologies in cars and LTVs.

```
        ELSE IF MY=1968 THEN DO; PS208E=.5; PV208E=.5; END; END;
/* LIVES SAVED AND EJECTIONS PREVENTED BY BACK-CENTER-SEAT LAP BELTS */
  OLDWTFA=WEIGHTFA;
  IF PS208E GT 0 OR PV208E GT 0 THEN DO;
```

The data from NHTSA's evaluation shows lap belts for rear-outboard passengers reduce fatality risk by 76 percent in rollovers and 39 percent in side impacts, but have little or no net benefit in frontals, rear impacts or other crashes. The same effectiveness estimates are assumed for the rear-center position.

```
    IF CRSH=1 THEN DO; E=.00; EJECT2=.26; END;
      ELSE IF CRSH=2 THEN DO; E=.39; EJECT2=.33; END;
      ELSE IF CRSH=3 THEN DO; E=.76; EJECT2=.64; END;
      ELSE IF CRSH=4 THEN DO; E=.00; EJECT2=.31; END;
    P=PS208E+PV208E;
    S=OLDWTFA*E*P / (1 - E*P);
    CLS208E=S*PS208E/P;
    CLV208E=S*PV208E/P;
    WEIGHTFA=OLDWTFA+CLS208E+CLV208E; END;
  ELSE DO; CLS208E=0; CLV208E=0; END;
  CL208E=CLS208E+CLV208E;
  CLS=CLS+CLS208E; CLV=CLV+CLV208E; CL=CL+CL208E;
```

As stated above, any rear-center passenger who was originally wearing 3-point belts was "transformed" to an unrestrained passenger back in the 208J step, and will have bypassed the 208E step.

```
/* ------------------------------------------------------------------ */
/*   105A DUAL MASTER CYLINDERS                                        */
/*   MEDIAN INSTALLATION YEAR: LATE 1966                               */
/* ------------------------------------------------------------------ */
```

As discussed in the analysis of non-occupant fatalities, dual master cylinders were gradually introduced into passenger cars during 1962-67 and they reduce fatal crash involvements of all types by an estimated 0.7 percent.

```
/* IMPLEMENTATION OF DUAL MASTER CYLINDERS */
  IF MY GE 1969 THEN DO; PS105A=1; PV105A=0; END;
    ELSE IF MY LE 1961 THEN DO; PS105A=0; PV105A=0; END;
    ELSE IF MY IN (1962,1963) THEN DO; PS105A=0; PV105A=.09; END;
    ELSE IF MY IN (1964,1965) THEN DO; PS105A=0; PV105A=.07; END;
    ELSE IF MY=1966 THEN DO; PS105A=0; PV105A=.54; END;
    ELSE IF MY=1967 THEN DO; PS105A=0; PV105A=1; END;
    ELSE IF MY=1968 THEN DO; PS105A=.5; PV105A=.5; END;
/* CAR OCCUPANT LIVES SAVED BY DUAL MASTER CYLINDERS */
```

```
OLDWTFA=WEIGHTFA;
IF PS105A GT 0 OR PV105A GT 0 THEN DO;
  E=.007;
  P=PS105A+PV105A;
  S=OLDWTFA*E*P / (1 - E*P);
  CLS105A=S*PS105A/P;
  CLV105A=S*PV105A/P;
  WEIGHTFA=OLDWTFA+CLS105A+CLV105A; END;
ELSE DO; CLS105A=0; CLV105A=0; END;
CL105A=CLS105A+CLV105A;
CLS=CLS+CLS105A; CLV=CLV+CLV105A; CL=CL+CL105A;

/* ---------------------------------------------------------------- */
/*   208D LAP BELTS FOR FRONT-SEAT CENTER OCCUPANTS                  */
/*   MEDIAN INSTALLATION YEAR: 1966                                  */
/* ---------------------------------------------------------------- */
```

A double-pair comparison analysis of lap belt effectiveness for front-outboard occupants, conducted in support of this study, found that lap belts reduce fatality risk by 22 percent in frontals, 43 percent in rollovers, and 21 percent in side impacts, rear impacts and other crashes. The same effectiveness is assumed for front-center occupants. However, the proportion of unrestrained fatalities that were ejected was derived specifically from data on front-center occupants.

```
/* IDENTIFIES FRONT-CENTER-SEAT OCCUPANTS WEARING LAP BELTS  */
PS208D=0; PV208D=0;
IF SEAT2=12 AND REST3=2 AND (5 LE AGE LE 99 OR AGE=.) THEN DO;
    IF MY GE 1969 THEN PS208D=1;
      ELSE IF MY LE 1967 THEN PV208D=1;
      ELSE IF MY=1968 THEN DO; PS208D=.5; PV208D=.5; END; END;
/* LIVES SAVED AND EJECTIONS PREVENTED BY FRONT-CENTER-SEAT LAP BELTS */
OLDWTFA=WEIGHTFA;
IF PS208D GT 0 OR PV208D GT 0 THEN DO;
    IF CRSH=1 THEN DO; E=.22; EJECT2=.20; END;
      ELSE IF CRSH=2 THEN DO; E=.21; EJECT2=.22; END;
      ELSE IF CRSH=3 THEN DO; E=.43; EJECT2=.58; END;
      ELSE IF CRSH=4 THEN DO; E=.21; EJECT2=.25; END;
    P=PS208D+PV208D;
    S=OLDWTFA*E*P / (1 - E*P);
    CLS208D=S*PS208D/P;
    CLV208D=S*PV208D/P;
    WEIGHTFA=OLDWTFA+CLS208D+CLV208D; END;
ELSE DO; CLS208D=0; CLV208D=0; END;
CL208D=CLS208D+CLV208D;
CLS=CLS+CLS208D; CLV=CLV+CLV208D; CL=CL+CL208D;
```

```
/* ------------------------------------------------------------------ */
/*   212 ADHESIVE WINDSHIELD BONDING                                   */
/*   MEDIAN INSTALLATION YEAR: EARLY 1966                              */
/* ------------------------------------------------------------------ */
```

We know exactly when each domestic make-model, and some imported make-models were equipped with adhesive windshield bonding. PS212 and PV212 can be set to 0 or 1 (or possibly to .5 in MY 1970 because FMVSS 212 took effect in the middle of the model year). Identification of these make-models is based on the VIN, if available, and on the FARS MAK_MOD code, otherwise. Note that some domestic models continued with rubber gasket installations after 1970 (meeting FMVSS 212, but looser than adhesive bonding); for these models, PS212 = 0 as long as the rubber gaskets continued.

For all other make-models, including most Japanese cars, we will conservatively assume that PS212 = 1 only from 1980, even though there is some evidence from NCSS that the rubber gasket installations of the 1970's were about as tight as adhesive bonding in domestic cars.

```
/* IDENTIFIES CARS WITH ADHESIVE WINDSHIELD BONDING           */
  IF MY GE 1980 THEN DO; PS212=1; PV212=0; GOTO SAVE212; END;
  IF MY LE 1962 THEN DO; PS212=0; PV212=0; GOTO SAVE212; END;
  IF MAKE IN (41,51) THEN DO;
    IF MY GE 1971 THEN DO; PS212=1; PV212=0; END;
      ELSE IF MY=1970 THEN DO; PS212=.5; PV212=.5; END;
      ELSE IF MY LE 1969 THEN DO; PS212=0; PV212=0; END; GOTO SAVE212; END;
  IF 1963 LE MY LE 1969 THEN PS212=0;
    ELSE IF 1971 LE MY LE 1979 THEN PV212=0;
  IF MAKE IN (1,6,7,8,9,12,13,14,18,19,20,21,22,30,32,35,49) THEN GOTO SOMEVOL2;
  PS212=0; PV212=0; GOTO SAVE212;
SOMEVOL2: V13=V1||V2||V3;
  IF V13 IN ('999','000','   ') THEN GOTO NOVIN2;
  IF MY=1963 THEN DO;
    IF MAKE=18 AND V1 IN ('A','B','C','0','1','3') THEN PV212=.5;
      ELSE IF MAKE=18 AND V1='7' THEN PV212=1;
      ELSE IF MAKE=21 AND V3 IN ('0','1') THEN PV212=.5;
      ELSE PV212=0; GOTO SAVE212; END;
  IF MY=1964 THEN DO;
    IF MAKE=18 AND V1 IN ('A','B','C','0','1','3','7') THEN PV212=1;
      ELSE IF MAKE=20 AND V2='5' THEN PV212=1;
      ELSE IF MAKE IN (21,22) AND V2 IN ('0','1','2') THEN PV212=1;
      ELSE PV212=0; GOTO SAVE212; END;
  IF MY=1965 THEN DO;
    IF MAKE IN (18,19,21,22) THEN PV212=1;
      ELSE IF MAKE=20 AND V2 IN ('1','9') THEN PV212=0;
      ELSE IF MAKE=20 THEN PV212=1;
      ELSE IF MAKE=12 AND V3 IN ('5','6','7') THEN PV212=1;
      ELSE IF MAKE=14 AND V3 IN ('4','5','6','7') THEN PV212=1;
      ELSE PV212=0; GOTO SAVE212; END;
  IF MY=1966 THEN DO;
    IF MAKE IN (13,14,18,19,21,22) THEN PV212=1;
      ELSE IF MAKE=20 AND V2 IN ('1','9') THEN PV212=0;
      ELSE IF MAKE=20 THEN PV212=1;
      ELSE IF MAKE=12 AND (V3='8' OR (V3='0' AND V4 IN ('7','8','9'))) THEN PV212=0;
      ELSE IF MAKE=12 THEN PV212=1;
      ELSE PV212=0; GOTO SAVE212; END;
  IF MY=1967 THEN DO;
    IF MAKE IN (13,18,19,21,22) THEN PV212=1;
      ELSE IF MAKE=20 AND V2 IN ('1','9') THEN PV212=0;
      ELSE IF MAKE=20 THEN PV212=1;
      ELSE IF MAKE=12 AND V3='0' THEN PV212=0;
      ELSE IF MAKE=12 THEN PV212=1;
      ELSE IF MAKE=14 AND V3='9' THEN PV212=0;
```

```
              ELSE IF MAKE=14 THEN PV212=1;
              ELSE IF MAKE=1 THEN GOTO NOVIN2;
              ELSE PV212=0; GOTO SAVE212; END;
        IF MY=1968 THEN DO;
           IF MAKE IN (13,18,19,20,21,22) THEN PV212=1;
              ELSE IF MAKE=12 AND V3='0' THEN PV212=0;
              ELSE IF MAKE=12 THEN PV212=1;
              ELSE IF MAKE=14 AND V3='9' THEN PV212=0;
              ELSE IF MAKE=14 THEN PV212=1;
              ELSE IF MAKE=1 AND V4='0' THEN PV212=0;
              ELSE IF MAKE=1 THEN PV212=1;
              ELSE PV212=0; GOTO SAVE212; END;
        IF MY=1969 THEN DO;
           IF MAKE IN (6,8,12,13,14,18,19,20,21,22) THEN PV212=1;
              ELSE IF MAKE=7 AND V1='D' THEN PV212=1;
              ELSE IF MAKE=9 AND V1='P' THEN PV212=1;
              ELSE IF MAKE=1 AND V4='0' THEN PV212=0;
              ELSE IF MAKE=1 THEN PV212=1;
              ELSE PV212=0; GOTO SAVE212; END;
        IF MY=1970 THEN DO;
           IF MAKE IN (1,6,8,13,14,18,19,20,21,22) THEN DO; PV212=.5; PS212=.5; END;
              ELSE IF MAKE=12 AND V3='9' THEN DO; PV212=0; PS212=0; END;
              ELSE IF MAKE=12 THEN DO; PV212=.5; PS212=.5; END;
              ELSE IF MAKE=7 AND V1 IN ('D','J') THEN DO; PV212=.5; PS212=.5; END;
              ELSE IF MAKE=9 AND V1 IN ('B','P') THEN DO; PV212=.5; PS212=.5; END;
              ELSE IF MAKE=30 THEN DO; PV212=0; PS212=.5; END;
              ELSE DO; PV212=0; PS212=0; END; GOTO SAVE212; END;
        IF MY IN (1971,1972,1973) THEN DO;
           IF MAKE IN (1,6,8,13,18,19,20,21,22,30) THEN PS212=1;
              ELSE IF MAKE=12 AND V3 IN ('1','9') THEN PS212=0;
              ELSE IF MAKE=12 THEN PS212=1;
              ELSE IF MAKE=14 AND V3='3' THEN PS212=0;
              ELSE IF MAKE=14 THEN PS212=1;
              ELSE IF MAKE=7 AND V1='L' THEN PS212=0;
              ELSE IF MAKE=7 THEN PS212=1;
              ELSE IF MAKE=9 AND V1='V' THEN PS212=0;
              ELSE IF MAKE=9 THEN PS212=1;
              ELSE PS212=0; GOTO SAVE212; END;
        IF MY IN (1974,1975,1976) THEN DO;
           IF MAKE IN (1,6,8,13,18,19,20,21,22,30) THEN PS212=1;
              ELSE IF MAKE=12 AND V3 IN ('1','9') THEN PS212=0;
              ELSE IF MAKE=12 THEN PS212=1;
              ELSE IF MAKE=14 AND V3 IN ('2','3') THEN PS212=0;
              ELSE IF MAKE=14 THEN PS212=1;
              ELSE IF MAKE=7 AND V1='L' THEN PS212=0;
              ELSE IF MAKE=7 THEN PS212=1;
              ELSE IF MAKE=9 AND V1='V' THEN PS212=0;
              ELSE IF MAKE=9 THEN PS212=1;
              ELSE IF MAKE=35 THEN GOTO NOVIN2;
              ELSE PS212=0;  GOTO SAVE212; END;
        IF MY=1977 THEN DO;
           IF MAKE IN (1,6,7,8,9,13,18,19,20,21,22,30) THEN PS212=1;
              ELSE IF MAKE=12 AND V3 IN ('1','9') THEN PS212=0;
              ELSE IF MAKE=12 THEN PS212=1;
              ELSE IF MAKE=14 AND V3 IN ('2','3') THEN PS212=0;
              ELSE IF MAKE=14 THEN PS212=1;
              ELSE IF MAKE=35 THEN GOTO NOVIN2;
              ELSE PS212=0;  GOTO SAVE212; END;
        IF MY=1978 THEN DO;
           IF MAKE IN (1,6,7,8,9,13,18,19,20,21,22,30,32) THEN PS212=1;
              ELSE IF MAKE=12 AND V3='1' THEN PS212=.5;
              ELSE IF MAKE=12 THEN PS212=1;
              ELSE IF MAKE=14 AND V3='2' THEN PS212=.5;
              ELSE IF MAKE=14 THEN PS212=1;
              ELSE IF MAKE=49 AND V2='A' THEN PS212=1;
```

```
           ELSE PS212=0;   GOTO SAVE212; END;
        IF MY=1979 THEN DO;
          IF MAKE IN (1,6,7,8,9,12,13,14,18,19,20,21,22,30,32) THEN PS212=1;
            ELSE IF MAKE=49 AND V2='A' AND V4 NE '6' THEN PS212=1;
            ELSE PS212=0;   GOTO SAVE212; END;
NOVIN2: IF MY=1963 THEN DO;
          IF MAK_MOD IN (1801,18001,2101,21001) THEN PV212=.5;
            ELSE IF MAK_MOD IN (1805,18005) THEN PV212=1;
            ELSE PV212=0; GOTO SAVE212; END;
        IF MY=1964 THEN DO;
          IF MAK_MOD IN (1801,18001,1805,18005,2001,20001,2101,21001,2201,22001) THEN PV212=1;
            ELSE PV212=0; GOTO SAVE212; END;
        IF MY=1965 THEN DO;
          IF MAKE IN (18,19,21,22) THEN PV212=1;
            ELSE IF MAK_MOD IN (2004,20004,2008,20008) THEN PV212=0;
            ELSE IF MAKE=20 THEN PV212=1;
            ELSE IF MAK_MOD IN (1206,12006,1406,14006) THEN PV212=1;
            ELSE PV212=0; GOTO SAVE212; END;
        IF MY=1966 THEN DO;
          IF MAKE IN (13,14,18,19,21,22) THEN PV212=1;
            ELSE IF MAK_MOD IN (2004,20004,2008,20008) THEN PV212=0;
            ELSE IF MAKE=20 THEN PV212=1;
            ELSE IF MAK_MOD IN (1203,12003,1204,12004) THEN PV212=0;
            ELSE IF MAKE=12 THEN PV212=1;
            ELSE PV212=0; GOTO SAVE212; END;
        IF MY=1967 THEN DO;
          IF MAKE IN (13,18,19,21,22) THEN PV212=1;
            ELSE IF MAK_MOD IN (2004,20004,2008,20008) THEN PV212=0;
            ELSE IF MAKE=20 THEN PV212=1;
            ELSE IF MAK_MOD IN (1203,12003,1404,14004) THEN PV212=0;
            ELSE IF MAKE IN (12,14) THEN PV212=1;
            ELSE IF MAK_MOD IN (101,1001) THEN PV212=0;
            ELSE IF MAKE=1 THEN PV212=1;
            ELSE PV212=0; GOTO SAVE212; END;
        IF MY=1968 THEN DO;
          IF MAKE IN (13,18,19,20,21,22) THEN PV212=1;
            ELSE IF MAK_MOD IN (1203,12003,1404,14004) THEN PV212=0;
            ELSE IF MAKE IN (12,14) THEN PV212=1;
            ELSE IF MAK_MOD IN (101,1001) THEN PV212=0;
            ELSE IF MAKE=1 THEN PV212=1;
            ELSE PV212=0; GOTO SAVE212; END;
        IF MY=1969 THEN DO;
          IF MAKE IN (6,8,12,13,14,18,19,20,21,22) THEN PV212=1;
            ELSE IF MAK_MOD IN (703,7003,704,7004,903,9003,904,9004) THEN PV212=1;
            ELSE IF MAK_MOD IN (101,1001) THEN PV212=0;
            ELSE IF MAKE=1 THEN PV212=1;
            ELSE PV212=0; GOTO SAVE212; END;
        IF MY=1970 THEN DO;
          IF MAKE IN (1,6,8,13,14,18,19,20,21,22) THEN DO; PV212=.5; PS212=.5; END;
            ELSE IF MAK_MOD IN (1208,12008) THEN DO; PV212=0; PS212=0; END;
            ELSE IF MAKE=12 THEN DO; PV212=.5; PS212=.5; END;
            ELSE IF MAK_MOD IN (703,7003,704,7004,705,7005,903,9003,904,9004,905,9005)
               THEN DO; PV212=.5; PS212=.5; END;
            ELSE IF MAKE=30 THEN DO; PV212=0; PS212=.5; END;
            ELSE DO; PV212=0; PS212=0; END; GOTO SAVE212; END;
        IF MY IN (1971,1972,1973) THEN DO;
          IF MAKE IN (1,6,8,13,18,19,20,21,22,30) THEN PS212=1;
            ELSE IF MAK_MOD IN (1208,12008,1209,12009,1408,14008) THEN PS212=0;
            ELSE IF MAKE IN (12,14) THEN PS212=1;
            ELSE IF MAK_MOD IN (701,7001,901,9001) THEN PS212=0;
            ELSE IF MAKE IN (7,9) THEN PS212=1;
            ELSE PS212=0; GOTO SAVE212; END;
        IF MY IN (1974,1975,1976) THEN DO;
          IF MAKE IN (1,6,8,13,18,19,20,21,22,30) THEN PS212=1;
            ELSE IF MAK_MOD IN (1208,12008,1209,12009,1408,14008,1409,14009) THEN PS212=0;
```

```
        ELSE IF MAKE IN (12,14) THEN PS212=1;
        ELSE IF MAK_MOD IN (701,7001,901,9001) THEN PS212=0;
        ELSE IF MAKE IN (7,9) THEN PS212=1;
        ELSE IF MAK_MOD IN (3538,35038,35008) THEN PS212=1;
        ELSE PS212=0; GOTO SAVE212; END;
   IF MY=1977 THEN DO;
      IF MAKE IN (1,6,7,8,9,13,18,19,20,21,22,30) THEN PS212=1;
        ELSE IF MAK_MOD IN (1208,12008,1209,12009,1408,14008,1409,14009) THEN PS212=0;
        ELSE IF MAKE IN (12,14) THEN PS212=1;
        ELSE IF MAK_MOD IN (3538,35038,35008) THEN PS212=1;
        ELSE PS212=0; GOTO SAVE212; END;
   IF MY=1978 THEN DO;
      IF MAKE IN (1,6,7,8,9,13,18,19,20,21,22,30,32) THEN PS212=1;
        ELSE IF MAK_MOD IN (1209,12009,1409,14009) THEN PS212=.5;
        ELSE IF MAKE IN (12,14) THEN PS212=1;
        ELSE IF MAK_MOD IN (4933,49033,49003) THEN PS212=1;
        ELSE PS212=0; GOTO SAVE212; END;
   IF MY=1979 THEN DO;
      IF MAKE IN (1,6,7,8,9,12,13,14,18,19,20,21,22,30,32) THEN PS212=1;
        ELSE IF MAK_MOD IN (4933,49033,49003) THEN PS212=1;
        ELSE PS212=0; GOTO SAVE212; END;
/* REDUCTION IN EJECTION FATALITY RISK WITH ADHESIVE WINDSHIELD BONDING */
SAVE212:  OLDWTFA=WEIGHTFA;
```

Adhesive windshield bonding is potentially beneficial for any ejected front-seat occupant [SEAT2 IN (11,12,13,18) AND EJECT2 NE 0]. In frontal impacts and rollovers of cars without adhesive bonding, 22 percent of ejections were through the windshield portal; in side and rear impacts, 5 percent were through the windshield portal.[3] Adhesive bonding saves 15 percent of the deaths of windshield ejectees.[4] Thus, adhesive bonding saves .22 x .15 = 3.3 percent of all ejection fatalities in frontals and rollovers; .05 x .15 = 0.75 percent of all ejection fatalities in side and rear impacts.

```
   IF (PS212 GT 0 OR PV212 GT 0) AND SEAT2 IN (11,12,13,18) AND EJECT2 NE 0 THEN DO;
      IF CRSH IN (1,3) THEN E=.033;
        ELSE IF CRSH IN (2,4) THEN E=.0075;
      P=PS212+PV212;
      REL_S=EJECT2*E*P / (1 - E*P);
      S=OLDWTFA*REL_S;
      CLS212=S*PS212/P;
      CLV212=S*PV212/P;
      WEIGHTFA=OLDWTFA+CLS212+CLV212;
```

Because adhesive bonding reduces ejection fatalities while leaving non-ejected fatalities unchanged, EJECT2 = (EJECT2 + REL_S)/(1 + REL_S) must be recomputed after this step.

```
      EJECT2=(EJECT2 + REL_S)/(1 + REL_S); END;
   ELSE DO; CLS212=0; CLV212=0; END;
   CL212=CLS212+CLV212;
   CLS=CLS+CLS212; CLV=CLV+CLV212; CL=CL+CL212;


/* -------------------------------------------------------------------- */
/*   208C LAP BELTS FOR BACK-SEAT OUTBOARD OCCUPANTS                     */
/*   MEDIAN INSTALLATION YEAR: LATE 1965                                 */
/* -------------------------------------------------------------------- */

/* IDENTIFIES BACK-OUTBOARD-SEAT OCCUPANTS WEARING LAP BELTS  */
   PS208C=0; PV208C=0;
   IF SEAT2=21 AND REST3=2 AND (5 LE AGE LE 99 OR AGE=.) THEN DO;
      IF MY GE 1969 THEN PS208C=1;
```

[3] Kahane, C.J., *An Evaluation of Windshield Glazing and Installation Methods for Passenger Cars*, NHTSA Technical Report No. DOT HS 806 693, Washington, 1985, p. 167.

[4] *Ibid.*, p. xxvii.

```
      ELSE IF MY LE 1967 THEN PV208C=1;
      ELSE IF MY=1968 THEN DO; PS208C=.5; PV208C=.5; END; END;
/* LIVES SAVED AND EJECTIONS PREVENTED BY BACK-OUTBOARD-SEAT LAP BELTS */
  OLDWTFA=WEIGHTFA;
  IF PS208C GT 0 OR PV208C GT 0 THEN DO;
```

The data from NHTSA's evaluation shows lap belts for rear-outboard passengers reduce fatality risk by 76 percent in rollovers and 39 percent in side impacts, but have little or no net benefit in frontals, rear impacts or other crashes. The same effectiveness estimates are assumed for the rear-center position.

```
  IF CRSH=1 THEN DO; E=.00; EJECT2=.23; END;
     ELSE IF CRSH=2 THEN DO; E=.39; EJECT2=.28; END;
     ELSE IF CRSH=3 THEN DO; E=.76; EJECT2=.61; END;
     ELSE IF CRSH=4 THEN DO; E=.00; EJECT2=.27; END;
  P=PS208C+PV208C;
  S=OLDWTFA*E*P / (1 - E*P);
  CLS208C=S*PS208C/P;
  CLV208C=S*PV208C/P;
  WEIGHTFA=OLDWTFA+CLS208C+CLV208C; END;
 ELSE DO; CLS208C=0; CLV208C=0; END;
 CL208C=CLS208C+CLV208C;
 CLS=CLS+CLS208C; CLV=CLV+CLV208C; CL=CL+CL208C;
```

```
/* ----------------------------------------------------------------- */
/*  206 DOOR LOCK IMPROVEMENTS                                        */
/*  MEDIAN INSTALLATION YEAR: 1965                                    */
/* ----------------------------------------------------------------- */

/* DOOR LOCKS IMPROVED GRADUALLY FROM 1962 TO 1968        */
  IF MY GE 1969 THEN DO; PS206=1; PV206=0; END;
    ELSE IF MY=1968 THEN DO; PS206=.5; PV206=.5; END;
    ELSE IF 1962 LE MY LE 1967 THEN DO; PS206=0; PV206=(MY-1961)/7; END;
    ELSE IF MY LE 1961 THEN DO; PS206=0; PV206=0; END;
/* REDUCTION IN EJECTION ROLLOVER FATALITY RISK WITH IMPROVED DOOR LOCKS */
  OLDWTFA=WEIGHTFA;
```

NHTSA's evaluation found that improved door locks would save 15.38 percent (400 of 2,600 in the baseline year for that study) of ejection fatalities (excluding occupants riding outside the passenger compartment).[5]

```
  IF (PS206 GT 0 OR PV206 GT 0) AND (CRSH=3 OR M_HARM=1)
      AND SEAT2 NE 52 AND EJECT2 NE 0 THEN DO;
    E=.1538;
    P=PS206+PV206;
    REL_S=EJECT2*E*P / (1 - E*P);
    S=OLDWTFA*REL_S;
    CLS206=S*PS206/P;
    CLV206=S*PV206/P;
    WEIGHTFA=OLDWTFA+CLS206+CLV206;
    EJECT2=(EJECT2 + REL_S)/(1 + REL_S); END;
  ELSE DO; CLS206=0; CLV206=0; END;
  CL206=CLS206+CLV206;
  CLS=CLS+CLS206; CLV=CLV+CLV206; CL=CL+CL206;


/* ----------------------------------------------------------------- */
/*  208B LAP BELT USE BY CHILDREN AGE 1-4                             */
/*  MEDIAN INSTALLATION YEAR FOR LAP BELTS USED BY CHILDREN: 1964     */
/* ----------------------------------------------------------------- */

/* IDENTIFIES CHILD PASSENGERS AGE 1-4 USING LAP BELTS */
  PS208B=0; PV208B=0;
  IF 1 LE AGE LE 4 AND REST3=2 THEN DO;
    IF MY GE 1969 THEN PS208B=1;
      ELSE IF MY LE 1967 THEN PV208B=1;
      ELSE IF MY=1968 THEN DO; PS208B=.5; PV208B=.5; END; END;
/* LIVES SAVED AND EJECTIONS PREVENTED BY LAP BELTS (AGE 1-4) */
  OLDWTFA=WEIGHTFA;
  IF PS208B GT 0 OR PV208B GT 0 THEN DO;
```

NHTSA's evaluation estimates that lap belts reduce fatality risk by 33 percent for toddlers. Effectiveness is not estimated separately by crash mode.[6] In general, 19 percent of unrestrained child passenger fatalities in frontal crashes were ejected, 26 percent in side impacts, 65 percent in rollovers and 34 percent in rear/other impacts.

```
    E=.33;
    IF CRSH=1 THEN EJECT2=.19;
      ELSE IF CRSH=2 THEN EJECT2=.26;
```

---

[5] Kahane, C.J., *An Evaluation of Door Locks and Roof Crush Resistance of Passenger Cars – Federal Motor Vehicle Safety Standards 206 and 216*, NHTSA Technical Report No. DOT HS 807 489, Washington, 1989, pp. 206-209 and 222-225.
[6] This evaluation is discussed in the "Child Safety Seats" section of the FMVSS 213 chapter of Part 1, not the FMVSS 208 chapter.

```
     ELSE IF CRSH=3 THEN EJECT2=.65;
     ELSE IF CRSH=4 THEN EJECT2=.34;
  P=PS208B+PV208B;
  S=OLDWTFA*E*P / (1 - E*P);
  CLS208B=S*PS208B/P;
  CLV208B=S*PV208B/P;
   WEIGHTFA=OLDWTFA+CLS208B+CLV208B; END;
 ELSE DO; CLS208B=0; CLV208B=0; END;
 CL208B=CLS208B+CLV208B;
 CLS=CLS+CLS208B; CLV=CLV+CLV208B; CL=CL+CL208B;

/* -------------------------------------------------------------- */
/*  208A LAP BELTS FOR FRONT-SEAT OUTBOARD OCCUPANTS              */
/*  MEDIAN INSTALLATION YEAR: 1962                               */
/* -------------------------------------------------------------- */
```

A double-pair comparison analysis of lap belt effectiveness for front-outboard occupants, conducted in support of this study, found that lap belts reduce fatality risk by 22 percent in frontals, 43 percent in rollovers, and 21 percent in side impacts, rear impacts and other crashes.

```
/* IDENTIFIES FRONT-OUTBOARD-SEAT OCCUPANTS WEARING LAP BELTS  */
  PS208A=0; PV208A=0;
  IF SEAT2 IN (11,13) AND REST3=2 AND (5 LE AGE LE 99 OR AGE=.) THEN DO;
     IF MY GE 1969 THEN PS208A=1;
        ELSE IF MY LE 1967 THEN PV208A=1;
        ELSE IF MY=1968 THEN DO; PS208A=.5; PV208A=.5; END; END;
/* LIVES SAVED AND EJECTIONS PREVENTED BY FRONT-OUTBOARD-SEAT LAP BELTS */
  OLDWTFA=WEIGHTFA;
  IF PS208A GT 0 OR PV208A GT 0 THEN DO;
    IF CRSH=1 THEN DO; E=.22; EJECT2=.21; END;
       ELSE IF CRSH=2 THEN DO; E=.21; EJECT2=.23; END;
       ELSE IF CRSH=3 THEN DO; E=.43; EJECT2=.66; END;
       ELSE IF CRSH=4 THEN DO; E=.21; EJECT2=.35; END;
    P=PS208A+PV208A;
    S=OLDWTFA*E*P / (1 - E*P);
    CLS208A=S*PS208A/P;
    CLV208A=S*PV208A/P;
     WEIGHTFA=OLDWTFA+CLS208A+CLV208A; END;
   ELSE DO; CLS208A=0; CLV208A=0; END;
  CL208A=CLS208A+CLV208A;
  CLS=CLS+CLS208A; CLV=CLV+CLV208A; CL=CL+CL208A;
  DROP V13;
RUN;
```

That concludes the model for passenger cars. It has estimated, on a case-by-case basis, by how much fatalities would increase if all the safety technologies were "removed" from vehicles. The next step is to tally up the lives saved over all the cases, by calendar year.

```
PROC MEANS SUM NOPRINT DATA=CAR2; BY CY;
  VAR ORIGWT WEIGHTFA
     CLV208J CLS208J CL208J
     CLV108 CLS108 CL108
     CLV208I CLS208I CL208I
     CLV214B CLS214B CL214B
     CLV208H CLS208H CL208H
     CLV208G CLS208G CL208G
     CLV213 CLS213 CL213
     CLVNCAP CLSNCAP CLNCAP
     CLV208F CLS208F CL208F
     CLV216 CLS216 CL216
     CLV214A CLS214A CL214A
     CLV105B CLS105B CL105B
     CLV201 CLS201 CL201
     CLV203 CLS203 CL203
     CLV208E CLS208E CL208E
     CLV105A CLS105A CL105A
     CLV208D CLS208D CL208D
     CLV212 CLS212 CL212
     CLV208C CLS208C CL208C
     CLV206 CLS206 CL206
     CLV208B CLS208B CL208B
     CLV208A CLS208A CL208A
     CABDRV CABRF CABKID
     CLV CLS CL;
  OUTPUT OUT=CAR3
```

C_ORIGWT = actual car occupant fatalities. C_WTFA = estimated number of fatalities if all safety technologies had been removed.

```
     SUM=C_ORIGWT C_WTFA
     CLV208J CLS208J CL208J
     CLV108 CLS108 CL108
     CLV208I CLS208I CL208I
     CLV214B CLS214B CL214B
     CLV208H CLS208H CL208H
     CLV208G CLS208G CL208G
     CLV213 CLS213 CL213
     CLVNCAP CLSNCAP CLNCAP
     CLV208F CLS208F CL208F
     CLV216 CLS216 CL216
     CLV214A CLS214A CL214A
     CLV105B CLS105B CL105B
     CLV201 CLS201 CL201
     CLV203 CLS203 CL203
     CLV208E CLS208E CL208E
     CLV105A CLS105A CL105A
     CLV208D CLS208D CL208D
     CLV212 CLS212 CL212
     CLV208C CLS208C CL208C
     CLV206 CLS206 CL206
     CLV208B CLS208B CL208B
     CLV208A CLS208A CL208A
     CABDRV CABRF CABKID
     CLV CLS CL;
```

Prints the number of lives saved by each technology in each calendar year, and the sum of lives saved in 1975-2002.

```
PROC PRINT DATA=CAR3;
  FORMAT CLV208J CLS208J CL208J CLV108 CLS108 CL108 CLV208I CLS208I CL208I 9.0;
  ID CY; VAR CLV208J CLS208J CL208J CLV108 CLS108 CL108 CLV208I CLS208I CL208I;
  SUM CLV208J CLS208J CL208J CLV108 CLS108 CL108 CLV208I CLS208I CL208I;
```

```
TITLE1 'CAR OCCUPANT LIVES SAVED BY BACK-CENTER 3-POINT BELTS, TRAILER CONSPICUITY TAPE, AND AIR
BAGS, 1975-2002';
TITLE2 ' ';
TITLE3 '...208J = 3-POINT BELTS FOR BACK-CENTER OCCUPANTS (MEDIAN INSTALLATION YEAR 2001, FMVSS
PHASE-IN TO BEGIN 9/2005)';
TITLE4 '...108 = TRAILER CONSPICUITY TAPE (ON-ROAD FLEET 50% EQUIPPED 1996, FMVSS 12/1/93,
RETROFIT 6/1/2001)';
TITLE5 '...208I = FRONTAL AIR BAGS (MEDIAN INSTALLATION YEAR 1994, FMVSS PHASE-IN BEGAN 9/1/86)';
TITLE6 '...V = VOLUNTARY INSTALLATIONS, BEFORE FMVSS EFFECTIVE DATE';
TITLE7 '...S = STANDARD INSTALLATIONS, CARS BUILT ON OR AFTER FMVSS EFFECTIVE DATE';

PROC PRINT DATA=CAR3;
   FORMAT CLV214B CL214B CLV208H CLS208H CL208H CLV208G CLS208G CL208G 9.0;
   ID CY; VAR CLV214B CL214B CLV208H CLS208H CL208H CLV208G CLS208G CL208G;
   SUM CLV214B CL214B CLV208H CLS208H CL208H CLV208G CLS208G CL208G;
TITLE1 'CAR OCCUPANT LIVES SAVED BY VOLUNTARY TTI(d) REDUCTIONS, AUTOMATIC 2-POINT BELTS, AND BACK-
OUTBOARD 3-POINT BELTS, 1975-2002';
TITLE2 ' ';
TITLE3 '...214B = VOLUNTARY TTI(d) REDUCTIONS IN 2-DOOR CARS (MEDIAN IMPLEMENTATION YEAR 1993)';
TITLE4 '...208H = AUTOMATIC 2-POINT BELTS (PEAK INSTALLATION YEAR 1991, FMVSS PHASE-IN BEGAN
9/1/86)';
TITLE5 '...208G = 3-POINT BELTS FOR BACK-OUTBOARD OCCUPANTS (MEDIAN INSTALLATION YEAR 1989, FMVSS
EFFECTIVE 12/11/89)';
TITLE6 '...V = VOLUNTARY INSTALLATIONS, BEFORE FMVSS EFFECTIVE DATE';
TITLE7 '...S = STANDARD INSTALLATIONS, CARS BUILT ON OR AFTER FMVSS EFFECTIVE DATE';

PROC PRINT DATA=CAR3;
   FORMAT CLV213 CLS213 CL213 CLVNCAP CLSNCAP CLNCAP CLV208F CLS208F CL208F 9.0;
   ID CY; VAR CLV213 CLS213 CL213 CLVNCAP CLNCAP CLV208F CLS208F CL208F;
   SUM CLV213 CLS213 CL213 CLVNCAP CLNCAP CLV208F CLS208F CL208F;
TITLE1 'CAR OCCUPANT LIVES SAVED BY CHILD SAFETY SEATS, VOLUNTARY NCAP IMPROVEMENTS, AND FRONT-SEAT
3-POINT BELTS, 1975-2002';
TITLE2 ' ';
TITLE3 '...213 = CHILD SAFETY SEATS (USE EXCEEDED 50% IN 1985, FMVSS EFFECTIVE 4/1/71, STATE LAWS
1978-85)';
TITLE4 '...NCAP = VOLUNTARY NCAP IMPROVEMENTS IN NON-AIR-BAG CARS (MEDIAN IMPLEMENTATION YEAR EARLY
1984)';
TITLE5 '...208F = 3-POINT BELTS FOR FRONT-SEAT OCCUPANTS (MEDIAN INSTALLATION YEAR 1974, FMVSS
EFFECTIVE 9/1/73)';
TITLE6 '...V = VOLUNTARY INSTALLATIONS, BEFORE FMVSS EFFECTIVE DATE';
TITLE7 '...S = STANDARD INSTALLATIONS, CARS BUILT ON OR AFTER FMVSS EFFECTIVE DATE';

PROC PRINT DATA=CAR3;
   FORMAT CLV216 CLS216 CL216 CLV214A CLS214A CL214A CLV105B CLS105B CL105B 9.0;
   ID CY; VAR CLV216 CLS216 CL216 CLV214A CLS214A CL214A CLV105B CLS105B CL105B;
   SUM CLV216 CLS216 CL216 CLV214A CLS214A CL214A CLV105B CLS105B CL105B;
TITLE1 'CAR OCCUPANT LIVES SAVED BY ROOF CRUSH STRENGTH, SIDE DOOR BEAMS, AND FRONT DISC BRAKES,
1975-2002';
TITLE2 ' ';
TITLE3 '...216 = ROOF CRUSH STRENGTH, B-PILLARS, ETC. (MEDIAN INSTALLATION YEAR 1973, FMVSS
EFFECTIVE 9/1/73)';
TITLE4 '...214A = SIDE DOOR BEAMS (MEDIAN INSTALLATION YEAR EARLY 1973, FMVSS EFFECTIVE 1/1/73)';
TITLE5 '...105B = FRONT DISC BRAKES (MEDIAN INSTALLATION YEAR 1971, FMVSS EFFECTIVE 1/1/76)';
TITLE6 '...V = VOLUNTARY INSTALLATIONS, BEFORE FMVSS EFFECTIVE DATE';
TITLE7 '...S = STANDARD INSTALLATIONS, CARS BUILT ON OR AFTER FMVSS EFFECTIVE DATE';

PROC PRINT DATA=CAR3;
   FORMAT CLV201 CL201 CLV203 CLS203 CL203 CLV208E CLS208E CL208E 9.0;
   ID CY; VAR CLV201 CL201 CLV203 CLS203 CL203 CLV208E CLS208E CL208E;
   SUM CLV201 CL201 CLV203 CLS203 CL203 CLV208E CLS208E CL208E;
TITLE1 'CAR OCCUPANT LIVES SAVED BY SAFER INSTRUMENT PANELS, EA STEERING ASSEMBLIES, AND BACK-
CENTER LAP BELTS, 1975-2002';
TITLE2 ' ';
TITLE3 '...201 = VOLUNTARY INSTRUMENT PANEL IMPROVEMENTS (MEDIAN IMPLEMENTATION YEAR 1968, FMVSS
EFFECTIVE 1/1/68)';
```

329

```
TITLE4 '...203  = ENERGY-ABSORBING STEERING ASSEMBLIES (MEDIAN IMPLEMENTATION YEAR 1967, FMVSS
EFFECTIVE 1/1/68)';
TITLE5 '...208E = LAP BELTS FOR BACK-CENTER OCCUPANTS (MEDIAN INSTALLATION YEAR EARLY 1967, FMVSS
EFFECTIVE 1/1/68)';
TITLE6 '...V = VOLUNTARY INSTALLATIONS, BEFORE FMVSS EFFECTIVE DATE';
TITLE7 '...S = STANDARD INSTALLATIONS, CARS BUILT ON OR AFTER FMVSS EFFECTIVE DATE';

PROC PRINT DATA=CAR3;
   FORMAT CLV105A CLS105A CL105A CLV208D CLS208D CL208D CLV212 CLS212 CL212 9.0;
   ID CY; VAR CLV105A CLS105A CL105A CLV208D CLS208D CL208D CLV212 CLS212 CL212;
   SUM CLV105A CLS105A CL105A CLV208D CLS208D CL208D CLV212 CLS212 CL212;
TITLE1 'CAR OCCUPANT LIVES SAVED BY DUAL MASTER CYLINDERS, FRONT-CENTER LAP BELTS, AND ADHESIVE
WINDSHIELD BONDING, 1975-2002';
TITLE2 ' ';
TITLE3 '...105A = DUAL MASTER CYLINDERS (MEDIAN INSTALLATION YEAR LATE 1966, FMVSS EFFECTIVE
1/1/68)';
TITLE4 '...208D = LAP BELTS FOR FRONT-CENTER OCCUPANTS (MEDIAN INSTALLATION YEAR 1966, FMVSS
EFFECTIVE 1/1/68)';
TITLE5 '...212  = ADHESIVE WINDSHIELD BONDING (MEDIAN INSTALLATION YEAR EARLY 1966, FMVSS EFFECTIVE
1/1/70)';
TITLE6 '...V = VOLUNTARY INSTALLATIONS, BEFORE FMVSS EFFECTIVE DATE';
TITLE7 '...S = STANDARD INSTALLATIONS, CARS BUILT ON OR AFTER FMVSS EFFECTIVE DATE';

PROC PRINT DATA=CAR3;
   FORMAT CLV208C CLS208C CL208C CLV206 CLS206 CL206 CLV208B CLS208B CL208B 9.0;
   ID CY; VAR CLV208C CLS208C CL208C CLV206 CLS206 CL206 CLV208B CLS208B CL208B;
   SUM CLV208C CLS208C CL208C CLV206 CLS206 CL206 CLV208B CLS208B CL208B;
TITLE1 'CAR OCCUPANT LIVES SAVED BY BACK-OUTBOARD LAP BELTS, IMPROVED DOOR LOCKS, AND LAP BELTS FOR
CHILDREN AGE 1-4, 1975-2002';
TITLE2 ' ';
TITLE3 '...208C = LAP BELTS FOR BACK-OUTBOARD OCCUPANTS (MEDIAN INSTALLATION YEAR LATE 1965, FMVSS
EFFECTIVE 1/1/68)';
TITLE4 '...206  = IMPROVED DOOR LOCKS (MEDIAN INSTALLATION YEAR 1965, FMVSS EFFECTIVE 1/1/68)';
TITLE5 '...208B = LAP BELT USE BY CHILDREN AGE 1-4 (MEDIAN INSTALLATION YEAR 1964, FMVSS EFFECTIVE
1/1/68)';
TITLE6 '...V = VOLUNTARY INSTALLATIONS, BEFORE FMVSS EFFECTIVE DATE';
TITLE7 '...S = STANDARD INSTALLATIONS, CARS BUILT ON OR AFTER FMVSS EFFECTIVE DATE';

PROC PRINT DATA=CAR3;
   FORMAT CLV208A CLS208A CL208A 9.0;
   ID CY; VAR CLV208A CLS208A CL208A;
   SUM CLV208A CLS208A CL208A;
TITLE1 'CAR OCCUPANT LIVES SAVED BY FRONT-OUTBOARD LAP BELTS, 1975-2002';
TITLE2 ' ';
TITLE3 '...208A = LAP BELTS FOR FRONT-OUTBOARD OCCUPANTS (MEDIAN INSTALLATION YEAR 1963, FMVSS
EFFECTIVE 1/1/68)';
TITLE4 '...V = VOLUNTARY INSTALLATIONS, BEFORE FMVSS EFFECTIVE DATE';
TITLE5 '...S = STANDARD INSTALLATIONS, CARS BUILT ON OR AFTER FMVSS EFFECTIVE DATE';
```

Summary results for cars. CL = total lives saved. PCT_SAVE = percentage of potential fatalities that were saved by the technologies.

```
DATA CAR4; SET CAR3;
PCT_SAVE=100*CL/C_WTFA;
PROC PRINT DATA=CAR4;
   FORMAT C_ORIGWT C_WTFA CLV CLS CL 9.0 PCT_SAVE 6.2;
   ID CY; VAR C_ORIGWT C_WTFA CLV CLS CL PCT_SAVE;
   SUM C_ORIGWT C_WTFA CLV CLS CL;
TITLE1 'OVERALL CAR OCCUPANT LIVES SAVED BY VEHICLE SAFETY STANDARDS AND TECHNOLOGIES, 1975-2002';
TITLE2 ' ';
TITLE3 'C_ORIGWT = ACTUAL CAR OCCUPANT FATALITIES';
TITLE4 'C_WTFA = FATALITIES THAT WOULD HAVE OCCURRED WITHOUT ANY VEHICLE SAFETY IMPROVEMENTS';
TITLE5 'CLV = OVERALL LIVES SAVED BY VOLUNTARY IMPROVEMENTS, BEFORE FMVSS EFFECTIVE DATE';
TITLE6 'CLS = OVERALL LIVES SAVED IN CARS BUILT ON OR AFTER FMVSS EFFECTIVE DATE';
```

330

```
TITLE7 'CL = OVERALL LIVES SAVED BY VEHICLE SAFETY IMPROVEMENTS (CLV + CLS)';
TITLE8 'PCT_SAVE = PERCENT OF WOULD-HAVE-BEEN FATALITIES SAVED BY SAFETY STANDARDS AND
TECHNOLOGIES';
RUN;


/* ------------------------------------------------------------------ */
/* ------------------------------------------------------------------ */
/*  LIVES SAVED BY FMVSS IN LIGHT TRUCKS                              */
/* ------------------------------------------------------------------ */
/* ------------------------------------------------------------------ */
```

The main model for LTVs begins here. The basic approach is the same as for cars, but the implementation dates and effectiveness of the various technologies may be different. In general, the notes on this section will be limited to the spots where LTVs differ from cars.

```
DATA TRK2; SET TRK1;
IF EJECTION IN (1,2) THEN EJECT2=1; ELSE EJECT2=0;


/* ------------------------------------------------------------------ */
/*  208J 3-POINT BELTS FOR BACK-SEAT CENTER OCCUPANTS (ANTON'S LAW)  */
/*  MEDIAN INSTALLATION YEAR: > 2001                                 */
/* ------------------------------------------------------------------ */

/* IDENTIFIES BACK-SEAT CENTER OCCUPANTS WEARING 3-POINT BELTS */
  PS208J=0; PV208J=0;
  IF SEAT2=22 AND REST3=3 THEN DO; IF MY LE 2008 THEN PV208J=1; ELSE PS208J=1; END;
/* LIVES SAVED AND EJECTIONS PREVENTED BY BACK-SEAT CENTER 3-POINT BELTS */
  OLDWTFA=WEIGHTFA;
  IF PS208J GT 0 OR PV208J GT 0 THEN DO;
```

In general, belts are more effective in LTVs than in cars; also, a higher percentage of unrestrained fatalities were ejected. The data in NHTSA's evaluation of 3-point belts in the rear-outboard seats indicate fatality reduction of 50 percent in frontals, 84 percent in rollovers and 70 percent in side, rear and other impacts. We will assume the same effectiveness for 3-point belts in the rear-center seat. In frontal crashes, 49 percent of unrestrained rear-center occupant fatalities were ejected, 59 percent in side impacts, 85 percent in rollovers and 58 percent in rear/other impacts.

TLV208J: "T" = car, "L" = lives saved, "V" = voluntary, "208J" = rear-center 3-point belt

```
    IF CRSH=1 THEN DO; E=.50; EJECT2=.49; END;
     ELSE IF CRSH=2 THEN DO; E=.70; EJECT2=.59; END;
     ELSE IF CRSH=3 THEN DO; E=.84; EJECT2=.85; END;
     ELSE IF CRSH=4 THEN DO; E=.70; EJECT2=.58; END;
  P=PS208J+PV208J;
  S=OLDWTFA*E*P / (1 - E*P);
  TLS208J=S*PS208J/P;
  TLV208J=S*PV208J/P;
   WEIGHTFA=OLDWTFA+TLS208J+TLV208J; END;
 ELSE DO; TLS208J=0; TLV208J=0; END;
 TL208J=TLS208J+TLV208J;
 TLS=TLS208J; TLV=TLV208J; TL=TL208J;


/* ------------------------------------------------------------------ */
/*  108 TRAILER CONSPICUITY TAPE - LIVES SAVED, LTRKS NOT HITTING THEM */
/*  50% OF TRAILERS ON THE ROAD HAD THEM IN: 1996                     */
/*  FMCSA RETROFIT REQUIREMENT ISSUED 3/31/1999, TOOK EFFECT 6/1/2001 */
/* ------------------------------------------------------------------ */

/* IDENTIFIES SIDE/REAR IMPACTS INTO TRAILERS IN THE DARK AND % OF TRAILERS WITH TAPE */
  PS108=0; PV108=0;
```

```
  IF CY GE 1991 AND (TAPECASE=1 OR (OVTYP=3 AND
    ((1991 LE CY LE 1994 AND OVCONFIG=5) OR (CY GE 1995 AND OVCONFIG=6))
    AND 2 LE LGT_COND LE 5 AND
      (OIMPACT2 IN (14,16) AND MAN_COLL IN (1,4,5) AND IMPACT2 IN (11,12,1))))
    THEN DO;
      IF CY=1991 THEN DO; PS108=0; PV108=.09; END;
        ELSE IF CY=1992 THEN DO; PS108=0; PV108=.18; END;
        ELSE IF CY=1993 THEN DO; PS108=0; PV108=.27; END;
        ELSE IF CY=1994 THEN DO; PS108=.09; PV108=.27; END;
        ELSE IF CY=1995 THEN DO; PS108=.18; PV108=.27; END;
        ELSE IF CY=1996 THEN DO; PS108=.27; PV108=.27; END;
        ELSE IF CY=1997 THEN DO; PS108=.36; PV108=.27; END;
        ELSE IF CY=1998 THEN DO; PS108=.45; PV108=.27; END;
        ELSE IF CY=1999 THEN DO; PS108=.54; PV108=.27; END;
        ELSE IF CY=2000 THEN DO; PS108=.63; PV108=.27; END;
        ELSE IF CY=2001 THEN DO; PS108=.8775; PV108=.1125; END;
        ELSE IF CY GE 2002 THEN DO; PS108=1; PV108=0; END; END;
/* LIGHT TRUCK OCCUPANT LIVES SAVED BY CONSPICUITY TAPE ON TRUCK TRAILERS */
  OLDWTFA=WEIGHTFA;
  IF PS108 GT 0 OR PV108 GT 0 THEN DO;
    E=.29;
    P=PS108+PV108;
    S=OLDWTFA*E*P / (1 - E*P);
    TLS108=S*PS108/P;
    TLV108=S*PV108/P;
    WEIGHTFA=OLDWTFA+TLS108+TLV108; END;
  ELSE DO; TLS108=0; TLV108=0; END;
  TL108=TLS108+TLV108;
  TLS=TLS+TLS108; TLV=TLV+TLV108; TL=TL+TL108;

/* ------------------------------------------------------------------ */
/*  2081 FRONTAL AIR BAGS FOR DRIVERS AND RF PASSENGERS AGE 13+       */
/*  ASSUMES 18 PERCENT OF ON-OFF SWITCHES ARE TURNED OFF FOR RF 13+   */
/*  MEDIAN INSTALLATION YEAR: 1995                                    */
/* ------------------------------------------------------------------ */

/* IDENTIFIES LIGHT TRUCKS WITH AIR BAGS   */
  PS2081=0; PV2081=0; UDAB=0; UPAB=0;
```

"PASSIVE," the variable that indicates the types of occupant protection at the front-outboard seats, is only defined if FARS has at least a partial VIN to decode. However, even when there is no VIN at all, we can often infer from the body type, the model year and the make if the LTV had driver and/or passenger air bags. All SUVs and minivans from MY 1998 onwards have had dual air bags; all compact pickups have had driver air bags from MY 1998 onwards. Some makes had driver air bags (UDAB = 1) or passenger air bags (UPAB = 1) on all LTVs of a certain body type before 1998. Because the FARS body types "large van" and "standard pickup" include vehicles with GVWR over 8,500 pounds, they may be without air bags even after 1998.

```
/* LTVs WITH UNKNOWN "PASSIVE" (BAD VINs) CLASSIFIED BY MAKE, BODY_TYP AND MODEL YEAR */

  IF CY GE 1991 AND MY GE 1992 AND (PASSIVE=. OR PASSIVE=9999) AND SEAT2 IN (11,13) THEN DO;
    IF SEAT2=11 THEN DO;
      IF (MAKE IN (6,9) AND MY GE 1992)
      OR (MAKE IN (7,12,49) AND BODY_TYP=20 AND MY GE 1992)
      OR (MAKE=41 AND BODY_TYP=20 AND MY GE 1993)
      OR (MAKE IN (14,21,22) AND MY GE 1994)
      OR (MAKE=7 AND BODY_TYP=30 AND MY GE 1994)
      OR (MAKE=12 AND BODY_TYP IN (15,16) AND MY GE 1994)
      OR (MAKE IN (20,23,35) AND BODY_TYP=20 AND MY GE 1994)
      OR (MAKE=49 AND BODY_TYP=31 AND MY GE 1994)
      OR (MAKE=30 AND MY GE 1995)
      OR (MAKE=7 AND BODY_TYP=21 AND MY GE 1995)
```

```
        OR (MAKE IN (12,20,23) AND BODY_TYP=30 AND MY GE 1995)
        OR (MAKE IN (12,23) AND BODY_TYP=14 AND MY GE 1995)
        OR (MAKE IN (20,23,49) AND BODY_TYP IN (15,16) AND MY GE 1995)
        OR (MAKE=12 AND BODY_TYP=21 AND MY GE 1995)
        OR (MAKE IN (2,37,53,54,59,62) AND MY GE 1996)
        OR (MAKE IN (20,38,49) AND BODY_TYP=14 AND MY GE 1996)
        OR (MAKE IN (35,38,41,49) AND BODY_TYP=30 AND MY GE 1996)
        OR (MAKE=38 AND BODY_TYP=20 AND MY GE 1996)
        OR (MAKE IN (52,58,63) AND MY GE 1997)
        OR (MAKE=35 AND BODY_TYP=14 AND MY GE 1997)
        OR (MAKE IN (13,18,19,34,42,48,55) AND MY GE 1998)
        OR (MAKE=7 AND BODY_TYP IN (15,16) AND MY GE 1998)
        OR (MAKE=41 AND BODY_TYP=14 AND MY GE 1998)
        OR (MAKE=12 AND BODY_TYP=31 AND MY GE 1999)
        OR (MAKE IN (20,23) AND BODY_TYP=31 AND MY GE 2001)
      THEN UDAB=1; END;

      ELSE IF SEAT2=13 THEN DO;
        IF (MAKE IN (6,9) AND MY GE 1994)
        OR (MAKE IN (7,49) AND BODY_TYP=20 AND MY GE 1994)
        OR (MAKE=12 AND BODY_TYP=14 AND MY GE 1995)
        OR (MAKE=49 AND BODY_TYP IN (15,16) AND MY GE 1995)
        OR (MAKE IN (14,37,53,54,59,62) AND MY GE 1996)
        OR (MAKE IN (23,35,38,41) AND BODY_TYP=20 AND MY GE 1996)
        OR (MAKE IN (38,49) AND BODY_TYP=14 AND MY GE 1996)
        OR (MAKE IN (2,22,30,52,58) AND MY GE 1997)
        OR (MAKE IN (12,20,23) AND BODY_TYP IN (15,16) AND MY GE 1997)
        OR (MAKE=12 AND BODY_TYP=21 AND MY GE 1997)
        OR (MAKE=20 AND BODY_TYP=20 AND MY GE 1997)
        OR (MAKE=35 AND BODY_TYP=14 AND MY GE 1997)
        OR (MAKE IN (13,18,19,21,34,42,48,55,63) AND MY GE 1998)
        OR (MAKE=7 AND BODY_TYP IN (15,16) AND MY GE 1998)
        OR (MAKE=7 AND BODY_TYP=21 AND MY GE 1998)
        OR (MAKE=12 AND BODY_TYP=20 AND MY GE 1998)
        OR (MAKE IN (20,23,41) AND BODY_TYP=14 AND MY GE 1998)
      THEN UPAB=1; END; END;
```

These are the codes of PASSIVE, based on VIN analysis, that indicate a car is equipped with
frontal air bags and the driver or right-front seats, respectively. PS208I includes any installation
from MY 1995 onwards, because the phase-in of automatic occupant protection began on
September 1, 1994.

```
IF CY GE 1990 AND MY GE 1991 AND
   (SEAT2=11 AND (PASSIVE IN (1,2,3,1090) OR UDAB=1))
   OR (SEAT2=13 AND (PASSIVE IN (2,3) OR UPAB=2)) THEN DO;
     IF 1991 LE MY LE 1994 THEN PV208I=1;
       ELSE IF MY GE 1995 THEN PS208I=1; END;

/* ------------------------------------------------------------------ */
/*  SPLITS CASES INTO SCENARIOS DEPENDING ON AIR BAG TYPE, SEAT POS,  */
/*  ON-OFF SWITCH PRESENCE, OCCUPANT AGE, RESTRAINT USE               */
/*  COMPUTES LIVES SAVED BY AIR BAGS IN EACH SCENARIO                 */
/* ------------------------------------------------------------------ */
```

For LTV occupants, there can be eight different effectiveness scenarios – belted drivers, unbelted
drivers, child passengers in LTVs (primarily pickup trucks) with on-off switches, child passengers
in LTVs (primarily SUVs and vans) without on-off switches, belted adult (age 13+) passengers
with switches, belted adult passengers without switches, unbelted adult passengers with switches
and unbelted adult passengers without switches – plus the default scenario, no air bag. Within each
scenario, there are several effectiveness estimates.

```
IF PS208I=0 AND PV208I=0 THEN GOTO TNOBAG;              /* NO AIR BAG */
   ELSE IF SEAT2=11 AND REST3 GT 0 THEN GOTO TDRVBELT;  /* BELTED DRIVER */
   ELSE IF SEAT2=11 THEN GOTO TDRVUNR;                  /* UNRESTRAINED DRIVER */
   ELSE IF PASSIVE=3 AND 0 LE AGE LE 12 THEN GOTO TKIDSW; /* CHILD RF WITH SWITCH */
   ELSE IF 0 LE AGE LE 12 THEN GOTO TKID;               /* CHILD RF NO SWITCH */
   ELSE IF PASSIVE=3 AND REST3 GT 0 THEN GOTO TRFBELSW; /* BELTED ADULT RF W SWITCH */
   ELSE IF PASSIVE=3 THEN GOTO TRFUNRSW;                /* UNBELTED ADULT RF W SWITCH */
   ELSE IF REST3 GT 0 THEN GOTO TRFBELT;                /* BELTED ADULT RF NO SWITCH */
   ELSE GOTO TRFUNR;                                    /* UNBELTED ADULT RF NO SWITCH */
```

In addition to the basic statistics, WEIGHTFA, TLS208I, TLV208I and TL208I, the model compiles more detailed information on air bags:

- TABDRV = LTV driver lives saved by air bags
- TABRF = adult (age 13+) passenger lives saved by air bags in LTVs without on-off switches
- TABKID = effect of air bags on child (age 0-12) passengers in LTVs without switches; if this is a fatality increase, it will show up as a negative number.
- SW0 = actual effect of air bags on adult passengers in pickup trucks with switches: number of lives saved by air bags because the switch was left "on" for the adult
- SW1 = potential number of adult passenger lives saved in these pickup trucks if the switch had always been "on" for adults (or if the trucks had passenger air bags but no switches)
- SW2 = SW0 – SW1 = potential savings not realized because the switches were "off" for some adults. This will be a negative number.
- SWKID0 = actual (harmful) effect of air bags on child passengers in pickup trucks with switches that were left "on" for the child. This will be a negative number, because a fatality increase is a negative number of lives saved.
- SWKID1 = potential (harmful) effect of air bags on child passengers in pickup trucks if those trucks had been equipped with dual air bags and no switches. This will be a more negative number than SWKID0.
- SWKID2 = SWKID0 – SWKID1 = child passengers avoiding potentially fatal injury from air bags because the switches were available and appropriately turned "off" for the children. This will be a positive number.

```
TNOBAG:    TLS208I=0; TLV208I=0; TL208I=0;
   TABDRV=0; TABRF=0; TABKID=0; SW0=0; SW1=0; SW2=0;
   SWKID0=0; SWKID1=0; SWKID2=0; GOTO T214;
```

NHTSA's evaluation found that air bags have about the same effect in LTVs as in cars.

```
TDRVBELT:  OLDWTFA=WEIGHTFA;
   IF 1 LE M_HARM LE 6 THEN E=0;
      ELSE IF IMPACT2=12 THEN E=.25258;
      ELSE IF IMPACT2 IN (1,11) THEN E=.13239;
      ELSE IF IMPACT2 IN (2,10) THEN E=.05052; ELSE E=0;
   P=PS208I+PV208I;
   S=OLDWTFA*E*P / (1 - E*P);
   TLS208I=S*PS208I/P;
   TLV208I=S*PV208I/P;
   WEIGHTFA=OLDWTFA+TLS208I+TLV208I;
   TL208I=TLS208I+TLV208I;
   TLS=TLS+TLS208I; TLV=TLV+TLV208I; TL=TL+TL208I;
   TABDRV=TL208I; TABRF=0; TABKID=0; SW0=0; SW1=0; SW2=0;
   SWKID0=0; SWKID1=0; SWKID2=0; GOTO T214;
```

```
TDRVUNR:    OLDWTFA=WEIGHTFA;
      IF 1 LE M_HARM LE 6 THEN E=0;
        ELSE IF IMPACT2=12 THEN E=.32742;
        ELSE IF IMPACT2 IN (1,11) THEN E=.17161;
        ELSE IF IMPACT2 IN (2,10) THEN E=.06548; ELSE E=0;
      P=PS208I+PV208I;
      S=OLDWTFA*E*P / (1 - E*P);
      TLS208I=S*PS208I/P;
      TLV208I=S*PV208I/P;
      WEIGHTFA=OLDWTFA+TLS208I+TLV208I;
      TL208I=TLS208I+TLV208I;
      TLS=TLS+TLS208I; TLV=TLV+TLV208I; TL=TL+TL208I;
      TABDRV=TL208I; TABRF=0; TABKID=0; SW0=0; SW1=0; SW2=0;
      SWKID0=0; SWKID1=0; SWKID2=0; GOTO T214;

TRFBELT:    OLDWTFA=WEIGHTFA;
      IF 1 LE M_HARM LE 6 THEN E=0;
        ELSE IF IMPACT2=12 THEN E=.27784;
        ELSE IF IMPACT2 IN (1,11) THEN E=.14563;
        ELSE IF IMPACT2 IN (2,10) THEN E=.05557; ELSE E=0;
      P=PS208I+PV208I;
      S=OLDWTFA*E*P / (1 - E*P);
      TLS208I=S*PS208I/P;
      TLV208I=S*PV208I/P;
      WEIGHTFA=OLDWTFA+TLS208I+TLV208I;
      TL208I=TLS208I+TLV208I;
      TLS=TLS+TLS208I; TLV=TLV+TLV208I; TL=TL+TL208I;
      TABDRV=0; TABRF=TL208I; TABKID=0; SW0=0; SW1=0; SW2=0;
      SWKID0=0; SWKID1=0; SWKID2=0; GOTO T214;

TRFUNR:    OLDWTFA=WEIGHTFA;
      IF 1 LE M_HARM LE 6 THEN E=0;
        ELSE IF IMPACT2=12 THEN E=.36016;
        ELSE IF IMPACT2 IN (1,11) THEN E=.18877;
        ELSE IF IMPACT2 IN (2,10) THEN E=.07203; ELSE E=0;
      P=PS208I+PV208I;
      S=OLDWTFA*E*P / (1 - E*P);
      TLS208I=S*PS208I/P;
      TLV208I=S*PV208I/P;
      WEIGHTFA=OLDWTFA+TLS208I+TLV208I;
      TL208I=TLS208I+TLV208I;
      TLS=TLS+TLS208I; TLV=TLV+TLV208I; TL=TL+TL208I;
      TABDRV=0; TABRF=TL208I; TABKID=0; SW0=0; SW1=0; SW2=0;
      SWKID0=0; SWKID1=0; SWKID2=0; GOTO T214;

TKID:       OLDWTFA=WEIGHTFA;
      IF KCRSH=1 AND RESTGP=1 THEN E=-4.5;
        ELSE IF KCRSH=1 AND RESTGP IN (2,3) THEN E=-1;
        ELSE IF KCRSH=1 AND RESTGP IN (4,5,6) THEN E=-.70;
        ELSE IF KCRSH IN (2,3) AND RESTGP=1 THEN E=-1.71;
        ELSE IF KCRSH IN (2,3) AND RESTGP IN (2,3) THEN E=-.38;
        ELSE IF KCRSH IN (2,3) AND RESTGP IN (4,5,6) THEN E=-.27;
        ELSE E=0;
      P=PS208I+PV208I;
      S=OLDWTFA*E*P / (1 - E*P);
      TLS208I=S*PS208I/P;
      TLV208I=S*PV208I/P;
      WEIGHTFA=OLDWTFA+TLS208I+TLV208I;
      TL208I=TLS208I+TLV208I;
      TLS=TLS+TLS208I; TLV=TLV+TLV208I; TL=TL+TL208I;
      TABDRV=0; TABRF=0; TABKID=TL208I; SW0=0; SW1=0; SW2=0;
      SWKID0=0; SWKID1=0; SWKID2=0; GOTO T214;

TRFBELSW:   OLDWTFA=WEIGHTFA;
      IF 1 LE M_HARM LE 6 THEN E=0;
```

```
ELSE IF IMPACT2=12 THEN E=.27784;
ELSE IF IMPACT2 IN (1,11) THEN E=.14563;
ELSE IF IMPACT2 IN (2,10) THEN E=.05557; ELSE E=0;
```

For the three scenarios involving on-off switches, the additional parameter U indicates the use rate of the switches – i.e., the proportion of air bags turned **_off_** – as observed in NHTSA's survey of pickup trucks. U ranges from .15 for adults age 20-59 up to .86 for infants under age 1. (Ideally, though, U should be 0 for those adults and 1 for the infants.) When the switch is "used," the air bag is off, and it has no effect on fatality risk. The effect of air bags is (1-U)E.

```
   /* USE RATES FOR ON-OFF SWITCHES BY PASSENGER AGE                    */
    IF 13 LE AGE LE 15 THEN U=.22;
       ELSE IF 16 LE AGE LE 19 THEN U=.17;
       ELSE IF 20 LE AGE LE 59 THEN U=.15;
       ELSE IF 60 LE AGE LE 69 THEN U=.19;
       ELSE IF AGE GE 70 THEN U=.56;

    P=PS2081+PV2081;
    X=OLDWTFA/(1 - E*P + U*E*P);
    S=X - OLDWTFA;
    TLS2081=S*PS2081/P;
    TLV2081=S*PV2081/P;
    WEIGHTFA=OLDWTFA+TLS2081+TLV2081;
    Y=X*(1-E);
    SW1=X-Y;
    SW2=Y-OLDWTFA;
    TL2081=TLS2081+TLV2081;
    TLS=TLS+TLS2081; TLV=TLV+TLV2081; TL=TL+TL2081;
    SW0=TL2081;
    SWKID0=0; SWKID1=0; SWKID2=0; TABDRV=0; TABRF=0; TABKID=0; GOTO T214;

TRFUNRSW:  OLDWTFA=WEIGHTFA;
    IF 1 LE M_HARM LE 6 THEN E=0;
       ELSE IF IMPACT2=12 THEN E=.36016;
       ELSE IF IMPACT2 IN (1,11) THEN E=.18877;
       ELSE IF IMPACT2 IN (2,10) THEN E=.07203; ELSE E=0;

   /* USE RATES FOR ON-OFF SWITCHES BY PASSENGER AGE                    */
    IF 13 LE AGE LE 15 THEN U=.22;
       ELSE IF 16 LE AGE LE 19 THEN U=.17;
       ELSE IF 20 LE AGE LE 59 THEN U=.15;
       ELSE IF 60 LE AGE LE 69 THEN U=.19;
       ELSE IF AGE GE 70 THEN U=.56;

    P=PS2081+PV2081;
    X=OLDWTFA/(1 - E*P + U*E*P);
    S=X - OLDWTFA;
    TLS2081=S*PS2081/P;
    TLV2081=S*PV2081/P;
    WEIGHTFA=OLDWTFA+TLS2081+TLV2081;
    Y=X*(1-E);
    SW1=X-Y;
    SW2=Y-OLDWTFA;
    TL2081=TLS2081+TLV2081;
    TLS=TLS+TLS2081; TLV=TLV+TLV2081; TL=TL+TL2081;
    SW0=TL2081;
    SWKID0=0; SWKID1=0; SWKID2=0; TABDRV=0; TABRF=0; TABKID=0; GOTO T214;

TKIDSW:    OLDWTFA=WEIGHTFA;
    IF KCRSH=1 AND RESTGP=1 THEN E=-4.5;
       ELSE IF KCRSH=1 AND RESTGP IN (2,3) THEN E=-1;
       ELSE IF KCRSH=1 AND RESTGP IN (4,5,6) THEN E=-.70;
       ELSE IF KCRSH IN (2,3) AND RESTGP=1 THEN E=-1.71;
       ELSE IF KCRSH IN (2,3) AND RESTGP IN (2,3) THEN E=-.38;
```

```
          ELSE IF KCRSH IN (2,3) AND RESTGP IN (4,5,6) THEN E=-.27;
          ELSE E=0;

   /* USE RATES FOR ON-OFF SWITCHES BY PASSENGER AGE                    */
      IF AGE=0 THEN U=.86;
        ELSE IF 1 LE AGE LE 6 THEN U=.74;
        ELSE IF 7 LE AGE LE 8 THEN U=.59;
        ELSE IF 9 LE AGE LE 10 THEN U=.47;
        ELSE IF 11 LE AGE LE 12 THEN U=.30;

      P=PS208I+PV208I;
      X=OLDWTFA/(1 - E*P + U*E*P);
      S=X - OLDWTFA;
      TLS208I=S*PS208I/P;
      TLV208I=S*PV208I/P;
      WEIGHTFA=OLDWTFA+TLS208I+TLV208I;
      Y=X*(1-E);
      SWKID1=X-Y;
      SWKID2=Y-OLDWTFA;
      TL208I=TLS208I+TLV208I;
      TLS=TLS+TLS208I; TLV=TLV+TLV208I; TL=TL+TL208I;
      SWKID0=TL208I;
      SW0=0; SW1=0; SW2=0; TABDRV=0; TABRF=0; TABKID=0;

/* ------------------------------------------------------------------ */
/*   214A SIDE DOOR BEAMS                                              */
/*   MEDIAN INSTALLATION YEAR: 1994                                   */
/* ------------------------------------------------------------------ */
```

Side door beams were introduced in a few LTV models in 1991-93. All MY 1994 LTVs were equipped with them.

```
/* IDENTIFIES LIGHT TRUCKS WITH SIDE DOOR BEAMS                  */
T214: PS214A=0; PV214A=0;
  IF MY LE 1990 THEN GOTO SAVE214T;
  IF MY GE 1994 THEN DO; PS214A=1; GOTO SAVE214T; END;
  IF CG IN (12303,12304,18405,49402) AND 1991 LE MY LE 1993 THEN DO;
      PV214A=1; GOTO SAVE214T; END;
  IF CG=12404 AND 1992 LE MY LE 1993 THEN DO; PV214A=1; GOTO SAVE214T; END;
  IF CG IN (12405,35203,35204,35301,49303) AND MY=1993 THEN DO;
      PV214A=1; GOTO SAVE214T; END;
  IF 1201 LE CG LE 63401 THEN GOTO SAVE214T;
  IF MAKE IN (12,14,20,21,22,35,41,49) THEN GOTO SOMEVOLT; GOTO SAVE214T;

SOMEVOLT: IF MAK_MOD IN (12401,41401,20442,21441,22441,49441)
     AND 1991 LE MY LE 1993 THEN DO; PV214A=1; GOTO SAVE214T; END;
IF MAK_MOD=12461 AND 1992 LE MY LE 1993 THEN DO; PV214A=1; GOTO SAVE214T; END;
IF MAK_MOD IN (14443,35443,35471,35401,49421) AND MY=1993 THEN DO;
    PV214A=1; GOTO SAVE214T; END;

/* REDUCTION IN SINGLE-VEHICLE SIDE-IMPACT FATALITIES WITH SIDE DOOR BEAMS   */
SAVE214T: OLDWTFA=WEIGHTFA;
```

NHTSA's evaluation showed side door beams are effective for outboard occupants in single-vehicle side-impact crashes. Fatality reduction is 26 percent for nearside occupants and 11 percent for farside occupants.

```
  IF (PS214A GT 0 OR PV214A GT 0) AND CRSH=2 AND
     SEAT2 IN (11,13,21) AND VE_FORMS=1 THEN DO;
  IF (SEAT_POS IN (11,21) AND IMPACT2 IN (8,9,10)) OR
     (SEAT_POS IN (13,23) AND IMPACT2 IN (2,3,4)) THEN E=.25; ELSE E=.11;
  P=PS214A+PV214A;
  S=OLDWTFA*E*P / (1 - E*P);
```

```
    TLS214A=S*PS214A/P;
    TLV214A=S*PV214A/P;
     WEIGHTFA=OLDWTFA+TLS214A+TLV214A; END;
  ELSE DO; TLS214A=0; TLV214A=0; END;
  TL214A=TLS214A+TLV214A;
  TLS=TLS+TLS214A; TLV=TLV+TLV214A; TL=TL+TL214A;


/* ---------------------------------------------------------------- */
/*  208G 3-POINT BELTS FOR BACK-SEAT OUTBOARD OCCUPANTS             */
/*  MEDIAN INSTALLATION YEAR: 1990                                 */
/* ---------------------------------------------------------------- */

/* IDENTIFIES BACK-SEAT OUTBOARD OCCUPANTS WEARING 3-POINT BELTS */
  PS208G=0; PV208G=0;
  IF SEAT2=21 AND REST3=3 THEN DO; IF MY LE 1991 THEN PV208G=1; ELSE PS208G=1; END;
/* LIVES SAVED AND EJECTIONS PREVENTED BY BACK-SEAT OUTBOARD 3-POINT BELTS */
  OLDWTFA=WEIGHTFA;
  IF PS208G GT 0 OR PV208G GT 0 THEN DO;
```

Three-point belts in the rear-outboard seats reduce fatality risk by 50 percent in frontals, 84 percent in rollovers and 70 percent in side, rear and other impacts.

```
    IF CRSH=1 THEN DO; E=.50; EJECT2=.48; END;
     ELSE IF CRSH=2 THEN DO; E=.70; EJECT2=.59; END;
     ELSE IF CRSH=3 THEN DO; E=.84; EJECT2=.85; END;
     ELSE IF CRSH=4 THEN DO; E=.70; EJECT2=.58; END;
    P=PS208G+PV208G;
    S=OLDWTFA*E*P / (1 - E*P);
    TLS208G=S*PS208G/P;
    TLV208G=S*PV208G/P;
     WEIGHTFA=OLDWTFA+TLS208G+TLV208G; END;
  ELSE DO; TLS208G=0; TLV208G=0; END;
  TL208G=TLS208G+TLV208G;
  TLS=TLS+TLS208G; TLV=TLV+TLV208G; TL=TL+TL208G;


/* ---------------------------------------------------------------- */
/*  213 CHILD SAFETY SEATS                                         */
/*  USE RATE WENT OVER 50% IN: 1985                                */
/* ---------------------------------------------------------------- */

/* IDENTIFIES CHILD PASSENGERS IN SAFETY SEATS  */
  PS213=0; PV213=0;
  IF REST3=4 THEN DO;
    IF STATE=47 AND CY GE 1978 AND AGE LE 3 THEN PS213=1;
     ELSE IF STATE=44 AND CY GE 1980 AND AGE LE 5 THEN PS213=1;
     ELSE IF STATE=54 AND CY GE 1981 AND AGE LE 2 THEN PS213=1;
     ELSE IF STATE=25 AND CY GE 1982 AND AGE LE 5 THEN PS213=1;
     ELSE IF STATE IN (9,10,20,21,26,36,37) AND CY GE 1982 AND AGE LE 3 THEN PS213=1;
     ELSE IF STATE IN (32,34,35) AND CY GE 1983 AND AGE LE 4 THEN PS213=1;
     ELSE IF STATE IN (1,6,11,12,17,28,33,39,45,51,55) AND CY GE 1983 AND AGE LE 3 THEN PS213=1;
     ELSE IF STATE=15 AND CY GE 1983 AND AGE LE 2 THEN PS213=1;
     ELSE IF STATE IN (4,5,46,50) AND CY GE 1984 AND AGE LE 4 THEN PS213=1;
     ELSE IF STATE IN (8,18,23,24,27,29,31,40,41,42) AND CY GE 1984 AND AGE LE 3 THEN PS213=1;
     ELSE IF STATE IN (13,38,53) AND CY GE 1984 AND AGE LE 2 THEN PS213=1;
     ELSE IF STATE IN (30,49) AND CY GE 1984 AND AGE LE 1 THEN PS213=1;
     ELSE IF STATE=56 AND CY GE 1985 AND AGE LE 4 THEN PS213=1;
     ELSE IF STATE IN (2,16) AND CY GE 1985 AND AGE LE 3 THEN PS213=1;
     ELSE IF STATE IN (19,22) AND CY GE 1985 AND AGE LE 2 THEN PS213=1;
     ELSE IF STATE=48 AND CY GE 1985 AND AGE LE 1 THEN PS213=1;
     ELSE PV213=1; END;
/* LIVES SAVED AND EJECTIONS PREVENTED BY CHILD SAFETY SEATS */
  OLDWTFA=WEIGHTFA;
  IF PS213 GT 0 OR PV213 GT 0 THEN DO;
```

NHTSA's evaluation estimates that safety seats reduce fatality risk by 58 percent for infants and by 59 percent for toddlers. Effectiveness is not estimated separately by crash mode. In general, 34 percent of unrestrained child passenger fatalities in frontal crashes were ejected, 48 percent in side impacts, 79 percent in rollovers and 42 percent in rear/other impacts.

```
IF AGE=0 THEN E=.58; ELSE E=.59;
IF CRSH=1 THEN EJECT2=.34;
  ELSE IF CRSH=2 THEN EJECT2=.48;
  ELSE IF CRSH=3 THEN EJECT2=.79;
  ELSE IF CRSH=4 THEN EJECT2=.42;
P=PS213+PV213;
S=OLDWTFA*E*P / (1 - E*P);
TLS213=S*PS213/P;
TLV213=S*PV213/P;
WEIGHTFA=OLDWTFA+TLS213+TLV213; END;
ELSE DO; TLS213=0; TLV213=0; END;
TL213=TLS213+TLV213;
TLS=TLS+TLS213; TLV=TLV+TLV213; TL=TL+TL213;


/* -------------------------------------------------------- */
/*  212 ADHESIVE WINDSHIELD BONDING                         */
/*  MEDIAN INSTALLATION YEAR: 1980                          */
/* -------------------------------------------------------- */

/* IDENTIFIES LIGHT TRUCKS WITH ADHESIVE WINDSHIELD BONDING     */
```

FMVSS 212 took effect in LTVs on September 1, 1978. However, many LTVs had rubber gasket installations that complied with the standard. The transition to adhesive bonding began shortly before the standard and continued for several years afterward.

```
IF MY LE 1977 THEN DO; PS212=0; PV212=0; END;
  ELSE IF MY GE 1985 THEN DO; PS212=1; PV212=0; END;
  ELSE IF MY=1978 THEN DO; PS212=0; PV212=.2; END;
  ELSE IF MY=1979 THEN DO; PS212=.4; PV212=0; END;
  ELSE IF MY=1980 THEN DO; PS212=.5; PV212=0; END;
  ELSE IF MY=1981 THEN DO; PS212=.6; PV212=0; END;
  ELSE IF MY=1982 THEN DO; PS212=.7; PV212=0; END;
  ELSE IF MY=1983 THEN DO; PS212=.8; PV212=0; END;
  ELSE IF MY=1984 THEN DO; PS212=.9; PV212=0; END;
/* REDUCTION IN EJECTION FATALITY RISK WITH ADHESIVE WINDSHIELD BONDING */
  OLDWTFA=WEIGHTFA;
```

NCSS data show that the proportion of ejections that are via the windshield portal is about the same in LTVs with rubber gaskets as in cars with rubber gaskets. We will assume that adhesive bonding has the same percentage effect on ejection fatalities in LTVs as in cars.

```
IF (PS212 GT 0 OR PV212 GT 0) AND SEAT2 IN (11,12,13,18) AND EJECT2 NE 0 THEN DO;
  IF CRSH IN (1,3) THEN E=.033;
    ELSE IF CRSH IN (2,4) THEN E=.0075;
  P=PS212+PV212;
  REL_S=EJECT2*E*P / (1 - E*P);
  S=OLDWTFA*REL_S;
  TLS212=S*PS212/P;
  TLV212=S*PV212/P;
  WEIGHTFA=OLDWTFA+TLS212+TLV212;
  EJECT2=(EJECT2 + REL_S)/(1 + REL_S); END;
ELSE DO; TLS212=0; TLV212=0; END;
TL212=TLS212+TLV212;
TLS=TLS+TLS212; TLV=TLV+TLV212; TL=TL+TL212;

/* -------------------------------------------------------- */
```

```
/*   208F 3-POINT BELTS FOR FRONT-SEAT OUTBOARD OCCUPANTS              */
/*   MEDIAN INSTALLATION YEAR: 1976                                    */
/* ----------------------------------------------------------------- */

/* IDENTIFIES FRONT-SEAT OCCUPANTS WEARING 3-POINT BELTS  */
  PS208F=0; PV208F=0;
   IF SEAT2 IN (11,13) AND REST3=3 THEN DO; IF MY LE 1976 THEN PV208F=1;
    ELSE IF 1977 LE MY LE 1981 AND NEWVTYP IN (12,13) THEN PV208F=1;
    ELSE PS208F=1; END;
/* LIVES SAVED AND EJECTIONS PREVENTED BY FRONT-OUTBOARD 3-POINT BELTS */
  OLDWTFA=WEIGHTFA;
   IF PS208F GT O OR PV208F GT O THEN DO;
```

In LTVs, 3-point belts reduce fatality risk by 64 percent in single-vehicle frontals, 40 percent in multivehicle frontals; 47 percent in single-vehicle nearside; 36 percent in multivehicle nearside; 61 percent in single-vehicle farside; 54 percent in multivehicle farside; 80 percent in rollovers; and 81 percent in rear/other impacts.

```
     IF CRSH=1 AND VE_FORMS=1 THEN DO; E=.64; EJECT2=.42; END;
       ELSE IF CRSH=1 THEN DO; E=.40; EJECT2=.24; END;
       ELSE IF CRSH=2 THEN DO;
         IF VE_FORMS=1 AND
           ((SEAT2=11 AND IMPACT2 IN (8,9,10)) OR (SEAT2=13 AND IMPACT2 IN (2,3,4)))
            THEN DO; E=.47; EJECT2=.44; END;
          ELSE IF ((SEAT2=11 AND IMPACT2 IN (8,9,10)) OR (SEAT2=13 AND IMPACT2 IN (2,3,4)))
            THEN DO; E=.36; EJECT2=.37; END;
          ELSE IF VE_FORMS=1 THEN DO; E=.61; EJECT2=.49; END;
          ELSE DO; E=.54; EJECT2=.41; END; END;
       ELSE IF CRSH=3 THEN DO; E=.80; EJECT2=.77; END;
       ELSE IF CRSH=4 THEN DO; E=.81; EJECT2=.51; END;
     P=PS208F+PV208F;
     S=OLDWTFA*E*P / (1 - E*P);
     TLS208F=S*PS208F/P;
     TLV208F=S*PV208F/P;
      WEIGHTFA=OLDWTFA+TLS208F+TLV208F; END;
    ELSE DO; TLS208F=0; TLV208F=0; END;
   TL208F=TLS208F+TLV208F;
   TLS=TLS+TLS208F; TLV=TLV+TLV208F; TL=TL+TL208F;


/* ----------------------------------------------------------------- */
/*   203 ENERGY-ABSORBING AND TELESCOPING STEERING ASSEMBLIES         */
/*   MEDIAN IMPLEMENTATION YEAR: EARLY 1976                           */
/* ----------------------------------------------------------------- */

/* IDENTIFIES LIGHT TRUCKS WITH ENERGY-ABSORBING STEERING ASSEMBLIES    */
  PS203=0; PV203=0;
   IF MY GE 1970 AND SEAT2=11 THEN DO;
```

The introduction date for energy-absorbing steering assemblies in LTVs depends on the manufacturer and the truck type (pickup, SUV or van).

```
     IF MY GE 1982 THEN PS203=1;
       ELSE IF MAKE=30 THEN PV203=1;
       ELSE IF MAKE IN (2,20,23) AND MY GE 1973 THEN PV203=1;
       ELSE IF MAKE IN (7,9) AND NEWVTYP IN (11,12) AND MY GE 1977 THEN PV203=1;
       ELSE IF MAKE IN (7,9) AND MY GE 1979 THEN PV203=1;
       ELSE IF MAKE=12 AND NEWVTYP IN (11,12) AND MY GE 1980 THEN PV203=1; END;
/* LIVES SAVED IN FRONTAL IMPACTS BY INSTRUMENT PANEL IMPROVEMENTS */
  OLDWTFA=WEIGHTFA;
   IF PS203 GT O OR PV203 GT O THEN DO;
```

NHTSA's evaluations suggest that energy-absorbing steering assemblies are about equally effective in LTVs and cars.

```
   IF IMPACT2 IN (11,12,1) THEN E=.121; ELSE E=0;
   P=PS203+PV203;
   S=OLDWTFA*E*P / (1 - E*P);
   TLS203=S*PS203/P;
   TLV203=S*PV203/P;
   WEIGHTFA=OLDWTFA+TLS203+TLV203; END;
 ELSE DO; TLS203=0; TLV203=0; END;
 TL203=TLS203+TLV203;
 TLS=TLS+TLS203; TLV=TLV+TLV203; TL=TL+TL203;


/* ------------------------------------------------------------------ */
/*  201 VOLUNTARY INSTRUMENT PANEL IMPROVEMENTS (FMVSS-201 INSPIRED)  */
/*  MEDIAN IMPLEMENTATION YEAR: 1972                                  */
/* ------------------------------------------------------------------ */

/* INSTRUMENT PANELS WERE GRADUALLY IMPROVED FROM 1969 TO 1977    */
 PS201=0; PV201=0;
```

NHTSA's evaluation suggests that LTVs received the same types of instrument panel modifications as cars, and that they were gradually introduced, perhaps over a 1969-77 timeframe. NHTSA has no details about specific make-models.

```
 IF MY GE 1969 AND SEAT2=13 THEN DO;
     IF MY GE 1977 THEN PV201=1;
        ELSE PV201=.125*(MY-1968); END;
/* LIVES SAVED IN FRONTAL IMPACTS BY INSTRUMENT PANEL IMPROVEMENTS */
 OLDWTFA=WEIGHTFA;
 IF PS201 GT 0 OR PV201 GT 0 THEN DO;
```

NHTSA's evaluation found the instrument panel improvements to be about equally effective in LTVs and cars.

```
   IF CRSH=1 AND REST3 NE 2 THEN E=.159; ELSE E=0;
   P=PS201+PV201;
   S=OLDWTFA*E*P / (1 - E*P);
   TLS201=S*PS201/P;
   TLV201=S*PV201/P;
   WEIGHTFA=OLDWTFA+TLS201+TLV201; END;
 ELSE DO; TLS201=0; TLV201=0; END;
 TL201=TLS201+TLV201;
 TLS=TLS+TLS201; TLV=TLV+TLV201; TL=TL+TL201;


/* ------------------------------------------------------------------ */
/*  105B FRONT DISC BRAKES                                            */
/*  MEDIAN INSTALLATION YEAR: 1971 (INSTALLATION COMPLETED 1977)      */
/* ------------------------------------------------------------------ */

/* IMPLEMENTATION OF FRONT DISC BRAKES */
```

We believe the timeframe for introducing front disc brakes was approximately the same for LTVs as cars; we are using the same implementation schedule as for cars. (However, FMVSS 105 was not extended to LTVs until September 1, 1983.)

```
 IF MY GE 1984 THEN DO; PS105B=1; PV105B=0; END;
  ELSE IF MY LE 1964 THEN DO; PS105B=0; PV105B=0; END;
  ELSE IF MY=1965 THEN DO; PS105B=0; PV105B=.02; END;
  ELSE IF MY=1966 THEN DO; PS105B=0; PV105B=.03; END;
  ELSE IF MY=1967 THEN DO; PS105B=0; PV105B=.06; END;
```

```
        ELSE IF MY=1968 THEN DO; PS105B=0; PV105B=.13; END;
        ELSE IF MY=1969 THEN DO; PS105B=0; PV105B=.28; END;
        ELSE IF MY=1970 THEN DO; PS105B=0; PV105B=.41; END;
        ELSE IF MY=1971 THEN DO; PS105B=0; PV105B=.63; END;
        ELSE IF MY=1972 THEN DO; PS105B=0; PV105B=.74; END;
        ELSE IF MY=1973 THEN DO; PS105B=0; PV105B=.86; END;
        ELSE IF MY=1974 THEN DO; PS105B=0; PV105B=.84; END;
        ELSE IF MY=1975 THEN DO; PS105B=0; PV105B=.93; END;
        ELSE IF MY=1976 THEN DO; PS105B=0; PV105B=.99; END;
        ELSE IF 1977 LE MY LE 1983 THEN DO; PS105B=0; PV105B=1; END;

/* LIGHT TRUCK OCCUPANT LIVES SAVED BY FRONT DISC BRAKES */
  OLDWTFA=WEIGHTFA;
  IF PS105B GT 0 OR PV105B GT 0 THEN DO;
    E=.0017;
    P=PS105B+PV105B;
    S=OLDWTFA*E*P / (1 - E*P);
    TLS105B=S*PS105B/P;
    TLV105B=S*PV105B/P;
     WEIGHTFA=OLDWTFA+TLS105B+TLV105B; END;
  ELSE DO; TLS105B=0; TLV105B=0; END;
  TL105B=TLS105B+TLV105B;
  TLS=TLS+TLS105B; TLV=TLV+TLV105B; TL=TL+TL105B;
/* ---------------------------------------------------------------- */
/*   208E LAP BELTS FOR BACK-SEAT CENTER OCCUPANTS                   */
/*   MEDIAN INSTALLATION YEAR: 1968                                 */
/* ---------------------------------------------------------------- */

/* IDENTIFIES BACK-CENTER-SEAT OCCUPANTS WEARING LAP BELTS  */
  PS208E=0; PV208E=0;
  IF SEAT2=22 AND REST3=2 AND (5 LE AGE LE 99 OR AGE=.) THEN DO;
```

The original FMVSS 208 requirement for lap belts at all designated seating positions was extended to LTVs effective July 1, 1971.

```
    IF MY GE 1972 THEN PS208E=1;
       ELSE IF MY LE 1970 THEN PV208E=1;
       ELSE IF MY=1971 THEN DO; PS208E=.17; PV208E=.83; END; END;
/* LIVES SAVED AND EJECTIONS PREVENTED BY BACK-CENTER-SEAT LAP BELTS */
  OLDWTFA=WEIGHTFA;
  IF PS208E GT 0 OR PV208E GT 0 THEN DO;
```

Lap belts in the rear-outboard seats reduce fatality risk by 44 percent in frontals, 80 percent in rollovers and 64 percent in side, rear and other impacts. The same effectiveness is assumed for the rear-center seat.

```
    IF CRSH=1 THEN DO; E=.44; EJECT2=.49; END;
      ELSE IF CRSH=2 THEN DO; E=.64; EJECT2=.59; END;
      ELSE IF CRSH=3 THEN DO; E=.80; EJECT2=.85; END;
      ELSE IF CRSH=4 THEN DO; E=.64; EJECT2=.58; END;
    P=PS208E+PV208E;
    S=OLDWTFA*E*P / (1 - E*P);
    TLS208E=S*PS208E/P;
    TLV208E=S*PV208E/P;
     WEIGHTFA=OLDWTFA+TLS208E+TLV208E; END;
  ELSE DO; TLS208E=0; TLV208E=0; END;
  TL208E=TLS208E+TLV208E;
  TLS=TLS+TLS208E; TLV=TLV+TLV208E; TL=TL+TL208E;
```

Even though lap belts for rear-center occupants and the next two technologies, lap belts for front-center occupants and lap belts for rear-outboard occupants have approximately the same median implementation year in LTVs, 1968, it makes no difference what their sequential order is in the

342

model. Only one of these technologies can apply to any occupant: a properly seated, belted person cannot be in two different seat positions at the same time.

```
/* ------------------------------------------------------------ */
/*   208D LAP BELTS FOR FRONT-SEAT CENTER OCCUPANTS             */
/*   MEDIAN INSTALLATION YEAR: 1968                             */
/* ------------------------------------------------------------ */

/* IDENTIFIES FRONT-CENTER-SEAT OCCUPANTS WEARING LAP BELTS  */
  PS208D=0; PV208D=0;
  IF SEAT2=12 AND REST3=2 AND (5 LE AGE LE 99 OR AGE=.) THEN DO;
     IF MY GE 1972 THEN PS208D=1;
        ELSE IF MY LE 1970 THEN PV208D=1;
        ELSE IF MY=1971 THEN DO; PS208D=.17; PV208D=.83; END; END;
/* LIVES SAVED AND EJECTIONS PREVENTED BY FRONT-CENTER-SEAT LAP BELTS */
  OLDWTFA=WEIGHTFA;
  IF PS208D GT 0 OR PV208D GT 0 THEN DO;
```

A double-pair comparison analysis of lap belt effectiveness for front-outboard occupants, conducted in support of this study, found that lap belts reduce fatality risk by 48 percent in LTVs. There were not enough data for separate effectiveness estimates by crash mode. This effectiveness will also be assumed for the front-center occupant.

```
  E=.48;
  IF CRSH=1 THEN EJECT2=.33; ELSE IF CRSH=2 THEN EJECT2=.38;
    ELSE IF CRSH=3 THEN EJECT2=.72; ELSE IF CRSH=4 THEN EJECT2=.53;
  P=PS208D+PV208D;
  S=OLDWTFA*E*P / (1 - E*P);
  TLS208D=S*PS208D/P;
  TLV208D=S*PV208D/P;
  WEIGHTFA=OLDWTFA+TLS208D+TLV208D; END;
  ELSE DO; TLS208D=0; TLV208D=0; END;
  TL208D=TLS208D+TLV208D;
  TLS=TLS+TLS208D; TLV=TLV+TLV208D; TL=TL+TL208D;

/* ------------------------------------------------------------ */
/*   208C LAP BELTS FOR BACK-SEAT OUTBOARD OCCUPANTS           */
/*   MEDIAN INSTALLATION YEAR: 1968                            */
/* ------------------------------------------------------------ */

/* IDENTIFIES BACK-OUTBOARD-SEAT OCCUPANTS WEARING LAP BELTS  */
  PS208C=0; PV208C=0;
  IF SEAT2=21 AND REST3=2 AND (5 LE AGE LE 99 OR AGE=.) THEN DO;
     IF MY GE 1972 THEN PS208C=1;
        ELSE IF MY LE 1970 THEN PV208C=1;
        ELSE IF MY=1971 THEN DO; PS208C=.17; PV208C=.83; END; END;
/* LIVES SAVED AND EJECTIONS PREVENTED BY BACK-OUTBOARD-SEAT LAP BELTS */
  OLDWTFA=WEIGHTFA;
  IF PS208C GT 0 OR PV208C GT 0 THEN DO;
```

Lap belts in the rear-outboard seats reduce fatality risk by 44 percent in frontals, 80 percent in rollovers and 64 percent in side, rear and other impacts.

```
  IF CRSH=1 THEN DO; E=.44; EJECT2=.48; END;
    ELSE IF CRSH=2 THEN DO; E=.64; EJECT2=.59; END;
    ELSE IF CRSH=3 THEN DO; E=.80; EJECT2=.85; END;
    ELSE IF CRSH=4 THEN DO; E=.64; EJECT2=.58; END;
  P=PS208C+PV208C;
  S=OLDWTFA*E*P / (1 - E*P);
  TLS208C=S*PS208C/P;
  TLV208C=S*PV208C/P;
  WEIGHTFA=OLDWTFA+TLS208C+TLV208C; END;
```

```
ELSE DO; TLS208C=0; TLV208C=0; END;
TL208C=TLS208C+TLV208C;
TLS=TLS+TLS208C; TLV=TLV+TLV208C; TL=TL+TL208C;
```

```
/* ------------------------------------------------------------------ */
/*   206 DOOR LOCK IMPROVEMENTS                                        */
/*   MEDIAN INSTALLATION YEAR: 1967                                    */
/* ------------------------------------------------------------------ */

/* DOOR LOCKS IMPROVED GRADUALLY FROM 1962 TO 1972          */
```

FMVSS 206 took effect on January 1, 1972 in LTVs (and on January 1, 1968 in cars). Door lock improvements were gradually introduced in cars from MY 1962 until MY 1968, the year the standard took effect. We will assume that LTV door locks received comparable improvements, also starting about 1962 but extending until the 1972 effective date.

```
IF MY GE 1973 THEN DO; PS206=1; PV206=0; END;
  ELSE IF MY=1972 THEN DO; PS206=.5; PV206=.5; END;
  ELSE IF 1962 LE MY LE 1971 THEN DO; PS206=0; PV206=(MY-1961)/11; END;
  ELSE IF MY LE 1961 THEN DO; PS206=0; PV206=0; END;
/* REDUCTION IN EJECTION ROLLOVER FATALITY RISK WITH IMPROVED DOOR LOCKS */
  OLDWTFA=WEIGHTFA;
  IF (PS206 GT 0 OR PV206 GT 0) AND (CRSH=3 OR M_HARM=1)
      AND SEAT2 NE 52 AND EJECT2 NE 0 THEN DO;
```

In passenger cars, the reduction of ejection fatalities in rollovers was 15.38 percent. However, in NCSS, 32.11 percent of ejection fatalities in rollovers in cars were through doors, but only 20.45 percent in LTVs. Thus, the effectiveness in LTVs is estimated to be (.2045/.3211) x .1538 = 9.8 percent (excluding occupants riding in the beds of pickups or elsewhere outside the passenger compartment).

```
    E=.098;
    P=PS206+PV206;
    REL_S=EJECT2*E*P / (1 - E*P);
    S=OLDWTFA*REL_S;
    TLS206=S*PS206/P;
    TLV206=S*PV206/P;
    WEIGHTFA=OLDWTFA+TLS206+TLV206;
    EJECT2=(EJECT2 + REL_S)/(1 + REL_S); END;
  ELSE DO; TLS206=0; TLV206=0; END;
  TL206=TLS206+TLV206;
  TLS=TLS+TLS206; TLV=TLV+TLV206; TL=TL+TL206;


/* ------------------------------------------------------------------ */
/*   105A DUAL MASTER CYLINDERS                                        */
/*   MEDIAN INSTALLATION YEAR: LATE 1966                               */
/* ------------------------------------------------------------------ */

/* IMPLEMENTATION OF DUAL MASTER CYLINDERS */
```

We believe the timeframe for introducing dual master cylinders was approximately the same for LTVs as cars; we are using the same implementation schedule as for cars. (However, FMVSS 105 was not extended to LTVs until September 1, 1983.)

```
IF MY GE 1984 THEN DO; PS105A=1; PV105A=0; END;
  ELSE IF MY LE 1961 THEN DO; PS105A=0; PV105A=0; END;
  ELSE IF MY IN (1962,1963) THEN DO; PS105A=0; PV105A=.09; END;
  ELSE IF MY IN (1964,1965) THEN DO; PS105A=0; PV105A=.07; END;
  ELSE IF MY=1966 THEN DO; PS105A=0; PV105A=.54; END;
  ELSE IF 1967 LE MY LE 1983 THEN DO; PS105A=0; PV105A=1; END;
/* LIGHT TRUCK OCCUPANT LIVES SAVED BY DUAL MASTER CYLINDERS */
  OLDWTFA=WEIGHTFA;
  IF PS105A GT 0 OR PV105A GT 0 THEN DO;
    E=.007;
    P=PS105A+PV105A;
```

```
    S=OLDWTFA*E*P / (1 - E*P);
    TLS105A=S*PS105A/P;
    TLV105A=S*PV105A/P;
    WEIGHTFA=OLDWTFA+TLS105A+TLV105A; END;
  ELSE DO; TLS105A=0; TLV105A=0; END;
  TL105A=TLS105A+TLV105A;
  TLS=TLS+TLS105A; TLV=TLV+TLV105A; TL=TL+TL105A;


/* ------------------------------------------------------------------ */
/*  208B LAP BELT USE BY CHILDREN AGE 1-4                             */
/*  MEDIAN INSTALLATION YEAR FOR LAP BELTS USED BY CHILDREN: 1966     */
/* ------------------------------------------------------------------ */

/* IDENTIFIES CHILD PASSENGERS AGE 1-4 USING LAP BELTS */
  PS208B=0; PV208B=0;
  IF 1 LE AGE LE 4 AND REST3=2 THEN DO;
    IF MY GE 1972 THEN PS208B=1;
      ELSE IF MY LE 1970 THEN PV208B=1;
      ELSE IF MY=1971 THEN DO; PS208B=.17; PV208B=.83; END; END;
/* LIVES SAVED AND EJECTIONS PREVENTED BY LAP BELTS (AGE 1-4) */
  OLDWTFA=WEIGHTFA;
  IF PS208B GT 0 OR PV208B GT 0 THEN DO;
```

NHTSA's evaluation estimates that lap belts reduce fatality risk by 48 percent for toddlers in LTVs. Effectiveness is not estimated separately by crash mode.[7]

```
    E=.48;
    IF CRSH=1 THEN EJECT2=.34; ELSE IF CRSH=2 THEN EJECT2=.48;
      ELSE IF CRSH=3 THEN EJECT2=.79; ELSE IF CRSH=4 THEN EJECT2=.42;
    P=PS208B+PV208B;
    S=OLDWTFA*E*P / (1 - E*P);
    TLS208B=S*PS208B/P;
    TLV208B=S*PV208B/P;
    WEIGHTFA=OLDWTFA+TLS208B+TLV208B; END;
  ELSE DO; TLS208B=0; TLV208B=0; END;
  TL208B=TLS208B+TLV208B;
  TLS=TLS+TLS208B; TLV=TLV+TLV208B; TL=TL+TL208B;


/* ------------------------------------------------------------------ */
/*  208A LAP BELTS FOR FRONT-SEAT OUTBOARD OCCUPANTS                  */
/*  MEDIAN INSTALLATION YEAR: 1964                                    */
/* ------------------------------------------------------------------ */

/* IDENTIFIES FRONT-OUTBOARD-SEAT OCCUPANTS WEARING LAP BELTS  */
  PS208A=0; PV208A=0;
  IF SEAT2 IN (11,13) AND REST3=2 AND (5 LE AGE LE 99 OR AGE=.) THEN DO;
    IF MY GE 1972 THEN PS208A=1;
      ELSE IF MY LE 1970 THEN PV208A=1;
      ELSE IF MY=1971 THEN DO; PS208A=.17; PV208A=.83; END; END;
/* LIVES SAVED AND EJECTIONS PREVENTED BY FRONT-OUTBOARD-SEAT LAP BELTS */
  OLDWTFA=WEIGHTFA;
  IF PS208A GT 0 OR PV208A GT 0 THEN DO;
```

A double-pair comparison analysis of lap belt effectiveness for front-outboard occupants, conducted in support of this study, found that lap belts reduce fatality risk by 48 percent in LTVs. There were not enough data for separate effectiveness estimates by crash mode.

```
    E=.48;
    IF CRSH=1 THEN EJECT2=.34; ELSE IF CRSH=2 THEN EJECT2=.42;
```

---

[7] This evaluation is discussed in the "Child Safety Seats" section of the FMVSS 213 chapter of Part 1, not the FMVSS 208 chapter.

```
      ELSE IF CRSH=3 THEN EJECT2=.77; ELSE IF CRSH=4 THEN EJECT2=.51;
    P=PS208A+PV208A;
    S=OLDWTFA*E*P / (1 - E*P);
    TLS208A=S*PS208A/P;
    TLV208A=S*PV208A/P;
    WEIGHTFA=OLDWTFA+TLS208A+TLV208A; END;
  ELSE DO; TLS208A=0; TLV208A=0; END;
  TL208A=TLS208A+TLV208A;
  TLS=TLS+TLS208A; TLV=TLV+TLV208A; TL=TL+TL208A;
RUN;
```

That concludes the model for LTVs. It has estimated, on a case-by-case basis, by how much fatalities would increase if all the safety technologies were "removed" from vehicles. The next step is to tally up the lives saved over all the cases, by calendar year.

```
PROC MEANS SUM NOPRINT DATA=TRK2; BY CY;
  VAR ORIGWT WEIGHTFA
    TLV208J TLS208J TL208J
    TLV108 TLS108 TL108
    TLV208I TLS208I TL208I
    TLV214A TLS214A TL214A
    TLV208G TLS208G TL208G
    TLV213 TLS213 TL213
    TLV212 TLS212 TL212
    TLV208F TLS208F TL208F
    TLV203 TLS203 TL203
    TLV201 TLS201 TL201
    TLV105B TLS105B TL105B
    TLV208E TLS208E TL208E
    TLV208D TLS208D TL208D
    TLV208C TLS208C TL208C
    TLV206 TLS206 TL206
    TLV105A TLS105A TL105A
    TLV208B TLS208B TL208B
    TLV208A TLS208A TL208A
    TABDRV TABRF TABKID SW0 SW1 SW2
    SWKID0 SWKID1 SWKID2
    TLV TLS TL;
  OUTPUT OUT=TRK3
    SUM=T_ORIGWT T_WTFA
    TLV208J TLS208J TL208J
    TLV108 TLS108 TL108
    TLV208I TLS208I TL208I
    TLV214A TLS214A TL214A
    TLV208G TLS208G TL208G
    TLV213 TLS213 TL213
    TLV212 TLS212 TL212
    TLV208F TLS208F TL208F
    TLV203 TLS203 TL203
    TLV201 TLS201 TL201
    TLV105B TLS105B TL105B
    TLV208E TLS208E TL208E
    TLV208D TLS208D TL208D
    TLV208C TLS208C TL208C
    TLV206 TLS206 TL206
    TLV105A TLS105A TL105A
    TLV208B TLS208B TL208B
    TLV208A TLS208A TL208A
    TABDRV TABRF TABKID SW0 SW1 SW2
    SWKID0 SWKID1 SWKID2
    TLV TLS TL;
```

347

Prints the number of lives saved by each technology in each calendar year, and the sum of lives saved in 1975-2002.

```
PROC PRINT DATA=TRK3;
   FORMAT TLV208J TLS208J TL208J TLV108 TLS108 TL108 TLV208I TLS208I TL208I 9.0;
   ID CY; VAR TLV208J TLS208J TL208J TLV108 TLS108 TL108 TLV208I TLS208I TL208I;
   SUM TLV208J TLS208J TL208J TLV108 TLS108 TL108 TLV208I TLS208I TL208I;
TITLE1 'LIGHT TRUCK OCCUPANT LIVES SAVED BY BACK-CENTER 3-POINT BELTS, TRAILER CONSPICUITY TAPE,
AND AIR BAGS, 1975-2002';
TITLE2 ' ';
TITLE3 '...208J = 3-POINT BELTS FOR BACK-CENTER OCCUPANTS (MEDIAN INSTALLATION YEAR > 2001, FMVSS
PHASE-IN TO BEGIN 9/2005)';
TITLE4 '...108  = TRAILER CONSPICUITY TAPE (ON-ROAD FLEET 50% EQUIPPED 1996, FMVSS 12/1/93,
RETROFIT 6/1/2001)';
TITLE5 '...208I = FRONTAL AIR BAGS (MEDIAN INSTALLATION YEAR 1995, FMVSS PHASE-IN BEGAN 9/1/94)';
TITLE6 '...V = VOLUNTARY INSTALLATIONS, BEFORE FMVSS EFFECTIVE DATE';
TITLE7 '...S = STANDARD INSTALLATIONS, LIGHT TRUCKS BUILT ON OR AFTER FMVSS EFFECTIVE DATE';

PROC PRINT DATA=TRK3;
   FORMAT TLV214A TLS214A TL214A TLV208G TLS208G TL208G TLV213 TLS213 TL213 9.0;
   ID CY; VAR TLV214A TLS214A TL214A TLV208G TLS208G TL208G TLV213 TLS213 TL213;
   SUM TLV214A TLS214A TL214A TLV208G TLS208G TL208G TLV213 TLS213 TL213;
TITLE1 'LIGHT TRUCK OCCUPANT LIVES SAVED BY SIDE DOOR BEAMS, BACK-OUTBOARD 3-POINT BELTS, AND CHILD
SAFETY SEATS, 1975-2002';
TITLE2 ' ';
TITLE3 '...214A = SIDE DOOR BEAMS (MEDIAN INSTALLATION YEAR 1994, FMVSS EFFECTIVE 9/1/93)';
TITLE4 '...208G = 3-POINT BELTS FOR BACK-SEAT OCCUPANTS (MEDIAN INSTALLATION YEAR 1990, FMVSS
EFFECTIVE 9/1/91)';
TITLE5 '...213  = CHILD SAFETY SEATS (USE EXCEEDED 50% IN 1985, FMVSS EFFECTIVE 4/1/71, STATE LAWS
1978-85)';
TITLE6 '...V = VOLUNTARY INSTALLATIONS, BEFORE FMVSS EFFECTIVE DATE';
TITLE7 '...S = STANDARD INSTALLATIONS, LIGHT TRUCKS BUILT ON OR AFTER FMVSS EFFECTIVE DATE';

PROC PRINT DATA=TRK3;
   FORMAT TLV212 TLS212 TL212 TLV208F TLS208F TL208F TLV203 TLS203 TL203 9.0;
   ID CY; VAR TLV212 TLS212 TL212 TLV208F TLS208F TL208F TLV203 TLS203 TL203;
   SUM TLV212 TLS212 TL212 TLV208F TLS208F TL208F TLV203 TLS203 TL203;
TITLE1 'LIGHT TRUCK OCC LIVES SAVED BY ADHESIVE WINDSHIELD BONDING, FRONT-SEAT 3-POINT BELTS, AND
EA STEERING ASSEMBLIES, 1975-2002';
TITLE2 ' ';
TITLE3 '...212  = ADHESIVE WINDSHIELD BONDING (MEDIAN INSTALLATION YEAR 1979, FMVSS EFFECTIVE
9/1/78)';
TITLE4 '...208F = 3-POINT BELTS FOR FRONT-SEAT OCCUPANTS (MEDIAN INSTALLATION YEAR 1976, FMVSS
EFFECTIVE 1/1/76)';
TITLE5 '...203  = ENERGY-ABSORBING STEERING ASSEMBLIES (MEDIAN IMPLEMENTATION YEAR EARLY 1976,
FMVSS EFFECTIVE 9/1/81)';
TITLE6 '...V = VOLUNTARY INSTALLATIONS, BEFORE FMVSS EFFECTIVE DATE';
TITLE7 '...S = STANDARD INSTALLATIONS, LIGHT TRUCKS BUILT ON OR AFTER FMVSS EFFECTIVE DATE';

PROC PRINT DATA=TRK3;
   FORMAT TLV201 TL201 TLV105B TLS105B TL105B TLV208E TLS208E TL208E 9.0;
   ID CY; VAR TLV201 TL201 TLV105B TLS105B TL105B TLV208E TLS208E TL208E;
   SUM TLV201 TL201 TLV105B TLS105B TL105B TLV208E TLS208E TL208E;
TITLE1 'LIGHT TRUCK OCCUPANT LIVES SAVED BY SAFER INSTRUMENT PANELS, FRONT DISC BRAKES, AND BACK-
CENTER LAP BELTS, 1975-2002';
TITLE2 ' ';
TITLE3 '...201  = VOLUNTARY INSTRUMENT PANEL IMPROVEMENTS (MEDIAN IMPLEMENTATION YEAR 1972, FMVSS
EFFECTIVE 9/1/81)';
TITLE4 '...105B = FRONT DISC BRAKES (MEDIAN INSTALLATION YEAR 1971, FMVSS EFFECTIVE 9/1/83)';
TITLE5 '...208E = LAP BELTS FOR BACK-CENTER OCCUPANTS (MEDIAN INSTALLATION YEAR 1968, FMVSS
EFFECTIVE 7/1/71)';
TITLE6 '...V = VOLUNTARY INSTALLATIONS, BEFORE FMVSS EFFECTIVE DATE';
TITLE7 '...S = STANDARD INSTALLATIONS, LIGHT TRUCKS BUILT ON OR AFTER FMVSS EFFECTIVE DATE';
```

```
PROC PRINT DATA=TRK3;
   FORMAT TLV208D TLS208D TL208D TLV208C TLS208C TL208C TLV206 TLS206 TL206 9.0;
   ID CY; VAR TLV208D TLS208D TL208D TLV208C TLS208C TL208C TLV206 TLS206 TL206;
   SUM TLV208D TLS208D TL208D TLV208C TLS208C TL208C TLV206 TLS206 TL206;
TITLE1 'LIGHT TRUCK OCCUPANT LIVES SAVED BY FRONT-CENTER LAP BELTS, BACK-OUTBOARD LAP BELTS, AND
IMPROVED DOOR LOCKS, 1975-2002';
TITLE2 ' ';
TITLE3 '...208D = LAP BELTS FOR FRONT-CENTER OCCUPANTS (MEDIAN INSTALLATION YEAR 1968, FMVSS
EFFECTIVE 7/1/71)';
TITLE4 '...208C = LAP BELTS FOR BACK-OUTBOARD OCCUPANTS (MEDIAN INSTALLATION YEAR 1968, FMVSS
EFFECTIVE 7/1/71)';
TITLE5 '...206  = IMPROVED DOOR LOCKS (MEDIAN INSTALLATION YEAR 1967, FMVSS EFFECTIVE 1/1/72)';
TITLE6 '...V = VOLUNTARY INSTALLATIONS, BEFORE FMVSS EFFECTIVE DATE';
TITLE7 '...S = STANDARD INSTALLATIONS, LIGHT TRUCKS BUILT ON OR AFTER FMVSS EFFECTIVE DATE';

PROC PRINT DATA=TRK3;
   FORMAT TLV105A TLS105A TL105A TLV208B TLS208B TL208B TLV208A TLS208A TL208A 9.0;
   ID CY; VAR TLV105A TLS105A TL105A TLV208B TLS208B TL208B TLV208A TLS208A TL208A;
   SUM TLV105A TLS105A TL105A TLV208B TLS208B TL208B TLV208A TLS208A TL208A;
TITLE1 'LIGHT TRUCK OCC LIVES SAVED BY DUAL MASTER CYLINDERS, LAP BELTS FOR CHILDREN AGE 1-4, AND
FRONT-OUTBOARD LAP BELTS, 1975-2002';
TITLE2 ' ';
TITLE3 '...105A = DUAL MASTER CYLINDERS (MEDIAN INSTALLATION YEAR LATE 1966, FMVSS EFFECTIVE
9/1/83)';
TITLE4 '...208B = LAP BELT USE BY CHILDREN AGE 1-4 (MEDIAN INSTALLATION YEAR 1966, FMVSS EFFECTIVE
7/1/71)';
TITLE5 '...208A = LAP BELTS FOR FRONT-OUTBOARD OCCUPANTS (MEDIAN INSTALLATION YEAR 1964, FMVSS
EFFECTIVE 7/1/71)';
TITLE6 '...V = VOLUNTARY INSTALLATIONS, BEFORE FMVSS EFFECTIVE DATE';
TITLE7 '...S = STANDARD INSTALLATIONS, LIGHT TRUCKS BUILT ON OR AFTER FMVSS EFFECTIVE DATE';

DATA TRK4; SET TRK3;
PCT_SAVE=100*TL/T_WTFA;
PROC PRINT DATA=TRK4;
   FORMAT T_ORIGWT T_WTFA TLV TLS TL 9.0 PCT_SAVE 6.2;
   ID CY; VAR T_ORIGWT T_WTFA TLV TLS TL PCT_SAVE;
   SUM T_ORIGWT T_WTFA TLV TLS TL;
TITLE1 'OVERALL LIGHT TRUCK OCCUPANT LIVES SAVED BY VEHICLE SAFETY STANDARDS AND TECHNOLOGIES,
1975-2002';
TITLE2 ' ';
TITLE3 'T_ORIGWT = ACTUAL LIGHT TRUCK OCCUPANT FATALITIES';
TITLE4 'T_WTFA = FATALITIES THAT WOULD HAVE OCCURRED WITHOUT ANY VEHICLE SAFETY IMPROVEMENTS';
TITLE5 'TLV = OVERALL LIVES SAVED BY VOLUNTARY IMPROVEMENTS, BEFORE FMVSS EFFECTIVE DATE';
TITLE6 'TLS = OVERALL LIVES SAVED IN LIGHT TRUCKS BUILT ON OR AFTER FMVSS EFFECTIVE DATE';
TITLE7 'TL = OVERALL LIVES SAVED BY VEHICLE SAFETY IMPROVEMENTS (TLV + TLS)';
TITLE8 'PCT_SAVE = PERCENT OF WOULD-HAVE-BEEN FATALITIES SAVED BY SAFETY STANDARDS AND
TECHNOLOGIES';
RUN;


/* ------------------------------------------------------------------- */
/* ------------------------------------------------------------------- */
/*  SUMMARY OF LIVES SAVED BY FMVSS FOR CARS, LIGHT TRUCKS & PEDS       */
/* ------------------------------------------------------------------- */
/* ------------------------------------------------------------------- */
```

Adds up the lives saved among car occupants, LTV occupants, non-occupants and motorcyclists.
PCT_SAVE is the percent of potential car + LTV occupant fatalities saved.

```
DATA CTP3; MERGE CAR3 TRK3 PED3 MC3; BY CY;
SAVED=CL+TL+PL+ML;
VOL_SAVE=CLV+TLV+PLV+MLV;
STD_SAVE=CLS+TLS+PLS+MLS;
F_ACTUAL=C_ORIGWT+T_ORIGWT;
```

```
F_POTNTL=C_WTFA+T_WTFA;
PV_SAVE=CL+TL;
PCT_SAVE=100*PV_SAVE/F_POTNTL;
BELTS=CL208J+CL208H+CL208G+CLNCAP+CL208F+CL208E+CL208D+CL208C+CL208B+CL208A;
  BELTS=BELTS+TL208J+TL208G+TL208F+TL208E+TL208D+TL208C+TL208B+TL208A;
AIR_BAGS=CL208I+TL208I;
KIDSEATS=CL213+TL213;
FMVSS105=ML105B+ML105A+PL105B+PL105A+CL105B+CL105A+TL105B+TL105A;
FMVSS108=CL108+TL108;
FMVSS201=CL201+TL201;
FMVSS203=CL203+TL203;
FMVSS206=CL206+TL206;
FMVSS212=CL212+TL212;
FMVSS214=CL214B+CL214A+TL214A;
FMVSS216=CL216;
STDS8=FMVSS105+FMVSS108+FMVSS201+FMVSS203+FMVSS206+FMVSS212+FMVSS214+FMVSS216;
AB_DRV=CABDRV+TABDRV;
AB_RF = CABRF+ TABRF;
AB_KID=CABKID+TABKID;
RUN;

PROC PRINT DATA=CTP3;
  FORMAT SAVED VOL_SAVE STD_SAVE CL TL PL ML COMMA11.0;
  ID CY; VAR SAVED VOL_SAVE STD_SAVE CL TL PL ML;
  SUM SAVED VOL_SAVE STD_SAVE CL TL PL ML;
TITLE1 'OVERALL CAR OCCUPANT, LIGHT TRUCK OCCUPANT, PEDESTRIAN AND MOTORCYCLIST LIVES SAVED, 1975-
2002';
TITLE2 ' ';
TITLE3 'SAVED = OVERALL LIVES SAVED BY VEHICLE SAFETY IMPROVEMENTS (VOL_SAVE + STD_SAVE)';
TITLE4 'VOL_SAVE = OVERALL LIVES SAVED BY VOLUNTARY IMPROVEMENTS, BEFORE FMVSS EFFECTIVE DATE';
TITLE5 'STD_SAVE = OVERALL LIVES SAVED BY VEHICLES BUILT ON OR AFTER FMVSS EFFECTIVE DATE';
TITLE6 'CL = CAR OCCUPANT LIVES SAVED';
TITLE7 'TL = LIGHT TRUCK OCCUPANT LIVES SAVED';
TITLE8 'PL = PEDESTRIAN/NONOCCUPANT LIVES SAVED BY CAR AND LIGHT TRUCK BRAKING IMPROVEMENTS';
TITLE9 'ML = MOTORCYCLIST LIVES SAVED BY CAR AND LIGHT TRUCK BRAKING IMPROVEMENTS';
RUN;

PROC PRINT DATA=CTP3;
  FORMAT F_ACTUAL F_POTNTL PV_SAVE COMMA11.0 PCT_SAVE 6.2;
  ID CY; VAR F_ACTUAL F_POTNTL PV_SAVE PCT_SAVE;
  SUM F_ACTUAL F_POTNTL PV_SAVE;
TITLE1 'OVERALL LIVES SAVED AND NET EFFECTIVENESS OF SAFETY IMPROVEMENTS FOR PASSENGER VEHICLES
(CARS + LIGHT TRUCKS), 1975-2002';
TITLE2 ' ';
TITLE3 'F_ACTUAL = ACTUAL CAR + LIGHT TRUCK OCCUPANT FATALITIES';
TITLE4 'F_POTNTL = FATALITIES THAT WOULD HAVE OCCURRED WITHOUT ANY VEHICLE SAFETY IMPROVEMENTS';
TITLE5 'PV_SAVE = CAR + LIGHT TRUCK OCCUPANT LIVES SAVED BY VEHICLE SAFETY IMPROVEMENTS';
TITLE6 'PCT_SAVE = PERCENT OF WOULD-HAVE-BEEN FATALITIES SAVED BY SAFETY STANDARDS AND
TECHNOLOGIES';
RUN;

PROC PRINT DATA=CTP3;
  FORMAT BELTS AIR_BAGS KIDSEATS COMMA11.0;
  ID CY; VAR BELTS AIR_BAGS KIDSEATS;
  SUM BELTS AIR_BAGS KIDSEATS;
TITLE1 'LIVES SAVED IN CARS + LIGHT TRUCKS BY SAFETY BELTS, AIR BAGS AND CHILD SAFETY SEATS, 1975-
2002';
TITLE2 ' ';
TITLE3 'BELTS = LIVES SAVED BY SAFETY BELTS (ALL TYPES, ALL SEAT POSITIONS)';
TITLE4 'AIR_BAGS = LIVES SAVED BY FRONTAL AIR BAGS, DRIVERS + RIGHT-FRONT PASSENGERS';
TITLE5 'KIDSEATS = LIVES SAVED BY CHILD SAFETY SEATS (ALL TYPES, ALL SEAT POSITIONS, ALL AGES)';
RUN;

PROC PRINT DATA=CTP3;
  FORMAT FMVSS105 FMVSS108 FMVSS201 FMVSS203 FMVSS206 FMVSS212 FMVSS214 FMVSS216 STDS8 9.0;
```

```
    ID CY; VAR FMVSS105 FMVSS108 FMVSS201 FMVSS203 FMVSS206 FMVSS212 FMVSS214 FMVSS216 STDS8;
    SUM FMVSS105 FMVSS108 FMVSS201 FMVSS203 FMVSS206 FMVSS212 FMVSS214 FMVSS216 STDS8;
TITLE1 'LIVES SAVED BY THE OTHER SAFETY STANDARDS, 1975-2002';
TITLE2 ' ';
TITLE3 'FMVSS105 = DUAL MASTER CYLINDERS + FRONT DISC BRAKES (INCLUDES OCCUPANT + PEDESTRIAN LIVES
SAVED)';
TITLE4 'FMVSS108 = TRAILER CONSPICUITY TAPE (LIVES SAVED BY CARS & LIGHT TRUCKS NOT HITTING THE
TRAILERS)';
TITLE5 'FMVSS201 = VOLUNTARY INSTRUMENT PANEL IMPROVEMENTS    FMVSS203 = ENERGY-ABSORBING STEERING
ASSEMBLIES (INCLUDES FMVSS 204)';
TITLE6 'FMVSS206 = IMPROVED DOOR LOCKS                        FMVSS212 = ADHESIVE WINDSHIELD
BONDING';
TITLE7 'FMVSS214 = SIDE DOOR BEAMS IN CARS AND LIGHT TRUCKS, VOLUNTARY TTI(d) REDUCTIONS IN 2-DOOR
CARS';
TITLE8 'FMVSS216 = ROOF CRUSH STRENGTH FOR PASSENGER CARS';
TITLE9 'STDS8    = LIVES SAVED BY THESE EIGHT STANDARDS (I.E., EVERTHING EXCEPT BELTS, AIR BAGS AND
SAFETY SEATS)';
RUN;
```

Five pages of more detailed summary for air bags: (1) driver lives saved; (2) adult passenger lives saved in vehicles without switches; (3) estimated net fatality increase for child passengers in vehicles without switches; (4) actual adult passenger lives saved in pickup trucks with switches, and potential lives saved if the trucks did not have the switches (or if the air bags were always "on" for adults); (5) estimated actual net increase in child passenger fatalities in pickup trucks with switches, potential number of child fatalities if these trucks had air bags and no switches, and lives saved because the switches were turned "off."

```
PROC PRINT DATA=CTP3;
    FORMAT AB_DRV CABDRV TABDRV COMMA11.0;
    ID CY; VAR AB_DRV CABDRV TABDRV;
    SUM AB_DRV CABDRV TABDRV;
TITLE1 'LIVES SAVED BY DRIVER AIR BAGS IN ALL CARS AND LIGHT TRUCKS, 1975-2002';
TITLE2 ' ';
TITLE3 'AB_DRV = LIVES SAVED BY DRIVER AIR BAGS IN CARS + LIGHT TRUCKS';
TITLE4 'CABDRV = LIVES SAVED BY DRIVER AIR BAGS IN CARS';
TITLE5 'TABDRV = LIVES SAVED BY DRIVER AIR BAGS IN LIGHT TRUCKS';
RUN;

PROC PRINT DATA=CTP3;
    FORMAT AB_RF CABRF TABRF COMMA11.0;
    ID CY; VAR AB_RF CABRF TABRF;
    SUM AB_RF CABRF TABRF;
TITLE1 'ADULT (AGE 13+) LIVES SAVED BY PASSENGER AIR BAGS IN CARS AND LIGHT TRUCKS WITHOUT ON-OFF
SWITCHES';
TITLE2 ' ';
TITLE3 'AB_RF = ADULT LIVES SAVED BY PASSENGER AIR BAGS IN CARS + LIGHT TRUCKS WITHOUT ON-OFF
SWITCHES';
TITLE4 'CABRF = ADULT LIVES SAVED BY PASSENGER AIR BAGS IN CARS';
TITLE5 'TABRF = ADULT LIVES SAVED BY PASSENGER AIR BAGS IN LIGHT TRUCKS WITHOUT ON-OFF SWITCHES';
RUN;

PROC PRINT DATA=CTP3;
    FORMAT AB_KID CABKID TABKID COMMA11.0;
    ID CY; VAR AB_KID CABKID TABKID;
    SUM AB_KID CABKID TABKID;
TITLE1 'CHILD (AGE 0-12) FATALITIES DUE TO AIR BAGS IN CARS AND LIGHT TRUCKS WITHOUT ON-OFF
SWITCHES';
TITLE2 ' ';
TITLE3 'AB_KID = CHILD FATALITIES DUE TO AIR BAGS IN CARS + LIGHT TRUCKS WITHOUT ON-OFF SWITCHES';
TITLE4 'CABKID = CHILD FATALITIES DUE TO AIR BAGS IN CARS';
TITLE5 'TABKID = CHILD FATALITIES DUE TO AIR BAGS IN LIGHT TRUCKS WITHOUT ON-OFF SWITCHES';
RUN;
```

```
PROC PRINT DATA=CTP3;
  FORMAT SW0 SW1 SW2 COMMA11.0;
  ID CY; VAR SW0 SW1 SW2;
  SUM SW0 SW1 SW2;
TITLE1 'ADULT (AGE 13+) LIVES SAVED BY PASSENGER AIR BAGS IN PICKUP TRUCKS WITH ON-OFF SWITCHES';
TITLE2 ' ';
TITLE3 'SW0 = ACTUAL ADULT LIVES SAVED BY AIR BAGS & SWITCHES LEFT ON';
TITLE4 'SW1 = THEORETICAL ADDITIONAL LIVES SAVED IF SWITCHES HAD NOT EXISTED';
TITLE5 'SW2 = LIVES LOST BECAUSE SWITCHES WERE TURNED OFF FOR ADULT PASSENGERS';
RUN;
PROC PRINT DATA=CTP3;
  FORMAT SWKID0 SWKID1 SWKID2 COMMA11.0;
  ID CY; VAR SWKID0 SWKID1 SWKID2;
  SUM SWKID0 SWKID1 SWKID2;
TITLE1 'CHILD (AGE 0-12) FATALITIES DUE TO AIR BAGS IN PICKUP TRUCKS WITH ON-OFF SWITCHES';
TITLE2 ' ';
TITLE3 'SWKID0 = ACTUAL ADDITIONAL CHILD PASSENGER FATALITIES DUE TO AIR BAGS & SWITCHES LEFT ON';
TITLE4 'SWKID1 = THEORETICAL ADDITIONAL FATALITIES IF SWITCHES HAD NOT EXISTED';
TITLE5 'SWKID2 = LIVES SAVED BECAUSE SWITCHES WERE TURNED OFF FOR CHILD PASSENGERS';
RUN;
```

The main model has been completed. However, subsets of the main databases are generated to describe the distribution of CY 1975-80 fatality cases on certain core variables. These subsets will be saved as *FARS7580*, *PED7580*, and *MC7580*. They will be used to set up the post-processor (lives saved in 1960-74).

```
/* ------------------------------------------------------------------ */
/*  CREATES CY 75-80 OCCUPANT, NON-OCC & MOTORCYCLIST DATABASES NEEDED TO RUN  */
/*  OLDFARS1 AND OLDFARS2 (LIVES SAVED IN CY 1960-74)                  */
/* ------------------------------------------------------------------ */

DATA FARS7580.FARS7580; SET CAR1 TRK1;
IF 1975 LE CY LE 1980;
IF EJECTION IN (1,2) THEN EJECT2=1; ELSE EJECT2=0;
VEHAGE=CY-MY;
KEEP AGE CRSH CY EJECT2 EJECTION IMPACT2 MY M_HARM ORIGWT REST3
  SEAT2 VEHAGE VE_FORMS VTYP WEIGHTFA;
RUN;

DATA PED7580.PED7580; SET PED1;
IF 1975 LE CY LE 1980;
VEHAGE=CY-MY;
KEEP CY MY VTYP ORIGWT VEHAGE WEIGHTFA;
RUN;

DATA MC7580.MC7580; SET MC1;
IF 1975 LE CY LE 1980;
VEHAGE=CY-OMY;
MY=OMY;
VTYP=OVTYP;
KEEP CY MY VTYP ORIGWT VEHAGE WEIGHTFA;
RUN;

/* ------------------------------------------------------------ */
/*  AVERAGE COMPLIANCE WITH STDS, BY MODEL YEAR, TO SET UP OLDFA24  */
/* ------------------------------------------------------------ */
```

Also needed for the post-processor are the percentage, by model year (1960-75), of cars with energy-absorbing columns, side door beams and adhesive bonding, and the percent of LTVs with energy-absorbing columns. For collisions of cars/LTVs with non-occupants or motorcyclists, we need the percentage, by model year of the car or LTV of post-standard vs. voluntary brake improvements (this percentage depends on the proportion of striking vehicles that are cars vs. the proportion that are LTVs, in certain model years).

```
DATA OLDCAR; SET CAR2;
IF 1960 LE MY LE 1975;
IF SEAT2=11;
KEEP MY PS203 PV203 PS214A PV214A PS212 PV212 ORIGWT;
PROC SORT; BY MY;
PROC MEANS MEAN NOPRINT; BY MY; WEIGHT ORIGWT;
   VAR PS203 PV203 PS214A PV214A PS212 PV212;
   OUTPUT OUT=OLDCAR2 MEAN=PS203 PV203 PS214A PV214A PS212 PV212;
PROC PRINT DATA=OLDCAR2; ID MY; VAR PS203 PV203 PS214A PV214A PS212 PV212;
TITLE 'CARS';
RUN;

DATA OLDTRK; SET TRK2;
IF 1960 LE MY LE 1975;
IF SEAT2=11;
KEEP MY PV203 ORIGWT;
PROC SORT; BY MY;
PROC MEANS MEAN NOPRINT; BY MY; WEIGHT ORIGWT; VAR PV203;
   OUTPUT OUT=OLDTRK2 MEAN=PV203;
PROC PRINT DATA=OLDTRK2; ID MY; VAR PV203;
TITLE 'LIGHT TRUCKS';
RUN;

DATA OLDPED; SET PED2;
IF 1960 LE MY LE 1975;
KEEP MY PS105A PV105A PS105B PV105B ORIGWT;
PROC SORT; BY MY;
PROC MEANS MEAN NOPRINT; BY MY; WEIGHT ORIGWT;
   VAR PS105A PV105A PS105B PV105B;
   OUTPUT OUT=OLDPED2 MEAN=PS105A PV105A PS105B PV105B;
PROC PRINT DATA=OLDPED2; ID MY; VAR PS105A PV105A PS105B PV105B;
TITLE 'NON-OCCUPANTS HIT BY CARS OR LTVs, BY MODEL YEAR OF THE CAR OR LTV';
RUN;

DATA OLDMC; SET MC2;
IF 1960 LE OMY LE 1975;
KEEP OMY MS105A MV105A MS105B MV105B ORIGWT;
PROC SORT; BY OMY;
PROC MEANS MEAN NOPRINT; BY OMY; WEIGHT ORIGWT;
   VAR MS105A MV105A MS105B MV105B;
   OUTPUT OUT=OLDMC2 MEAN=MS105A MV105A MS105B MV105B;
PROC PRINT DATA=OLDMC2; ID OMY; VAR MS105A MV105A MS105B MV105B;
TITLE 'MOTORCYCLISTS HIT BY CARS OR LTVs, BY MODEL YEAR OF THE CAR OR LTV';
RUN;

/* ------------------------------------------------------------------ */
/*  AVERAGE COMPLIANCE WITH STDS, BY MY, TO FIND CHRONOLOGICAL ORDER  */
/* ------------------------------------------------------------------ */
```

Also computes the percent of cars and LTVs, by model year, equipped with the various safety technologies (except belts, child safety seats and conspicuity tape). This information is used to check that the models "remove" the technologies in the correct (reverse-chronological) order.

The first set of technologies is counted "per driver" (IF SEAT2=11), because they are installed at the driver's seat (e.g., energy-absorbing columns), or are vehicle-wide (e.g., brakes).

```
DATA ALLCAR1; SET CAR2;
IF 1960 LE MY LE 2002;
IF SEAT2=11;
KEEP MY PVNCAP PS216 PV216 PS105B PV105B PS203 PV203 PS214A PV214A
   PS105A PV105A PS206 PV206 PS212 PV212 ORIGWT;
PROC SORT; BY MY;
PROC MEANS MEAN NOPRINT; BY MY; WEIGHT ORIGWT;
```

```
    VAR PVNCAP PS216 PV216 PS105B PV105B PS203 PV203 PS214A PV214A
       PS105A PV105A PS206 PV206 PS212 PV212;
    OUTPUT OUT=ALLCAR1A
          MEAN=PVNCAP PS216 PV216 PS105B PV105B PS203 PV203 PS214A
               PV214A PS105A PV105A PS206 PV206 PS212 PV212;
RUN;
```

Air bags and automatic belts are counted "per front-seat occupant."

```
DATA ALLCAR2; SET CAR2;
IF 1960 LE MY LE 2002;
IF SEAT2 IN (11,13);
KEEP MY PS208I PV208I PS208H PV208H ORIGWT;
PROC SORT; BY MY;
PROC MEANS MEAN NOPRINT; BY MY; WEIGHT ORIGWT;
  VAR PS208I PV208I PS208H PV208H;
  OUTPUT OUT=ALLCAR2A MEAN=PS208I PV208I PS208H PV208H;
RUN;
```

Voluntary TTI(d) improvements are counted "per front-seat occupant of 2-door cars."

```
DATA ALLCAR3; SET CAR2;
IF 1960 LE MY LE 2002;
IF SEAT2 IN (11,13) AND BOD2 IN (1,2,3);
KEEP MY PV214B ORIGWT;
PROC SORT; BY MY;
PROC MEANS MEAN NOPRINT; BY MY; WEIGHT ORIGWT;
  VAR PV214B;
  OUTPUT OUT=ALLCAR3A MEAN=PV214B;
RUN;
```

Instrument panel improvements are counted "per RF passenger" because, in CAR2, PV201 can equal 1 only for RF passengers.

```
DATA ALLCAR4; SET CAR2;
IF 1960 LE MY LE 2002;
IF SEAT2=13;
KEEP MY PV201 ORIGWT;
PROC SORT; BY MY;
PROC MEANS MEAN NOPRINT; BY MY; WEIGHT ORIGWT;
  VAR PV201;
  OUTPUT OUT=ALLCAR4A MEAN=PV201;
RUN;

DATA ALLCAR5; MERGE ALLCAR1A ALLCAR2A ALLCAR3A ALLCAR4A; BY MY;
PROC PRINT DATA=ALLCAR5;
  FORMAT PS208I PV208I PV214B PS208H PV208H PVNCAP PS216 PV216 PS105B PV105B PS214A
         PV214A PV201 PS203 PV203 PS105A PV105A PS206 PV206 PS212 PV212 6.4;
  ID MY;
  VAR PS208I PV208I PV214B PS208H PV208H PVNCAP PS216 PV216 PS105B PV105B PS214A
      PV214A PV201 PS203 PV203 PS105A PV105A PS206 PV206 PS212 PV212;
TITLE 'CARS: PERCENT OF NEW VEHICLES WITH SAFETY TECHNOLOGY';
RUN;

DATA ALLTRK1; SET TRK2;
IF 1960 LE MY LE 2002;
IF SEAT2=11;
KEEP MY PS105B PV105B PS203 PV203 PS214A PV214A
   PS105A PV105A PS206 PV206 PS212 PV212 ORIGWT;
PROC SORT; BY MY;
PROC MEANS MEAN NOPRINT; BY MY; WEIGHT ORIGWT;
  VAR PS105B PV105B PS203 PV203 PS214A PV214A
      PS105A PV105A PS206 PV206 PS212 PV212;
  OUTPUT OUT=ALLTRK1A
```

```
            MEAN=PS105B PV105B PS203 PV203 PS214A
                  PV214A PS105A PV105A PS206 PV206 PS212 PV212;
RUN;

DATA ALLTRK2; SET TRK2;
IF 1960 LE MY LE 2002;
IF SEAT2 IN (11,13);
KEEP MY PS208I PV208I ORIGWT;
PROC SORT; BY MY;
PROC MEANS MEAN NOPRINT; BY MY; WEIGHT ORIGWT;
  VAR PS208I PV208I;
  OUTPUT OUT=ALLTRK2A MEAN=PS208I PV208I;
RUN;

DATA ALLTRK4; SET TRK2;
IF 1960 LE MY LE 2002;
IF SEAT2=13;
KEEP MY PV201 ORIGWT;
PROC SORT; BY MY;
PROC MEANS MEAN NOPRINT; BY MY; WEIGHT ORIGWT;
  VAR PV201;
  OUTPUT OUT=ALLTRK4A MEAN=PV201;
RUN;

DATA ALLTRK5; MERGE ALLTRK1A ALLTRK2A ALLTRK4A; BY MY;
PROC PRINT DATA=ALLTRK5;
  FORMAT PS208I PV208I PS105B PV105B PS214A
      PV214A PV201 PS203 PV203 PS105A PV105A PS206 PV206 PS212 PV212 6.4;
  ID MY;
  VAR PS208I PV208I PS105B PV105B PS214A
      PV214A PV201 PS203 PV203 PS105A PV105A PS206 PV206 PS212 PV212;
TITLE 'LTVs: PERCENT OF NEW VEHICLES WITH SAFETY TECHNOLOGY';
RUN;
```

DESCRIPTION OF THE POSTPROCESSOR PROGRAM OLDFA24
(LIVES SAVED IN CY 1960-74)

```
OPTION NOCENTER NOFMTERR OBS=5000000 LS=153 PAGESIZE=47;
/*  PROGRAM NAME: OLDFA24.SAS           UPDATED: 04/14/2004 */
/*  LIVES SAVED BY FMVSS IN 1960-74                         */
```

Codes for impact location/crash type (CRSH):

```
PROC FORMAT; VALUE CRSH 1='FRONTAL' 2='SIDE IMPACT' 2.1='SIDE FIXED OBJ'
  2.2='SIDE MULTIVEH' 3='ROLLOVER (PRIMARY)' 4='REAR & OTHER' 9='UNKNOWN';
```

Codes for vehicle type:

```
PROC FORMAT; VALUE VTYP 1='PASSENGER CAR' 2='LIGHT TRUCK' 3='BIG TRUCK, BUS'
  4='MOTORCYCLE' 5='OTHER' 9='UNKNOWN';
PROC FORMAT; VALUE MY 1900-1954='PRE-1955' 1955-1959='1955-59' 1960-1961='1960-61';
```

Codes for streamlined seat position variable (SEAT2):

```
PROC FORMAT; VALUE NEWSEAT 11='DRIVER' 12='CENTER FRONT' 13='RIGHT FRONT' 18='OTHER FRONT'
  19='UNK FRONT' 21='OUTBOARD REAR' 22='CENTER REAR' 28='OTHER REAR'
  29='UNK REAR' 51='OTHER ENCLOSED' 52='UNENCLOSED' 99='UNK PASSENGER';
PROC FORMAT; VALUE NEWVTYP 1='CONVERTIBLE' 2='2-DOOR CAR' 4='4-DOOR CAR'
  6='STATION WAGON' 9='CAR, UNK STYLE'
  11='PICKUP TRUCK' 12='S U V' 13='VAN' 19='TRUCK, UNK STYLE';
```

Codes for streamlined restraint use variables (REST2 and REST3):

```
PROC FORMAT; VALUE NEWREST 0='UNRESTRAINED' 2='LAP BELT ONLY'
  3='LAP+SHOULDER BELT' 4='CHILD SAFETY SEAT' 7='2-PT AUTOMATIC BELT';
proc format; value vehage 25-99='ge 25';
RUN;
```

OLDFA24 begins by computing the ratio of certain fatality counts on FARS to counts in *Accident Facts* for CY 1975-80.[1] By applying these ratios to *Accident Facts* counts for any year before 1975, it estimates what the fatality count "would have been on FARS" if FARS had existed in that year. For example, *Accident Facts* counts of non-occupant fatalities total up to 60,100 for 1975-80. We do not know what vehicles struck the non-occupants: they could have been cars, LTVs, heavy trucks, or some other type. The 1975-80 FARS data suggest that 47,250 non-occupants were fatally injured in collisions by passenger cars or LTVs under conditions that qualify as "fatal motor-vehicle traffic crashes" as defined in FARS. This number, 47,250, is produced by another program, OLDFA41, that analyzes the 1975-80 data files produced by LS2004. Thus, the *Accident Facts* counts of non-occupant fatalities for any preceding year is multiplied by 47,250/60,100 to obtain an estimate of how many non-occupant fatalities in collisions with cars or LTVs would have been on FARS, if FARS had existed that year.

---

[1] For example, *Accident Facts 1978 Edition*, National Safety Council, Chicago, 1978, pp. 56 and 58.

```
/* -------------------------------------------------------------------------- */
/*  1960-74 FATALITY COUNTS IMPUTED FROM ACCIDENT FACTS                        */
/*                                                                             */
/*  USING 1975-80 FARS AND ACCIDENT FACTS FATALITY COUNTS,                     */
/*  ADJUSTS THE 1960-74 ACCIDENT FACTS COUNTS (AFPED, AFCAR, AFTRK, AFMC)      */
/*  TO MATCH WHAT WOULD HAVE BEEN ON FARS (FARSPED, FARSCAR, FARSTRK, FARSMC)  */
/*                                                                             */
/*  FOR EXAMPLE:                                                               */
/*  60100=ACC FACTS 1975-80 NONOCCS                                            */
/*  47250=FARS 75-80 NONOCCS STRUCK BY CARS OR LIGHT TRUCKS (OLDFA14, P. 3)    */
/*  FARSPED = 47250/60100 * AFPED IN EACH YEAR BEFORE 1975                     */
/*                                                                             */
/*  168100=ACC FACTS 1975-80 CAR OCCUPANT FATALITIES                          */
/*  35950=ACC FACTS 1975-80 TRUCK OCCUPANT FATALITIES                         */
/*                                                                             */
/*  ADDITIONAL ADJUSTMENT BEFORE 1965, BASED ON 1965                          */
/*  37900=ACC FACTS 65 TOTAL FATALITIES MINUS NONOCC + MCYCLE, ETC. IN EARLIER YEARS */
/*  31990.47 = FARSCAR FOR 65 (32600*164957/168100)                           */
/*   4527.41 = FARSTRK FOR 65 (4400*36991/35950)                              */
/* -------------------------------------------------------------------------- */
```

ACCFAC1 reads in the *Accident Facts* counts for total fatalities, non-occupants, car occupants, "truck" occupants and motorcyclists in each year, 1960-74. The FARS – *Accident Facts* fatality ratios for 1975-80 are:

- 47,250 FARS non-occupants in collisions with cars/LTVs to 60,100 *Accident Facts* non-occupants (any collision type).
- 11,787 FARS motorcyclists in collisions with cars/LTVs to 22,800 *Accident Facts* motorcyclists (any collision type).
- 164,957 FARS car occupant fatalities to 168,100 *Accident Facts* car occupant fatalities.
- 36,991 FARS LTV occupant fatalities to 35,950 *Accident Facts* "truck" occupant fatalities.

These ratios are used to estimate the numbers of fatalities that would have been on FARS, if it had existed, in 1960-74: FARSPED, FARSMC, FARSCAR, FARSTRK.

An additional calculation (as shown below) is needed for 1960-64, when *Accident Facts* did not separately indicate car and truck occupant fatalities, only total, non-occupant and motorcyclist fatalities.

```
DATA ACCFAC1;
INPUT CY 1-4 AFTOT 6-10 AFPED 12-16 AFCAR 18-22 AFTRK 24-27 AFMC 29-32;
FARSPED=AFPED*47250/60100;                                   /* 47250 FROM OLDFA14, P. 3 */
FARSMC=AFMC*11787/22800;                                     /* 11787 FROM OLDFA14, P. 4 */
IF AFCAR=. THEN FARSCAR=(AFTOT-(AFPED+AFMC))*(31990.47/37900);
  ELSE FARSCAR=AFCAR*164957/168100;                          /* 164957 FROM OLDFA14, P. 5 */
IF AFTRK=. THEN FARSTRK=(AFTOT-(AFPED+AFMC))*( 4527.41/37900);
  ELSE FARSTRK=AFTRK*36991/35950;                            /* 36991 FROM OLDFA14, P. 6 */
CARDS;
1960 38200  8200         .    .  750
1961 38000  8150         .    .  700
1962 40900  8400         .    .  800
1963 43600  8800         .    .  900
1964 47700  9650         .    . 1100
1965 49000  9500      32600 4400 1600
1966 53000 10050      35000 4650 2150
1967 53100 10100      35200 4650 2000
1968 55200 10600      36400 5150 1900
1969 56400 10600      37100 5550 1950
```

358

```
1970 54800 11200 34700 5350 2300
1971 54700 11450 34100 5300 2200
1972 56600 11800 35100 5500 2700
1973 55800 11650 33500 5700 3100
1974 46200  9700 26600 4900 3200
RUN;

PROC PRINT; ID CY;
TITLE 'FATALITIES REPORTED IN ACCIDENT FACTS AND IMPUTED IN FARS, BASED ON 1975-80 RATIOS';
RUN;
```

Here are the resulting estimates of FARSPED, FARSMC, FARSCAR and FARSTRK:

CY	FARSPED	FARSMC	FARSCAR	FARSTRK
1960	6446.76	387.73	24689.21	3494.11
1961	6407.45	361.88	24604.81	3482.16
1962	6603.99	413.58	26757.20	3786.78
1963	6918.47	465.28	28614.17	4049.58
1964	7586.73	568.67	31188.60	4413.93
1965	7468.80	827.16	31990.47	4527.41
1966	7901.21	1111.49	34345.60	4784.65
1967	7940.52	1033.95	34541.86	4784.65
1968	8333.61	982.25	35719.42	5299.13
1969	8333.61	1008.10	36406.33	5710.71
1970	8805.32	1189.04	34051.21	5504.92
1971	9001.87	1137.34	33462.43	5453.47
1972	9277.04	1395.83	34443.73	5659.26
1973	9159.11	1602.62	32873.64	5865.05
1974	7626.04	1654.32	26102.65	5041.89

```
/* -------------------------------------------------------------- */
/* VEHICLE AGE DISTRIBUTIONS IN 1975-80 FARS FATALITIES:          */
/* CARS/LIGHT TRUCKS HITTING PEDS, CAR OCCUPANTS, LIGHT TRUCK OCCS */
/*                                                                 */
/* TYPE IN FIRST COLUMNS ON PP. 3-6 OF OLDFA14                     */
/* -------------------------------------------------------------- */
```

Working with the files *FARS7580, PED7580,* and *MC7580* generated by LS2004, OLDFA14 computes the vehicle-age distribution, in 1975-80, of the cars/LTVs that hit non-occupants, the car occupant fatalities, the LTV occupant fatalities, and the cars/LTVs that hit motorcyclists.

```
DATA VEHAGE1; INPUT VEHAGE 4-5 NPED 10-17 NCAR 22-29 NTRK NMC;
CASEFLAG=1;
CARDS;
    -1    162.4817    605.2072    109.1351     21.3214
     0    3364.35     11344.2     3351.701    796.6358
     1    4819.269    15132.48    5172.744   1213.928
     2    4603.945    13773.37    4285.246   1119.965
     3    4382.934    13421.52    3597.582    977.4574
     4    3983.189    13338.52    3032.132    945.1746
     5    3867.286    13428.7     2596.071    964.3163
     6    4034.862    14013.89    2346.24     985.7485
     7    3787.245    13593.19    2068.892    936.6506
     8    3224.628    12302.93    1682.43     860.7186
     9    2942.550    10758.34    1374.067    670.5461
```

```
        10     2298.511      9071.043     1304.867     618.9576
        11     1829.233      7177.197     1109.991     493.7239
        12     1298.031      5443.185      924.8238    340.0405
        13      876.5308     4058.674      768.6906    291.7421
        14      578.49       2736.983      609.642     157.1982
        15      385.4057     1688.982      442.6576    100.5024
        16      212.7371     1010.59       360.1974     78.11997
        17      138.6325      587.4548     287.0335     49.64865
        18       97.32384     348.0791     247.7068     29.4394
        19       70.10814     220.0954     185.4293     20.31554
        20       64.19948     226.6397     175.3913     21.4539
        21       51.34826     155.1664     139.663      11.14231
        22       33.84781     129.4678     139.7618     10.18185
        23       34.0774      102.6976     136.4589     13.26287
        24       30.53487      66.08272    113.1879      9.18104
ge      25       78.29803     222.098      429.6281     50.01732
RUN;
```

The above are actual (weighted) fatality counts. This code transforms them to percentages of the total:

```
PROC MEANS SUM NOPRINT DATA=VEHAGE1; VAR NPED NCAR NTRK NMC;
  OUTPUT OUT=VEHAGE2 SUM=TOTPED TOTCAR TOTTRK TOTMC;
DATA VEHAGE3; SET VEHAGE2; CASEFLAG=1;
DATA VEHAGE4; MERGE VEHAGE1 VEHAGE3; BY CASEFLAG;
PCTPED=NPED/TOTPED;
PCTCAR=NCAR/TOTCAR;
PCTTRK=NTRK/TOTTRK;
PCTMC =NMC /TOTMC;
KEEP VEHAGE PCTPED PCTCAR PCTTRK PCTMC;
```

First is the analysis of the effect of car/LTV brake improvements on non-occupant fatalities. To perform this analysis, we need to know, in each CY, how many vehicles of each MY hit non-occupants. Based on the MY, we know the probability they had dual master cylinders and/or front disc brakes.

```
/* ------------------------------------------------------------------ */
/* ------------------------------------------------------------------ */
/*   ANALYZES EFFECT OF BRAKE IMPROVEMENTS ON NONOCCUPANT FATALITIES   */
/*                                                                    */
/*   FIRST STEP: CREATE FILE WITH ALL POSSIBLE CY-MY COMBINATIONS      */
/*   WITH ORIGWT INDICATING THE NUMBER OF PEDESTRIAN FATALITIES        */
/*   FOR THAT COMBINATION (MY=MODEL YEAR OF THE VEH THAT HIT THE PED)  */
/* ------------------------------------------------------------------ */
/* ------------------------------------------------------------------ */
```

The file ACCFAC1 has one record for each CY, 1960-74, with FARSPED showing the total number of non-occupant fatalities in collisions with a car or LTV. This data step creates another file, PED5, that is a matrix with CY running from 1960 to 1974 and VEHAGE (vehicle age) running from −1, 0, 1, 2, … , 24, 25+. So far, it just has FARSPED, the CY grand total, on each record.

```
DATA PED5(KEEP=CY FARSPED VEHAGE);
SET ACCFAC1;
DO VEHAGE = -1 TO 25;
  OUTPUT;
END;
RUN;
PROC SORT DATA=PED5; BY VEHAGE;
```

360

To create PED6, we merge onto PED5 the proportion of the non-occupant fatalities (PCTPED) at each value of VEHAGE. ORIGWT = PCTPED * FARSPED is the number of fatalities, in that CY, at that VEHAGE. Because MY = CY – VEHAGE, PED6 is also a matrix of non-occupant fatalities by CY and MY (of the striking car/LTV).

PED6 looks like this:

CY	MY	ORIGWT
1960	≤ 1935	10.683
1960	1936	4.166
1960	1937	4.649
...		
1960	1959	657.537
1960	1960	459.029
1960	1961	22.169
1961	≤ 1936	10.618
1961	1937	4.141
...		
1961	1961	456.230
1961	1962	22.034
...		
...		
1974	≤ 1949	12.637
1974	1950	4.928
...		
1974	1973	777.818
1974	1974	542.998
1974	1975	26.224

```
DATA PED6; MERGE PED5 VEHAGE4; BY VEHAGE;
ORIGWT=PCTPED*FARSPED;
WEIGHTFA=ORIGWT;
MY=CY-VEHAGE;
KEEP CY MY ORIGWT WEIGHTFA;
PROC SORT DATA=PED6; BY CY MY;
RUN;

/* ---------------------------------------------------------------- */
/*   REPEATS THE NONOCCUPANT ANALYSIS OF LS2004.SAS                  */
/* ---------------------------------------------------------------- */
```

The main analysis sections in the post-processor OLDFA24 are similar to the corresponding sections of the main processor LS2004. The principal difference is that LS2004 worked through the individual FARS cases, one case at a time. OLDFA24 works on cells rather than cases, one cell at a time. In the post-processor for non-occupant fatalities, two safety technologies, dual master cylinders (105A) and front disc brakes (105B) saved non-occupant lives because they enabled car/LTV drivers to avoid hitting the non-occupants. The model operates on the file PED6 that, as shown above, has one cell for each CY, MY combination. For example, the cell with CY = 1974 and MY = 1973 originally comprises 777.818 fatalities (ORIGWT). The model assumes that 86

361

percent of MY 1973 cars are equipped with front disc brakes and 100 percent with dual master cylinders. Disc brakes reduce pedestrian crashes by 0.17 percent, dual master cylinders by 0.7 percent. Thus, "removing" the disc brakes from the MY 1973 cars will increase fatalities in that cell to 777.818 / (1 – (.86x.0017)) = 778.957 and removing the dual master cylinders will further increase them to 778.957 / (1 – .007) = 784.448 (the final WEIGHTFA). In this cell, braking technologies saved 784.448 – 777.818 = 6.63 lives.

For more details on the computation methods, see the notes on the corresponding sections of LS2004. In general, the notes on this program will be limited to the spots where it differs from LS2004.

```
DATA PED7; SET PED6;
/* IMPLEMENTATION OF FRONT DISC BRAKES IN THE VEHICLES THAT HIT THESE PEDESTRIANS */
 IF MY LE 1964 THEN DO; PS105B=0; PV105B=0; END;
   ELSE IF MY=1965 THEN DO; PS105B=0; PV105B=.02; END;
   ELSE IF MY=1966 THEN DO; PS105B=0; PV105B=.03; END;
   ELSE IF MY=1967 THEN DO; PS105B=0; PV105B=.06; END;
   ELSE IF MY=1968 THEN DO; PS105B=0; PV105B=.13; END;
   ELSE IF MY=1969 THEN DO; PS105B=0; PV105B=.28; END;
   ELSE IF MY=1970 THEN DO; PS105B=0; PV105B=.41; END;
   ELSE IF MY=1971 THEN DO; PS105B=0; PV105B=.63; END;
   ELSE IF MY=1972 THEN DO; PS105B=0; PV105B=.74; END;
   ELSE IF MY=1973 THEN DO; PS105B=0; PV105B=.86; END;
   ELSE IF MY=1974 THEN DO; PS105B=0; PV105B=.84; END;
   ELSE IF MY=1975 THEN DO; PS105B=0; PV105B=.93; END;
/* IMPLEMENTATION OF DUAL MASTER CYLINDERS IN THE VEHICLES THAT HIT THESE PEDESTRIANS */
IF MY LE 1961 THEN DO; PS105A=0; PV105A=0; END;
   ELSE IF MY IN (1962,1963) THEN DO; PS105A=0; PV105A=.09; END;
   ELSE IF MY IN (1964,1965) THEN DO; PS105A=0; PV105A=.07; END;
   ELSE IF MY=1966 THEN DO; PS105A=0; PV105A=.54; END;
   ELSE IF MY=1967 THEN DO; PS105A=0; PV105A=1; END;
   ELSE IF MY=1968 AND CY=1967 THEN DO; PS105A=0; PV105A=1; END;
```

In MY 1968-75, after January 1, 1968, the mix of post-standard and voluntary installations depends on the mix of cars and LTVs hitting non-occupants, because FMVSS 105 took effect on cars on January 1, 1968 but on LTVs not until September 1, 1983. (But all vehicles had dual master cylinders.) LS2004 prints these percentages on a page titled, "non-occupants hit by cars or LTVs, by model year of the car or LTV."

```
   ELSE IF MY=1968 THEN DO; PS105A=.42; PV105A=.58; END;
   ELSE IF MY=1969 THEN DO; PS105A=.81; PV105A=.19; END;
   ELSE IF MY=1970 THEN DO; PS105A=.82; PV105A=.18; END;
   ELSE IF MY=1971 THEN DO; PS105A=.81; PV105A=.19; END;
   ELSE IF MY=1972 THEN DO; PS105A=.80; PV105A=.20; END;
   ELSE IF MY=1973 THEN DO; PS105A=.78; PV105A=.22; END;
   ELSE IF MY=1974 THEN DO; PS105A=.74; PV105A=.26; END;
   ELSE IF MY=1975 THEN DO; PS105A=.76; PV105A=.24; END;
/* PEDESTRIAN LIVES SAVED BY FRONT DISC BRAKES */
  OLDWTFA=WEIGHTFA;
  IF PS105B GT 0 OR PV105B GT 0 THEN DO;
   E=.0017;
   P=PS105B+PV105B;
   S=OLDWTFA*E*P / (1 - E*P);
   PLS105B=S*PS105B/P;
   PLV105B=S*PV105B/P;
   WEIGHTFA=OLDWTFA+PLS105B+PLV105B; END;
  ELSE DO; PLS105B=0; PLV105B=0; END;
/* PEDESTRIAN LIVES SAVED BY DUAL MASTER CYLINDERS */
  OLDWTFA=WEIGHTFA;
```

```
IF PS105A GT O OR PV105A GT O THEN DO;
  E=.007;
  P=PS105A+PV105A;
  S=OLDWTFA*E*P / (1 - E*P);
  PLS105A=S*PS105A/P;
  PLV105A=S*PV105A/P;
  WEIGHTFA=OLDWTFA+PLS105A+PLV105A; END;
ELSE DO; PLS105A=0; PLV105A=0; END;
PLS=PLS105A+PLS105B; PLV=PLV105A+PLV105B;
PL105A=PLS105A+PLV105A; PL105B=PLS105B+PLV105B; PL=PLS+PLV;
RUN;
```

Adds up and prints out the totals for each calendar year, and the sum for all years, 1960-74.

```
PROC MEANS SUM NOPRINT DATA=PED7; BY CY;
  VAR ORIGWT WEIGHTFA PLV105B PLS105B PL105B PLV105A PLS105A PL105A PLV PLS PL;
  OUTPUT OUT=PED8
    SUM=P_ORIGWT P_WTFA PLV105B PLS105B PL105B PLV105A PLS105A PL105A PLV PLS PL;
PROC PRINT DATA=PED8;
  FORMAT P_ORIGWT P_WTFA PLV105B PLS105B PL105B PLV105A PLS105A PL105A PLV PLS PL 9.0;
  ID CY; VAR P_ORIGWT P_WTFA PLV105B PLS105B PL105B PLV105A PLS105A PL105A PLV PLS PL;
  SUM P_ORIGWT P_WTFA PLV105B PLS105B PL105B PLV105A PLS105A PL105A PLV PLS PL;
TITLE1 'PEDESTRIANS SAVED BY CAR/LIGHT TRUCK DISC BRAKES AND DUAL MASTER CYLINDERS, 1960-74';
TITLE2 ' ';
TITLE3 '...105B = FRONT DISC BRAKES (MEDIAN INSTALLATION YEAR 1971, FMVSS EFFECTIVE 1/1/76 OR
9/1/83)';
TITLE4 '...105A = DUAL MASTER CYLINDERS (MEDIAN INSTALLATION YEAR LATE 1966, FMVSS EFFECTIVE 1/1/68
OR 9/1/83)';
TITLE5 '...V = VOLUNTARY INSTALLATIONS, BEFORE EFFECTIVE DATE';
TITLE6 '...S = STANDARD INSTALLATIONS, AFTER EFFECTIVE DATE';
TITLE7 'P_ORIGWT = ACTUAL PED/BIKE/NONMOTORIST FATALITIES';
TITLE8 'P_WTFA = FATALITIES THAT WOULD HAVE OCCURRED WITHOUT THESE SAFETY IMPROVEMENTS';
TITLE9 'PL = OVERALL PED/BIKE/NONMOTORIST LIVES SAVED BY BRAKE IMPROVEMENTS IN CARS/LIGHT TRUCKS';
RUN;
```

The analysis of the effect of car/LTV brake improvements on motorcyclist fatalities is quite similar to the non-occupant analysis.

```
/* ----------------------------------------------------------------- */
/* ----------------------------------------------------------------- */
/*   ANALYZES EFFECT OF BRAKE IMPROVEMENTS ON MOTORCYCLIST FATALITIES */
/*                                                                    */
/*   FIRST STEP: CREATE FILE WITH ALL POSSIBLE CY-MY COMBINATIONS     */
/*   WITH ORIGWT INDICATING THE NUMBER OF MOTORCYCLIST FATALITIES     */
/*   FOR THAT COMBINATION (MY=MODEL YEAR OF THE VEH THAT HIT THE MC)  */
/* ----------------------------------------------------------------- */
/* ----------------------------------------------------------------- */

DATA MC5(KEEP=CY FARSMC VEHAGE);
SET ACCFAC1;
DO VEHAGE = -1 TO 25;
  OUTPUT;
END;
RUN;
PROC SORT DATA=MC5; BY VEHAGE;
DATA MC6; MERGE MC5 VEHAGE4; BY VEHAGE;
ORIGWT=PCTMC*FARSMC;
WEIGHTFA=ORIGWT;
MY=CY-VEHAGE;
KEEP CY MY ORIGWT WEIGHTFA;
PROC SORT DATA=MC6; BY CY MY;
RUN;


/* ----------------------------------------------------------------- */
/*   REPEATS THE MOTORCYCLIST ANALYSIS OF LS2004.SAS                  */
```

```
/* ---------------------------------------------------------------- */

DATA MC7; SET MC6;
/* IMPLEMENTATION OF FRONT DISC BRAKES IN THE VEHICLES THAT HIT THESE MOTORCYCLISTS */
IF MY LE 1964 THEN DO; MS105B=0; MV105B=0; END;
    ELSE IF MY=1965 THEN DO; MS105B=0; MV105B=.02; END;
    ELSE IF MY=1966 THEN DO; MS105B=0; MV105B=.03; END;
    ELSE IF MY=1967 THEN DO; MS105B=0; MV105B=.06; END;
    ELSE IF MY=1968 THEN DO; MS105B=0; MV105B=.13; END;
    ELSE IF MY=1969 THEN DO; MS105B=0; MV105B=.28; END;
    ELSE IF MY=1970 THEN DO; MS105B=0; MV105B=.41; END;
    ELSE IF MY=1971 THEN DO; MS105B=0; MV105B=.63; END;
    ELSE IF MY=1972 THEN DO; MS105B=0; MV105B=.74; END;
    ELSE IF MY=1973 THEN DO; MS105B=0; MV105B=.86; END;
    ELSE IF MY=1974 THEN DO; MS105B=0; MV105B=.84; END;
    ELSE IF MY=1975 THEN DO; MS105B=0; MV105B=.93; END;
/* IMPLEMENTATION OF DUAL MASTER CYLINDERS IN THE VEHICLES THAT HIT THESE MOTORCYCLISTS */
  IF MY LE 1961 THEN DO; MS105A=0; MV105A=0; END;
    ELSE IF MY IN (1962,1963) THEN DO; MS105A=0; MV105A=.09; END;
    ELSE IF MY IN (1964,1965) THEN DO; MS105A=0; MV105A=.07; END;
    ELSE IF MY=1966 THEN DO; MS105A=0; MV105A=.54; END;
    ELSE IF MY=1967 THEN DO; MS105A=0; MV105A=1; END;
    ELSE IF MY=1968 AND CY=1967 THEN DO; MS105A=0; MV105A=1; END;
```

In MY 1968-75, after January 1, 1968, the mix of post-standard and voluntary installations depends on the mix of cars and LTVs hitting non-occupants, because FMVSS 105 took effect on cars on January 1, 1968 but on LTVs not until September 1, 1983. (But all vehicles had dual master cylinders.) LS2004 prints these percentages on a page titled, "motorcyclists hit by cars or LTVs, by model year of the car or LTV." These percentages are slightly different for motorcyclists and non-occupants, because motorcycles have proportionately more collisions with LTVs and fewer with cars.

```
    ELSE IF MY=1968 THEN DO; MS105A=.40; MV105A=.60; END;
    ELSE IF MY=1969 THEN DO; MS105A=.74; MV105A=.26; END;
    ELSE IF MY=1970 THEN DO; MS105A=.76; MV105A=.24; END;
    ELSE IF MY=1971 THEN DO; MS105A=.72; MV105A=.28; END;
    ELSE IF MY=1972 THEN DO; MS105A=.72; MV105A=.28; END;
    ELSE IF MY=1973 THEN DO; MS105A=.72; MV105A=.28; END;
    ELSE IF MY=1974 THEN DO; MS105A=.68; MV105A=.32; END;
    ELSE IF MY=1975 THEN DO; MS105A=.69; MV105A=.31; END;
/* MOTORCYCLIST LIVES SAVED BY FRONT DISC BRAKES */
  OLDWTFA=WEIGHTFA;
  IF MS105B GT 0 OR MV105B GT 0 THEN DO;
    E=.0017;
    P=MS105B+MV105B;
    S=OLDWTFA*E*P / (1 - E*P);
    MLS105B=S*MS105B/P;
    MLV105B=S*MV105B/P;
    WEIGHTFA=OLDWTFA+MLS105B+MLV105B; END;
  ELSE DO; MLS105B=0; MLV105B=0; END;
/* MOTORCYCLIST LIVES SAVED BY DUAL MASTER CYLINDERS */
  OLDWTFA=WEIGHTFA;
  IF MS105A GT 0 OR MV105A GT 0 THEN DO;
    E=.007;
    P=MS105A+MV105A;
    S=OLDWTFA*E*P / (1 - E*P);
    MLS105A=S*MS105A/P;
    MLV105A=S*MV105A/P;
    WEIGHTFA=OLDWTFA+MLS105A+MLV105A; END;
  ELSE DO; MLS105A=0; MLV105A=0; END;
  MLS=MLS105A+MLS105B; MLV=MLV105A+MLV105B;
  ML105A=MLS105A+MLV105A; ML105B=MLS105B+MLV105B; ML=MLS+MLV;
```

```
RUN;

PROC MEANS SUM NOPRINT DATA=MC7; BY CY;
  VAR ORIGWT WEIGHTFA MLV105B MLS105B ML105B MLV105A MLS105A ML105A MLV MLS ML;
  OUTPUT OUT=MC8
    SUM=M_ORIGWT M_WTFA MLV105B MLS105B ML105B MLV105A MLS105A ML105A MLV MLS ML;
PROC PRINT DATA=MC8;
  FORMAT M_ORIGWT M_WTFA MLV105B MLS105B ML105B MLV105A MLS105A ML105A MLV MLS ML 9.0;
  ID CY; VAR M_ORIGWT M_WTFA MLV105B MLS105B ML105B MLV105A MLS105A ML105A MLV MLS ML;
  SUM M_ORIGWT M_WTFA MLV105B MLS105B ML105B MLV105A MLS105A ML105A MLV MLS ML;
TITLE1 'MOTORCYCLISTS SAVED BY CAR/LIGHT TRUCK DISC BRAKES AND DUAL MASTER CYLINDERS, 1960-74';
TITLE2 ' ';
TITLE3 '...105B = FRONT DISC BRAKES (MEDIAN INSTALLATION YEAR 1971, FMVSS EFFECTIVE 1/1/76 OR
9/1/83)';
TITLE4 '...105A = DUAL MASTER CYLINDERS (MEDIAN INSTALLATION YEAR LATE 1966, FMVSS EFFECTIVE 1/1/68
OR 9/1/83)';
TITLE5 '...V = VOLUNTARY INSTALLATIONS, BEFORE EFFECTIVE DATE';
TITLE6 '...S = STANDARD INSTALLATIONS, AFTER EFFECTIVE DATE';
TITLE7 'M_ORIGWT = ACTUAL MOTORCYCLIST FATALITIES';
TITLE8 'M_WTFA = FATALITIES THAT WOULD HAVE OCCURRED WITHOUT THESE SAFETY IMPROVEMENTS';
TITLE9 'ML = OVERALL MOTORCYCLIST LIVES SAVED BY BRAKE IMPROVEMENTS IN CARS/LIGHT TRUCKS';
RUN;


/* ------------------------------------------------------------------ */
/* ------------------------------------------------------------------ */
/*   THE ANALYSIS OF PASSENGER CAR OCCUPANTS BEGINS HERE              */
/*                                                                    */
/*   CREATES A FILE WITH ALL POSSIBLE COMBINATIONS OF                 */
/*   CY, MY, CRASH MODE, OCC AGE GP, SEAT POSITION, RESTRAINT USE     */
/*   ORIGWT IS THE NUMBER OF OCC FATALITIES FOR EACH COMBINATION      */
/* ------------------------------------------------------------------ */
/* ------------------------------------------------------------------ */
```

The basic approach for car occupant fatalities resembles the analysis of non-occupants: first create a matrix of cells showing the distribution of fatalities by CY, MY and other factors, then analyze the matrix one cell at a time, estimating the fatality increase as the safety technologies are "removed" one by one. The difference is that a more complex matrix is needed here. In order to find out what technologies apply and how effective they are, we also need to know the impact location/crash type, and the occupant's age group, seat position, and restraint use. We will need a joint distribution of all these variables.

```
/* ------------------------------------------------------------------ */
/*   FIRST STEP: ALL CY-MY COMBINATIONS                              */
/*   BASED ON VEHICLE AGE DISTRIBUTION IN 1975-80 FARS               */
/* ------------------------------------------------------------------ */
```

The initial matrix of fatalities by CY and MY (CAR6) is generated by the same code as used for non-occupants and motorcyclists.

```
DATA CAR5(KEEP=CY FARSCAR VEHAGE);
SET ACCFAC1;
DO VEHAGE = -1 TO 25;
  OUTPUT;
END;
RUN;
PROC SORT DATA=CAR5; BY VEHAGE;
DATA CAR6; MERGE CAR5 VEHAGE4; BY VEHAGE;
ORIGWT=PCTCAR*FARSCAR;
MY=CY-VEHAGE;
KEEP CY MY ORIGWT;
PROC SORT DATA=CAR6; BY CY MY;
```

```
RUN;

/* ------------------------------------------------------------------ */
/*   ALL PERMITTED COMBINATIONS OF CRASH MODE, AGE GROUP, SEAT POS     */
/*   AGE GROUPS ARE INFANT, TODDLER 1-4, ALL OTHERS 5 AND OLDER        */
/*   WEIGHT FACTORS BASED ON DISTRIBUTIONS IN 1975-80 FARS             */
/* ------------------------------------------------------------------ */
```

CAR7 is a matrix distributing fatalities not only by CY and MY but also by impact location/crash type (1 = frontal, 2.1 = side, single-vehicle, 2.2 = side, multivehicle, 3 = rollover, 4 = rear/other), seat position (see the NEWSEAT format at the beginning of the program) and occupant age group (0 = infant, 1 = toddler age 1-4, 5 = all other occupants, age 5+). The model assumes (for simplicity) that the crash mode is independent from the seat position and occupant age. Of course, seat position is not independent from occupant age (e.g., there are no infants or toddlers driving cars), and a joint distribution is needed for those two parameters. All distributions are based on 1975-80 FARS data, specifically the analysis by OLDFA14 of the file *FARS7580* generated by LS2004.

```
DATA CAR7(KEEP=CY MY ORIGWT CRSH SEAT2 AGEGP);
SET CAR6;
```

There were 120,401.5 weighted cases in the file analyzed by OLDFA14.
Thus, CRASHFA * ASFA / X is the proportion of cases for each crash mode-seat position-occupant age combination.

```
X=120401.5*120401.5;                                        /* FROM OLDFA14, P. 7 */
OLDWT=ORIGWT;
DO ICRSH=1 TO 5;
DO IAGESEAT=1 TO 25;
IF ICRSH=1 THEN DO; CRSH=1; CRSHFA=61393.68; END;           /* FROM OLDFA14, P. 7 */
  ELSE IF ICRSH=2 THEN DO; CRSH=2.1; CRSHFA=12942.7; END;
  ELSE IF ICRSH=3 THEN DO; CRSH=2.2; CRSHFA=22710.49; END;
  ELSE IF ICRSH=4 THEN DO; CRSH=3;   CRSHFA=18427.09; END;
  ELSE IF ICRSH=5 THEN DO; CRSH=4;   CRSHFA=4927.502; END;
IF IAGESEAT=1 THEN DO; SEAT2=11; AGEGP=5; ASFA=76173; END;   /* FROM OLDFA14, P. 7 */
  ELSE IF IAGESEAT= 2 THEN DO; SEAT2=12; AGEGP=0; ASFA=230.42; END;
  ELSE IF IAGESEAT= 3 THEN DO; SEAT2=12; AGEGP=1; ASFA=451.80; END;
  ELSE IF IAGESEAT= 4 THEN DO; SEAT2=12; AGEGP=5; ASFA=2458.9; END;
  ELSE IF IAGESEAT= 5 THEN DO; SEAT2=13; AGEGP=0; ASFA=302.37; END;
  ELSE IF IAGESEAT= 6 THEN DO; SEAT2=13; AGEGP=1; ASFA=622.66; END;
  ELSE IF IAGESEAT= 7 THEN DO; SEAT2=13; AGEGP=5; ASFA= 29280; END;
  ELSE IF IAGESEAT= 8 THEN DO; SEAT2=18; AGEGP=0; ASFA=33.569; END;
  ELSE IF IAGESEAT= 9 THEN DO; SEAT2=18; AGEGP=1; ASFA=39.501; END;
  ELSE IF IAGESEAT=10 THEN DO; SEAT2=18; AGEGP=5; ASFA=42.877; END;
  ELSE IF IAGESEAT=11 THEN DO; SEAT2=21; AGEGP=0; ASFA=85.037; END;
  ELSE IF IAGESEAT=12 THEN DO; SEAT2=21; AGEGP=1; ASFA=519.94; END;
  ELSE IF IAGESEAT=13 THEN DO; SEAT2=21; AGEGP=5; ASFA=7661.6; END;
  ELSE IF IAGESEAT=14 THEN DO; SEAT2=22; AGEGP=0; ASFA=63.575; END;
  ELSE IF IAGESEAT=15 THEN DO; SEAT2=22; AGEGP=1; ASFA=218.39; END;
  ELSE IF IAGESEAT=16 THEN DO; SEAT2=22; AGEGP=5; ASFA=1625.6; END;
  ELSE IF IAGESEAT=17 THEN DO; SEAT2=28; AGEGP=0; ASFA=4.8798; END;
  ELSE IF IAGESEAT=18 THEN DO; SEAT2=28; AGEGP=1; ASFA=13.574; END;
  ELSE IF IAGESEAT=19 THEN DO; SEAT2=28; AGEGP=5; ASFA=84.192; END;
  ELSE IF IAGESEAT=20 THEN DO; SEAT2=51; AGEGP=0; ASFA=5.8478; END;
  ELSE IF IAGESEAT=21 THEN DO; SEAT2=51; AGEGP=1; ASFA= 37.321; END;
  ELSE IF IAGESEAT=22 THEN DO; SEAT2=51; AGEGP=5; ASFA= 204.76; END;
  ELSE IF IAGESEAT=23 THEN DO; SEAT2=52; AGEGP=0; ASFA=  1.1127; END;
  ELSE IF IAGESEAT=24 THEN DO; SEAT2=52; AGEGP=1; ASFA=2.0335; END;
  ELSE IF IAGESEAT=25 THEN DO; SEAT2=52; AGEGP=5; ASFA=238.91; END;
ORIGWT=OLDWT*CRSHFA*ASFA/X;
```

```
OUTPUT;
END;
END;
RUN;
```

```
/* ---------------------------------------------------------------- */
/* ALL PERMITTED CONFIGURATIONS OF RESTRAINT USE                    */
/*                                                                  */
/* SEPARATE ANALYSES BY SEAT POSITION, MODEL YEAR & AGE GROUP       */
/* DEPENDING ON WHAT TYPES AND HOW MANY SYSTEMS ARE AVAILABLE       */
/* USE RATES BASED ON 1975-80 AVGS, EXCEPT WHERE 1975 IS HIGHER     */
/* FIRST MODERN SAFETY SEATS: CY 1967 (TODDLER) AND 1970 (INFANT)   */
/* ---------------------------------------------------------------- */
```

Each of the cells in CAR7 is now further subdivided into a theoretical maximum of four cells, based on the occupant's restraint use (REST3): 0 = unrestrained, 2 = lap belt, 3 = lap plus shoulder belt, 4 = child safety seat. OLDFA14 generally tabulates the belt use of fatally injured occupants in CY 1975 (i.e., before the temporary decline of belt use in the late 1970's). Only if there are insufficient cases from CY 1975 does OLDFA14 also include data from later calendar years. Eight scenarios are considered:

- RES0, cells where REST3 is always set to zero, either because restraints do not exist (e.g., non-designated seat positions, or infants before the existence of infant seats) or because belt use is so rare that there are no cases of restrained occupant fatalities on *FARS7580* (e.g., center-front occupants before MY 1967).
- INF1, infants in designated seat positions, after the existence of infant seats. REST3 can be 0 or 4.
- TOD11, toddlers after the existence of child seats. REST3 can be 0, 2 or 4.
- TOD21, toddlers before the existence of child seats. REST3 can be 0 or 2.
- FOB11, front-outboard occupants age 5+ in MY 1974-75, when all cars had 3-point belts. REST3 can be 0 or 3.
- FOB21, front-outboard occupants age 5+ in MY 1968-73, when most cars had separate lap and shoulder belts (and some had 3-point belts). REST3 can be 0, 2 or 3.
- FOB31, front-outboard occupants age 5+ in MY ≤ 1967, when most cars lap belts only (although a few had 3-point belts, there is not a single fatality case in 1975-80). REST3 can be 0 or 2.
- LAP1, occupants age 5+ at all other seat positions where there is some lap belt use. REST3 can be 0 or 2.

```
DATA RES0(KEEP=CY MY ORIGWT CRSH SEAT2 AGEGP REST3)
     INF1(KEEP=CY MY ORIGWT CRSH SEAT2 AGEGP)
     TOD11(KEEP=CY MY ORIGWT CRSH SEAT2 AGEGP)
     TOD21(KEEP=CY MY ORIGWT CRSH SEAT2 AGEGP)
     FOB11(KEEP=CY MY ORIGWT CRSH SEAT2 AGEGP)
     FOB21(KEEP=CY MY ORIGWT CRSH SEAT2 AGEGP)
     FOB31(KEEP=CY MY ORIGWT CRSH SEAT2 AGEGP)
     LAP1(KEEP=CY MY ORIGWT CRSH SEAT2 AGEGP);
SET CAR7;
IF SEAT2 IN (18,28,51,52) OR (AGEGP=0 AND CY LT 1970) OR
   (SEAT2=12 AND MY LT 1967) OR (SEAT2=21 AND MY LT 1964) OR
   (SEAT2=22 AND MY LT 1967) THEN DO; REST3=0; OUTPUT RES0; END;
 ELSE IF AGEGP=0 THEN OUTPUT INF1;
 ELSE IF AGEGP=1 AND 1967 LE CY LE 1974 THEN OUTPUT TOD11;
 ELSE IF AGEGP=1 THEN OUTPUT TOD21;
 ELSE IF SEAT2 IN (11,13) AND 1974 LE MY LE 1975 THEN OUTPUT FOB11;
 ELSE IF SEAT2 IN (11,13) AND 1968 LE MY LE 1973 THEN OUTPUT FOB21;
 ELSE IF SEAT2 IN (11,13) THEN OUTPUT FOB31;
 ELSE OUTPUT LAP1;
RUN;
```

```
DATA INF2(KEEP=CY MY ORIGWT CRSH SEAT2 AGEGP REST3);
SET INF1;
OLDWT=ORIGWT;
```

OLDFA14 shows 4.34 percent of infant fatality cases were restrained in a safety seat in CY 1975-78 (i.e., before child passenger safety laws took effect). Given that infant seats first became available in CY 1970, we will assume a gradual increase in use from 1970, reaching 4.34 percent in 1974. Because the safety seat is not part of the car's original equipment, the MY of the car is not as important a factor as the CY of the crash (and for simplicity, only the CY is considered here).

```
CRDUSE=.0434*(CY-1969)/5;                              /* FROM OLDFA14, P. 1 */
```

The original cell is split into two cells, one for restrained infants, one unrestrained. For example, if there were 100 infant fatalities in MY 1974 (ORIGWT = 100), we will create one cell of restrained infants with ORIGWT = 4.34 and one cell of unrestrained infants with ORIGWT = 95.66.

```
REST3=4; ORIGWT=   CRDUSE *OLDWT; OUTPUT;
REST3=0; ORIGWT=(1-CRDUSE)*OLDWT; OUTPUT;
RUN;

DATA TOD12(KEEP=CY MY ORIGWT CRSH SEAT2 AGEGP REST3);
SET TOD11;
OLDWT=ORIGWT;
```

1.59 percent of toddler fatalities were restrained in safety seats in CY 1975-78. Given that toddler seats first became available in CY 1967, but more widely starting in 1970, we will assume a slow increase in 1967-69, and a faster increase in use from 1970, reaching 1.59 percent in 1974.

```
IF 1970 LE CY LE 1974 THEN CRDUSE=.0159*(CY-1969)/5;     /* FROM OLDFA14, P. 2 */
  ELSE IF 1967 LE CY LE 1969 THEN CRDUSE=.0159*(CY-1966)/20;
```

Lap belt use by toddlers was close to 1 percent, and depends more on the model year of the vehicle than the calendar year of the crash.

```
IF MY GE 1955 THEN LAPUSE=.0118; ELSE LAPUSE=.0099;     /* .0118 FROM OLDFA14, P. 2 */
REST3=4; ORIGWT=             CRDUSE  *OLDWT; OUTPUT;     /* .0099 FROM OLDFA14, P. 11 */
REST3=2; ORIGWT=             LAPUSE  *OLDWT; OUTPUT;
REST3=0; ORIGWT=(1-(CRDUSE+LAPUSE))*OLDWT; OUTPUT;
RUN;

DATA TOD22(KEEP=CY MY ORIGWT CRSH SEAT2 AGEGP REST3);
SET TOD21;
OLDWT=ORIGWT;
IF MY GE 1955 THEN LAPUSE=.0118; ELSE LAPUSE=.0099;     /* .0118 FROM OLDFA14, P. 2 */
REST3=2; ORIGWT=   LAPUSE *OLDWT; OUTPUT;                /* .0099 FROM OLDFA14, P. 11 */
REST3=0; ORIGWT=(1-LAPUSE)*OLDWT; OUTPUT;
RUN;

DATA FOB12(KEEP=CY MY ORIGWT CRSH SEAT2 AGEGP REST3);
SET FOB11;
OLDWT=ORIGWT;
```

In CY 1975, 10.61 percent of front-outboard fatalities age 5+ in MY 1974-75 cars were restrained by 3-point belts.

```
REST3=3; ORIGWT=.1061*OLDWT; OUTPUT;                    /* FROM OLDFA14, P. 12 */
REST3=0; ORIGWT=.8939*OLDWT; OUTPUT;
RUN;
```

```
DATA FOB22(KEEP=CY MY ORIGWT CRSH SEAT2 AGEGP REST3);
SET FOB21;
OLDWT=ORIGWT;
```

In CY 1975, depending on the model year, the percentage of front-outboard fatalities wearing lap belts only ranged from 2.47 to 6.42 percent (REST_USE = 2 or 8 on FARS). Far fewer were explicitly coded as also using the shoulder belt (REST_USE = 3).

```
IF MY=1968 THEN LS_USE=.0011; ELSE IF 1969 LE MY LE 1970 THEN LS_USE=.0051;
  ELSE IF MY=1971 THEN LS_USE=.0081; ELSE IF MY=1972 THEN LS_USE=.0099;
  ELSE IF MY=1973 THEN LS_USE=.0112;
IF MY=1968 THEN LAPUSE=.0247; ELSE IF 1969 LE MY LE 1971 THEN LAPUSE=.0366;
  ELSE IF 1972 LE MY LE 1973 THEN LAPUSE=.0642;              /* FROM OLDFA14, P. 12 */
REST3=3; ORIGWT=          LS_USE *OLDWT; OUTPUT;
REST3=2; ORIGWT=          LAPUSE *OLDWT; OUTPUT;
REST3=0; ORIGWT=(1-(LS_USE+LAPUSE))*OLDWT; OUTPUT;
RUN;

DATA FOB32(KEEP=CY MY ORIGWT CRSH SEAT2 AGEGP REST3);
SET FOB31;
OLDWT=ORIGWT;
```

In pre-1968 cars, 1 to 2 percent of front-outboard fatalities were lap-belted. Belt use is lower in model years before 1964, because many cars did not have any safety belts installed. These rates are based on CY 1975-80 FARS data because there are not enough cases in CY 1975 alone for meaningful rates.

```
IF 1964 LE MY LE 1967 THEN LAPUSE=.0195;                    /* FROM OLDFA14, P. 11 */
  ELSE IF 1955 LE MY LE 1963 THEN LAPUSE=.0131;
  ELSE LAPUSE=.0099;
REST3=2; ORIGWT=   LAPUSE *OLDWT; OUTPUT;
REST3=0; ORIGWT=(1-LAPUSE)*OLDWT; OUTPUT;
RUN;

DATA LAP2(KEEP=CY MY ORIGWT CRSH SEAT2 AGEGP REST3);
SET LAP1;
OLDWT=ORIGWT;
```

Lap belt use of fatally injured occupants in the rear-outboard seats, MY 1964-74 and the center seats, MY 1967-74.

```
IF SEAT2=12 THEN LAPUSE=.0145;                              /* FROM OLDFA14, P. 13 */
  ELSE IF SEAT2=21 AND MY GE 1972 THEN LAPUSE=.0374;        /* FROM OLDFA14, P. 15 */
  ELSE IF SEAT2=21 THEN LAPUSE=.0109;                       /* FROM OLDFA14, P. 14 */
  ELSE IF SEAT2=22 THEN LAPUSE=.0049;                       /* FROM OLDFA14, P. 16 */
REST3=2; ORIGWT=   LAPUSE *OLDWT; OUTPUT;
REST3=0; ORIGWT=(1-LAPUSE)*OLDWT; OUTPUT;
RUN;

/* ---------------------------------------------------------------- */
/* CAR8 IS THE FULL ANALYSIS FILE                                   */
/* EJECT2 IS PROPORTION OF FATALITIES EJECTED IN EACH CELL          */
/* ---------------------------------------------------------------- */

DATA CAR8; SET RESO INF2 TOD12 TOD22 FOB12 FOB22 FOB32 LAP2;
WEIGHTFA=ORIGWT;
```

CAR8 has a separate record (cell) for each permissible combination of CY, MY, CRSH, SEAT2, AGEGP and REST3. As in LS2004, we also need to know EJECT2, the probability that a fatality was ejected. Unlike LS2004, we do not know whether specific individuals were ejected (the

370

EJECTION variable on FARS) because, more generally, CAR8 does not have information on specific crashes. We only have an overall proportion ejected (EJECT2). It is based on the OLDFA14 analysis of *FARS7580*. For simplicity, we will use one set of EJECT2 values, by crash mode, for all unrestrained fatalities, and another set for all restrained fatalities; we will not further subdivide by seat position or occupant age group. CY, MY, CRSH, SEAT2, AGEGP, REST3 and EJECT2 are all the core variables needed to estimate the lives saved by technologies available in 1960-74.

ORIGWT is the number of car-occupant fatalities in that cell.

```
IF REST3=0 THEN DO;
  IF CRSH=1 THEN EJECT2=.1549; ELSE IF CRSH=2.1 THEN EJECT2=.2779;   /* FROM OLDFA14, P. 9 */
    ELSE IF CRSH=2.2 THEN EJECT2=.1499; ELSE IF CRSH=3 THEN EJECT2=.5556;
    ELSE IF CRSH=4 THEN EJECT2=.2273; END;
ELSE DO;
  IF CRSH=1 THEN EJECT2=.0562; ELSE IF CRSH=2.1 THEN EJECT2=.1537;   /* FROM OLDFA14, P. 10 */
    ELSE IF CRSH=2.2 THEN EJECT2=.0627; ELSE IF CRSH=3 THEN EJECT2=.2387;
    ELSE IF CRSH=4 THEN EJECT2=.0923; END;
RUN;

PROC SORT DATA=CAR8; BY CY;
RUN;

/* --------------------------------------------------------------------- */
/* --------------------------------------------------------------------- */
/*  REPEATS THE CAR OCCUPANT ANALYSIS OF LS2004.SAS                       */
/*                                                                        */
/*  HAS TO BE SIMPLIFIED FOR 203, 212, 214 SINCE VIN/MAKE-MODEL UNK       */
/* --------------------------------------------------------------------- */
/* --------------------------------------------------------------------- */
```

The estimation of lives saved by the various safety technologies closely parallels the "DATA CAR2; SET CAR1;" step of LS2004.

Three "tags" of information accompany each cell as it processes through the model:

- ORIGWT, the original number of fatalities in the cell, remains unchanged.
- WEIGHTFA, the inflated cell fatality count, grows as safety technologies are "removed" from the vehicle. The existence of ORIGWT fatalities in a cell containing vehicles equipped certain safety technologies implies that there would have been WEIGHTFA fatalities if the vehicle had no safety technologies at all, in the sense that the combined effectiveness of these technologies would reduce WEIGHTFA fatalities to ORIGWT fatalities.
- EJECT2, the probability that an occupant in this cell was ejected. Belts reduce the probability of ejection, whereas other technologies, such as improved door locks, are effective only if the occupant was ejected. Thus, it is necessary to track the changing probability of ejection as technologies are removed.

```
DATA CAR9; SET CAR8;

/* --------------------------------------------------------------------- */
/*  ZERO BENEFITS FOR TECHNOLOGIES NOT AVAILABLE IN 1960-74               */
/* --------------------------------------------------------------------- */
```

The first difference from LS2004 is that seven of the later technologies in LS2004 did not exist in any car on the road in CY 1960-74. We may skip the analyses and set the lives saved to zero.

```
PS208J=0; PV208J=0; CLS208J=0; CLV208J=0; CL208J=0;
PS108=0; PV108=0; CLS108=0; CLV108=0; CL108=0;
PS208I=0; PV208I=0; CLS208I=0; CLV208I=0; CL208I=0;
PS214B=0; PV214B=0; CLS214B=0; CLV214B=0; CL214B=0;
PS208H=0; PV208H=0; CLS208H=0; CLV208H=0; CL208H=0;
PS208G=0; PV208G=0; CLS208G=0; CLV208G=0; CL208G=0;
PSNCAP=0; PVNCAP=0; CLSNCAP=0; CLVNCAP=0; CLNCAP=0;
CLS=0; CLV=0; CL=0;
```

```
/* ------------------------------------------------------------- */
/*   213 CHILD SAFETY SEATS                                      */
/*   USE RATE WENT OVER 50% IN: 1985                             */
/* ------------------------------------------------------------- */
```

Even though the "median installation year" is 1985, some children were in safety seats long before that, and well before 1974. We must proceed to estimate lives saved.

```
/* IDENTIFIES CHILD PASSENGERS IN SAFETY SEATS */
```

Benefits are estimated for those cells where every child was in a safety seat (REST3 = 4). The sizes of those cells (ORIGWT), as discussed above, depends on the use rate for safety seats.

```
  PS213=0; IF REST3=4 THEN PV213=1; ELSE PV213=0;
/* LIVES SAVED AND EJECTIONS PREVENTED BY CHILD SAFETY SEATS */
  OLDWTFA=WEIGHTFA;
  IF PS213 GT 0 OR PV213 GT 0 THEN DO;
    IF AGEGP=0 THEN E=.71; ELSE E=.54;
```

When we "remove" a restraint system from the vehicle, the occupant becomes unrestrained and, in general, 15.49 percent of unrestrained fatalities in frontal crashes were ejected, 27.79 percent in single-vehicle side impacts, etc. (These probabilities are the same as for the originally unrestrained occupants in the DATA CAR8 step above, and are generated by OLDFA14. The same probabilities will be used for all subsequent groups of transformed-to-unrestrained car occupants.) One of these becomes the new values of EJECT2 as this now-unrestrained occupant fatality cell proceeds through the rest of the model.

```
    IF CRSH=1 THEN EJECT2=.1549; ELSE IF CRSH=2.1 THEN EJECT2=.2779;   /* FROM OLDFA14, P. 9 */
      ELSE IF CRSH=2.2 THEN EJECT2=.1499; ELSE IF CRSH=3 THEN EJECT2=.5556;
      ELSE IF CRSH=4 THEN EJECT2=.2273;
    P=PS213+PV213;
    S=OLDWTFA*E*P / (1 - E*P);
    CLS213=S*PS213/P;
    CLV213=S*PV213/P;
    WEIGHTFA=OLDWTFA+CLS213+CLV213; END;
  ELSE DO; CLS213=0; CLV213=0; END;
  CL213=CLS213+CLV213;
  CLS=CLS+CLS213; CLV=CLV+CLV213; CL=CL+CL213;
```

```
/* ------------------------------------------------------------- */
/*   208F 3-POINT BELTS FOR FRONT-SEAT OUTBOARD OCCUPANTS        */
/* ------------------------------------------------------------- */
```

Includes occupants who buckle both belts in 1968-73 cars with separate lap and shoulder belts.

```
/*   MEDIAN INSTALLATION YEAR: 1974                             */
/* ------------------------------------------------------------- */
```

```
/* IDENTIFIES FRONT-SEAT OCCUPANTS WEARING 3-POINT BELTS */
```

```
PS208F=0; PV208F=0;
IF SEAT2 IN (11,13) AND REST3=3 THEN DO; IF MY LE 1973 THEN PV208F=1; ELSE PS208F=1; END;
/* LIVES SAVED AND EJECTIONS PREVENTED BY FRONT-OUTBOARD 3-POINT BELTS */
OLDWTFA=WEIGHTFA;
IF PS208F GT O OR PV208F GT O THEN DO;
```

Unlike LS2004, which has many different effectiveness values depending on the crash mode, we will for simplicity use the overall average fatality reduction by lap/shoulder belts: 45 percent.

```
E=.45;
IF CRSH=1 THEN EJECT2=.1549; ELSE IF CRSH=2.1 THEN EJECT2=.2779;    /* FROM OLDFA14, P. 9 */
  ELSE IF CRSH=2.2 THEN EJECT2=.1499; ELSE IF CRSH=3 THEN EJECT2=.5556;
  ELSE IF CRSH=4 THEN EJECT2=.2273;
P=PS208F+PV208F;
S=OLDWTFA*E*P / (1 - E*P);
CLS208F=S*PS208F/P;
CLV208F=S*PV208F/P;
  WEIGHTFA=OLDWTFA+CLS208F+CLV208F; END;
ELSE DO; CLS208F=0; CLV208F=0; END;
CL208F=CLS208F+CLV208F;
CLS=CLS+CLS208F; CLV=CLV+CLV208F; CL=CL+CL208F;

/* ---------------------------------------------------------------- */
/*  216 B-PILLARS FOR HARDTOP CARS AND OTHER ROOF STRENGTHENING      */
/*  MEDIAN INSTALLATION YEAR: 1973                                   */
/* ---------------------------------------------------------------- */
```

Same code as LS2004. For the remaining technologies, if the code is also the same as in LS2004, notes will be omitted.

```
/* ROOF STRENGTH IMPROVED GRADUALLY FROM 1970 TO 1977         */
IF MY LE 1969 THEN DO; PS216=0; PV216=0; END;
  ELSE IF 1970 LE MY LE 1973 THEN DO; PS216=0; PV216=.125*(MY-1969); END;
  ELSE IF 1974 LE MY LE 1976 THEN DO; PS216=.125*(MY-1969); PV216=0; END;
/* REDUCTION IN NONEJECTION ROLLOVER FATALITY RISK WITH IMPROVED ROOF STRENGTH */
OLDWTFA=WEIGHTFA;
IF (PS216 GT O OR PV216 GT O) AND CRSH=3
   AND SEAT2 NE 52 AND EJECT2 NE 1 THEN DO;
  E=.074;
  P=PS216+PV216;
  REL_S=(1-EJECT2)*E*P / (1 - E*P);
  S=OLDWTFA*REL_S;
  CLS216=S*PS216/P;
  CLV216=S*PV216/P;
  WEIGHTFA=OLDWTFA+CLS216+CLV216;
  EJECT2=EJECT2/(1+ REL_S); END;
ELSE DO; CLS216=0; CLV216=0; END;
CL216=CLS216+CLV216;
CLS=CLS+CLS216; CLV=CLV+CLV216; CL=CL+CL216;

/* ---------------------------------------------------------------- */
/*  214A SIDE DOOR BEAMS                                             */
/*  MEDIAN INSTALLATION YEAR: EARLY 1973                            */
/* ---------------------------------------------------------------- */

/* IDENTIFIES CARS WITH SIDE DOOR BEAMS                    */
```

In LS2004, we could identify the specific vehicles equipped with side door beams, based on the VIN; PS214A and PV214A were either 1 or 0. Here, we can only find the proportion of vehicles in each cell with side door beams, based on the model year. A tabulation near the end of LS2004 indicates that side door beams were in 17 percent of MY 1969 cars, 35 percent in 1970, 44 percent in 1971, 49 percent in 1972 and 85 percent in 1973 (the standard took effect in mid-MY 1973).

```
       IF MY GE 1974 THEN DO; PS214A=1; PV214A=0; END;
         ELSE IF MY LE 1968 THEN DO; PS214A=0; PV214A=0; END;
         ELSE IF MY=1969 THEN DO; PS214A=0; PV214A=.17; END;                    /* FROM LS2004 */
         ELSE IF MY=1970 THEN DO; PS214A=0; PV214A=.35; END;
         ELSE IF MY=1971 THEN DO; PS214A=0; PV214A=.44; END;
         ELSE IF MY=1972 THEN DO; PS214A=0; PV214A=.49; END;
         ELSE IF MY=1973 AND CY=1972 THEN DO; PS214A=0; PV214A=.85; END;
         ELSE IF MY=1973 THEN DO; PS214A=.50; PV214A=.35; END;
  /* REDUCTION IN SINGLE-VEHICLE SIDE-IMPACT FATALITIES WITH SIDE DOOR BEAMS     */
    OLDWTFA=WEIGHTFA;
    IF (PS214A GT 0 OR PV214A GT 0) AND CRSH=2.1 AND SEAT2 IN (11,13,21) THEN DO;
      E=.14;
      P=PS214A+PV214A;
      S=OLDWTFA*E*P / (1 - E*P);
      CLS214A=S*PS214A/P;
      CLV214A=S*PV214A/P;
      WEIGHTFA=OLDWTFA+CLS214A+CLV214A; END;
    ELSE DO; CLS214A=0; CLV214A=0; END;
    CL214A=CLS214A+CLV214A;
    CLS=CLS+CLS214A; CLV=CLV+CLV214A; CL=CL+CL214A;
```

```
/* ------------------------------------------------------------------ */
/*   105B FRONT DISC BRAKES                                           */
/*   MEDIAN INSTALLATION YEAR: 1971 (INSTALLATION COMPLETED 1977)     */
/* ------------------------------------------------------------------ */

/* IMPLEMENTATION OF FRONT DISC BRAKES */
  IF MY GE 1977 THEN DO; PS105B=1; PV105B=0; END;
    ELSE IF MY LE 1964 THEN DO; PS105B=0; PV105B=0; END;
    ELSE IF MY=1965 THEN DO; PS105B=0; PV105B=.02; END;
    ELSE IF MY=1966 THEN DO; PS105B=0; PV105B=.03; END;
    ELSE IF MY=1967 THEN DO; PS105B=0; PV105B=.06; END;
    ELSE IF MY=1968 THEN DO; PS105B=0; PV105B=.13; END;
    ELSE IF MY=1969 THEN DO; PS105B=0; PV105B=.28; END;
    ELSE IF MY=1970 THEN DO; PS105B=0; PV105B=.41; END;
    ELSE IF MY=1971 THEN DO; PS105B=0; PV105B=.63; END;
    ELSE IF MY=1972 THEN DO; PS105B=0; PV105B=.74; END;
    ELSE IF MY=1973 THEN DO; PS105B=0; PV105B=.86; END;
    ELSE IF MY=1974 THEN DO; PS105B=0; PV105B=.84; END;
    ELSE IF MY=1975 THEN DO; PS105B=0; PV105B=.93; END;
/* CAR OCCUPANT LIVES SAVED BY FRONT DISC BRAKES */
  OLDWTFA=WEIGHTFA;
  IF PS105B GT 0 OR PV105B GT 0 THEN DO;
    E=.0017;
    P=PS105B+PV105B;
    S=OLDWTFA*E*P / (1 - E*P);
    CLS105B=S*PS105B/P;
    CLV105B=S*PV105B/P;
    WEIGHTFA=OLDWTFA+CLS105B+CLV105B; END;
  ELSE DO; CLS105B=0; CLV105B=0; END;
  CL105B=CLS105B+CLV105B;
  CLS=CLS+CLS105B; CLV=CLV+CLV105B; CL=CL+CL105B;

/* ------------------------------------------------------------------ */
/*   201 VOLUNTARY INSTRUMENT PANEL IMPROVEMENTS (FMVSS-201 INSPIRED) */
/*   MEDIAN IMPLEMENTATION YEAR: 1968                                 */
/* ------------------------------------------------------------------ */

/* IDENTIFIES CARS WITH VOLUNTARY INSTRUMENT PANEL IMPROVEMENTS     */
  PS201=0; PV201=0;
  IF MY GE 1967 AND SEAT2=13 THEN DO;
      IF MY GE 1973 THEN PV201=1;
        ELSE IF MY=1967 THEN PV201=.25; ELSE IF MY=1968 THEN PV201=.5;
        ELSE IF MY=1969 THEN PV201=.6;  ELSE IF MY=1970 THEN PV201=.7;
        ELSE IF MY=1971 THEN PV201=.8;  ELSE IF MY=1972 THEN PV201=.9; END;
/* LIVES SAVED IN FRONTAL IMPACTS BY INSTRUMENT PANEL IMPROVEMENTS */
  OLDWTFA=WEIGHTFA;
  IF PS201 GT 0 OR PV201 GT 0 THEN DO;
    IF CRSH=1 AND REST3 NE 2 THEN E=.159; ELSE E=0;
    P=PS201+PV201;
    S=OLDWTFA*E*P / (1 - E*P);
    CLS201=S*PS201/P;
    CLV201=S*PV201/P;
    WEIGHTFA=OLDWTFA+CLS201+CLV201; END;
  ELSE DO; CLS201=0; CLV201=0; END;
  CL201=CLS201+CLV201;
  CLS=CLS+CLS201; CLV=CLV+CLV201; CL=CL+CL201;

/* ------------------------------------------------------------------ */
/*   203 ENERGY-ABSORBING AND TELESCOPING STEERING ASSEMBLIES        */
/*   MEDIAN IMPLEMENTATION YEAR: 1967                                 */
/* ------------------------------------------------------------------ */

/* IDENTIFIES CARS WITH ENERGY-ABSORBING STEERING ASSEMBLIES       */
  PS203=0; PV203=0;
  IF MY GE 1967 AND SEAT2=11 THEN DO;
```

LS2004 indicates that 62 percent of MY 1967 cars (and all MY 1968 cars) were equipped with energy-absorbing steering assemblies.

```
     IF MY GE 1969 THEN PS203=1;
        ELSE IF MY=1968 AND CY=1967 THEN PV203=1;
        ELSE IF MY=1968 THEN DO; PS203=.5; PV203=.5; END;              /* FROM LS2004 */
        ELSE IF MY=1967 THEN PV203=.62; END;
/* LIVES SAVED IN FRONTAL IMPACTS BY INSTRUMENT PANEL IMPROVEMENTS */
 OLDWTFA=WEIGHTFA;
 IF PS203 GT 0 OR PV203 GT 0 THEN DO;
   IF CRSH=1 THEN E=.121; ELSE E=0;
   P=PS203+PV203;
   S=OLDWTFA*E*P / (1 - E*P);
   CLS203=S*PS203/P;
   CLV203=S*PV203/P;
   WEIGHTFA=OLDWTFA+CLS203+CLV203; END;
 ELSE DO; CLS203=0; CLV203=0; END;
 CL203=CLS203+CLV203;
 CLS=CLS+CLS203; CLV=CLV+CLV203; CL=CL+CL203;


/* ------------------------------------------------------------------ */
/*  208E LAP BELTS FOR BACK-SEAT CENTER OCCUPANTS                      */
/*  MEDIAN INSTALLATION YEAR: EARLY 1967                               */
/* ------------------------------------------------------------------ */

/* IDENTIFIES BACK-CENTER-SEAT OCCUPANTS WEARING LAP BELTS  */
 PS208E=0; PV208E=0;
 IF SEAT2=22 AND REST3=2 AND AGEGP=5 THEN DO;
    IF MY GE 1969 THEN PS208E=1;
       ELSE IF MY LE 1967 THEN PV208E=1;
       ELSE IF MY=1968 AND CY=1967 THEN PV208E=1;
       ELSE IF MY=1968 THEN DO; PS208E=.5; PV208E=.5; END; END;
/* LIVES SAVED AND EJECTIONS PREVENTED BY BACK-CENTER-SEAT LAP BELTS */
 OLDWTFA=WEIGHTFA;
 IF PS208E GT 0 OR PV208E GT 0 THEN DO;
```

NHTSA's evaluation estimates an overall 32 percent fatality reduction by lap belts for rear-outboard occupants. The same effectiveness is assumed for the rear-center occupant.

```
   E=.32;
   IF CRSH=1 THEN EJECT2=.1549; ELSE IF CRSH=2.1 THEN EJECT2=.2779;   /* FROM OLDFA14, P. 9 */
      ELSE IF CRSH=2.2 THEN EJECT2=.1499; ELSE IF CRSH=3 THEN EJECT2=.5556;
      ELSE IF CRSH=4 THEN EJECT2=.2273;
   P=PS208E+PV208E;
   S=OLDWTFA*E*P / (1 - E*P);
   CLS208E=S*PS208E/P;
   CLV208E=S*PV208E/P;
   WEIGHTFA=OLDWTFA+CLS208E+CLV208E; END;
 ELSE DO; CLS208E=0; CLV208E=0; END;
 CL208E=CLS208E+CLV208E;
 CLS=CLS+CLS208E; CLV=CLV+CLV208E; CL=CL+CL208E;


/* ------------------------------------------------------------------ */
/*  105A DUAL MASTER CYLINDERS                                         */
/*  MEDIAN INSTALLATION YEAR: LATE 1966                                */
/* ------------------------------------------------------------------ */

/* IMPLEMENTATION OF DUAL MASTER CYLINDERS */
 IF MY GE 1969 THEN DO; PS105A=1; PV105A=0; END;
    ELSE IF MY LE 1961 THEN DO; PS105A=0; PV105A=0; END;
    ELSE IF MY IN (1962,1963) THEN DO; PS105A=0; PV105A=.09; END;
    ELSE IF MY IN (1964,1965) THEN DO; PS105A=0; PV105A=.07; END;
    ELSE IF MY=1966 THEN DO; PS105A=0; PV105A=.54; END;
```

```
          ELSE IF MY=1967 THEN DO; PS105A=0; PV105A=1; END;
          ELSE IF MY=1968 AND CY=1967 THEN DO; PS105A=0; PV105A=1; END;
          ELSE IF MY=1968 THEN DO; PS105A=.5; PV105A=.5; END;
/* CAR OCCUPANT LIVES SAVED BY DUAL MASTER CYLINDERS */
   OLDWTFA=WEIGHTFA;
   IF PS105A GT 0 OR PV105A GT 0 THEN DO;
     E=.007;
     P=PS105A+PV105A;
     S=OLDWTFA*E*P / (1 - E*P);
     CLS105A=S*PS105A/P;
     CLV105A=S*PV105A/P;
     WEIGHTFA=OLDWTFA+CLS105A+CLV105A; END;
   ELSE DO; CLS105A=0; CLV105A=0; END;
   CL105A=CLS105A+CLV105A;
   CLS=CLS+CLS105A; CLV=CLV+CLV105A; CL=CL+CL105A;

/* ------------------------------------------------------------------ */
/*   208D LAP BELTS FOR FRONT-SEAT CENTER OCCUPANTS                    */
/*   MEDIAN INSTALLATION YEAR: 1966                                   */
/* ------------------------------------------------------------------ */

/* IDENTIFIES FRONT-CENTER-SEAT OCCUPANTS WEARING LAP BELTS  */
   PS208D=0; PV208D=0;
   IF SEAT2=12 AND REST3=2 AND AGEGP=5 THEN DO;
       IF MY GE 1969 THEN PS208D=1;
         ELSE IF MY LE 1967 THEN PV208D=1;
         ELSE IF MY=1968 AND CY=1967 THEN PV208D=1;
         ELSE IF MY=1968 THEN DO; PS208D=.5; PV208D=.5; END; END;
/* LIVES SAVED AND EJECTIONS PREVENTED BY FRONT-CENTER-SEAT LAP BELTS */
   OLDWTFA=WEIGHTFA;
   IF PS208D GT 0 OR PV208D GT 0 THEN DO;
```

A double-pair comparison analysis of lap belt effectiveness for front-outboard occupants, conducted in support of this study, found 27 percent overall fatality reduction. The same effectiveness is assumed for front-center occupants.

```
     E=.27;
     IF CRSH=1 THEN EJECT2=.1549; ELSE IF CRSH=2.1 THEN EJECT2=.2779;   /* FROM OLDFA14, P. 9 */
       ELSE IF CRSH=2.2 THEN EJECT2=.1499; ELSE IF CRSH=3 THEN EJECT2=.5556;
       ELSE IF CRSH=4 THEN EJECT2=.2273;
     P=PS208D+PV208D;
     S=OLDWTFA*E*P / (1 - E*P);
     CLS208D=S*PS208D/P;
     CLV208D=S*PV208D/P;
     WEIGHTFA=OLDWTFA+CLS208D+CLV208D; END;
   ELSE DO; CLS208D=0; CLV208D=0; END;
   CL208D=CLS208D+CLV208D;
   CLS=CLS+CLS208D; CLV=CLV+CLV208D; CL=CL+CL208D;

/* ------------------------------------------------------------------ */
/*   212 ADHESIVE WINDSHIELD BONDING                                  */
/*   MEDIAN INSTALLATION YEAR: EARLY 1966                             */
/* ------------------------------------------------------------------ */

/* IDENTIFIES CARS WITH ADHESIVE WINDSHIELD BONDING         */
   IF MY LE 1962 THEN DO; PS212=0; PV212=0; END;
```

LS2004 tabulates the proportion of cars with adhesive windshield bonding, by model year.

```
     ELSE IF MY=1963 THEN DO; PS212=0; PV212=.02; END;      /* FROM LS2004 */
     ELSE IF MY=1964 THEN DO; PS212=0; PV212=.11; END;
     ELSE IF MY=1965 THEN DO; PS212=0; PV212=.48; END;
     ELSE IF MY=1966 THEN DO; PS212=0; PV212=.56; END;
     ELSE IF MY=1967 THEN DO; PS212=0; PV212=.60; END;
```

```
      ELSE IF MY=1968 THEN DO; PS212=0; PV212=.65; END;
      ELSE IF MY=1969 THEN DO; PS212=0; PV212=.75; END;
      ELSE IF MY=1970 AND CY=1969 THEN DO; PS212=0; PV212=.73; END;
      ELSE IF MY=1970 THEN DO; PS212=.39; PV212=.34; END;
      ELSE IF MY=1971 THEN DO; PS212=.74; PV212=0; END;
      ELSE IF MY=1972 THEN DO; PS212=.74; PV212=0; END;
      ELSE IF MY=1973 THEN DO; PS212=.77; PV212=0; END;
      ELSE IF MY=1974 THEN DO; PS212=.69; PV212=0; END;
      ELSE IF MY=1975 THEN DO; PS212=.71; PV212=0; END;
/* REDUCTION IN EJECTION FATALITY RISK WITH ADHESIVE WINDSHIELD BONDING */
  OLDWTFA=WEIGHTFA;
  IF (PS212 GT 0 OR PV212 GT 0) AND SEAT2 IN (11,12,13,18) AND EJECT2 NE 0 THEN DO;
    IF CRSH IN (1,3) THEN E=.033;
      ELSE IF CRSH IN (2.1,2.2,4) THEN E=.0075;
    P=PS212+PV212;
    REL_S=EJECT2*E*P / (1 - E*P);
    S=OLDWTFA*REL_S;
    CLS212=S*PS212/P;
    CLV212=S*PV212/P;
    WEIGHTFA=OLDWTFA+CLS212+CLV212;
    EJECT2=(EJECT2 + REL_S)/(1 + REL_S); END;
  ELSE DO; CLS212=0; CLV212=0; END;
  CL212=CLS212+CLV212;
  CLS=CLS+CLS212; CLV=CLV+CLV212; CL=CL+CL212;


/* ------------------------------------------------------------------ */
/*   208C LAP BELTS FOR BACK-SEAT OUTBOARD OCCUPANTS                   */
/*   MEDIAN INSTALLATION YEAR: LATE 1965                              */
/* ------------------------------------------------------------------ */

/* IDENTIFIES BACK-OUTBOARD-SEAT OCCUPANTS WEARING LAP BELTS   */
  PS208C=0; PV208C=0;
  IF SEAT2=21 AND REST3=2 AND AGEGP=5 THEN DO;
    IF MY GE 1969 THEN PS208C=1;
      ELSE IF MY LE 1967 THEN PV208C=1;
      ELSE IF MY=1968 AND CY=1967 THEN PV208C=1;
      ELSE IF MY=1968 THEN DO; PS208C=.5; PV208C=.5; END; END;
/* LIVES SAVED AND EJECTIONS PREVENTED BY BACK-OUTBOARD-SEAT LAP BELTS */
  OLDWTFA=WEIGHTFA;
  IF PS208C GT 0 OR PV208C GT 0 THEN DO;
```

NHTSA's evaluation estimates an overall 32 percent fatality reduction by lap belts for rear-outboard occupants.

```
    E=.32;
    IF CRSH=1 THEN EJECT2=.1549; ELSE IF CRSH=2.1 THEN EJECT2=.2779;    /* FROM OLDFA14, P. 9 */
      ELSE IF CRSH=2.2 THEN EJECT2=.1499; ELSE IF CRSH=3 THEN EJECT2=.5556;
      ELSE IF CRSH=4 THEN EJECT2=.2273;
    P=PS208C+PV208C;
    S=OLDWTFA*E*P / (1 - E*P);
    CLS208C=S*PS208C/P;
    CLV208C=S*PV208C/P;
    WEIGHTFA=OLDWTFA+CLS208C+CLV208C; END;
  ELSE DO; CLS208C=0; CLV208C=0; END;
  CL208C=CLS208C+CLV208C;
  CLS=CLS+CLS208C; CLV=CLV+CLV208C; CL=CL+CL208C;


/* ------------------------------------------------------------------ */
/*   206 DOOR LOCK IMPROVEMENTS                                        */
/*   MEDIAN INSTALLATION YEAR: 1965                                   */
/* ------------------------------------------------------------------ */

/* DOOR LOCKS IMPROVED GRADUALLY FROM 1962 TO 1968         */
  IF MY GE 1969 THEN DO; PS206=1; PV206=0; END;
```

```
      ELSE IF MY=1968 AND CY=1967 THEN DO; PS206=0; PV206=1; END;
      ELSE IF MY=1968 THEN DO; PS206=.5; PV206=.5; END;
      ELSE IF 1962 LE MY LE 1967 THEN DO; PS206=0; PV206=(MY-1961)/7; END;
      ELSE IF MY LE 1961 THEN DO; PS206=0; PV206=0; END;
/* REDUCTION IN EJECTION ROLLOVER FATALITY RISK WITH IMPROVED DOOR LOCKS */
  OLDWTFA=WEIGHTFA;
  IF (PS206 GT 0 OR PV206 GT 0) AND CRSH=3
      AND SEAT2 NE 52 AND EJECT2 NE 0 THEN DO;
    E=.1538;
    P=PS206+PV206;
    REL_S=EJECT2*E*P / (1 - E*P);
    S=OLDWTFA*REL_S;
    CLS206=S*PS206/P;
    CLV206=S*PV206/P;
    WEIGHTFA=OLDWTFA+CLS206+CLV206;
    EJECT2=(EJECT2 + REL_S)/(1 + REL_S); END;
  ELSE DO; CLS206=0; CLV206=0; END;
  CL206=CLS206+CLV206;
  CLS=CLS+CLS206; CLV=CLV+CLV206; CL=CL+CL206;


/* ------------------------------------------------------------------ */
/*  208B LAP BELT USE BY CHILDREN AGE 1-4                             */
/*  MEDIAN INSTALLATION YEAR FOR LAP BELTS USED BY CHILDREN: 1964     */
/* ------------------------------------------------------------------ */

/* IDENTIFIES CHILD PASSENGERS AGE 1-4 USING LAP BELTS */
  PS208B=0; PV208B=0;
  IF AGEGP=1 AND REST3=2 THEN DO;
      IF MY GE 1969 THEN PS208B=1;
        ELSE IF MY LE 1967 THEN PV208B=1;
        ELSE IF MY=1968 AND CY=1967 THEN PV208B=1;
        ELSE IF MY=1968 THEN DO; PS208B=.5; PV208B=.5; END; END;
/* LIVES SAVED AND EJECTIONS PREVENTED BY LAP BELTS (AGE 1-4) */
  OLDWTFA=WEIGHTFA;
  IF PS208B GT 0 OR PV208B GT 0 THEN DO;
    E=.33;
    IF CRSH=1 THEN EJECT2=.1549; ELSE IF CRSH=2.1 THEN EJECT2=.2779;   /* FROM OLDFA14, P. 9 */
      ELSE IF CRSH=2.2 THEN EJECT2=.1499; ELSE IF CRSH=3 THEN EJECT2=.5556;
      ELSE IF CRSH=4 THEN EJECT2=.2273;
    P=PS208B+PV208B;
    S=OLDWTFA*E*P / (1 - E*P);
    CLS208B=S*PS208B/P;
    CLV208B=S*PV208B/P;
    WEIGHTFA=OLDWTFA+CLS208B+CLV208B; END;
  ELSE DO; CLS208B=0; CLV208B=0; END;
  CL208B=CLS208B+CLV208B;
  CLS=CLS+CLS208B; CLV=CLV+CLV208B; CL=CL+CL208B;


/* ------------------------------------------------------------------ */
/*  208A LAP BELTS FOR FRONT-SEAT OUTBOARD OCCUPANTS                  */
/*  MEDIAN INSTALLATION YEAR: 1962                                    */
/* ------------------------------------------------------------------ */

/* IDENTIFIES FRONT-OUTBOARD-SEAT OCCUPANTS WEARING LAP BELTS  */
  PS208A=0; PV208A=0;
  IF SEAT2 IN (11,13) AND REST3=2 AND AGEGP=5 THEN DO;
      IF MY GE 1969 THEN PS208A=1;
        ELSE IF MY LE 1967 THEN PV208A=1;
        ELSE IF MY=1968 AND CY=1967 THEN PV208A=1;
        ELSE IF MY=1968 THEN DO; PS208A=.5; PV208A=.5; END; END;
/* LIVES SAVED AND EJECTIONS PREVENTED BY FRONT-OUTBOARD-SEAT LAP BELTS */
  OLDWTFA=WEIGHTFA;
  IF PS208A GT 0 OR PV208A GT 0 THEN DO;
```

A double-pair comparison analysis of lap belt effectiveness for front-outboard occupants, conducted in support of this study, found 27 percent overall fatality reduction.

```
E=.27;
   IF CRSH=1 THEN EJECT2=.1549; ELSE IF CRSH=2.1 THEN EJECT2=.2779;   /* FROM OLDFA14, P. 9 */
      ELSE IF CRSH=2.2 THEN EJECT2=.1499; ELSE IF CRSH=3 THEN EJECT2=.5556;
      ELSE IF CRSH=4 THEN EJECT2=.2273;
   P=PS208A+PV208A;
   S=OLDWTFA*E*P / (1 - E*P);
   CLS208A=S*PS208A/P;
   CLV208A=S*PV208A/P;
   WEIGHTFA=OLDWTFA+CLS208A+CLV208A; END;
  ELSE DO; CLS208A=0; CLV208A=0; END;
  CL208A=CLS208A+CLV208A;
  CLS=CLS+CLS208A; CLV=CLV+CLV208A; CL=CL+CL208A;
RUN;
```

That concludes the model for passenger cars. It has estimated, on a cell-by-cell basis, by how much fatalities would increase if all the safety technologies were "removed." The next steps are to tally up the lives saved over all the cells, by calendar year, and print the results.

```
PROC MEANS SUM NOPRINT DATA=CAR9; BY CY;
  VAR ORIGWT WEIGHTFA
     CLV208J CLS208J CL208J
     CLV108 CLS108 CL108
     CLV208I CLS208I CL208I
     CLV214B CLS214B CL214B
     CLV208H CLS208H CL208H
     CLV208G CLS208G CL208G
     CLV213 CLS213 CL213
     CLVNCAP CLSNCAP CLNCAP
     CLV208F CLS208F CL208F
     CLV216 CLS216 CL216
     CLV105B CLS105B CL105B
     CLV214A CLS214A CL214A
     CLV208E CLS208E CL208E
     CLV208D CLS208D CL208D
     CLV201 CLS201 CL201
     CLV203 CLS203 CL203
     CLV105A CLS105A CL105A
     CLV208C CLS208C CL208C
     CLV206 CLS206 CL206
     CLV208B CLS208B CL208B
     CLV212 CLS212 CL212
     CLV208A CLS208A CL208A
     CLV CLS CL;
  OUTPUT OUT=CAR10
     SUM=C_ORIGWT C_WTFA
     CLV208J CLS208J CL208J
     CLV108 CLS108 CL108
     CLV208I CLS208I CL208I
     CLV214B CLS214B CL214B
     CLV208H CLS208H CL208H
     CLV208G CLS208G CL208G
     CLV213 CLS213 CL213
     CLVNCAP CLSNCAP CLNCAP
     CLV208F CLS208F CL208F
     CLV216 CLS216 CL216
     CLV105B CLS105B CL105B
     CLV214A CLS214A CL214A
     CLV208E CLS208E CL208E
     CLV208D CLS208D CL208D
     CLV201 CLS201 CL201
```

```
      CLV203 CLS203 CL203
      CLV105A CLS105A CL105A
      CLV208C CLS208C CL208C
      CLV206 CLS206 CL206
      CLV208B CLS208B CL208B
      CLV212 CLS212 CL212
      CLV208A CLS208A CL208A
      CLV CLS CL;

PROC PRINT DATA=CAR10;
   FORMAT CLV208H CLS208H CL208H CLV208G CLS208G CL208G CLV213 CLS213 CL213 9.0;
   ID CY; VAR CLV208H CLS208H CL208H CLV208G CLS208G CL208G CLV213 CLS213 CL213;
   SUM CLV208H CLS208H CL208H CLV208G CLS208G CL208G CLV213 CLS213 CL213;
TITLE1 'CAR OCCUPANT LIVES SAVED BY AUTOMATIC 2-POINT BELTS, BACK-SEAT 3-POINT BELTS, AND CHILD
SAFETY SEATS, 1960-74';
TITLE2 ' ';
TITLE3 '...208H = AUTOMATIC 2-POINT BELTS (PEAK INSTALLATION YEAR 1991, FMVSS PHASE-IN BEGAN
9/1/86)';
TITLE4 '...208G = 3-POINT BELTS FOR BACK-SEAT OCCUPANTS (MEDIAN INSTALLATION YEAR 1989, FMVSS
EFFECTIVE 12/11/89)';
TITLE5 '...213 = CHILD SAFETY SEATS (USE EXCEEDED 50% IN 1985, FMVSS EFFECTIVE 4/1/71, STATE LAWS
1978-85)';
TITLE6 '...V = VOLUNTARY INSTALLATIONS, BEFORE FMVSS EFFECTIVE DATE';
TITLE7 '...S = STANDARD INSTALLATIONS, CARS BUILT ON OR AFTER FMVSS EFFECTIVE DATE';

PROC PRINT DATA=CAR10;
   FORMAT CLVNCAP CLSNCAP CLNCAP CLV208F CLS208F CL208F CLV216 CLS216 CL216 9.0;
   ID CY; VAR CLVNCAP CLNCAP CLV208F CLS208F CL208F CLV216 CLS216 CL216;
   SUM CLVNCAP CLNCAP CLV208F CLS208F CL208F CLV216 CLS216 CL216;
TITLE1 'CAR OCCUPANT LIVES SAVED BY VOLUNTARY NCAP IMPROVEMENTS, FRONT-SEAT 3-POINT BELTS, AND ROOF
CRUSH STRENGTH, 1960-74';
TITLE2 ' ';
TITLE3 '...NCAP = VOLUNTARY NCAP IMPROVEMENTS IN NON-AIR-BAG CARS (MEDIAN IMPLEMENTATION YEAR EARLY
1984)';
TITLE4 '...208F = 3-POINT BELTS FOR FRONT-SEAT OCCUPANTS (MEDIAN INSTALLATION YEAR 1974, FMVSS
EFFECTIVE 9/1/73)';
TITLE5 '...216 = ROOF CRUSH STRENGTH, B-PILLARS, ETC. (MEDIAN INSTALLATION YEAR 1973, FMVSS
EFFECTIVE 9/1/73)';
TITLE6 '...V = VOLUNTARY INSTALLATIONS, BEFORE FMVSS EFFECTIVE DATE';
TITLE7 '...S = STANDARD INSTALLATIONS, CARS BUILT ON OR AFTER FMVSS EFFECTIVE DATE';

PROC PRINT DATA=CAR10;
   FORMAT CLV105B CLS105B CL105B CLV214A CLS214A CL214A CLV208E CLS208E CL208E 9.0;
   ID CY; VAR CLV105B CLS105B CL105B CLV214A CLS214A CL214A CLV208E CLS208E CL208E;
   SUM CLV105B CLS105B CL105B CLV214A CLS214A CL214A CLV208E CLS208E CL208E;
TITLE1 'CAR OCCUPANT LIVES SAVED BY FRONT DISC BRAKES, SIDE DOOR BEAMS, AND BACK-CENTER LAP BELTS,
1960-74';
TITLE2 ' ';
TITLE3 '...105B = FRONT DISC BRAKES (MEDIAN INSTALLATION YEAR 1971, FMVSS EFFECTIVE 1/1/76)';
TITLE4 '...214A = SIDE DOOR BEAMS (MEDIAN INSTALLATION YEAR EARLY 1973, FMVSS EFFECTIVE 1/1/73)';
TITLE5 '...208E = LAP BELTS FOR BACK-CENTER OCCUPANTS (MEDIAN INSTALLATION YEAR EARLY 1967, FMVSS
EFFECTIVE 1/1/68)';
TITLE6 '...V = VOLUNTARY INSTALLATIONS, BEFORE FMVSS EFFECTIVE DATE';
TITLE7 '...S = STANDARD INSTALLATIONS, CARS BUILT ON OR AFTER FMVSS EFFECTIVE DATE';

PROC PRINT DATA=CAR10;
   FORMAT CLV208D CLS208D CL208D CLV201 CL201 CLV203 CLS203 CL203 9.0;
   ID CY; VAR CLV208D CLS208D CL208D CLV201 CL201 CLV203 CLS203 CL203;
   SUM CLV208D CLS208D CL208D CLV201 CL201 CLV203 CLS203 CL203;
TITLE1 'CAR OCCUPANT LIVES SAVED BY FRONT-CENTER LAP BELTS, SAFER INSTRUMENT PANELS, AND EA
STEERING ASSEMBLIES, 1960-74';
TITLE2 ' ';
TITLE3 '...208D = LAP BELTS FOR FRONT-CENTER OCCUPANTS (MEDIAN INSTALLATION YEAR 1966, FMVSS
EFFECTIVE 1/1/68)';
```

```
TITLE4 '...201  = VOLUNTARY INSTRUMENT PANEL IMPROVEMENTS (MEDIAN IMPLEMENTATION YEAR 1968, FMVSS
EFFECTIVE 1/1/68)';
TITLE5 '...203  = ENERGY-ABSORBING STEERING ASSEMBLIES (MEDIAN IMPLEMENTATION YEAR 1967, FMVSS
EFFECTIVE 1/1/68)';
TITLE6 '...V = VOLUNTARY INSTALLATIONS, BEFORE FMVSS EFFECTIVE DATE';
TITLE7 '...S = STANDARD INSTALLATIONS, CARS BUILT ON OR AFTER FMVSS EFFECTIVE DATE';

PROC PRINT DATA=CAR10;
   FORMAT CLV105A CLS105A CL105A CLV208C CLS208C CL208C CLV206 CLS206 CL206 9.0;
   ID CY; VAR CLV105A CLS105A CL105A CLV208C CLS208C CL208C CLV206 CLS206 CL206;
   SUM CLV105A CLS105A CL105A CLV208C CLS208C CL208C CLV206 CLS206 CL206;
TITLE1 'CAR OCCUPANT LIVES SAVED BY DUAL MASTER CYLINDERS, BACK-OUTBOARD LAP BELTS, AND IMPROVED
DOOR LOCKS, 1960-74';
TITLE2 ' ';
TITLE3 '...105A = DUAL MASTER CYLINDERS (MEDIAN INSTALLATION YEAR LATE 1966, FMVSS EFFECTIVE
1/1/68)';
TITLE4 '...208C = LAP BELTS FOR BACK-OUTBOARD OCCUPANTS (MEDIAN INSTALLATION YEAR LATE 1965, FMVSS
EFFECTIVE 1/1/68)';
TITLE5 '...206  = IMPROVED DOOR LOCKS (MEDIAN INSTALLATION YEAR 1965, FMVSS EFFECTIVE 1/1/68)';
TITLE6 '...V = VOLUNTARY INSTALLATIONS, BEFORE FMVSS EFFECTIVE DATE';
TITLE7 '...S = STANDARD INSTALLATIONS, CARS BUILT ON OR AFTER FMVSS EFFECTIVE DATE';

PROC PRINT DATA=CAR10;
   FORMAT CLV208B CLS208B CL208B CLV212 CLS212 CL212 CLV208A CLS208A CL208A 9.0;
   ID CY; VAR CLV208B CLS208B CL208B CLV212 CLS212 CL212 CLV208A CLS208A CL208A;
   SUM CLV208B CLS208B CL208B CLV212 CLS212 CL212 CLV208A CLS208A CL208A;
TITLE1 'CAR OCCUPANT LIVES SAVED BY LAP BELTS FOR CHILDREN AGE 1-4, ADHESIVE WINDSHIELD BONDING,
AND FRONT-OUTBOARD LAP BELTS, 1960-74';
TITLE2 ' ';
TITLE3 '...208B = LAP BELT USE BY CHILDREN AGE 1-4 (MEDIAN INSTALLATION YEAR 1964, FMVSS EFFECTIVE
1/1/68)';
TITLE4 '...212  = ADHESIVE WINDSHIELD BONDING (MEDIAN INSTALLATION YEAR EARLY 1966, FMVSS EFFECTIVE
1/1/70)';
TITLE5 '...208A = LAP BELTS FOR FRONT-OUTBOARD OCCUPANTS (MEDIAN INSTALLATION YEAR 1962, FMVSS
EFFECTIVE 1/1/68)';
TITLE6 '...V = VOLUNTARY INSTALLATIONS, BEFORE FMVSS EFFECTIVE DATE';
TITLE7 '...S = STANDARD INSTALLATIONS, CARS BUILT ON OR AFTER FMVSS EFFECTIVE DATE';

DATA CAR11; SET CAR10;
PCT_SAVE=100*CL/C_WTFA;
PROC PRINT DATA=CAR11;
   FORMAT C_ORIGWT C_WTFA CLV CLS CL 9.0 PCT_SAVE 6.2;
   ID CY; VAR C_ORIGWT C_WTFA CLV CLS CL PCT_SAVE;
   SUM C_ORIGWT C_WTFA CLV CLS CL;
TITLE1 'OVERALL CAR OCCUPANT LIVES SAVED BY VEHICLE SAFETY STANDARDS AND TECHNOLOGIES, 1960-74';
TITLE2 ' ';
TITLE3 'C_ORIGWT = ACTUAL CAR OCCUPANT FATALITIES (ESTIMATED)';
TITLE4 'C_WTFA = FATALITIES THAT WOULD HAVE OCCURRED WITHOUT ANY VEHICLE SAFETY IMPROVEMENTS';
TITLE5 'CLV = OVERALL LIVES SAVED BY VOLUNTARY IMPROVEMENTS, BEFORE FMVSS EFFECTIVE DATE';
TITLE6 'CLS = OVERALL LIVES SAVED IN CARS BUILT ON OR AFTER FMVSS EFFECTIVE DATE';
TITLE7 'CL = OVERALL LIVES SAVED BY VEHICLE SAFETY IMPROVEMENTS (CLV + CLS)';
TITLE8 'PCT_SAVE = PERCENT OF WOULD-HAVE-BEEN FATALITIES SAVED BY SAFETY STANDARDS AND
TECHNOLOGIES';
RUN;
```

```
/* ------------------------------------------------------------- */
/* ------------------------------------------------------------- */
/*   THE ANALYSIS OF LIGHT TRUCK OCCUPANTS BEGINS HERE           */
/*                                                               */
/*   CREATES A FILE WITH ALL POSSIBLE COMBINATIONS OF            */
/*   CY, MY, CRASH MODE, OCC AGE GP, SEAT POSITION, RESTRAINT USE */
/*   ORIGWT IS THE NUMBER OF OCC FATALITIES FOR EACH COMBINATION */
/* ------------------------------------------------------------- */
/* ------------------------------------------------------------- */


/* ------------------------------------------------------------- */
/*   FIRST STEP: ALL CY-MY COMBINATIONS                          */
/*   BASED ON VEHICLE AGE DISTRIBUTION IN 1975-80 FARS           */
/* ------------------------------------------------------------- */
```

The model for LTVs begins here. The basic approach is the same as for cars, but the implementation dates and effectiveness of the various technologies may be different. Belt use is also different. In general, the notes on this section will be limited to the spots where LTVs differ from cars.

```
DATA TRK5(KEEP=CY FARSTRK VEHAGE);
SET ACCFAC1;
DO VEHAGE = -1 TO 25;
  OUTPUT;
END;
RUN;
PROC SORT DATA=TRK5; BY VEHAGE;
```

TRK6 is a matrix of LTV occupant fatalities by CY and MY.

```
DATA TRK6; MERGE TRK5 VEHAGE4; BY VEHAGE;
ORIGWT=PCTTRK*FARSTRK;
MY=CY-VEHAGE;
KEEP CY MY ORIGWT;
PROC SORT DATA=TRK6; BY CY MY;
RUN;
```

```
/* ------------------------------------------------------------- */
/*   ALL PERMITTED COMBINATIONS OF CRASH MODE, AGE GROUP, SEAT POS */
/*   AGE GROUPS ARE INFANT, TODDLER 1-4, ALL OTHERS 5 AND OLDER  */
/*   WEIGHT FACTORS BASED ON DISTRIBUTIONS IN 1975-80 FARS       */
/* ------------------------------------------------------------- */
```

TRK7 is a matrix of LTV occupant fatalities by CY, MY, CRSH, SEAT2 and AGEGP. All distributions are based on 1975-80 FARS data, specifically the analysis by OLDFA14 of the file *FARS7580* generated by LS2004. There were 22,676.59 weighted cases of LTV occupant fatalities in the file analyzed by OLDFA14.

```
DATA TRK7(KEEP=CY MY ORIGWT CRSH SEAT2 AGEGP);
SET TRK6;
X=22676.59*22676.59;                                    /* FROM OLDFA14, P. 7 */
OLDWT=ORIGWT;
DO ICRSH=1 TO 5;
DO IAGESEAT=1 TO 25;                                     /* FROM OLDFA14, P. 7 */
IF ICRSH=1 THEN DO; CRSH=1; CRSHFA=10937.08; END;
  ELSE IF ICRSH=2 THEN DO; CRSH=2.1; CRSHFA=1912.215; END;
  ELSE IF ICRSH=3 THEN DO; CRSH=2.2; CRSHFA=2233.393; END;
  ELSE IF ICRSH=4 THEN DO; CRSH=3;   CRSHFA=6184.181; END;
  ELSE IF ICRSH=5 THEN DO; CRSH=4;   CRSHFA=1409.718; END;
IF IAGESEAT=1 THEN DO; SEAT2=11; AGEGP=5; ASFA=15269; END;  /* FROM OLDFA14, P. 7 */
  ELSE IF IAGESEAT= 2 THEN DO; SEAT2=12; AGEGP=0; ASFA=30.883; END;
  ELSE IF IAGESEAT= 3 THEN DO; SEAT2=12; AGEGP=1; ASFA=116.04; END;
```

383

```
      ELSE IF IAGESEAT= 4 THEN DO; SEAT2=12; AGEGP=5; ASFA=827.00; END;
      ELSE IF IAGESEAT= 5 THEN DO; SEAT2=13; AGEGP=0; ASFA=32.045; END;
      ELSE IF IAGESEAT= 6 THEN DO; SEAT2=13; AGEGP=1; ASFA=130.29; END;
      ELSE IF IAGESEAT= 7 THEN DO; SEAT2=13; AGEGP=5; ASFA=4389.6; END;
      ELSE IF IAGESEAT= 8 THEN DO; SEAT2=18; AGEGP=0; ASFA=7.7778; END;
      ELSE IF IAGESEAT= 9 THEN DO; SEAT2=18; AGEGP=1; ASFA=13.863; END;
      ELSE IF IAGESEAT=10 THEN DO; SEAT2=18; AGEGP=5; ASFA=36.997; END;
      ELSE IF IAGESEAT=11 THEN DO; SEAT2=21; AGEGP=0; ASFA=4.7558; END;
      ELSE IF IAGESEAT=12 THEN DO; SEAT2=21; AGEGP=1; ASFA=28.356; END;
      ELSE IF IAGESEAT=13 THEN DO; SEAT2=21; AGEGP=5; ASFA=313.38; END;
      ELSE IF IAGESEAT=14 THEN DO; SEAT2=22; AGEGP=0; ASFA=3.981; END;
      ELSE IF IAGESEAT=15 THEN DO; SEAT2=22; AGEGP=1; ASFA=17.394; END;
      ELSE IF IAGESEAT=16 THEN DO; SEAT2=22; AGEGP=5; ASFA=138.31; END;
      ELSE IF IAGESEAT=17 THEN DO; SEAT2=28; AGEGP=0; ASFA=0.0666; END;
      ELSE IF IAGESEAT=18 THEN DO; SEAT2=28; AGEGP=1; ASFA=0.5587; END;
      ELSE IF IAGESEAT=19 THEN DO; SEAT2=28; AGEGP=5; ASFA= 11.949; END;
      ELSE IF IAGESEAT=20 THEN DO; SEAT2=51; AGEGP=0; ASFA=6.0551; END;
      ELSE IF IAGESEAT=21 THEN DO; SEAT2=51; AGEGP=1; ASFA=56.304; END;
      ELSE IF IAGESEAT=22 THEN DO; SEAT2=51; AGEGP=5; ASFA=1016.3; END;
      ELSE IF IAGESEAT=23 THEN DO; SEAT2=52; AGEGP=0; ASFA=0.0641; END;
      ELSE IF IAGESEAT=24 THEN DO; SEAT2=52; AGEGP=1; ASFA=3.4122; END;
      ELSE IF IAGESEAT=25 THEN DO; SEAT2=52; AGEGP=5; ASFA= 222.19; END;
ORIGWT=OLDWT*CRSHFA*ASFA/X;
OUTPUT;
END;
END;
RUN;


/* ----------------------------------------------------------------- */
/*  ALL PERMITTED CONFIGURATIONS OF RESTRAINT USE                    */
/*                                                                   */
/*  SEPARATE ANALYSES BY SEAT POSITION, MODEL YEAR & AGE GROUP       */
/*  DEPENDING ON WHAT TYPES AND HOW MANY SYSTEMS ARE AVAILABLE       */
/*  USE RATES BASED ON 1975-80 AVGS, EXCEPT WHERE 1975 IS HIGHER     */
/*  FIRST MODERN SAFETY SEATS: CY 1967 (TODDLER) AND 1970 (INFANT)   */
/* ----------------------------------------------------------------- */
```

Similar scenarios for analysis of restraint use as in the car model, but the model years in the scenarios may differ, and so do the belt use rates.

```
DATA RESO(KEEP=CY MY ORIGWT CRSH SEAT2 AGEGP REST3)
     INF1(KEEP=CY MY ORIGWT CRSH SEAT2 AGEGP)
     TOD11(KEEP=CY MY ORIGWT CRSH SEAT2 AGEGP)
     TOD21(KEEP=CY MY ORIGWT CRSH SEAT2 AGEGP)
     FOB11(KEEP=CY MY ORIGWT CRSH SEAT2 AGEGP)
     FOB21(KEEP=CY MY ORIGWT CRSH SEAT2 AGEGP)
     FOB31(KEEP=CY MY ORIGWT CRSH SEAT2 AGEGP)
     LAP1(KEEP=CY MY ORIGWT CRSH SEAT2 AGEGP);
SET TRK7;
IF SEAT2 IN (18,28,51,52) OR (AGEGP=0 AND CY LT 1970) OR
    (SEAT2=12 AND MY LT 1968) OR (SEAT2=21 AND MY LT 1968) OR
    (SEAT2=22 AND MY LT 1968) THEN DO; REST3=0; OUTPUT RESO; END;
  ELSE IF AGEGP=0 THEN OUTPUT INF1;
  ELSE IF AGEGP=1 AND 1967 LE CY LE 1974 THEN OUTPUT TOD11;
  ELSE IF AGEGP=1 THEN OUTPUT TOD21;
  ELSE IF SEAT2 IN (11,13) AND 1969 LE MY LE 1975 THEN OUTPUT FOB11;
  ELSE IF SEAT2 IN (11,13) THEN OUTPUT FOB21;
  ELSE OUTPUT LAP1;
RUN;

DATA INF2(KEEP=CY MY ORIGWT CRSH SEAT2 AGEGP REST3);
SET INF1;
OLDWT=ORIGWT;
```

Use of infant seats among fatally injured passengers gradually increased from CY 1970, reaching 3.82 percent in 1974.

```
CRDUSE=.0382*(CY-1969)/5;                              /* FROM OLDFA14, P. 1 */
REST3=4; ORIGWT=   CRDUSE *OLDWT; OUTPUT;
REST3=0; ORIGWT=(1-CRDUSE)*OLDWT; OUTPUT;
RUN;

DATA TOD12(KEEP=CY MY ORIGWT CRSH SEAT2 AGEGP REST3);
SET TOD11;
OLDWT=ORIGWT;
```

The percentage of toddler fatalities restrained in safety seats increased from CY 1967, reaching 0.71 percent in 1974. Lap belt use by child passenger fatalities age 1-4 was 0.39 percent.

```
IF 1970 LE CY LE 1974 THEN CRDUSE=.0071*(CY-1969)/5;       /* FROM OLDFA14, P. 2 */
  ELSE IF 1967 LE CY LE 1969 THEN CRDUSE=.0071*(CY-1966)/20;
LAPUSE=.0039;
REST3=4; ORIGWT=            CRDUSE *OLDWT; OUTPUT;
REST3=2; ORIGWT=            LAPUSE *OLDWT; OUTPUT;
REST3=0; ORIGWT=(1-(CRDUSE+LAPUSE))*OLDWT; OUTPUT;
RUN;

DATA TOD22(KEEP=CY MY ORIGWT CRSH SEAT2 AGEGP REST3);
SET TOD21;
OLDWT=ORIGWT;
```

Before CY 1967, lap belt use by child passenger fatalities age 1-4 was 0.39 percent.

```
REST3=2; ORIGWT=.0039*OLDWT; OUTPUT;                       /* FROM OLDFA14, P. 7 */
REST3=0; ORIGWT=.9961*OLDWT; OUTPUT;
RUN;

DATA FOB12(KEEP=CY MY ORIGWT CRSH SEAT2 AGEGP REST3);
SET FOB11;
OLDWT=ORIGWT;
```

For front-outboard fatalities age 5+, lap belt use was 2.85 percent in MY 1969-75 LTVs. A few LTVs were equipped with 3-point belts; overall use among fatalities was 0.13 percent in 1969-73 and 0.7 percent in 1974-75.

```
IF 1969 LE MY LE 1973 THEN LS_USE=.0013;                   /* FROM OLDFA14, P. 18 */
  ELSE IF 1974 LE MY LE 1975 THEN LS_USE=.0070;
REST3=3; ORIGWT=           LS_USE  *OLDWT; OUTPUT;
REST3=2; ORIGWT=            .0285  *OLDWT; OUTPUT;
REST3=0; ORIGWT=(1-(LS_USE+ .0285))*OLDWT; OUTPUT;
RUN;

DATA FOB22(KEEP=CY MY ORIGWT CRSH SEAT2 AGEGP REST3);
SET FOB21;
OLDWT=ORIGWT;
```

Lap belt use by front-outboard fatalities was 2.93 percent in MY 1968 and 0.79 percent before that.

```
IF MY=1968 THEN LAPUSE=.0293;                              /* FROM OLDFA14, P. 18 */
  ELSE LAPUSE=.0079;                                       /* FROM OLDFA14, P. 17 */
REST3=2; ORIGWT=   LAPUSE *OLDWT; OUTPUT;
REST3=0; ORIGWT=(1-LAPUSE)*OLDWT; OUTPUT;
RUN;

DATA LAP2(KEEP=CY MY ORIGWT CRSH SEAT2 AGEGP REST3);
SET LAP1;
```

```
OLDWT=ORIGWT;
```

In MY 1968-75, 1.56 percent of rear-seat fatalities and 0.45 percent of front-center fatalities were lap-belted.

```
IF SEAT2=12 THEN LAPUSE=.0045;                              /* FROM OLDFA14, P. 19 */
  ELSE IF SEAT2=21 THEN LAPUSE=.0156;                      /* FROM OLDFA14, P. 20 */
  ELSE IF SEAT2=22 THEN LAPUSE=.0156;                      /* FROM OLDFA14, P. 20 */
REST3=2; ORIGWT=   LAPUSE *OLDWT; OUTPUT;
REST3=0; ORIGWT=(1-LAPUSE)*OLDWT; OUTPUT;
RUN;


/* -------------------------------------------------------------- */
/*  TRK8 IS THE FULL ANALYSIS FILE                                */
/*  EJECT2 IS PROPORTION OF FATALITIES EJECTED IN EACH CELL       */
/* -------------------------------------------------------------- */

DATA TRK8; SET RESO INF2 TOD12 TOD22 FOB12 FOB22 LAP2;
```

TRK8, the full analysis file, is a matrix of LTV occupant fatalities by CY, MY, CRSH, SEAT2, AGEGP and REST3.

```
WEIGHTFA=ORIGWT;
```

Proportions of unrestrained and restrained fatalities that were ejected, by crash mode.

```
IF REST3=0 THEN DO;
  IF CRSH=1 THEN EJECT2=.2331; ELSE IF CRSH=2.1 THEN EJECT2=.3664;   /* FROM OLDFA14, P. 9 */
    ELSE IF CRSH=2.2 THEN EJECT2=.3153; ELSE IF CRSH=3 THEN EJECT2=.6424;
    ELSE IF CRSH=4 THEN EJECT2=.4091; END;
ELSE DO;
  IF CRSH=1 THEN EJECT2=.1151; ELSE IF CRSH=2.1 THEN EJECT2=.2626;   /* FROM OLDFA14, P. 10 */
    ELSE IF CRSH=2.2 THEN EJECT2=.2315; ELSE IF CRSH=3 THEN EJECT2=.2483;
    ELSE IF CRSH=4 THEN EJECT2=.3106; END;
RUN;
PROC SORT DATA=TRK8; BY CY;
RUN;


/* -------------------------------------------------------------- */
/* -------------------------------------------------------------- */
/*  REPEATS THE TRK OCCUPANT ANALYSIS OF LS2004.SAS               */
/*                                                                */
/*  HAS TO BE SIMPLIFIED FOR 203 SINCE VIN/MAKE-MODEL UNK         */
/* -------------------------------------------------------------- */
/* -------------------------------------------------------------- */

DATA TRK9; SET TRK8;

/* -------------------------------------------------------------- */
/*  ZERO BENEFITS FOR TECHNOLOGIES NOT AVAILABLE IN 1960-74       */
/* -------------------------------------------------------------- */
```

Six of the later technologies in LS2004 did not exist in any LTV on the road in CY 1960-74.

```
PS208J=0;  PV208J=0; TLS208J=0; TLV208J=0; TL208J=0;
PS108=0;  PV108=0; TLS108=0; TLV108=0; TL108=0;
PS208I=0; PV208I=0; TLS208I=0; TLV208I=0; TL208I=0;
PS214A=0; PV214A=0; TLS214A=0; TLV214A=0; TL214A=0;
PS208G=0; PV208G=0; TLS208G=0; TLV208G=0; TL208G=0;
PS212=0; PV212=0; TLS212=0; TLV212=0; TL212=0;
TLS=0; TLV=0; TL=0;

/* -------------------------------------------------------------- */
```

```
/*   213 CHILD SAFETY SEATS                                            */
/*   USE RATE WENT OVER 50% IN: 1985                                   */
/* ------------------------------------------------------------------- */

/* IDENTIFIES CHILD PASSENGERS IN SAFETY SEATS   */
  PS213=0;  IF REST3=4 THEN PV213=1;  ELSE PV213=0;
/* LIVES SAVED AND EJECTIONS PREVENTED BY CHILD SAFETY SEATS */
  OLDWTFA=WEIGHTFA;
  IF PS213 GT 0 OR PV213 GT 0 THEN DO;
    IF AGEGP=0 THEN E=.58; ELSE E=.59;
```

"Removing" the safety seats makes the occupants unrestrained. This is the probability that an unrestrained occupant was ejected:

```
    IF CRSH=1 THEN EJECT2=.2331; ELSE IF CRSH=2.1 THEN EJECT2=.3664;   /* FROM OLDFA14, P. 9 */
      ELSE IF CRSH=2.2 THEN EJECT2=.3153; ELSE IF CRSH=3 THEN EJECT2=.6424;
      ELSE IF CRSH=4 THEN EJECT2=.4091;
    P=PS213+PV213;
    S=OLDWTFA*E*P / (1 - E*P);
    TLS213=S*PS213/P;
    TLV213=S*PV213/P;
    WEIGHTFA=OLDWTFA+TLS213+TLV213; END;
  ELSE DO; TLS213=0; TLV213=0; END;
  TL213=TLS213+TLV213;
  TLS=TLS+TLS213; TLV=TLV+TLV213; TL=TL+TL213;
```

```
/* -------------------------------------------------------------------- */
/*   208F 3-POINT BELTS FOR FRONT-SEAT OUTBOARD OCCUPANTS               */
/*   MEDIAN INSTALLATION YEAR: 1976                                    */
/* -------------------------------------------------------------------- */

/* IDENTIFIES FRONT-SEAT OCCUPANTS WEARING 3-POINT BELTS */
  PS208F=0; PV208F=0;
  IF SEAT2 IN (11,13) AND REST3=3 THEN PV208F=1;
/* LIVES SAVED AND EJECTIONS PREVENTED BY FRONT-OUTBOARD 3-POINT BELTS */
  OLDWTFA=WEIGHTFA;
  IF PS208F GT 0 OR PV208F GT 0 THEN DO;
```

Unlike LS2004, which has many different effectiveness values depending on the crash mode, we will for simplicity use the overall average fatality reduction by 3-point belts in LTVs: 60 percent.

```
  E=.60;
  IF CRSH=1 THEN EJECT2=.2331; ELSE IF CRSH=2.1 THEN EJECT2=.3664;   /* FROM OLDFA14, P. 9 */
    ELSE IF CRSH=2.2 THEN EJECT2=.3153; ELSE IF CRSH=3 THEN EJECT2=.6424;
    ELSE IF CRSH=4 THEN EJECT2=.4091;
  P=PS208F+PV208F;
  S=OLDWTFA*E*P / (1 - E*P);
  TLS208F=S*PS208F/P;
  TLV208F=S*PV208F/P;
  WEIGHTFA=OLDWTFA+TLS208F+TLV208F; END;
ELSE DO; TLS208F=0; TLV208F=0; END;
TL208F=TLS208F+TLV208F;
TLS=TLS+TLS208F; TLV=TLV+TLV208F; TL=TL+TL208F;


/* -------------------------------------------------------------------- */
/*   203 ENERGY-ABSORBING AND TELESCOPING STEERING ASSEMBLIES          */
/*   MEDIAN IMPLEMENTATION YEAR: EARLY 1976                            */
/* -------------------------------------------------------------------- */

/* IDENTIFIES LIGHT TRUCKS WITH ENERGY-ABSORBING STEERING ASSEMBLIES    */
  PS203=0; PV203=0;
  IF MY GE 1970 AND SEAT2=11 THEN DO;
```

LS2004 indicates that 3 percent of MY 1970 and 1971 LTVs were equipped with energy-absorbing steering assemblies. That had increased to 48 percent by MY 1974-75.

```
    IF MY=1970 THEN PV203=.03;                     /* FROM LS2004 */
      ELSE IF MY=1971 THEN PV203=.03;
      ELSE IF MY=1972 THEN PV203=.02;
      ELSE IF MY=1973 THEN PV203=.47;
      ELSE IF MY=1974 THEN PV203=.48;
      ELSE IF MY=1975 THEN PV203=.48; END;
/* LIVES SAVED IN FRONTAL IMPACTS BY INSTRUMENT PANEL IMPROVEMENTS */
  OLDWTFA=WEIGHTFA;
  IF PS203 GT 0 OR PV203 GT 0 THEN DO;
    IF CRSH=1 THEN E=.121; ELSE E=0;
    P=PS203+PV203;
    S=OLDWTFA*E*P / (1 - E*P);
    TLS203=S*PS203/P;
    TLV203=S*PV203/P;
    WEIGHTFA=OLDWTFA+TLS203+TLV203; END;
  ELSE DO; TLS203=0; TLV203=0; END;
  TL203=TLS203+TLV203;
  TLS=TLS+TLS203; TLV=TLV+TLV203; TL=TL+TL203;


/* -------------------------------------------------------------------- */
/*   201 VOLUNTARY INSTRUMENT PANEL IMPROVEMENTS (FMVSS-201 INSPIRED)   */
/*   MEDIAN IMPLEMENTATION YEAR: 1972                                  */
/* -------------------------------------------------------------------- */
```

```
/* INSTRUMENT PANELS WERE GRADUALLY IMPROVED FROM 1969 TO 1977    */
   PS201=0;  PV201=0;
   IF MY GE 1969 AND SEAT2=13 THEN PV201=.125*(MY-1968);
/* LIVES SAVED IN FRONTAL IMPACTS BY INSTRUMENT PANEL IMPROVEMENTS */
   OLDWTFA=WEIGHTFA;
   IF PS201 GT 0 OR PV201 GT 0 THEN DO;
     IF CRSH=1 AND REST3 NE 2 THEN E=.159; ELSE E=0;
     P=PS201+PV201;
     S=OLDWTFA*E*P / (1 - E*P);
     TLS201=S*PS201/P;
     TLV201=S*PV201/P;
     WEIGHTFA=OLDWTFA+TLS201+TLV201; END;
   ELSE DO; TLS201=0; TLV201=0; END;
   TL201=TLS201+TLV201;
   TLS=TLS+TLS201; TLV=TLV+TLV201; TL=TL+TL201;


/* ------------------------------------------------------------------ */
/*   105B FRONT DISC BRAKES                                           */
/*   MEDIAN INSTALLATION YEAR: 1971 (INSTALLATION COMPLETED 1977)     */
/* ------------------------------------------------------------------ */

/* IMPLEMENTATION OF FRONT DISC BRAKES */
   IF MY LE 1964 THEN DO; PS105B=0; PV105B=0; END;
     ELSE IF MY=1965 THEN DO; PS105B=0; PV105B=.02; END;
     ELSE IF MY=1966 THEN DO; PS105B=0; PV105B=.03; END;
     ELSE IF MY=1967 THEN DO; PS105B=0; PV105B=.06; END;
     ELSE IF MY=1968 THEN DO; PS105B=0; PV105B=.13; END;
     ELSE IF MY=1969 THEN DO; PS105B=0; PV105B=.28; END;
     ELSE IF MY=1970 THEN DO; PS105B=0; PV105B=.41; END;
     ELSE IF MY=1971 THEN DO; PS105B=0; PV105B=.63; END;
     ELSE IF MY=1972 THEN DO; PS105B=0; PV105B=.74; END;
     ELSE IF MY=1973 THEN DO; PS105B=0; PV105B=.86; END;
     ELSE IF MY=1974 THEN DO; PS105B=0; PV105B=.84; END;
     ELSE IF MY=1975 THEN DO; PS105B=0; PV105B=.93; END;
/* LIGHT TRUCK OCCUPANT LIVES SAVED BY FRONT DISC BRAKES */
   OLDWTFA=WEIGHTFA;
   IF PS105B GT 0 OR PV105B GT 0 THEN DO;
     E=.0017;
     P=PS105B+PV105B;
     S=OLDWTFA*E*P / (1 - E*P);
     TLS105B=S*PS105B/P;
     TLV105B=S*PV105B/P;
     WEIGHTFA=OLDWTFA+TLS105B+TLV105B; END;
   ELSE DO; TLS105B=0; TLV105B=0; END;
   TL105B=TLS105B+TLV105B;
   TLS=TLS+TLS105B; TLV=TLV+TLV105B; TL=TL+TL105B;
```

```
/* ------------------------------------------------------------------ */
/*   208E LAP BELTS FOR BACK-SEAT CENTER OCCUPANTS                     */
/*   MEDIAN INSTALLATION YEAR: 1968                                    */
/* ------------------------------------------------------------------ */

/* IDENTIFIES BACK-CENTER-SEAT OCCUPANTS WEARING LAP BELTS   */
  PS208E=0; PV208E=0;
  IF SEAT2=22 AND REST3=2 AND AGEGP=5 THEN DO;
```

The original requirements of FMVSS 208 took effect for LTVs on July 1, 1971.

```
    IF MY GE 1972 THEN PS208E=1;
      ELSE IF MY LE 1970 THEN PV208E=1;
      ELSE IF MY=1971 AND CY=1970 THEN PV208E=1;
      ELSE IF MY=1971 THEN DO; PS208E=.17; PV208E=.83; END; END;
/* LIVES SAVED AND EJECTIONS PREVENTED BY BACK-CENTER-SEAT LAP BELTS */
  OLDWTFA=WEIGHTFA;
  IF PS208E GT 0 OR PV208E GT 0 THEN DO;
```

NHTSA's evaluation estimates an overall 63 percent fatality reduction by lap belts for rear-outboard occupants in LTVs. The same effectiveness is assumed for the rear-center occupant.

```
    E=.63;
    IF CRSH=1 THEN EJECT2=.2331; ELSE IF CRSH=2.1 THEN EJECT2=.3664;   /* FROM OLDFA14, P. 9 */
      ELSE IF CRSH=2.2 THEN EJECT2=.3153; ELSE IF CRSH=3 THEN EJECT2=.6424;
      ELSE IF CRSH=4 THEN EJECT2=.4091;
    P=PS208E+PV208E;
    S=OLDWTFA*E*P / (1 - E*P);
    TLS208E=S*PS208E/P;
    TLV208E=S*PV208E/P;
    WEIGHTFA=OLDWTFA+TLS208E+TLV208E; END;
  ELSE DO; TLS208E=0; TLV208E=0; END;
  TL208E=TLS208E+TLV208E;
  TLS=TLS+TLS208E; TLV=TLV+TLV208E; TL=TL+TL208E;

/* ------------------------------------------------------------------ */
/*   208D LAP BELTS FOR FRONT-SEAT CENTER OCCUPANTS                    */
/*   MEDIAN INSTALLATION YEAR: 1968                                    */
/* ------------------------------------------------------------------ */

/* IDENTIFIES FRONT-CENTER-SEAT OCCUPANTS WEARING LAP BELTS   */
  PS208D=0; PV208D=0;
  IF SEAT2=12 AND REST3=2 AND AGEGP=5 THEN DO;
    IF MY GE 1972 THEN PS208D=1;
      ELSE IF MY LE 1970 THEN PV208D=1;
      ELSE IF MY=1971 AND CY=1970 THEN PV208D=1;
      ELSE IF MY=1971 THEN DO; PS208D=.17; PV208D=.83; END; END;
/* LIVES SAVED AND EJECTIONS PREVENTED BY FRONT-CENTER-SEAT LAP BELTS */
  OLDWTFA=WEIGHTFA;
  IF PS208D GT 0 OR PV208D GT 0 THEN DO;
```

A double-pair comparison analysis of lap belt effectiveness for front-outboard occupants, conducted in support of this study, found that lap belts reduce fatality risk by 48 percent in LTVs. This effectiveness will also be assumed for the front-center occupant.

```
    E=.48;
    IF CRSH=1 THEN EJECT2=.2331; ELSE IF CRSH=2.1 THEN EJECT2=.3664;   /* FROM OLDFA14, P. 9 */
      ELSE IF CRSH=2.2 THEN EJECT2=.3153; ELSE IF CRSH=3 THEN EJECT2=.6424;
      ELSE IF CRSH=4 THEN EJECT2=.4091;
    P=PS208D+PV208D;
    S=OLDWTFA*E*P / (1 - E*P);
    TLS208D=S*PS208D/P;
    TLV208D=S*PV208D/P;
```

```
     WEIGHTFA=OLDWTFA+TLS208D+TLV208D; END;
   ELSE DO; TLS208D=0; TLV208D=0; END;
   TL208D=TLS208D+TLV208D;
   TLS=TLS+TLS208D; TLV=TLV+TLV208D; TL=TL+TL208D;


/* ------------------------------------------------------------ */
/*   208C LAP BELTS FOR BACK-SEAT OUTBOARD OCCUPANTS            */
/*   MEDIAN INSTALLATION YEAR: 1968                            */
/* ------------------------------------------------------------ */

/* IDENTIFIES BACK-OUTBOARD-SEAT OCCUPANTS WEARING LAP BELTS  */
   PS208C=0; PV208C=0;
   IF SEAT2=21 AND REST3=2 AND AGEGP=5 THEN DO;
      IF MY GE 1972 THEN PS208C=1;
        ELSE IF MY LE 1970 THEN PV208C=1;
        ELSE IF MY=1971 AND CY=1970 THEN PV208C=1;
        ELSE IF MY=1971 THEN DO; PS208C=.17; PV208C=.83; END; END;
/* LIVES SAVED AND EJECTIONS PREVENTED BY BACK-OUTBOARD-SEAT LAP BELTS */
   OLDWTFA=WEIGHTFA;
   IF PS208C GT 0 OR PV208C GT 0 THEN DO;
```

NHTSA's evaluation estimates an overall 63 percent fatality reduction by lap belts for rear-outboard occupants in LTVs.

```
   E=.63;
   IF CRSH=1 THEN EJECT2=.2331; ELSE IF CRSH=2.1 THEN EJECT2=.3664;   /* FROM OLDFA14, P. 9 */
     ELSE IF CRSH=2.2 THEN EJECT2=.3153; ELSE IF CRSH=3 THEN EJECT2=.6424;
     ELSE IF CRSH=4 THEN EJECT2=.4091;
   P=PS208C+PV208C;
   S=OLDWTFA*E*P / (1 - E*P);
   TLS208C=S*PS208C/P;
   TLV208C=S*PV208C/P;
    WEIGHTFA=OLDWTFA+TLS208C+TLV208C; END;
   ELSE DO; TLS208C=0; TLV208C=0; END;
   TL208C=TLS208C+TLV208C;
   TLS=TLS+TLS208C; TLV=TLV+TLV208C; TL=TL+TL208C;


/* ------------------------------------------------------------ */
/*   206 DOOR LOCK IMPROVEMENTS                                */
/*   MEDIAN INSTALLATION YEAR: 1967                            */
/* ------------------------------------------------------------ */

/* DOOR LOCKS IMPROVED GRADUALLY FROM 1962 TO 1972          */
   IF MY GE 1973 THEN DO; PS206=1; PV206=0; END;
     ELSE IF MY=1972 AND CY=1971 THEN DO; PS206=0; PV206=1; END;
     ELSE IF MY=1972 THEN DO; PS206=.5; PV206=.5; END;
     ELSE IF 1962 LE MY LE 1971 THEN DO; PS206=0; PV206=(MY-1961)/11; END;
     ELSE IF MY LE 1961 THEN DO; PS206=0; PV206=0; END;
/* REDUCTION IN EJECTION ROLLOVER FATALITY RISK WITH IMPROVED DOOR LOCKS */
   OLDWTFA=WEIGHTFA;
   IF (PS206 GT 0 OR PV206 GT 0) AND CRSH=3
       AND SEAT2 NE 52 AND EJECT2 NE 0 THEN DO;
     E=.098;
     P=PS206+PV206;
     REL_S=EJECT2*E*P / (1 - E*P);
     S=OLDWTFA*REL_S;
     TLS206=S*PS206/P;
     TLV206=S*PV206/P;
     WEIGHTFA=OLDWTFA+TLS206+TLV206;
     EJECT2=(EJECT2 + REL_S)/(1 + REL_S); END;
   ELSE DO; TLS206=0; TLV206=0; END;
   TL206=TLS206+TLV206;
   TLS=TLS+TLS206; TLV=TLV+TLV206; TL=TL+TL206;
```

```
/* ------------------------------------------------------------------ */
/*  105A DUAL MASTER CYLINDERS                                         */
/*  MEDIAN INSTALLATION YEAR: LATE 1966                               */
/* ------------------------------------------------------------------ */

/* IMPLEMENTATION OF DUAL MASTER CYLINDERS */
  IF MY GE 1967 THEN DO; PS105A=0; PV105A=1; END;
    ELSE IF MY LE 1961 THEN DO; PS105A=0; PV105A=0; END;
    ELSE IF MY IN (1962,1963) THEN DO; PS105A=0; PV105A=.09; END;
    ELSE IF MY IN (1964,1965) THEN DO; PS105A=0; PV105A=.07; END;
    ELSE IF MY=1966 THEN DO; PS105A=0; PV105A=.54; END;
/* LIGHT TRUCK OCCUPANT LIVES SAVED BY DUAL MASTER CYLINDERS */
  OLDWTFA=WEIGHTFA;
  IF PS105A GT 0 OR PV105A GT 0 THEN DO;
   E=.007;
   P=PS105A+PV105A;
   S=OLDWTFA*E*P / (1 - E*P);
   TLS105A=S*PS105A/P;
   TLV105A=S*PV105A/P;
   WEIGHTFA=OLDWTFA+TLS105A+TLV105A; END;
  ELSE DO; TLS105A=0; TLV105A=0; END;
  TL105A=TLS105A+TLV105A;
  TLS=TLS+TLS105A; TLV=TLV+TLV105A; TL=TL+TL105A;


/* ------------------------------------------------------------------ */
/*  208B LAP BELT USE BY CHILDREN AGE 1-4                             */
/*  MEDIAN INSTALLATION YEAR FOR LAP BELTS USED BY CHILDREN: 1966     */
/* ------------------------------------------------------------------ */

/* IDENTIFIES CHILD PASSENGERS AGE 1-4 USING LAP BELTS */
  PS208B=0; PV208B=0;
  IF AGEGP=1 AND REST3=2 THEN DO;
    IF MY GE 1972 THEN PS208B=1;
    ELSE IF MY LE 1970 THEN PV208B=1;
    ELSE IF MY=1971 AND CY=1970 THEN PV208B=1;
    ELSE IF MY=1971 THEN DO; PS208B=.17; PV208B=.83; END; END;
/* LIVES SAVED AND EJECTIONS PREVENTED BY LAP BELTS (AGE 1-4) */
  OLDWTFA=WEIGHTFA;
  IF PS208B GT 0 OR PV208B GT 0 THEN DO;
```

NHTSA's evaluation estimates that lap belts reduce fatality risk by 48 percent for toddlers in LTVs.[2]

```
   E=.48;
   IF CRSH=1 THEN EJECT2=.2331; ELSE IF CRSH=2.1 THEN EJECT2=.3664;   /* FROM OLDFA14, P. 9 */
     ELSE IF CRSH=2.2 THEN EJECT2=.3153; ELSE IF CRSH=3 THEN EJECT2=.6424;
     ELSE IF CRSH=4 THEN EJECT2=.4091;
   P=PS208B+PV208B;
   S=OLDWTFA*E*P / (1 - E*P);
   TLS208B=S*PS208B/P;
   TLV208B=S*PV208B/P;
   WEIGHTFA=OLDWTFA+TLS208B+TLV208B; END;
  ELSE DO; TLS208B=0; TLV208B=0; END;
  TL208B=TLS208B+TLV208B;
  TLS=TLS+TLS208B; TLV=TLV+TLV208B; TL=TL+TL208B;


/* ------------------------------------------------------------------ */
/*  208A LAP BELTS FOR FRONT-SEAT OUTBOARD OCCUPANTS                  */
/*  MEDIAN INSTALLATION YEAR: 1964                                    */
/* ------------------------------------------------------------------ */
```

---

[2] This evaluation is discussed in the "Child Safety Seats" section of the FMVSS 213 chapter of Part 1, not the FMVSS 208 chapter.

```
/* IDENTIFIES FRONT-OUTBOARD-SEAT OCCUPANTS WEARING LAP BELTS   */
  PS208A=0; PV208A=0;
  IF SEAT2 IN (11,13) AND REST3=2 AND AGEGP=5 THEN DO;
     IF MY GE 1972 THEN PS208A=1;
        ELSE IF MY LE 1970 THEN PV208A=1;
        ELSE IF MY=1971 AND CY=1970 THEN PV208A=1;
        ELSE IF MY=1971 THEN DO; PS208A=.17; PV208A=.83; END; END;
/* LIVES SAVED AND EJECTIONS PREVENTED BY FRONT-OUTBOARD-SEAT LAP BELTS */
  OLDWTFA=WEIGHTFA;
  IF PS208A GT 0 OR PV208A GT 0 THEN DO;
```

A double-pair comparison analysis of lap belt effectiveness for front-outboard occupants, conducted in support of this study, found that lap belts reduce fatality risk by 48 percent in LTVs.

```
     E=.48;
     IF CRSH=1 THEN EJECT2=.2331; ELSE IF CRSH=2.1 THEN EJECT2=.3664;    /* FROM OLDFA14, P. 9 */
       ELSE IF CRSH=2.2 THEN EJECT2=.3153; ELSE IF CRSH=3 THEN EJECT2=.6424;
       ELSE IF CRSH=4 THEN EJECT2=.4091;
     P=PS208A+PV208A;
     S=OLDWTFA*E*P / (1 - E*P);
     TLS208A=S*PS208A/P;
     TLV208A=S*PV208A/P;
     WEIGHTFA=OLDWTFA+TLS208A+TLV208A; END;
   ELSE DO; TLS208A=0; TLV208A=0; END;
   TL208A=TLS208A+TLV208A;
   TLS=TLS+TLS208A; TLV=TLV+TLV208A; TL=TL+TL208A;
RUN;
```

That concludes the model for LTVs. It has estimated, on a cell-by-cell basis, by how much fatalities would increase if all the safety technologies were "removed." The last steps are to tally up the lives saved over all the cells, by calendar year, and print the results.

```
PROC MEANS SUM NOPRINT DATA=TRK9; BY CY;
  VAR ORIGWT WEIGHTFA
     TLV208J TLS208J TL208J
     TLV108 TLS108 TL108
     TLV208I TLS208I TL208I
     TLV214A TLS214A TL214A
     TLV208G TLS208G TL208G
     TLV213 TLS213 TL213
     TLV212 TLS212 TL212
     TLV208F TLS208F TL208F
     TLV203 TLS203 TL203
     TLV201 TLS201 TL201
     TLV105B TLS105B TL105B
     TLV208E TLS208E TL208E
     TLV208D TLS208D TL208D
     TLV208C TLS208C TL208C
     TLV206 TLS206 TL206
     TLV105A TLS105A TL105A
     TLV208B TLS208B TL208B
     TLV208A TLS208A TL208A
     TLV TLS TL;
  OUTPUT OUT=TRK10
     SUM=T_ORIGWT T_WTFA TLV108 TLS108 TL108
     TLV208J TLS208J TL208J
     TLV208I TLS208I TL208I
     TLV214A TLS214A TL214A
     TLV208G TLS208G TL208G
     TLV213 TLS213 TL213
     TLV212 TLS212 TL212
     TLV208F TLS208F TL208F
     TLV203 TLS203 TL203
     TLV201 TLS201 TL201
```

```
      TLV105B TLS105B TL105B
      TLV208E TLS208E TL208E
      TLV208D TLS208D TL208D
      TLV208C TLS208C TL208C
      TLV206 TLS206 TL206
      TLV105A TLS105A TL105A
      TLV208B TLS208B TL208B
      TLV208A TLS208A TL208A
      TLV TLS TL;

PROC PRINT DATA=TRK10;
   FORMAT TLV213 TLS213 TL213 TLV212 TLS212 TL212 TLV208F TLS208F TL208F 9.0;
   ID CY; VAR TLV213 TLS213 TL213 TLV212 TLS212 TL212 TLV208F TLS208F TL208F;
   SUM TLV213 TLS213 TL213 TLV212 TLS212 TL212 TLV208F TLS208F TL208F;
TITLE1 'LIGHT TRUCK OCCUPANT LIVES SAVED BY CHILD SAFETY SEATS, ADHESIVE WINDSHIELD BONDING, AND
FRONT-SEAT 3-POINT BELTS, 1960-74';
TITLE2 ' ';
TITLE3 '...213  = CHILD SAFETY SEATS (USE EXCEEDED 50% IN 1985, FMVSS EFFECTIVE 4/1/71, STATE LAWS
1978-85)';
TITLE4 '...212  = ADHESIVE WINDSHIELD BONDING (MEDIAN INSTALLATION YEAR 1979, FMVSS EFFECTIVE
9/1/78)';
TITLE5 '...208F = 3-POINT BELTS FOR FRONT-SEAT OCCUPANTS (MEDIAN INSTALLATION YEAR 1976, FMVSS
EFFECTIVE 9/1/76)';
TITLE6 '...V = VOLUNTARY INSTALLATIONS, BEFORE FMVSS EFFECTIVE DATE';
TITLE7 '...S = STANDARD INSTALLATIONS, LIGHT TRUCKS BUILT ON OR AFTER FMVSS EFFECTIVE DATE';
PROC PRINT DATA=TRK10;
   FORMAT TLV203 TLS203 TL203 TLV201 TL201 TLV105B TLS105B TL105B 9.0;
   ID CY; VAR TLV203 TLS203 TL203 TLV201 TL201 TLV105B TLS105B TL105B;
   SUM TLV203 TLS203 TL203 TLV201 TL201 TLV105B TLS105B TL105B;
TITLE1 'LIGHT TRUCK OCCUPANT LIVES SAVED BY EA STEERING ASSEMBLIES, SAFER INSTRUMENT PANELS, AND
FRONT DISC BRAKES, 1960-74';
TITLE2 ' ';
TITLE3 '...203  = ENERGY-ABSORBING STEERING ASSEMBLIES (MEDIAN IMPLEMENTATION YEAR EARLY 1976,
FMVSS EFFECTIVE 9/1/81)';
TITLE4 '...201  = VOLUNTARY INSTRUMENT PANEL IMPROVEMENTS (MEDIAN IMPLEMENTATION YEAR 1972, FMVSS
EFFECTIVE 9/1/81)';
TITLE5 '...105B = FRONT DISC BRAKES (MEDIAN INSTALLATION YEAR 1971, FMVSS EFFECTIVE 9/1/83)';
TITLE6 '...V = VOLUNTARY INSTALLATIONS, BEFORE FMVSS EFFECTIVE DATE';
TITLE7 '...S = STANDARD INSTALLATIONS, LIGHT TRUCKS BUILT ON OR AFTER FMVSS EFFECTIVE DATE';

PROC PRINT DATA=TRK10;
   FORMAT TLV208E TLS208E TL208E TLV208D TLS208D TL208D TLV208C TLS208C TL208C 9.0;
   ID CY; VAR TLV208E TLS208E TL208E TLV208D TLS208D TL208D TLV208C TLS208C TL208C;
   SUM TLV208E TLS208E TL208E TLV208D TLS208D TL208D TLV208C TLS208C TL208C;
TITLE1 'LIGHT TRUCK OCCUPANT LIVES SAVED BY BACK-CENTER LAP BELTS, FRONT-CENTER LAP BELTS, AND
BACK-OUTBOARD LAP BELTS, 1960-74';
TITLE2 ' ';
TITLE3 '...208E = LAP BELTS FOR BACK-CENTER OCCUPANTS (MEDIAN INSTALLATION YEAR 1968, FMVSS
EFFECTIVE 7/1/71)';
TITLE4 '...208D = LAP BELTS FOR FRONT-CENTER OCCUPANTS (MEDIAN INSTALLATION YEAR 1968, FMVSS
EFFECTIVE 7/1/71)';
TITLE5 '...208C = LAP BELTS FOR BACK-OUTBOARD OCCUPANTS (MEDIAN INSTALLATION YEAR 1968, FMVSS
EFFECTIVE 7/1/71)';
TITLE6 '...V = VOLUNTARY INSTALLATIONS, BEFORE FMVSS EFFECTIVE DATE';
TITLE7 '...S = STANDARD INSTALLATIONS, LIGHT TRUCKS BUILT ON OR AFTER FMVSS EFFECTIVE DATE';

PROC PRINT DATA=TRK10;
   FORMAT TLV206 TLS206 TL206 TLV105A TLS105A TL105A TLV208B TLS208B TL208B 9.0;
   ID CY; VAR TLV206 TLS206 TL206 TLV105A TLS105A TL105A TLV208B TLS208B TL208B;
   SUM TLV206 TLS206 TL206 TLV105A TLS105A TL105A TLV208B TLS208B TL208B;
TITLE1 'LIGHT TRUCK OCCUPANT LIVES SAVED BY IMPROVED DOOR LOCKS, DUAL MASTER CYLINDERS, AND LAP
BELTS FOR CHILDREN AGE 1-4, 1960-74';
TITLE2 ' ';
TITLE3 '...206  = IMPROVED DOOR LOCKS (MEDIAN INSTALLATION YEAR 1967, FMVSS EFFECTIVE 1/1/72)';
```

```
TITLE4 '...105A = DUAL MASTER CYLINDERS (MEDIAN INSTALLATION YEAR LATE 1966, FMVSS EFFECTIVE
9/1/83)';
TITLE5 '...208B = LAP BELT USE BY CHILDREN AGE 1-4 (MEDIAN INSTALLATION YEAR 1966, FMVSS EFFECTIVE
7/1/71)';
TITLE6 '...V = VOLUNTARY INSTALLATIONS, BEFORE FMVSS EFFECTIVE DATE';
TITLE7 '...S = STANDARD INSTALLATIONS, LIGHT TRUCKS BUILT ON OR AFTER FMVSS EFFECTIVE DATE';

PROC PRINT DATA=TRK10;
   FORMAT TLV208A TLS208A TL208A 9.0;
   ID CY; VAR TLV208A TLS208A TL208A;
   SUM TLV208A TLS208A TL208A;
TITLE1 'LIGHT TRUCK OCCUPANT LIVES SAVED BY FRONT-OUTBOARD LAP BELTS, 1960-74';
TITLE2 ' ';
TITLE3 '...208A = LAP BELTS FOR FRONT-OUTBOARD OCCUPANTS (MEDIAN INSTALLATION YEAR 1964, FMVSS
EFFECTIVE 7/1/71)';
TITLE4 '...V = VOLUNTARY INSTALLATIONS, BEFORE FMVSS EFFECTIVE DATE';
TITLE5 '...S = STANDARD INSTALLATIONS, LIGHT TRUCKS BUILT ON OR AFTER FMVSS EFFECTIVE DATE';

DATA TRK11; SET TRK10;
PCT_SAVE=100*TL/T_WTFA;
PROC PRINT DATA=TRK11;
   FORMAT T_ORIGWT T_WTFA TLV TLS TL 9.0 PCT_SAVE 6.2;
   ID CY; VAR T_ORIGWT T_WTFA TLV TLS TL PCT_SAVE;
   SUM T_ORIGWT T_WTFA TLV TLS TL;
TITLE1 'OVERALL LIGHT TRUCK OCCUPANT LIVES SAVED BY VEHICLE SAFETY STANDARDS AND TECHNOLOGIES,
1960-74';
TITLE2 ' ';
TITLE3 'T_ORIGWT = ACTUAL LIGHT TRUCK OCCUPANT FATALITIES';
TITLE4 'T_WTFA = FATALITIES THAT WOULD HAVE OCCURRED WITHOUT ANY VEHICLE SAFETY IMPROVEMENTS';
TITLE5 'TLV = OVERALL LIVES SAVED BY VOLUNTARY IMPROVEMENTS, BEFORE FMVSS EFFECTIVE DATE';
TITLE6 'TLS = OVERALL LIVES SAVED IN LIGHT TRUCKS BUILT ON OR AFTER FMVSS EFFECTIVE DATE';
TITLE7 'TL = OVERALL LIVES SAVED BY VEHICLE SAFETY IMPROVEMENTS (TLV + TLS)';
TITLE8 'PCT_SAVE = PERCENT OF WOULD-HAVE-BEEN FATALITIES SAVED BY SAFETY STANDARDS AND
TECHNOLOGIES';
RUN;
```

APPENDIX B: SUMMARIES OF PUBLISHED EVALUATION REPORTS

A systematic program to evaluate the effectiveness of the Federal Motor Vehicle Safety Standards (FMVSS) was initiated in 1975, when NHTSA was just beginning to establish its own crash databases. The first "preliminary" evaluation of a standard was published in 1979 (side door strength) and the first "final" evaluations in 1981 (energy-absorbing steering assemblies, bumpers). Since 1979, 48 comprehensive evaluations of regulations, safety programs, consumer information programs, or safety technologies have been published. Here is a list of the 48 studies including summaries of principal findings [except where findings were superseded in a follow-up evaluation]:

2004

Preliminary Results Analyzing the Effectiveness of Electronic Stability Control (ESC) Systems (NHTSA Publication DOT HS 809 790)

ESC systems detect when a vehicle is about to go out of control and automatically intervene by applying the brakes to individual wheels and possibly reducing engine torque to help the driver stay on course. Preliminary analyses found statistically significant crash reductions in certain luxury passenger cars and SUVs currently equipped with ESC systems. Single vehicle crashes were reduced by 35 percent in passenger cars and by 67 percent in SUVs. Fatal single vehicle crashes were reduced by 30 percent in cars and by 63 percent in SUVs.

Evaluation of Rear Window Defrosting and Defogging Systems (NHTSA Publication DOT HS 809 724)

Almost all new cars, minivans and SUVs have rear window defoggers, even though Federal standards do not require them. Analyses of crashes where drivers were backing up or changing lanes during rain or snow, early morning hours, or in the winter did not show a statistically significant reduction with defoggers. Nevertheless, NHTSA would expect consumers to continue wanting rear window defoggers for their vehicles because they conveniently clear condensation, frost, ice, and/or snow from the back window.

Evaluation of FMVSS 214 Side Impact Protection for Light Trucks: Crush Resistance Requirements for Side Doors (NHTSA Publication DOT HS 809 719)

Light trucks (pickup trucks, vans, and sport utility vehicles) were required to meet a crush resistance standard for side doors beginning September 1, 1993. Side door beams were installed to reduce the velocity and depth of door intrusion in side impact crashes. The beams are estimated to reduce fatalities by 19 percent in single vehicle side impacts. When all light trucks on the road have head restraints, they will save an estimated 151 lives per year. Little or no fatality reduction was found in multivehicle crashes.

2003

Results of the Survey on the Use of Passenger Air Bag On-Off Switches (NHTSA Publication DOT HS 809 689)

On-off switches allow drivers to temporarily deactivate air bags when children must ride in the front seat of pickup trucks and other vehicles that cannot accommodate rear-facing child safety seats in the back seat. NHTSA recommends that passenger air bag be turned off when a child age 12 or younger must ride in the front seat, and turned on if all front-seat occupants are age 13 or older. In a 2000 survey, switches were left on for 14 percent of infants and 26 percent of child passengers age 1-6, but turned off for 17 percent of the adult passengers.

Vehicle Weight, Fatality Risk and Crash Compatibility of Model Year 1991-99 Passenger Cars and Light Trucks (NHTSA Publication DOT HS 809 662)

There is little association between vehicle weight and fatal-crash rates in the heavier light trucks and vans. However, in other groups of model year 1991-99 vehicles, fatality rates increased as weights decreased. Pickup trucks and SUVs of these model years had, on the average, higher fatality rates than passenger cars or minivans of comparable weight. Model year 1991-99 light trucks and vans, especially those with high, rigid frontal structures, were more aggressive than cars when they struck other vehicles.

NCAP Test Improvements with Pretensioners and Load Limiters (NHTSA Publication DOT HS 809 562)

Safety belt pretensioners pull belts snug as a crash begins. Load limiters allow belts to yield slightly during a crash to reduce the force on the wearer's chest. In New Car Assessment Program (NCAP) frontal barrier crashes at 35 mph, the combination of pretensioners and load limiters reduced average Head Injury Criterion (HIC) by 232, chest acceleration by 6.6 g's and chest deflection by 10.6 mm, for driver and right front passenger dummies, relative to cars and light trucks of the same make-models without these features.

2002

Evaluation of Child Safety Seat Registration (NHTSA Publication DOT HS 809 518)

Since March 1993, manufacturers of child safety seats have been required to provide a postage-paid registration form with each new child safety seat. Seat registration has increased from 3 percent prior to 1993 to 27 percent in 1996-2000. The repair rate for recalled child safety seats increased from 13.8 percent prior to 1993 to 21.5 percent.

Preliminary Report: The Incidence Rate of Odometer Fraud (NHTSA Publication DOT HS 809 441)

There are an estimated 452,000 cases of odometer rollback per year in the United States. The difference between the inflated prices that consumers paid for rolled-back vehicles and the prices they would have been willing to pay if they had known the true mileage average $2,336 per case of odometer rollback, amounting to $1,056 million per year in the United States.

2001

The Effectiveness of Head Restraints in Light Trucks (NHTSA Publication DOT HS 809 247)

The purpose of a head restraint is to prevent whiplash injuries in rear-impact crashes. Head restraints reduce overall injury risk in light trucks in rear impacts by a statistically significant 6 percent. When all light trucks on the road have head restraints, they will be preventing approximately 15,000 nonfatal injuries per year. (See also the 1982 evaluation of head restraints in passenger cars.)

The Effectiveness of Retroreflective Tape on Heavy Trailers (NHTSA Publication DOT HS 809 222)

Retroreflective tape enhances the visibility of heavy trailers in the dark. The tape reduces side and rear impacts by other vehicles into trailers by 29 percent in dark conditions (including dark-not-lighted, dark-lighted, dawn and dusk). In dark-not-lighted conditions, the tape reduces side and rear impacts by 41 percent. When all heavy trailers have the tape, it will prevent an estimated 191-350 fatalities, 3,100-5,000 injuries and 7,800 crashes per year.

Evaluation of the American Automobile Labeling Act (NHTSA Publication DOT HS 809 208)

In a survey of 646 recent or imminent new-vehicle buyers, over 75 percent were unaware of the existence of automobile parts content labels. Among those who had read the labels, many said they used the country-of-assembly information, but none said they used the numerical U.S./Canadian parts content score. Overall U.S./Canadian parts content in new cars and light trucks dropped from an average of 70 percent in model year 1995 to 67.6 percent in 1998. However, it increased from 47 to 59 percent in transplants while dropping from 89 to 84 percent in Big 3 vehicles: trends undoubtedly influenced by the 1995 U.S.-Japan Agreement on Autos and Auto Parts and the North American Free Trade Agreement (NAFTA).

2000

Fatality Reduction by Safety Belts for Front-Seat Occupants of Cars and Light Trucks: Updated and Expanded Estimates Based on 1986-99 FARS Data (NHTSA Publication DOT HS 809 199)

Manual three-point belts reduce fatality risk, relative to the unrestrained front-seat occupant, by 45 percent in passenger cars and by 60 percent in pickup trucks, vans and sport utility vehicles. The analyses reconfirm the agency's earlier (1984-89) estimates of fatality reduction.

1999

Evaluation of FMVSS 214 - Side Impact Protection: Dynamic Performance Requirement; Phase 1: Correlation of TTI(d) with Fatality Risk in Actual Side Impact Collisions of Model Year 1981-1993 Passenger Cars (NHTSA Publication DOT HS 809 004)

The test injury criterion TTI(d) has a statistically significant association with fatality risk in actual side-impact crashes on the highway. In model year 1981-93 cars, make-models with low TTI(d) on the FMVSS 214 test tend to have low fatality risk. The relationship is stronger in 2-door than 4-door cars. Reducing TTI(d) by one unit is associated with an estimated 0.927 percent reduction of fatality risk in side impacts of 2-door cars. The association in the corresponding analysis of 4-door cars was not statistically significant.

Effectiveness of Lap/Shoulder Belts in the Back Outboard Seating Positions (NHTSA Publication DOT HS 808 945)

Lap/shoulder belts reduce fatality risk by 44 percent relative to unrestrained back-seat occupants of passenger cars, and by 15 percent relative to lap-belted occupants. Lap belts reduce fatality risk by 32 percent relative to unrestrained occupants. Lap/shoulder belts are effective in all crashes, but lap belts only in nonfrontal crashes. Lap-belted occupants have substantially higher abdominal-injury risk than unrestrained back-seat occupants in frontal crashes, but lap/shoulder belts reduce abdominal injuries by 52 percent and head injuries by 47 percent relative to lap belts.

1998

Highway Safety Assessment: A Summary of Findings in Ten States (NHTSA Publication DOT HS 808 796)

Assessment of 1980-1993 safety programs in ten States showed that Federal grants and technology were used to address safety priorities as intended by Congress. Federal grants, amounting to less than two percent of total safety spending by States and communities, have acted as seed money to resolve important highway safety problems. Programs started with Federal funds were often extended or replicated elsewhere with State funds. Occupant protection programs, however, remain heavily dependent on Federal funds.

Auto Theft and Recovery - Effects of the Anti Car Theft Act of 1992 and the Motor Vehicle Theft Law Enforcement Act of 1984 - Report to the Congress (NHTSA Publication DOT HS 808 761)

Theft rates, which had increased during the 1980's, declined from 714 per million in 1990 to 597 in 1995. Parts marking and factory-installed anti-theft devices have had beneficial and complementary effects on auto thefts and/or recoveries. The Acts have given law enforcement tools to deter thefts, trace stolen vehicles and parts, and apprehend and convict thieves.

The Long-Term Effectiveness of Center High Mounted Stop Lamps in Passenger Cars and Light Trucks (NHTSA Publication DOT HS 808 696)

Throughout 1989-95, cars equipped with Center High Mounted Stop Lamps were 4.3 percent less likely to be struck in the rear than cars without the lamps. (In 1987, when the lamps were first introduced, the reduction was 8.5 percent.) The effectiveness of CHMSL in light trucks is about the same as in cars. At the 1989-95 effectiveness level, when all cars and light trucks on the road have the lamps, they would prevent 194,000-239,000 crashes, 58,000-70,000 nonfatal injuries and $655 million in property damage per year.

1997

Relationship of Vehicle Weight to Fatality and Injury Risk in Model Year 1985-93 Passenger Cars and Light Trucks (NHTSA Publication DOT HS 808 569); *Relationships between Vehicle Size and Fatality Risk in Model Year 1985-93 Passenger Cars and Light Trucks* (NHTSA Publication DOT HS 808 570)

[Findings have been superseded by the 2003 evaluation - see above.]

1996

Fatality Reduction by Air Bags: Analyses of Accident Data through Early 1996 (NHTSA Publication DOT HS 808 470)

Driver air bags reduce overall fatality risk by an estimated 11 percent in passenger cars and light trucks (essentially unchanged from the 1994 and 1992 NHTSA analyses). Passenger air bags are beneficial for right-front passengers age 13 or older. Air bags provide a life-saving benefit for belted as well as unbelted drivers. The fatality risk for child passengers age 0-12 in cars with passenger air bags is currently higher than in cars without them. Current air bags are significantly less effective for drivers age 70 or older than for younger drivers.

1995

Preliminary Evaluation of the Effectiveness of Antilock Brake Systems for Passenger Cars
(NHTSA Publication DOT HS 808 206)

> ABS significantly reduced multivehicle crashes on wet roads: fatal crashes by 24 percent, and nonfatal crashes by 14 percent. Fatal collisions with pedestrians and bicyclists were down a significant 27 percent. However, these reductions were offset by statistically significant increases in single vehicle, run-off-road crashes (rollovers or impacts with fixed objects). Fatal run-off-road crashes were up by 28 percent, and nonfatal crashes by 19 percent in the ABS-equipped cars, as compared to similar cars without ABS.

1994

Fatality Reduction by Automatic Occupant Protection in the United States (Proceedings of the 14th Conference on Enhanced Safety of Vehicles)

> The fatality risk of front-outboard occupants in cars with motorized 2-point belts (without disconnect) is 6 percent lower than in cars with manual belts; the risk in cars with non-motorized 3-point belts is the same as in cars with manual belts. [This report's findings on air bags have been superseded by the 1996 evaluation - see above.]

An Evaluation of the Effects of Glass-Plastic Windshield Glazing in Passenger Cars (NHTSA Publication DOT HS 808 062)

> Following an amendment to the glazing standard (FMVSS 205) in 1983, two manufacturers equipped some of their cars with glass-plastic windshields. Crash data indicate the injury reduction potential of these windshields is less than predicted. Fleet and warranty data show that durability problems are greater than anticipated. While glass-plastic windshields add $65 to the cost of a new car, their replacement costs are estimated to exceed $1,700.

Correlation of NCAP Performance with Fatality Risk in Actual Head-On Collisions (NHTSA Publication DOT HS 808 061)

> There is a statistically significant correlation between the performance of passenger cars on the NCAP test and the fatality risk of belted drivers in actual head-on collisions. In a head-on collision between a car with "good" NCAP performance and a car of equal mass with "poor" performance, the driver of the "good" car has, on the average, about 15-25 percent lower fatality risk. The steady improvement in NCAP scores during 1979-91 was paralleled by a 20-25 percent reduction of fatality risk for belted drivers in actual head-on collisions.

1993

Preliminary Evaluation of the Effectiveness of Rear-Wheel Antilock Brake Systems for Light Trucks (Submitted to NHTSA Docket No. 70-27-GR-026)

Rear-wheel ABS significantly reduced the risk of nonfatal run-off-road crashes in light trucks: rollovers by about 30-40 percent, side impacts with fixed objects by 15-30 percent and frontal impacts with fixed objects by 5-20 percent. The reductions mostly did not carry over to fatal run-off-road crashes. Collisions with pedestrians and bicyclists were reduced by 5-15 percent. Involvements in multivehicle crashes were not reduced, and may even have increased with rear-wheel ABS.

1992

Evaluation of the Effectiveness of Occupant Protection - Federal Motor Vehicle Safety FMVSS 208 - Interim Report (NHTSA Publication DOT HS 807 843)

Air bags and automatic belts have significantly reduced the risk of nonfatal injury and occupant ejection. [This report's findings on fatality reduction for air bags have been superseded by the 1996 evaluation; for automatic belts - by the 1994 evaluation.]

An Evaluation of the Uniform Tire Quality Grading Standards and Other Tire Labeling Requirements (NHTSA Publication DOT HS 807 805)

Consumers and tire dealers were surveyed about their knowledge and utilization of tire quality grades and other tire information supplied in response to Federal regulations. The ratings for treadwear were viewed as "important" by 29 percent of consumers who had recently purchased tires, and the ratings for traction, by 27 percent. The majority of consumers are not aware that these ratings are printed on the tires.

1991

Auto Theft and Recovery - Effects of the Motor Vehicle Theft Law Enforcement Act of 1984 - Report to the Congress (NHTSA Publication DOT HS 807 703)

[Findings have been superseded by the 1998 evaluation - see above.]

Effect of Car Size on Fatality and Injury Risk

[Findings have been superseded by the 2003 evaluation - see above.]

1990

Motor Vehicle Fires in Traffic Crashes and the Effects of the Fuel System Integrity Standard (NHTSA Publication DOT HS 807 675)

Modifications to fuel systems in response to FMVSS 301 reduced the frequency of fires in nonfatal crashes of passenger cars by an estimated 14 percent; fatalities in cars and light trucks, however, were not affected. During 1975-88, the number of fire-related fatalities has increased from 1,300 to 1,800, primarily due to an aging vehicle fleet.

1989

An Evaluation of Door Locks and Roof Crush Resistance of Passenger Cars - Federal Motor Vehicle Safety Standards 206 and 216 (NHTSA Publication DOT HS 807 489)

Door latch improvements implemented during 1963-68 (preceding or responding to FMVSS 206) save an estimated 400 lives per year, reducing the risk of ejection in rollover crashes by 15 percent. The shift from hardtops to pillared cars with stronger roof support, in response to FMVSS 216, saves an estimated 110 lives per year.

An Evaluation of Center High Mounted Stop Lamps Based on 1987 Data (NHTSA Publication DOT HS 807 442)

[Findings have been superseded by the 1998 evaluation - see above.]

1988

An Evaluation of Occupant Protection in Frontal Interior Impact for Unrestrained Front Seat Occupants of Cars and Light Trucks (NHTSA Publication DOT HS 807 203)

During the 1960's and early 1970's, the manufacturers modified instrument panels of cars and light trucks, installing padding, reducing the rigidity of structures and extending the panel downward and toward the passenger. The improvements reduced fatality risk and serious injury risk by nearly 25 percent for unrestrained right front passengers of cars in frontal crashes, saving up to 700 lives per year.

1987

An Evaluation of the Bumper Standard - As Modified in 1982 (NHTSA Publication DOT HS 807 072)

To reduce regulatory burden on manufacturers, damage resistance requirements for bumpers were relaxed in model year 1983: the impact test speed was lowered from 5 to 2.5 mph. The net costs to consumers did not significantly change. A small increase in the repair cost over the lifetime of the car is offset by a reduction in the initial cost of the lighter bumpers. (See also the 1981 evaluation of bumpers.)

A Preliminary Evaluation of Seat Back Locks for Two-Door Passenger Cars with Folding Front Seatbacks (NHTSA Publication DOT HS 807 067)

FMVSS 207 requires a locking device for front seats with folding seatbacks, designed to limit the forward motion of the seatback in a collision. These locks or other seat components often separate at moderate crash speeds when they are impacted by back-seat occupants. No statistically significant injury or fatality reductions were found for seat back locks in any of the crash data files or in sled tests.

Fatality and Injury Reducing Effectiveness of Lap Belts for Back Seat Occupants (SAE Paper 870486)

[Findings have been superseded by the 1999 evaluation - see above.]

The Effectiveness of Center High Mounted Stop Lamps - A Preliminary Evaluation (NHTSA Publication DOT HS 807 076)

[Findings have been superseded by the 1998 evaluation - see above.]

1986

Fuel Economy and Annual Travel for Passenger Cars and Light Trucks: National On-Road Survey (NHTSA Publication DOT HS 806 971)

The actual fuel economy of model year 1978-81 vehicles was measured by a national survey in which drivers maintained log books of mileage and fuel purchases. On-road fuel economy of cars increased by 41 percent during model years 1977-81; the fuel economy of light trucks increased by 17-26 percent. However, the actual on-road fuel economy is consistently 15-20 percent below laboratory (EPA) ratings.

An Evaluation of Child Passenger Safety: The Effectiveness and Benefits of Safety Seats (NHTSA Publication DOT HS 806 890)

A correctly used safety seat reduces fatality risk by an estimated 71 percent and serious injury risk by 67 percent. But misuse can partially or completely nullify this effect. In 1984, when 39 percent of safety seats were correctly used and 61 percent were misused, the average overall fatality reduction for safety seats (correct users plus misusers) was 46 percent. In all, 192 children were saved by safety seats and lap belts in 1984.

1985

An Evaluation of Windshield Glazing and Installation Methods for Passenger Cars (NHTSA Publication DOT HS 806 693)

The High Penetration Resistant windshield doubled the impact velocity needed for the occupant's head to penetrate the windshield, reducing serious facial lacerations by 74 percent, preventing 39,000 serious lacerations and 8,000 facial fractures per year. Adhesive bonding of the windshield halved the incidence of bond separation and occupant ejection through the windshield portal in crashes, saving 105 lives per year.

1984

Effectiveness - Manual Lap and Lap/Shoulder Belts (Chapter IV-A of "Final Regulatory Impact Analysis - Amendment to Federal Motor Vehicle Safety Standard 208 - Passenger Car Front Seat Occupant Protection," NHTSA Publication DOT HS 806 572)

Manual lap-shoulder belts are estimated to reduce the fatality risk of drivers and right-front passengers by 40-50 percent [reconfirmed and superseded by the 2000 evaluation - see above], and serious injury risk by 45-55 percent, relative to an unrestrained occupant. The manual lap belt, alone, is estimated to reduce fatality risk by 30-40 percent and serious injury risk by 25-35 percent.

1983

An Evaluation of Side Marker Lamps for Cars, Trucks and Buses (NHTSA Publication DOT HS 806 430)

Side marker lamps were installed in response to FMVSS 108 to enable a driver to see another vehicle that is approaching at an angle at night. The lamps reduced nonfatal nighttime angle collisions by 16 percent, preventing 106,000 crashes, 93,000 injuries and $347 million in property damage per year. The lamps have not been effective in reducing fatalities.

A Preliminary Evaluation of Two Braking Improvements for Passenger Cars - Dual Master Cylinders and Front Disc Brakes (NHTSA Publication DOT HS 806 359)

Dual master cylinders, by providing a backup braking system in case of certain types of brake failure, prevent 40,000 crashes, 260 fatalities, 24,000 injuries and $132 million in property damage per year. Front disc brakes, which improve vehicle handling under various braking conditions, are estimated to prevent 10,000 crashes, 64 fatalities, 5,700 injuries and $32 million in property damage per year.

Evaluation of Federal Motor Vehicle Safety Standard 301-75, Fuel System Integrity: Passenger Cars (NHTSA Publication DOT HS 806 335)

[Findings have been superseded by the 1990 evaluation - see above.]

1982

An Evaluation of Side Structure Improvements in Response to Federal Motor Vehicle Safety Standard 214 (NHTSA Publication DOT HS 806 314)

Side door beams were installed in passenger cars to reduce the velocity and depth of door intrusion in side impact crashes. The beams are especially effective in side impacts with fixed objects, preventing 480 fatalities and 4,500 hospitalizations per year. In vehicle-to-vehicle side impacts, they prevent 4,900 nonfatal hospitalizations per year, but have not reduced fatality risk.

An Evaluation of Head Restraints - Federal Motor Vehicle Safety Standard 202 (NHTSA Publication DOT HS 806 108)

The purpose of a head restraint is to prevent whiplash injury in rear-impact crashes. There are integral (fixed) and adjustable head restraints; 75 percent of adjustable restraints are left in the "down" position by occupants. In 1982, integral head restraints reduced injury risk in rear impacts by 17 percent; adjustable restraints by 10 percent. The 1982 mix of head restraints prevented 64,000 whiplash injuries per year. [Subsequently, manufacturers have enlarged adjustable restraints to provide better protection, even in the "down" position. See also the 2001 evaluation of head restraints in light trucks.]

1981

An Evaluation of the Bumper Standard (NHTSA Publication DOT HS 805 866)

In order to reduce car repair costs for consumers, damage resistance tests were established for bumpers in model year 1973 and upgraded in 1974 and 1979. The bumper standards did not significantly change net costs for consumers: the savings in repair costs over the lifetime of the car are almost equal to the increase in the initial cost of the bumpers. (See also the 1987 evaluation of bumpers.)

An Evaluation of Federal Motor Vehicle Safety Standards for Passenger Car Steering Assemblies: Standard 203 - Impact Protection for the Driver; Standard 204 - Rearward Column Displacement (NHTSA Publication DOT HS 805 705)

Energy-absorbing, telescoping steering columns reduced the risk of serious injury due to steering-assembly contact by 38 percent. Rearward column displacement was reduced by 81 percent. The standards prevent 1,300 fatalities and 23,000 hospitalizations per year. The performance of energy-absorbing steering assemblies is degraded under nonaxial impact conditions.

1979

An Evaluation of Standard 214 (NHTSA Publication DOT HS 804 858)

[Findings have been superseded by the 1982 evaluation - see above.]